Mastering MATLAB®

Mastering MATLAB®

Duane Hanselman

Bruce Littlefield

Department of Electrical and Computer Engineering

University of Maine

International Edition contributions by
Arindam Chatterjee
Salil Joshi

PEARSON

Upper Saddle River   Boston   Columbus   San Francisco
New York   Indianapolis   London   Toronto   Sydney   Singapore   Tokyo   Montreal
Dubai   Madrid   Hong Kong   Mexico City   Munich   Paris   Amsterdam   Cape Town

Vice President and Editorial Director, ECS:
  Marcia J. Horton
Senior Editor: Andrew Gilfillan
Associate Editor: Alice Dworkin
Editorial Assistant: William Opaluch
Marketing Manager: Tim Galligan
Production Manager: Pat Brown
Publisher, International Edition:
  Angshuman Chakraborty
Acquisitions Editor, International Edition:
  Somnath Basu
Publishing Assistant, International Edition:
  Shokhi Shah

Print and Media Editor, International Edition:
  Ashwitha Jayakumar
Project Editor, International Edition:
  Jayashree Arunachalam
Art Director: Jayne Conte
Cover Designer: Bruce Kenselaar
Full-Service Project Management/
Composition: Kiruthiga Anand/Integra
  Software Services Pvt. Ltd.
Cover Printer: R.R. Donnelley—
  Crawfordsville

Pearson Education Limited
Edinburgh Gate
Harlow
Essex CM20 2JE
England

and Associated Companies throughout the world

Visit us on the World Wide Web at:
www.pearsoninternationaleditions.com

© Pearson Education Limited 2012.

British Library Cataloguing-in-Publication Data

A Catalogue record for this book is available from the British Library

10 9 8 7 6 5 4 3 2 1
14 13 12 11

Typeset in Times Ten-Roman by Kiruthiga Anand/Integra Software Services Pvt. Ltd.
Printed and bound by Courier Westford in The United States of America

The publisher's policy is to use paper manufactured from sustainable forests.

PEARSON

ISBN 10: 0-273-75213-8
ISBN 13: 978-0-273-75213-4

Contents

Preface

This book is about MATLAB®. If you use MATLAB or consider using it, *Mastering MATLAB* is for you. The text represents an alternative to learning MATLAB on your own, with or without the help of the documentation that comes with the software. The informal style of this book makes it easy to read, and, as the title suggests, it provides the tools you need to master MATLAB. As a programming language and data-visualization tool, MATLAB offers a rich set of capabilities for solving problems in engineering, scientific, computing, and mathematical disciplines. The fundamental goal of this text is to help you increase your productivity by showing you how to use these capabilities efficiently. To optimize the interactive nature of MATLAB, the material is generally presented in the form of examples that you can duplicate by running MATLAB as you read this book.

This text covers only topics that are of use to a general audience. The material presented here generally applies to all computer platforms. None of the *Toolboxes, Blocksets,* or other *Libraries* that are available for additional cost are discussed, although some are referred to in appropriate places. There are simply too many additional product items to consider in one book.

Since MATLAB continues to evolve as a software tool, this text focuses on MATLAB version 7.12. For the most part, the material applies to versions 6.x through 7.11 of MATLAB, as well as later versions. When appropriate, distinctions between versions are made.

We, the authors, encourage you to give us feedback on this book. What are the best features of the text? What areas need more work? What topics should be left out? What topics should be added? We can be reached at the e-mail address *masteringmatlab@gmail.com*. In addition, errata, all examples in the text, and other related material can be found at *http://www.masteringmatlab.com*

Instructor Resources for the International Edition can be accessed at www.pearsoninternationaleditions.com/hanselman.

NEW TO THIS EDITION

Major changes from *Mastering MATLAB 7* (based on MATLAB version 7.0, Release 14, June 2004) were prompted by user feedback, changes and enhancements to MATLAB itself, increased availability of books on specialized aspects of MATLAB such as programming interfaces and Windows integration, and the desire for more examples in the text.

1. All chapters have been revised and updated for MATLAB version 7.12, Release 2011a, April 2011.

2. Chapter 32 ("Examples, Examples, Examples") is revised and greatly expanded to provide even more helpful examples.

3. Many topics are rewritten to take advantage of new features available in MATLAB 7.12 and beyond.

4. Depreciated or obsolete functions have been removed or de-emphasized.

5. Appendices have been expanded to include changes in MATLAB functions and functionality from version 5.0 through version 7.12. These appendices help the user identify incompatibilities between versions so that MATLAB code can be written that runs seamlessly across many MATLAB versions.

6. Chapters on programming in C, Fortran, and Java along with Windows integration were dropped to make room for expanded examples and coverage of the new features of MATLAB 7.12.

DUANE HANSELMAN
BRUCE LITTLEFIELD

The publishers would like to thank Dr Sudhir Bhide and Sourav Sen Gupta for reviewing the content of the International Edition.

Mastering MATLAB®

1

Getting Started

1.1 INTRODUCTION

This text assumes that you have some familiarity with matrices and computer programming. Matrices and arrays, in general, are at the heart of MATLAB, since all data in MATLAB are stored as arrays. Besides offering common matrix algebra operations, MATLAB offers array operations that allow you to quickly manipulate sets of data in a wide variety of ways. In addition to its matrix orientation, MATLAB offers programming features that are similar to those of other computer programming languages. Finally, MATLAB offers graphical user interface (GUI) tools that enable you to use MATLAB as an application-development tool. This combination of array data structures, programming features, and GUI tools makes MATLAB an extremely powerful tool for solving problems in many fields. In this text, each of these aspects of MATLAB is discussed in detail. To facilitate learning, detailed examples are presented.

1.2 TYPOGRAPHICAL CONVENTIONS

The following conventions are used throughout this book:

Bold italics	New terms or important facts
Boxed text	Important terms and facts
Bold Initial Caps	Keyboard key names, menu names, and menu items
`Constant width`	Computer user input, function and file names, commands, and screen displays

| Boxed constant width | Contents of a script, function, or data file |

Constant width italics   User input that is to be replaced and not taken
literally, such as » help *functionname*

Italics   Window names, object names, book titles, toolbox
names, company names, example text, and
mathematical notations

1.3 WHAT'S NEW IN MATLAB

The previous edition of this text, *Mastering MATLAB 7*, was based on MATLAB version 7.0. Over time, MATLAB has evolved through a number of versions. As of Release 2011a of the MATLAB/Simulink software suite, MATLAB has progressed to version 7.12.

The latest version is yet another evolutionary change in MATLAB. The *Command* window remains the primary user interface, *Figure* windows are used to display graphical information and to create graphical user interfaces (GUIs), and a text editor/debugger is provided for writing, editing, and debugging MATLAB code. The *MATLAB desktop* coordinates the position and visibility of a number of other windows, such as *Workspace*, *Editor*, *Help*, and *Command History*.

Under the surface, several numerical and operational changes have been made to the software. In addition to having added or modified several internal numerical algorithms, MATLAB 7 and above supports mathematical operations on data types other than double-precision arrays, which have always been at the heart of MATLAB. Perhaps even more importantly, the MATLAB command interpreter now includes acceleration features, collectively called *The MATLAB JIT-accelerator*. This accelerator, which first appeared in MATLAB 6.5, dramatically increases the speed of loop operations by interpreting and executing code within a loop as a whole, rather than line by line, as was done previously. This acceleration eliminates the need to *vectorize* some code, which requires you to create and maximize the use of arrays and array mathematics to achieve optimal performance. To make use of the JIT-accelerator, loop operation code must follow specific guidelines, which are covered in this text. When these guidelines are not followed, loop operation code is interpreted at the much slower line-by-line rate.

In sum, version 7.12 represents an evolutionary change in MATLAB. The basic operation of MATLAB and its capabilities haven't changed in any dramatic way. In almost all cases, MATLAB code written for MATLAB 6.0–7.11 will run without change in MATLAB 7.12 and above. For the most part, the new and changed features of MATLAB can increase your productivity in solving problems with MATLAB.

1.4 WHAT'S IN *MASTERING MATLAB*

MATLAB documentation, in both hard copy and electronic format, exceeds 5000 pages of information and help text. Given this exhaustive documentation and help, *Mastering MATLAB* does not attempt to be a comprehensive tutorial or reference, which just isn't possible in a single book containing less than 1000 pages. *Mastering MATLAB* does not even attempt to document all functions within the base product of MATLAB. This too is impossible, as there are more than 1500 operators and functions that are part of the base product of MATLAB. In light of this documentation set, the goals of *Mastering MATLAB* include the following: (1) introduce MATLAB to the novice user, (2) illustrate all key features and capabilities of MATLAB, and (3) demonstrate by example how to write efficient MATLAB code.

This text is intended to be a valuable resource when the MATLAB documentation set is unavailable. It follows the rule of providing 80 percent of the information needed in 20 percent of the space required to cover everything. The book is also intended to be a valuable resource when the MATLAB documentation set is available. In this case, it supplies numerous examples of efficient MATLAB coding that demonstrate how the many features of MATLAB come together to solve real problems.

Equally important is what's not included in *Mastering MATLAB*. The book does not discuss all of the windows, menus, menu items, submenus, and dialog boxes of the MATLAB user interface. While these user-interface aspects of MATLAB are very important, there is no room for them in this text. Moreover, they are difficult to cover effectively in a text and are best learned by hands-on exploration. Rather, this text focuses on the mathematical, programming, and graphical features that facilitate problem solution.

In addition to updating the second edition to reflect changes and additions to MATLAB since version 7.0, revisions include a major expansion of Chapter 32 ("Examples, Examples, Examples"), consolidation and revision of some of the other chapters, and dropping the specialized GUI and API (application programming interface) chapters. For those interested in building graphical user interfaces, the MATLAB online documentation is quite extensive. In addition to the MATLAB documentation, a number of specialized texts on integrating MATLAB with C, FORTRAN, Java, and Microsoft Windows applications are also available.

This book was written using information about MATLAB version 7.12. As MATLAB evolves, some things are bound to change. As a result, there may be isolated areas in this book where information about new features is missing and, worse yet, where information is incorrect. The authors have no control over MATLAB. We also cannot rewrite the text to reflect minor MATLAB releases. We are thankful that the makers of MATLAB are very careful when introducing new features and when changing old features. Historically, old features are grandfathered for one major release, and sometimes for additional releases. As a result, even though the book reflects MATLAB version 7.12, it will undoubtedly be useful for all future MATLAB versions as well.

To support this text, the authors maintain the Mastering MATLAB website at *http://www.masteringmatlab.com*. At the site, you can find errata for the text, as well as MATLAB script M-files for creating all of the figures in the book. The authors also encourage constructive feedback about the text at the e-mail address *masteringmatlab@gmail.com*.

2

Basic Features

Running MATLAB creates one or more windows on your computer monitor. One of these windows, entitled *MATLAB*, is commonly called the ***MATLAB desktop***. This window is the primary graphical user interface for MATLAB. Within the *MATLAB* window, there is a window called the ***Command*** window, which is the primary place wherein you interact with MATLAB. The prompt >> is displayed in the *Command* window, and when the *Command* window is active, a blinking cursor appears to the right of the prompt. This cursor and prompt signify that MATLAB is waiting to perform a mathematical operation.

2.1 SIMPLE MATH

Just like a calculator, MATLAB can do basic math. Consider the following simple example: Mary goes to the office supply store and buys five pens at 30 cents each, seven notebooks at 60 cents each, and one pair of scissors for 70 cents. How many items did Mary buy, and how much did they cost?

To solve this problem with a calculator, you enter

```
5 + 7 + 1 = 13 items
5 × 30 + 7 × 60 + 1 × 70 = 640 cents
```

In MATLAB, this problem can be solved in a number of different ways. First, the calculator approach can be taken:

```
>> 5+7+1
ans =
    13
```

```
>> 5*30 + 7*60 + 1*70
ans =
    640
```

Note that MATLAB doesn't care about spaces, for the most part, and that multiplication takes precedence over addition. Note also that MATLAB calls the result ans, which is short for *answer* for both computations.

As an alternative, the problem can be solved by storing information in **MATLAB variables**:

```
>> pens = 5
pens =
    5
>> notebooks = 7
notebooks =
        7
>> scissors = 1;
>> items = pens + notebooks + scissors
items =
    13
>> cost = pens*30 + notebooks*60 + scissors*70
cost =
    640
```

Here, we created three MATLAB variables—pens, notebooks, and scissors—to store the number of each item. After entering each statement, MATLAB displayed the results, except in the case of scissors. The semicolon at the end of the line tells MATLAB to evaluate the line, but not to display the answer. Finally, rather than calling the results ans, we told MATLAB to call the number of items purchased items and the total price paid cost. At each step, MATLAB remembered past information. Because MATLAB remembers things, let's ask what the average cost per item was:

```
>> average_cost = cost/items
average_cost =
    49.2308
```

Since the term *average cost* is two words and MATLAB variable names must be one word, an underscore was used to create the single MATLAB variable average_cost.

In all, MATLAB offers the following basic arithmetic operations:

Operation	Symbol	Example
Addition	+	3 + 22
Subtraction	–	54.4 – 16.5
Multiplication	*	3.14 * 6
Division	/ or \	19.54/7 or 7\19.54
Exponentiation	^	2^8

The order in which these operations are evaluated in a given expression is determined by the usual rules of precedence, summarized as follows:

> Expressions are evaluated from left to right, with the exponentiation operation having the highest precedence, followed by multiplication and division, having equal precedence, and then by addition and subtraction, having equal precedence.
>
>   Parentheses can be used to alter this ordering, in which case these rules of precedence are applied within each set of parentheses, by starting with the innermost set and proceeding outward.

More information about precedence rules can be displayed by typing `help precedence` at the MATLAB prompt.

2.2 THE MATLAB WORKSPACE

As you work in the *Command* window, MATLAB remembers the commands you enter, as well as the values of any variables you create. These commands and variables are said to reside in the ***MATLAB workspace*** or ***base workspace***, and they can be recalled whenever you wish. For example, to check the value of `scissors`, all you have to do is ask MATLAB for it by entering its name at the prompt:

```
>> scissors

scissors =

        1
```

If you can't remember the name of a variable, you can ask MATLAB for a list of the variables it knows by using the MATLAB command `who`:

```
>> who

Your variables are:

ans            items     scissors

average_cost   notebooks

cost           pens
```

Note that MATLAB doesn't tell you the value of all of the variables; it merely gives you their names. To find their values, you must enter their names individually at the MATLAB prompt.

To recall previous commands, MATLAB responds to the **Cursor** keys on your keyboard. For example, pressing the ↑ key once recalls the most recent command to the MATLAB prompt. Repeatedly pressing ↑ scrolls back through prior commands, one at a time. In a similar manner, pressing the ↓ key scrolls forward through commands. By pressing the → or ← keys, you can maneuver within a given command at the MATLAB prompt, thereby editing the command in much the same way that you edit text in a word-processing program. Other standard editing keys, such as **Delete** or **Backspace**, **Home**, and **End**, perform their commonly assigned tasks. The **Tab** key is useful for variable-name completion. Once a scrolled command is acceptable, pressing the **Return** key with the cursor *anywhere* in the command tells MATLAB to process it. Finally, and perhaps most useful, the **Escape** key erases the current command at the prompt. For those of you familiar with the EMACS editor, MATLAB also accepts common EMACS editing control-character sequences, such as **Control-U** to erase the current command.

2.3 ABOUT VARIABLES

Like any other computer language, MATLAB has rules about variable names. Earlier, it was noted that variable names must be a single word containing no spaces. More specifically, MATLAB variable-naming rules are listed as follows.

Variable-Naming Rules	Comments/Examples
Variable names are case sensitive.	Cost, cost, CoSt, and COST are all different MATLAB variables.
Variable names can contain up to `namelengthmax` characters (63 as of version 7.12). Any characters beyond the 63rd are ignored.	Howaboutthisvariablename
Variable names must start with a letter, followed by any number of letters, digits, or underscores.	how_about_this_variable_nameX51483
Punctuation characters are not allowed, because many of them have special meanings in MATLAB.	a_b_c_d_e

There are some specific exceptions to these naming rules. MATLAB has several names that cannot be used for variables. These names are *keywords* and form a *reserved word list* for MATLAB:

Reserved Word List
for end if while function return elseif case otherwise classdef switch continue else try catch global persistent break parfor spmd

This list is returned as an output of the iskeyword function. MATLAB will report an error if you try to use a reserved word as a variable. However, you can use words similar to keywords by capitalizing one or more letters. The function isvarname('teststring') returns True (1) if the character-string argument 'teststring' is a valid variable name; otherwise, it returns False (0).

In addition, just as your calculator stores constants such as π, MATLAB has a number of special variables:

Special Variables	Description
ans	Default variable name used for results
beep	Makes computer sound a beep
computer	Computer type
version	MATLAB version string
ver	MATLAB, computer OS, and Java version information
pi	Ratio of the circumference of a circle to its diameter
eps	Smallest number that, when added to 1, creates a number greater than 1 on the computer
inf	Stands for infinity (e.g., 1/0)
NaN or nan	Stands for Not-a-Number (e.g., 0/0)
i or j	Stands for $\sqrt{-1}$
nargin	Number of function input arguments
nargout	Number of function output arguments
intmin	Largest usable negative integer
intmax	Largest usable positive integer
realmin	Smallest usable positive real number
realmax	Largest usable positive real number
bitmax	Largest usable positive integer stored in double-precision format
varargin	Variable number of function input arguments
varargout	Variable number of function output arguments

If you reuse a variable (such as `scissors` in the earlier example) or assign a value to one of the special variables from the preceding list, the prior value of the variable is overwritten and lost. However, any other expressions computed with the prior values do not change. Consider the following example:

```
>> pens = 5;
>> notebooks = 7;
>> scissors = 1;
>> items = pens + notebooks + scissors
items =
    13
>> pens = 9
pens =
    9
>> items
items =
    13
```

Here, using the first example again, we found the number of items Mary purchased. Afterward, we changed the number of pens to nine, overwriting its prior value of 5. In doing so, the value of items has not changed. Unlike a common spreadsheet program, MATLAB does not recalculate the number of items based on the new value of pens. *When MATLAB performs a calculation, it does so using the values that it knows at the time the requested command is evaluated.* In the preceding example, if you wish to recalculate the number of items, the total cost, and the average cost, it is necessary to recall the appropriate MATLAB commands and ask MATLAB to evaluate them again.

The special variables given earlier follow this guideline, also, with the exception of the fact that the special values can be restored. When you start MATLAB, the variables have their original values; when you change their values, the original special values are lost. To restore a special value, all you have to do is *clear* the overwritten value. For example,

```
>> intmax
ans =
    4294967296
>> intmax = 1.23e-4
intmax =
        0.000123
```

```
>> clear intmax
>> intmax

ans =

    4294967296
```

shows that intmax has the special value of 4294967296, to ten significant digits; is overwritten with the value of 1.23e-4; and then, after being cleared using the clear function has its original special value once again. Remember that intmax is a platform-dependent constant.

2.4 COMMENTS, PUNCTUATION, AND ABORTING EXECUTION

As we saw earlier, placing a semicolon at the end of a command suppresses printing of the computed results. This feature is especially useful for suppressing the results of intermediate calculations. For instance,

```
>> pens = 9

pens =

     9

>> items = pens + notebooks + scissors;
>> cost = pens*30 + notebooks*60 + scissors*70;
>> average_cost = cost/items

average_cost =

    44.7059
```

displays the average cost of the items that Mary bought when she purchased nine pens, rather than the original five. The intermediate results items and cost were not printed, because semicolons appear at the ends of the commands defining them.

In addition to semicolons, MATLAB uses other punctuation symbols. For example, all text after a percent sign (%) is taken as a comment statement:

```
>> scissors = 1 % number of pairs of scissors purchased
```

The variable scissors is given the value of 1, and MATLAB simply ignores the percent sign and all text following it.

If they are separated by commas or semicolons, multiple commands can be placed on one line:

```
>> pens = 6, notebooks = 6; scissors = 2

pens =

     6

scissors =

     2
```

Commas tell MATLAB to display results; semicolons suppress printing.

Sometimes, expressions or commands are so long that it is convenient to continue them onto additional lines. In MATLAB, statement continuation is denoted by three periods in succession, as shown in the following code:

```
>> average_cost = cost/items % command as done earlier
average_cost =
        44.7059
>> average_cost = cost/...% command with valid continuation
items
average_cost =
        44.7059
>> average_cost = cost...% command with valid continuation
/items
average_cost =
        44.7059
>> average_cost = cost...command with valid continuation (no % needed)
/items
average_cost =
        44.7059
>> average_cost = cost/it...% command with Invalid continuation
ems
??? ems
    |
Error: Missing MATLAB operator.
```

Note that statement continuation works if the three periods appear between variable names and mathematical operators, but not in the middle of a variable name. That is, variable names cannot be split between two lines. Furthermore, all text after the three periods is considered to be a comment, so no percent symbol is needed. In addition, since comment lines are ignored, they cannot be continued either, as the following example shows:

```
>> % Comments cannot be continued...
>> either
??? Undefined function or variable 'either'.
```

In this case, the . . . in the comment line is part of the comment and is not processed by MATLAB.

Finally, MATLAB processing can be interrupted at any time by pressing **Control-C** (i.e., pressing the **Ctrl** and **C** keys simultaneously).

2.5 COMPLEX NUMBERS

One of the most powerful features of MATLAB is that it does not require any special handling for complex numbers. Complex numbers are formed in MATLAB in several ways. Examples of complex numbers include the following:

```
>> c1 = 1-2i % the appended i signifies the imaginary part
c1 =
   1.0000 - 2.0000i
>> c1 = 1-2j % j also works
c1 =
   1.0000 - 2.0000i
>> c1 = complex(1,-2) % a function that creates complex numbers
c1 =
   1.0000 - 2.0000i
>> c2 = 3*(2-sqrt(-1)*3)
c2 =
   6.0000 - 9.0000i
>> c3 = sqrt(-2)
c3 =
        0 + 1.4142i
>> c4 = 6+sin(.5)*1i
c4 =
   6.0000 + 0.4794i
>> c5 = 6+sin(.5)*1j
c5 =
   6.0000 + 0.4794i
```

In the last two examples, multiplication by 1i and 1j are used to form the imaginary part. Multiplication by 1i or 1j is required in these cases, since sin(.5)i and sin(.5)j have no meaning in MATLAB. Termination with the characters i and j, as shown in the first two examples above, works only with numbers, and not with expressions.

Some programming languages require special handling for complex numbers, wherever they appear. In MATLAB, no special handling is required. Mathematical operations on complex numbers are written the same way as those on real numbers:

```
>> c6 = (c1+c2)/c3 % from the above data
c6 =
   -7.7782 - 4.9497i
>> c6r = real(c6)
c6r =
    -7.7782
>> c6i = imag(c6)
c6i =
    -4.9497
>> check_it_out = 1i^2 % sqrt(-1) squared must be -1!
check_it_out =
   -1
```

In general, operations on complex numbers lead to complex number results. However, in the last case, MATLAB is smart enough to drop the zero imaginary part of the result. In addition, the foregoing shows that the functions `real` and `imag` extract the real and imaginary parts of a complex number, respectively.

As a final example of complex arithmetic, consider the Euler identity that relates the polar form of a complex number to its rectangular form $M\angle\theta = Me^{j\theta} = a + bj$, where the polar form is given by a magnitude M and an angle θ, and the rectangular form is given by $a + bj$. The relationships among these forms are $M = \sqrt{a^2 + b^2}$, $\theta = \tan^{-1}(b/a)$, $a = M\cos(\theta)$, $b = M\sin(\theta)$.

In MATLAB, the conversion between polar and rectangular forms makes use of the functions `real`, `imag`, `abs`, and `angle`:

```
>> c1
c1 =
   1.0000 - 2.0000i
>> mag_c1 = abs(c1) % magnitude
mag_c1 =
    2.2361
>> angle_c1 = angle(c1) % angle in radians
angle_c1 =
   -1.1071
>> deg_c1 = angle_c1*180/pi % angle in degrees
```

```
deg_c1 =
   -63.4349
>> real_c1 = real(c1) % real part
real_c1 =
   1
>> imag_c1 = imag(c1) % imaginary part
imag_c1 =
   -2
```

The MATLAB function abs computes the magnitude of complex numbers or the absolute value of real numbers, depending on which one you assign it to compute. Likewise, the MATLAB function angle computes the angle of a complex number in radians. ***MATLAB does not natively perform trigonometric operations with units of degrees; however, basic trigonometric functions supporting angles in degrees are provided in MATLAB.***

2.6 FLOATING-POINT ARITHMETIC

In almost all cases, numerical values in MATLAB are represented in double-precision arithmetic using a binary (base 2) representation internally. It is the most common representation used by computers and is a native format for numerical coprocessors. Because of this representation, not all numbers can be represented exactly. There are limiting values that can be represented, and there is a recognizable lower limit for addition.

The largest positive real number that can be represented is

```
>> format long % tell MATLAB to display more precision
>> realmax
ans =
   1.797693134862316e+308
```

The smallest positive number that can be represented is

```
>> realmin
ans =
   2.225073858507201e-308
```

The smallest number that can be added to 1 to produce a number larger than 1 in double precision is

```
>> eps
ans =
    2.220446049250313e-16
```

Generalizing this concept, eps(x) is the smallest increment that can be added to x to produce a number larger than x, as the following examples show:

```
> eps(1) % same as eps by itself
ans =
    2.220446049250313e-16
>> eps(10)
ans =
    1.776356839400251e-15
>> eps(1e10)
ans =
    1.907348632812500e-06
```

As the magnitude of a number increases, the distance between values that can be represented in finite precision (which is what eps(x) returns) increases as well.

The consequences of the limitations of finite-precision arithmetic are sometimes strange. For example, as shown here, addition is not *exactly* commutative:

```
>> 0.21 - 0.25 + 0.04
ans =
    -6.9389e-018
>> 0.04 - 0.25 + 0.21 % rearrange order
ans =
    0
>> 0.04 + 0.21 - 0.25 % rearrange order again
ans =
    0
```

All three of these results should be zero, but they are not. In each case, the arithmetic was performed from left to right. The issue here is that not all of the numbers can be represented exactly in double-precision arithmetic. In fact, only 0.25 has an exact representation. When numbers cannot be represented exactly, they are approximated with as much precision as possible—leading to inevitable errors in computed results. For the most part, these errors are minor; otherwise, double-precision arithmetic wouldn't be used in modern computers. In practice, the problems with double-precision arithmetic occur most often when asking MATLAB to compare two numbers for equality or inequality. Clearly, within MATLAB, $0.21 - 0.25 + 0.04$ does not equal $0.04 - 0.25 + 0.21$, even though our brains can do the exact arithmetic to show that it does.

A second drawback of finite-precision arithmetic appears in function evaluation. Not only is finite-precision arithmetic unable to always represent function arguments exactly, but also most functions cannot themselves be represented exactly, as the following code shows:

```
>> sin(0)
ans =
    0
>> sin(pi)
ans =
    1.224646799147353e-16
```

Here, both results should be zero, but $\sin(\pi)$ is not. It is interesting to note that the error here and the error in the previous example are both less than eps.

Finally, when using double-precision, floating-point arithmetic to represent integers, only integers up to a limit can be represented *exactly*. The limiting value is $2^{53} - 1$, which is represented by MATLAB as follows:

```
>> bitmax
ans =
    9.007199254740991e+15
```

2.7 MATHEMATICAL FUNCTIONS

Lists of the common functions that MATLAB support are shown in the tables at the end of this chapter. Most of these functions are used in the same way you would write them mathematically:

```
>> x = sqrt(3)/3
x =
    0.5773
>> y = asin(x)
y =
    0.6154
>> y_deg = y*180/pi
y_deg =
    35.2608
```

These commands find the angle where the sine function has a value of $\sqrt{3}/3$. Note, again, that MATLAB uses radians, not degrees, in trigonometric functions. Other examples include the following:

```
>> y = sqrt(3^2 + 4^2) % show 3-4-5 right triangle relationship
y =
    5
>> y = rem(23,4) % remainder function, 23/4 has a remainder of 3
y =
    3
>> x = 2.6, y1 = fix(x), y2 = floor(x), y3 = ceil(x), y4 = round(x)
x =
    2.6000
y1 =
    2
y2 =
    2
y3 =
    3
y4 =
    3
```

Trigonometric Function	Description
acos	Inverse cosine returning radians
acosd	Inverse cosine returning degrees
acosh	Inverse hyperbolic cosine returning radians
acot	Inverse cotangent returning radians
acotd	Inverse cotangent returning degrees
acoth	Inverse hyperbolic cotangent returning radians
acsc	Inverse cosecant returning radians
acscd	Inverse cosecant returning degrees
acsch	Inverse hyperbolic cosecant returning radians
asec	Inverse secant returning radians
asecd	Inverse secant returning degrees

Trigonometric Function	Description
asech	Inverse hyperbolic secant returning radians
asin	Inverse sine returning radians
asind	Inverse sine returning degrees
asinh	Inverse hyperbolic sine returning radians
atan	Inverse tangent returning radians
atand	Inverse tangent returning degrees
atanh	Inverse hyperbolic tangent returning radians
atan2	Four-quadrant inverse tangent returning radians
cos	Cosine returning radians
cosd	Cosine of argument in degrees
cosh	Hyperbolic cosine returning radians
cot	Cotangent returning radians
cotd	Cotangent of argument in degrees
coth	Hyperbolic cotangent returning radians
csc	Cosecant returning radians
cscd	Cosecant of argument in degrees
csch	Hyperbolic cosecant returning radians
hypot	Square root of sum of squares
sec	Secant returning radians
secd	Secant of argument in degrees
sech	Hyperbolic secant returning radians
sin	Sine returning radians
sind	Sine returning degrees
sinh	Hyperbolic sine returning radians
tan	Tangent returning radians
tand	Tangent returning degrees
tanh	Hyperbolic tangent returning radians

Exponential Function	Description
^	Power
exp	Exponential
expm1	Exponential minus 1 [i.e., $\exp(x) - 1$]
log	Natural logarithm
log10	Base 10 logarithm
log1p	Natural logarithm of $x + 1$ [i.e., $\log(x + 1)$]
log2	Base 2 logarithm and floating-point number dissection
nthroot	n^{th} real root of real numbers
pow2	Base 2 power and floating-point number scaling
reallog	Natural logarithm limited to real nonnegative values
realpow	Power limited to real-valued arguments
realsqrt	Square root limited to real-valued values
sqrt	Square root
nextpow2	Next higher power of 2

Complex Function	Description
abs	Absolute value or magnitude
angle	Phase angle in radians
conj	Complex conjugate
imag	Imaginary part
real	Real part
unwrap	Unwraps phase angle
isreal	True for real values
cplxpair	Sorts vector into complex conjugate pairs
complex	Forms complex number from real and imaginary parts
sign	Signum function

Rounding and Remainder Function	Description
fix	Rounds toward zero
floor	Rounds toward negative infinity
ceil	Rounds toward positive infinity
round	Rounds toward nearest integer
mod	Modulus or signed remainder
rem	Remainder after division
idivide	Integer division with rounding option
sign	Signum function

Coordinate Transformation Function	Description
cart2sph	Cartesian to spherical
cart2pol	Cartesian to cylindrical or polar
pol2cart	Cylindrical or polar to Cartesian
sph2cart	Spherical to Cartesian

Number Theoretic Function	Description
factor	Prime factors
isprime	True for prime numbers
primes	Generates list of prime numbers
gcd	Greatest common divisor
lcm	Least common multiple
rat	Rational approximation
rats	Rational output
perms	All possible combinations
nchoosek	All combinations of N elements taken K at a time
factorial	Factorial function

Specialized Function	Description
airy	Airy function
besselj	Bessel function of the first kind
bessely	Bessel function of the second kind
besselh	Bessel function of the third kind
besseli	Modified Bessel function of the first kind
besselk	Modified Bessel function of the second kind
beta	Beta function
betainc	Incomplete beta function
betaincinv	Inverse incomplete beta function
betaln	Logarithm of beta function
ellipj	Jacobi elliptic function
ellipke	Complete elliptic integral
erf	Error function
erfc	Complementary error function
erfcinv	Inverse complementary error function
erfcx	Scaled complementary error function
erfinv	Inverse error function
expint	Exponential error function
gamma	Gamma function
gammainc	Incomplete gamma function
gammaincinv	Inverse incomplete gamma function
gammaln	Logarithm of gamma function
legendre	Associated Legendre functions
psi	Psi (polygamma) function
cross	Vector cross product
dot	Vector dot product

3

The MATLAB Desktop

As stated in the preceding chapter, running MATLAB creates one or more windows on your computer monitor. One of these windows, entitled *MATLAB*, is commonly called the *MATLAB* desktop. This window contains or manages all other windows that are part of MATLAB. Depending on how you set up MATLAB, some windows associated with the desktop may or may not be visible, and some may or may not reside within (i.e., be ***docked*** in) the *MATLAB* window. Management of the *MATLAB* desktop window and its associated windows is not discussed in this text. If you are familiar with other window-based programs, manipulation of windows in MATLAB will be familiar to you. The *MATLAB* desktop menu items change, depending on which window is active. In addition, there are many helpful contextual menus (which are accessed by pressing the right mouse button over an item). If you are not familiar with window-based programs, it may be beneficial to seek general assistance in this area from other sources. In any case, to find out which windows are associated with the desktop, investigate the items on the **Desktop** menu within the *MATLAB* desktop window.

3.1 MATLAB WINDOWS

The windows used in MATLAB include the (1) *Command*, (2) *Command History*, (3) *Current Folder* (browser), (4) *Workspace* (browser), (5) *Help* (browser), (6) *Profiler*, (7) *File Exchange*, (8) *Editor*, (9) *Figures*, (10) *Web Browser*, (11) *Variable Editor*, and (12) *Comparison Tool*. The following table gives an overview of the purpose of each of these windows:

Window	Description
Command	Issues commands to MATLAB for processing
Command History	Running history of prior commands issued in the *Command* window
Current Folder	GUI for directory and file manipulation in MATLAB
Workspace	GUI for viewing, editing, loading, and saving MATLAB variables
Help	GUI for finding and viewing online documentation
Profiler	Tool for optimizing M-file performance
File Exchange	GUI for accessing the MATLAB File Exchange
Editor	Text editor for creating, editing, and debugging M-files
Figures	Creates, views, and modifies plots and other figures
Web Browser	MATLAB HTML viewer
Variable Editor	Tool for viewing and editing arrays in table format
Comparison Tool	GUI for comparing text files line by line

3.2 MANAGING THE MATLAB WORKSPACE

Within the *MATLAB* desktop, actions taken in all windows support computations performed in the *Command* window. As a result, the rest of this chapter provides more detailed information about the *Command* window.

The data and variables created in the *Command* window reside in what is called the **MATLAB workspace** or **base workspace**. In addition to viewing variables in the *Workspace* window, you can see what variable names exist in the MATLAB workspace by issuing the command who:

```
>> who
Your variables are:
angle_c1      c4           cost       notebooks
ans           c5           deg_c1     real_c1
average_cost  c6           pens       scissors
c1            c6i          imag_c1
c2            c6r          items
c3            check_it_out mag_c1
```

The variables you see may differ from those just listed, depending on what you've asked MATLAB to do since you opened the program. For more detailed information, use the command whos:

```
>> whos
  Name            Size        Bytes     Class       Attributes
  angle_c1        1x1             8     double
  ans             1x1             8     double
  average_cost    1x1             8     double
  c1              1x1            16     double      (complex)
  c2              1x1            16     double      (complex)
  c3              1x1            16     double      (complex)
  c4              1x1            16     double      (complex)
  c5              1x1            16     double      (complex)
  c6              1x1            16     double      (complex)
  c6i             1x1             8     double
  c6r             1x1             8     double
  check_it_out    1x1             8     double
  cost            1x1             8     double
  deg_c1          1x1             8     double
  pens            1x1             8     double
  imag_c1         1x1             8     double
  items           1x1             8     double
  mag_c1          1x1             8     double
  notebooks       1x1             8     double
  real_c1         1x1             8     double
  scissors        1x1             8     double
```

Here, each variable is listed, along with its size, the number of bytes used, its class, and other attributes, if any. Since MATLAB is array oriented, all of the variables belong to the class of double-precision arrays, even though all of the variables are scalars. Later, as other data types or classes are introduced, the information in this last column will become more useful. The preceding list is also displayed in the *Workspace* window, which can be viewed by typing workspace at the MATLAB prompt or by choosing the **Workspace** menu item on the **Desktop** menu of the *MATLAB* desktop.

The command clear deletes variables from the MATLAB workspace. For example,

```
>> clear real_c1 imag_c1 c*
>> who
Your variables are:
angle_c1        deg_c1      mag_c1
ans             pens        notebooks
average_cost    items       scissors
```

deletes the variables real_c1, imag_c1, and all variables starting with the letter c. Other options for the clear function can be identified by asking for information with the command help or helpwin:

```
>> help clear
 CLEAR Clear variables and functions from memory.
    CLEAR removes all variables from the workspace.
    CLEAR VARIABLES does the same thing.
    CLEAR GLOBAL removes all global variables.
    CLEAR FUNCTIONS removes all compiled M- and MEX-functions.

    CLEAR ALL removes all variables, globals, functions and MEX links.
    CLEAR ALL at the command prompt also removes the Java packages import
    list.

    CLEAR IMPORT removes the Java packages import list at the command
    prompt. It cannot be used in a function.

    CLEAR CLASSES is the same as CLEAR ALL except that class definitions
    are also cleared. If any objects exist outside the workspace (say in
    userdata or persistent in a locked m-file) a warning will be issued and
    the class definition will not be cleared. CLEAR CLASSES must be used if
    the number or names of fields in a class are changed.

    CLEAR JAVA is the same as CLEAR ALL except that java classes on the
    dynamic java path (defined using JAVACLASSPATH) are also cleared.

    CLEAR VAR1 VAR2 ... clears the variables specified. The wildcard
    character '*' can be used to clear variables that match a pattern. For
```

instance, CLEAR X* clears all the variables in the current workspace that start with X.

CLEAR -REGEXP PAT1 PAT2 can be used to match all patterns using regular expressions. This option only clears variables. For more information on using regular expressions, type "doc regexp" at the command prompt.

If X is global, CLEAR X removes X from the current workspace, but leaves it accessible to any functions declaring it global.
CLEAR GLOBAL X completely removes the global variable X.
CLEAR GLOBAL -REGEXP PAT removes global variables that match regular expression patterns.
Note that to clear specific global variables, the GLOBAL option must come first. Otherwise, all global variables will be cleared.

CLEAR FUN clears the function specified. If FUN has been locked by MLOCK it will remain in memory. Use a partial path (see PARTIALPATH) to distinguish between different overloaded versions of FUN. For instance, 'clear inline/display' clears only the INLINE method for DISPLAY, leaving any other implementations in memory.

CLEAR ALL, CLEAR FUN, or CLEAR FUNCTIONS also have the side effect of removing debugging breakpoints and reinitializing persistent variables since the breakpoints for a function and persistent variables are cleared whenever the m-file changes or is cleared.

Use the functional form of CLEAR, such as CLEAR('name'), when the variable name or function name is stored in a string.

Examples for pattern matching:
```
    clear a*                    % Clear variables starting with "a"
    clear -regexp ^b\d{3}$      % Clear variables starting with "b" and
                                %    followed by 3 digits
    clear -regexp \d            % Clear variables containing any digits
```

See also clearvars, who, whos, mlock, munlock, persistent.

```
Reference page in Help browser
doc clear
```

Obviously, the `clear` command does more than just delete variables. Its other uses will become apparent as you become familiar with more of MATLAB's features.

3.3 MEMORY MANAGEMENT

MATLAB allocates memory for variables as they are created and for M-file functions as they are used. Depending on the computer on which the program is installed, it is possible for MATLAB to run out of memory, making it impossible to do any further work. When you eliminate variables by using the `clear` command, MATLAB frees up the memory used by the variables. Over time, however, it is possible for memory to become fragmented, leaving MATLAB's memory space populated by variables surrounded by numerous small fragments of free memory. Since MATLAB always stores variables in contiguous chunks of memory, these fragments of free memory may not be reusable. To alleviate this problem, the `pack` command performs memory garbage collection. The command saves all MATLAB workspace variables to disk, clears all variables from the workspace, and then reloads the variables back into the workspace. On completion, all fragments of free memory are consolidated into one large, usable block. Depending on how much memory is allocated to MATLAB on your computer, how long you've been running a particular MATLAB session, and how many variables you've created, you may or may not ever need to use the `pack` command.

3.4 NUMBER DISPLAY FORMATS

When MATLAB displays numerical results, it follows several rules. By default, if a result is an integer, MATLAB displays it as an integer. Likewise, when a result is a real number, MATLAB displays it with approximately four digits to the right of the decimal point. If the significant digits in the result are outside of this range, MATLAB displays the result in scientific notation, similar to the display of a scientific calculator. You can override this default behavior by specifying a different numerical format. From the *Command* window **File** menu, choose the **Preferences** menu. Alternatively, type the appropriate MATLAB `format` command at the prompt. With the special variable `pi`, the numerical display formats produced by different `format` selections are as follows:

MATLAB Command	pi	Comments
format short	3.1416	5 digits
format long	3.141592653589793	16 digits

`format short e` `format shorte`	3.1416e+00	5 digits plus a 2-digit exponent
`format long e` `format longe`	3.141592653589793e+00	16 digits plus a 2-digit exponent
`format short eng` `format shorteng`	3.1416e+000	At least 5 digits plus a 3-digit exponent that is a multiple of three
`format long eng` `format longeng`	3.14159265358979e+000	15 significant digits plus a 3-digit exponent that is a multiple of three
`format short g`	3.1416	Best of `format short` or `format short e` with 5 digits
`format long g`	3.14159265358979	Best of `format long` or `format long e` with 15 digits
`format hex`	400921fb54442d18	Hexadecimal, floating point
`format bank`	3.14	2 decimal digits
`format +`	+	Positive (+), negative (−), or blank ( )
`format rat`	355/113	Rational approximation

`format debug`	`Structure address = 1214830`
	
	`m = 1`
	`n = 1`
	`pr = 11d60d0`
	`pi = 0`
	`    3.1416`
		Internal storage information in addition to `short g`

Note: In MATLAB, the internal representation of a number does not change when different display formats are chosen; only the display changes. All calculations using double-precision numbers are performed using double-precision arithmetic.

3.5 SYSTEM INFORMATION

MATLAB provides a number of commands that provide information about the computer in use, as well as about the MATLAB version in use. The command `computer` returns a character string identifying the computer in use:

```
>> computer

ans =

GLNX86
```

In this case, the computer is a PC that is running 32-bit *Linux*. The command version returns a character string identifying the MATLAB version:

```
>> version

ans =

7.12.0.635  (R2011a)
```

Note that the command ver returns information about MATLAB, as well as installed toolboxes:

```
>> ver
--------------------------------------------------------------------------------
MATLAB Version 7.12.0.635 (R2011a)
MATLAB License Number: 123456
Operating System: Linux 2.6.38.8-35.fc15.i686 #1 SMP Wed Jul 6 14:46:26 UTC 2011 i686
Java VM Version: Java 1.6.0_17-b04 with Sun Microsystems Inc. Java HotSpot(TM) Client
VM mixed mode
--------------------------------------------------------------------------------
MATLAB                          Version 7.12            (R2011a)
Mastering MATLAB Toolbox        Version 6.0
```

MATLAB licensing information can be found by using the commands license and hostid:

```
>> license

ans =

123456

>> hostid
   'no_file'
```

Of course, your results from entering these commands will most likely be different from those shown above since your computer and MATLAB version differ from those used to produce this text. The hostid command returns information about your license server. In this example, the license is node-locked and does not require a separate license server.

3.6 THE MATLAB SEARCH PATH

MATLAB uses a ***search path*** to find information stored in files on disk. MATLAB's files are organized into numerous directories and subdirectories.

> The list of all directories where MATLAB's files are found is called the MATLAB search path or simply MATLAB path.

Use of the MATLAB search path is described next. When you enter `cow` at the MATLAB prompt by typing `>> cow`, MATLAB does the following:

1. It checks to see if `cow` is a ***variable*** in the MATLAB workspace.
2. If not, checks to see if `cow` is a built-in function.
3. If not, searches for a file named `cow.m` in the ***current directory***.
4. If none exists, checks to see if `cow.m` exists anywhere on the ***MATLAB search path*** by searching in the order in which the path is specified.
5. If `cow` isn't found at this point, MATLAB reports an error.

MATLAB takes appropriate action according to this search strategy. If `cow` is a variable, MATLAB uses the variable. If `cow` is a built-in function or if `cow.m` is a file either in the current directory or anywhere on the MATLAB search path, the built-in function is executed or the file `cow.m` is opened, and MATLAB acts on what is found in the file. As is documented in Chapters 4 and 12, MATLAB has two basic file types that are of common use; both are simple text files containing MATLAB commands. (See Chapters 4 and 12 for further information regarding these ***M-files***.)

In reality, the search procedure is more complicated than just described, because of advanced features in MATLAB. For the most part, however, this search procedure provides sufficient detail for basic MATLAB work. (More detailed information regarding the MATLAB search path can be found in Chapter 12.)

When MATLAB starts up, it defines a default MATLAB search path that points to all of the directories in which MATLAB stores its files. This search path can be displayed and modified in several ways. The easiest way is to use the ***Path Browser***, which is a graphical user interface designed for viewing and modifying the MATLAB search path. The path browser is made available by choosing **Set Path . . .** from the **File** menu on the *MATLAB* desktop window. Since the MATLAB search path already points to all directories where MATLAB stores its files, the primary purpose for accessing the path browser is to add your own MATLAB file storage directories to the search path.

To display the MATLAB search path in the *Command* window, MATLAB provides the functions `path` and `matlabpath`. In addition, the features of the path browser can be duplicated in the *Command* window by using the functions `path`, `addpath`, and `rmpath`. For more information regarding these functions, see the online documentation.

4

Script M-files

For simple problems, entering your requests at the MATLAB prompt in the *Command* window is fast and efficient. However, as the number of commands increases, or when you wish to change the value of one or more variables and reevaluate a number of commands, typing at the MATLAB prompt quickly becomes tedious. MATLAB provides a logical solution to this problem. It allows you to place MATLAB commands in a simple text file and then tells MATLAB to open the file and evaluate the commands exactly as it would have if you had typed the commands at the MATLAB prompt. These files are called *script files* or *M-files*. The term *script* signifies that MATLAB simply reads from the script found in the file. The term *M-file* means that script filenames must end with the extension '.m', as in, for example, example1.m.

4.1 SCRIPT M-FILE USE

To create a script M-file, click on the blank page icon on the MATLAB desktop toolbar, or choose **New** from the **File** menu and select **M-file**. This procedure brings up a text editor window wherein you can enter MATLAB commands. The following script M-file shows the commands from an example considered earlier:

```
% script M-file example1.m
  pens = 5; % number of each item
  notebooks = 7;
  scissors = 1;
```

```
items = pens + notebooks + scissors
cost = pens*30 + notebooks*60 + scissors*70
average_cost = cost/items
```

This file can be saved to disk and executed immediately by (1) choosing **Save and Run** from the **Debug** menu, (2) pressing the **Save and Run** button on the *Editor* toolbar, or (3) simply pressing the function key **F5**. Alternatively, you can save this file as the M-file example1.m on your disk by choosing **Save** from the **File** menu; then, at the MATLAB prompt, it's just a matter of typing the name of the script file without the .m extension:

```
>> example1
items =
     13
cost =
   640
average_cost =
       49.2308
```

When MATLAB interprets the example1 statement, it follows the search path described in Chapter 3. In brief, MATLAB prioritizes current MATLAB variables ahead of M-file names. If example1 is not a current MATLAB variable or a built-in function name, MATLAB opens the file example1.m (if it can find it) and evaluates the commands found there just as if they had been entered directly at the *Command* window prompt. As a result, commands within the M-file have access to all of the variables in the MATLAB workspace, and all of the variables created by the M-file become part of the workspace. Normally, the M-file commands are not displayed as they are evaluated. The echo on command tells MATLAB to display, or echo, commands to the *Command* window as they are read and evaluated. You can probably guess what the echo off command does. Similarly, the command echo by itself toggles the echo state.

This ability to create script M-files makes it simple to answer "what-if?" questions. For example, you could repeatedly open the example1.m M-file, change the number of pens, notebooks, or scissors, and then save and execute the file.

The utility of MATLAB comments is readily apparent when you use script files, as shown in example1.m. Comments allow you to document the commands found in a script file, so that you do not forget the commands when you view them in the future. In addition, the use of semicolons at the ends of lines to suppress the display of results allows you to control script-file output, so that only important results are shown.

Because of the utility of script files, MATLAB provides several functions that are particularly helpful when used in M-files:

Function	Description
beep	Makes computer sound a beep
disp(variablename)	Displays results without identifying variable names
echo	Controls *Command* window echoing of script file contents as they are executed
input	Prompts user for input
keyboard	Temporarily gives control to keyboard. (Type return to return control to the executing script M-file.)
pause or pause(n)	Pauses until user presses any keyboard key, or pauses for n seconds and then continues
waitforbuttonpress	Pauses until user presses any keyboard key or any mouse button over a figure window

When a MATLAB command is not terminated by a semicolon, the results are displayed in the *Command* window, with the variable name identified. For a prettier display, it is sometimes convenient to suppress the variable name. In MATLAB, this is accomplished with the command disp:

```
>> items
items =
     13
>> disp(items)
     13
```

Rather than repeatedly edit a script file for computations for a variety of cases, you can employ the input command to prompt for input as a script file is executed. For example, reconsider the example1.m script file, with the following modifications:

```
% script M-file example1.m

pens = 5;  % Number of each item
notebooks = 7;
scissors = input('Enter the number of pairs of scissors purchased > ');
items = pens + notebooks + scissors
```

```
cost = pens*30 + notebooks*60 + scissors*70
average_cost = cost/items
```

Running this script M-file produces this result:

```
>> example1
Enter the number of pairs of scissors purchased > 1
items =
     13
cost =
    640
average_cost =
        49.2308
```

In response to the prompt, the number 1 was entered and the **Return** or **Enter** key was pressed. The remaining commands were evaluated as before. The function input accepts any valid MATLAB expression for input. For example, running the script file again and providing different input gives the following result:

```
>> example1
Enter the number of pairs of scissors purchased > round(sqrt(5))-1
items =
     13
cost =
    640
average_cost =
        49.2308
```

In this case, the number of pairs of scissors was set equal to the result of evaluating the expression

```
round(sqrt(5))-1.
```

To see the effect of the echo command, add it to the script file and execute it:

```
% script M-file example1.m
echo on
```

```
pens = 5;  % Number of each item
notebooks = 7;
scissors = input('Enter the number of pairs of scissors purchased > ');
items = pens + notebooks + scissors
cost = pens*40 + notebooks*50 + scissors*30
average_cost = cost/items
echo off
```

```
>> example1
pens = 5; % Number of each item
notebooks = 7;
scissors = input('Enter the number of pairs of scissors purchased > ');
Enter the number of pairs of scissors purchased > 3
items = pens + notebooks + scissors
items =
    13
cost = pens*40 + notebooks*50 + scissors*30
cost =
   640
average_cost = cost/items
average_cost =
     42.667
echo off
```

As you can see, the echo command made the result much harder to read. On the other hand, the echo command can be very helpful when debugging more complicated script files.

4.2 BLOCK COMMENTS AND CODE CELLS

In prior versions of MATLAB, comments were line oriented. That is, comments began with an initial percent sign and continued to the end of the current line. To continue the comments on the next line required another initial percent sign.

Therefore, a block of comments would all start with initial percent signs, as shown here:

```
% This is an example of multiple line comments
% in an M-file. Each line requires an initial % sign, or MATLAB
% assumes the line contains code to be executed.
```

While the block of comments is visually simple, it can become cumbersome to manage later, when comment text is augmented or edited. In this case, the percent signs must remain at the beginning of the lines, and the comment text must flow after the initial percent signs on each line. In the past, to make it less cumbersome, the MATLAB editor included commands for adding or removing the initial percent signs on a highlighted block of lines. Now, MATLAB 7 and above supports *block comments* through the use of the syntax %{ and %}. These symbols mark the beginning and end, respectively, of a block of text to be treated as comments by MATLAB. For example, using block comment syntax on the previous example produces

```
%{
This is an example of multiple line comments
in an M-file. Each line requires an initial % sign or MATLAB
assumes the line contains code to be executed.
(Now lines can be added and edited as desired without having to
place percent signs at the beginning of each line.)
%}
```

In addition to their utility in composing multiline comments, block comments will allow you to rapidly turn on and off the interpretation and execution of any number of lines of code in an M-file. This feature is particularly helpful in creating and debugging large M-files. Simply adding %{ before and %} after a block of MATLAB code turns the enclosed code into comments that are not processed by MATLAB. When this feature is used, different sections of a script file can be executed at different times during the editing and debugging processes.

In the past, the MATLAB editor offered commands for executing a block of highlighted code in an editor window. In MATLAB 7 and above, the editor now supports the selective execution of M-file code through the use of **code cells**. A code cell is simply a block of M-file code that starts with a comment line containing two percent signs followed by a space (i.e., % %). The code cell continues to the end of the M-file or to the beginning of another code cell. From within the MATLAB editor, cells can be created, individually executed, and sequentially executed, thereby enabling effective M-file debugging. The **Cell** menu in the *Editor* window facilitates these operations. It is important to note that the syntax for code cells is interpreted by the editor, not by the MATLAB command interpreter. As a result, when an M-file is executed after its name has been entered in the *Command* window, code-cell syntax is ignored, and all executable lines in the file are processed.

4.3 SETTING EXECUTION TIME

When the name of an M-file is typed at the *Command* window prompt, the M-file is executed immediately. In some circumstances (such as large programming projects, or situations wherein data to be manipulated becomes available only over a long period of time while MATLAB is running), it is convenient to specify execution times for M-files. In MATLAB, this capability is accomplished by using *timer objects*. A timer object is created using the function `timer`. For example,

```
>> my_timer = timer('TimerFcn','MfileName','StartDelay',100)
```

creates the timer object stored in the variable `my_timer`, which executes the M-file *MfileName* 100 seconds after the timer has been started by using the `start` function:

```
>> start(my_timer) % start the timer in the variable my_timer
```

In general terms, timer function syntax has the form

```
>> t=timer('PropertyName1',PropertyValue1,'PropertyName2',PropertyValue2,...)
```

where the arguments identify property names as character strings paired with corresponding property values.

Neither MATLAB nor the computer operating system is held up while a timer is running, but not executing code. MATLAB and other programs can be used during this idle time. When the timer object initiates code execution, the timer takes control of MATLAB and executes the code, just as if the code had been typed at the MATLAB prompt. When this execution ends, MATLAB returns control to the *Command* window prompt.

The timer object has many more features. For example, `'MfileName'` can be any statement executable at the MATLAB prompt. It can be a script M-file, a function handle or function M-file, or a set of MATLAB commands. You also can specify that timer code be executed on a periodic basis or executed a specified number of times. You can specify four different M-files or code sequences to be executed under different conditions. For example,

```
>> my_timer = timer('TimerFcn','Mfile1',...
                    'StartFcn','Mfile2',...
                    'StopFcn,, 'Mfile3',...
                    'ErrorFcn','Mfile4');
```

creates a timer function that executes (1) `'Mfile1'` as the primary timer code that may be repeatedly executed, (2) `'Mfile2'` when the timer is started with the `start` function, (3) `'Mfile3'` when the timer is stopped with the `stop` function, and (4) `'Mfile4'` if a MATLAB error occurs while any of these functions are executing. (See the MATLAB documentation for more complete information about timer objects.)

4.4 STARTUP AND FINISH

When MATLAB starts up, it executes two script M-files, `matlabrc.m` and `startup.m`. The first, `matlabrc.m`, comes with MATLAB and generally should not be modified. The commands in this M-file set the default *Figure* window size and placement, as well as a number of other default features. The default MATLAB search path is set by retrieving the script file `pathdef.m` from `matlabrc.m`. The *Path Browser* and *Command* window functions for editing the MATLAB search path maintain the file `pathdef.m`, so there is no need to edit it with a text editor.

Commands in `matlabrc.m` check for the existence of the script M-file `startup.m` on the MATLAB search path. If the M-file exists, the commands in it are executed. This optional M-file `startup.m` typically contains commands that add personal default features to MATLAB. For example, it is common to put one or more `addpath` or `path` commands in `startup.m` to append additional directories to the MATLAB search path. Similarly, the default number display format can be changed (e.g., `format compact`). Since `startup.m` is a standard script M-file, there are no restrictions as to what commands can be placed in it. (However, it's probably not wise to include the command `quit` in `startup.m`!) On single-user installations, `startup.m` is commonly stored in the `toolbox/local` subdirectory on the MATLAB path. On network installations, a convenient location for your `startup.m` file is the default directory where you start MATLAB sessions.

When you terminate MATLAB via the **Exit MATLAB** item on the **File** menu in the *MATLAB* desktop window (or by typing `exit` or `quit` at the MATLAB prompt), MATLAB searches the MATLAB path for a script M-file named `finish.m`. If one is found, the commands in it are executed before MATLAB terminates. For example, the following `finish.m` prompts the user for confirmation using a dialog box before quitting, and the command `quit cancel` provides a way to cancel quitting:

```
%FINISH Confirm Desire for Quitting MATLAB

question = 'Are You Sure You Want To Quit?';
button = questdlg(question,'Exit Request','Yes','No','No');

switch button
case 'No'
    quit cancel; % how to cancel quitting!
end
% 'Yes' lets script and MATLAB end.
```

5

Arrays and Array Operations

All of the computations considered to this point have involved single numbers called scalars. Operations involving scalars are the basis of mathematics. At the same time, when we wish to perform the same operation on more than one number at a time, performing repeated scalar operations is time consuming and cumbersome. To solve this problem, MATLAB defines operations on data arrays.

5.1 SIMPLE ARRAYS

Consider the problem of computing values of the sine function over one half of its period, namely, $y = \sin(x)$ over $0 \leq x \leq \pi$. Since it is impossible to compute $\sin(x)$ at all points over this range (there are an infinite number of them), we must choose a finite number of points. In doing so, we sample the function. To pick a number, let's evaluate $\sin(x)$ every 0.1π in this range; that is, let $x = 0, 0.1\pi, 0.2\pi, \ldots, 1.0\pi$. If you were using a scientific calculator to compute these values, you would start by making a list, or an array, of the values of x. Then, you would enter each value of x into your calculator, find its sine, and write down the result as the second array y. Perhaps you would write the arrays in an organized fashion, as follows:

x	0	0.1π	0.2π	0.3π	0.4π	0.5π	0.6π	0.7π	0.8π	0.9π	π
y	0	0.31	0.59	0.81	0.95	1	0.95	0.81	0.59	0.31	0

As shown, x and y are ordered lists of numbers; that is, the first value or element in y is associated with the first value or element in x, the second element in y is associated with the second element in x, and so on. Because of this ordering, it is common to refer to individual values or elements in x and y by using subscripts; for example, x_1 is the first element in x, y_5 is the fifth element in y, and x_n is the nth element in x.

MATLAB handles arrays in a straightforward, intuitive way. Creating arrays is easy—just follow the preceding visual organization to create the following array:

```
>> x = [0 .1*pi .2*pi .3*pi .4*pi .5*pi .6*pi .7*pi .8*pi .9*pi pi]
x =
  Columns 1 through 7
        0    0.3142    0.6283    0.9425    1.2566    1.5708    1.8850
  Columns 8 through 11
   2.1991    2.5133    2.8274    3.1416
>> y = sin(x)
y =
  Columns 1 through 7
        0    0.3090    0.5878    0.8090    0.9511    1.0000    0.9511
  Columns 8 through 11
   0.8090    0.5878    0.3090    0.0000
```

To create an array in MATLAB, all you have to do is start with a left bracket, enter the desired values separated by spaces (or commas), and then close the array with a right bracket. Notice that finding the sine of the values in x follows naturally. MATLAB understands that you want to find the sine of each element in x and place the results in an associated array called y. This fundamental capability makes MATLAB different from other computer languages.

Since spaces separate array values, complex numbers entered as array values cannot have embedded spaces, unless the expressions are enclosed in parentheses. For example, [1 -2i 3 4 5+6i] contains five elements, whereas the identical arrays [(1-2i) 3 4 5+6i] and [1-2i 3 4 5+6i] contain four.

5.2 ARRAY ADDRESSING OR INDEXING

In the previous example, since x has more than one element (it has 11 values separated into columns), MATLAB gives you the result with the columns identified. As shown, x is an array having 1 row and 11 columns; or in mathematical jargon, it is a row vector, a 1-by-11 array, or simply an array of length 11.

In MATLAB, individual array elements are accessed by using subscripts; for example, x(1) is the first element in x, x(2) is the second element in x, and so on. The following code is illustrative:

```
>> x(3) % The third element of x
ans =
    0.6283
>> y(5) % The fifth element of y
ans =
    0.9511
```

To access a block of elements at one time, MATLAB provides *colon notation*:

```
>> x(1:5)
ans =
         0    0.3142    0.6283    0.9425    1.2566
```

These are the first through fifth elements in x. The notation 1:5 says, start with 1 and count up to 5. The code

```
>> x(7:end)
ans =
      1.8850        2.1991        2.5133        2.8274        3.1416
```

starts with the seventh element and continues to the last element. Here, the word end signifies the last element in the array x. In the code

```
>> y(3:-1:1)
ans =
    0.5878      0.3090         0
```

the results contain the third, second, and first elements in reverse order. The notation 3:-1:1 says, start with 3, count down by 1, and stop at 1. Similarly, the results in the code

```
>> x(2:2:7)
ans =
    0.3142    0.9425    1.5708
```

consists of the second, fourth, and sixth elements in x. The notation 2:2:7 says, start with 2, count up by 2, and stop when you get to 7. (In this case adding 2 to 6 gives 8, which is greater than 7, and so the eighth element is not included.) The code

```
>> y([8 2 9 1])
ans =
    0.8090    0.3090    0.5878    0
```

uses another array, [8 2 9 1], to extract the elements of the array y in the order we wanted them! The first element taken is the eighth, the second is the second, the third is the ninth, and the fourth is the first. In reality, [8 2 9 1] itself is an array that addresses the desired elements of y. Note that in the code

```
>> y([1 1 3 4 2 2])
ans =
    0    0    0.5878    0.8090    0.3090    0.3090
```

there is no requirement that the array used as an index contain unique elements. This allows you to rearrange and duplicate array elements arbitrarily. Using this feature leads to efficient MATLAB coding.

Addressing one array with another works as long as the addressing array contains integers between 1 and the length of the array. In contrast, consider the following code:

```
>> y(3.2)
??? Subscript indices must either be real positive integers or logicals.
>> y(11.6)
??? Subscript indices must either be real positive integers or logicals.
>> y(12)
??? Index exceeds matrix dimensions.
```

In these examples, MATLAB simply returns an error when the index is noninteger and returns a statement that no such value exists when the index exceeds the dimensions of the variable. In all cases, no numerical output appears.

5.3 ARRAY CONSTRUCTION

Earlier, we entered the values of x by typing each individual element in x. While this is fine when there are only 11 values in x, what if there were 111 values? Two other ways of entering x are as follows:

```
>> x = (0:0.1:1)*pi
x =
```

```
  Columns 1 through 7
            0    0.3142    0.6283    0.9425    1.2566    1.5708    1.8850
  Columns 8 through 11
     2.1991    2.5133    2.8274    3.1416
>> x = linspace(0,pi,11)

x =
  Columns 1 through 7
            0    0.3142    0.6283    0.9425    1.2566    1.5708    1.8850
  Columns 8 through 11
     2.1991    2.5133    2.8274    3.1416
```

In the first case, the colon notation (0:0.1:1) creates an array that starts at 0, increments (or counts) by 0.1, and ends at 1. Each element in the array is then multiplied by π to create the desired values in x. In the second case, the MATLAB function linspace is used to create x. This function's arguments are described by

```
linspace(first_value,last_value,number_of_values)
```

Both of these array creation forms are common in MATLAB. The colon notation form allows you to directly specify the increment between data points, but not the number of data points. Using linspace, on the other hand, allows you to directly specify the number of data points, but not the increment between the data points.

Both of these array creation forms result in arrays in which the individual elements are linearly spaced with respect to each other. For the special case where a logarithmically spaced array is desired, MATLAB provides the logspace function:

```
>> logspace(0,2,11)

ans =
  Columns 1 through 7
     1.0000    1.5849    2.5119    3.9811    6.3096    10.0000    15.8489
  Columns 8 through 11
    25.1189   39.8107   63.0957  100.0000
```

Here, we created an array starting at 10^0, ending at 10^2, and containing 11 values. The function arguments are described by

```
logspace(first_exponent,last_exponent,number_of_values)
```

Although it is common to begin and end at integer powers of 10, logspace works equally well when nonintegers are used as the first two input arguments.

When using colon notation, or the functions linspace or logspace, there is often a temptation to enclose expressions in brackets:

```
>> a = [1:7]
a =
     1     2     3     4     5     6     7
>> b = [linspace(1,7,5)]
b =
     1.0000      2.5000      4.0000      5.5000      7.0000
```

Although using brackets does not change the results and may add clarity to the statements, the added brackets force MATLAB to do more work and take more time, because brackets signify a concatenation operation. In the preceding examples, no concatenation is performed, and so there's no need to ask MATLAB to take the time to consider that possibility.

Parentheses do not signify concatenation and therefore do not slow MATLAB down. As a result, parentheses can be used as needed:

```
>> a = (1:7)' % change row to column
a =
     1
     2
     3
     4
     5
     6
     7
```

Sometimes, an array is required that is not conveniently described by a linearly or logarithmically spaced element relationship. There is no uniform way to create these arrays. However, array addressing and the ability to combine expressions can help eliminate the need to enter individual elements, one at a time. For example,

```
>> a = 1:5,  b = 1:2:9
a =
     1     2     3     4     5
b =
     1     3     5     7     9
```

creates two arrays. Remember that multiple statements can appear on a single line if they are separated by commas or semicolons. The code

```
>> c = [b a]
c =
    1    3    5    7    9    1    2    3    4    5
```

creates an array c composed of the elements of b, followed by those of a, while the code

```
>> d = [a(1:2:5) 1 0 1]
d =
    1    3    5    1    0    1
```

creates an array d composed of the first, third, and fifth elements of a, followed by three additional elements.

The simple array construction features of MATLAB are summarized in the following table:

Array Construction Technique	Description
x=[2 2*pi sqrt(2) 2-3j]	Creates row vector x containing arbitrary elements
x=first:last	Creates row vector x starting with first, counting by 1, and ending at or before last (Note that x=[first:last] produces the same result, but takes longer, since MATLAB considers both bracket and colon array-creation forms.)
x=first:increment:last	Creates row vector x starting with first, counting by increment, and ending at or before last
x=linspace(first,last,n)	Creates linearly spaced row vector x starting with first, ending at last, having n elements
x=logspace(first,last,n)	Creates logarithmically spaced row vector x starting with 10^{first}, ending at 10^{last}, and having n elements

5.4 ARRAY ORIENTATION

In the preceding examples, arrays contained one row and multiple columns. As a result of this row orientation, the arrays are commonly called row vectors. It is also possible for an array to be a column vector, having one column and multiple rows. In this case, all of the previous array manipulation and mathematics apply without change. The only difference is that results are displayed as columns, rather than as rows.

Since the array creation functions previously illustrated all create row vectors, there must be some way to create column vectors. The most straightforward way to create a column vector is to specify it, element by element, and by using *semicolons* to separate values:

```
>> c = [1;2;3;4;5]
c =
      1
      2
      3
      4
      5
```

According to this example, separating elements by spaces or commas specifies elements in different columns, whereas separating elements by semicolons specifies elements in different rows.

To create a column vector using the colon notation start:increment:end or the functions linspace and logspace, you must *transpose* the resulting row into a column by using the MATLAB transpose operator ('). For example,

```
>> a = 1:5
a =
      1     2     3     4     5
```

creates a row vector using the colon notation format. The code

```
>> b = a'
b =
      1
      2
      3
      4
      5
```

uses the transpose operator to change the row vector a into the column vector b:

```
>> w = b'
w =
      1     2     3     4     5
```

This statement applies the transpose again and changes the column back into a row.

In addition to performing the simple transpose operations just described, MATLAB also offers a transpose operator with a preceding dot. In this case, the **dot-transpose operator** is interpreted as the noncomplex conjugate transpose. When an array is complex, the transpose (') gives the complex conjugate transpose; that is, the sign on the imaginary part is changed, as part of the transpose operation. On the other hand, the dot-transpose (. ') transposes the array, but does not conjugate it. The code

```
>> c = a.'
c =
      1
      2
      3
      4
      5
```

shows that . ' and ' are identical for real data, while the code

```
>> d = complex(a,a)
d =
  Columns 1 through 4
   1.0000 + 1.0000i  2.0000 + 2.0000i  3.0000 + 3.0000i  4.0000 + 4.0000i
  Column 5
   5.0000 + 5.0000i
```

creates a simple complex row vector from the array a by using the function complex. The code

```
>> e = d'
e =
    1.0000 - 1.0000i
    2.0000 - 2.0000i
    3.0000 - 3.0000i
    4.0000 - 4.0000i
    5.0000 - 5.0000i
```

creates a column vector e that is the complex conjugate transpose of d, while the code

```
>> f = d.'
f =
   1.0000 + 1.0000i
   2.0000 + 2.0000i
   3.0000 + 3.0000i
   4.0000 + 4.0000i
   5.0000 + 5.0000i
```

creates a column vector f that is the transpose of d.

  If an array can be a row vector or a column vector, it makes intuitive sense that arrays can have both multiple rows and multiple columns. That is, arrays can also be in the form of matrices. The creation of matrices follows the creation of row and column vectors. ***Commas or spaces are used to separate elements in a specific row, and semicolons are used to separate individual rows:***

```
>> g = [1 2 3 4;5 6 7 8]
g =
   1   2   3   4
   5   6   7   8
```

Here, g is an array or matrix having 2 rows and 4 columns; that is, it is a 2-by-4 matrix, or it is a matrix of dimension 2 by 4. The semicolon tells MATLAB to start a new row between the 4 and the 5. Note the use of line breaks in the following code:

```
>> g = [1 2 3 4
5 6 7 8
9 10 11 12]
g =
   1    2    3    4
   5    6    7    8
   9   10   11   12
```

Thus, in addition to using semicolons, pressing the **Return** or **Enter** key while entering an array tells MATLAB to start a new row. MATLAB strictly enforces the fact that all rows must contain the same number of columns:

```
>> h = [1 2 3;4 5 6 7]
??? Error using ==> vertcat
CAT arguments dimensions are not consistent.
```

5.5 SCALAR–ARRAY MATHEMATICS

In the first array example given, the array x is multiplied by the scalar π. Other simple mathematical operations between scalars and arrays follow the same natural interpretation. Addition, subtraction, multiplication, and division by a scalar simply apply the operation to all elements of the array. For example, the code

```
>> g-2
ans =
    -1    0    1    2
     3    4    5    6
     7    8    9   10
```

subtracts 2 from each element in g, while the code

```
>> 2*g - 1
ans =
     1    3    5    7
     9   11   13   15
    17   19   21   23
```

multiplies each element in g by 2 and subtracts 1 from each element of the result. Finally, the code

```
>> 2*g/5 + 1
ans =
    1.4000    1.8000    2.2000    2.6000
    3.0000    3.4000    3.8000    4.2000
    4.6000    5.0000    5.4000    5.8000
```

multiplies each element of g by 2, then divides each element of the result by 5, and finally adds 1 to each element.

Note that scalar–array mathematics uses the same order of precedence used in scalar expressions to determine the order of evaluation.

5.6 ARRAY–ARRAY MATHEMATICS

Mathematical operations between arrays are not quite as simple as those between scalars and arrays. Clearly, array operations between arrays of different sizes or dimensions are difficult to define and are of even more dubious value. However, when two arrays have the same dimensions, addition, subtraction, multiplication, and division apply on an element-by-element basis in MATLAB, as in the following example:

```
>> g % recall previous array
g =
   1    2    3    4
   5    6    7    8
   9   10   11   12
>> h = [1 1 1 1;2 2 2 2;3 3 3 3] % create new array
h =
   1    1    1    1
   2    2    2    2
   3    3    3    3
>> g + h % add h to g on an element-by-element basis
ans =
    2    3    4    5
    7    8    9   10
   12   13   14   15
>> ans - h % subtract h from the previous answer to get g back
ans =
   1    2    3    4
   5    6    7    8
   9   10   11   12
>> 2*g - h % multiplies g by 2 and subtracts h from the result
ans =
    1    3    5    7
    8   10   12   14
   15   17   19   21
>> 2*(g-h) % use parentheses to change order of operation
ans =
    0    2    4    6
    6    8   10   12
   12   14   16   18
```

Note that array–array mathematics uses the same order of precedence used in scalar expressions to determine the order of evaluation. Note also that parentheses can be used as desired to change the order of operation.

Element-by-element multiplication and division work similarly, but use slightly unconventional notation:

```
>> g.*h
ans =
      1      2      3      4
     10     12     14     16
     27     30     33     36
```

Here, we multiplied the arrays g and h, element by element by using the dot multiplication symbol .*.

> The dot preceding the standard asterisk multiplication symbol tells MATLAB to perform element-by-element array multiplication. Multiplication without the dot signifies matrix multiplication, which is discussed later.

For this particular example, matrix multiplication is not defined:

```
>> g*h
??? Error using ==> mtimes
Inner matrix dimensions must agree.
```

Element-by-element division, or dot division, also requires use of the dot symbol as follows:

```
>> g./h
ans =
1.0000    2.0000    3.0000    4.0000
2.5000    3.0000    3.5000    4.0000
3.0000    3.3333    3.6667    4.0000
>> h.\g
ans =
1.0000    2.0000    3.0000    4.0000
2.5000    3.0000    3.5000    4.0000
3.0000    3.3333    3.6667    4.0000
```

As with scalars, division is defined by using both forward and backward slashes. In both cases, the array *below* the slash is divided into the array *above* the slash.

> The dot preceding the forward or backward slash symbol tells MATLAB to perform element-by-element array division. Division without the dot signifies matrix inversion, which is discussed later.

Array or dot division also applies if the numerator is a scalar. Consider the following example:

```
>> 1./g
ans =
      1.0000     0.5000     0.3333     0.2500
      0.2000     0.1667     0.1429     0.1250
      0.1111     0.1000     0.0909     0.0833
```

In this case, the scalar 1 in the numerator is expanded to an array the same size as the denominator, and then element-by-element division is performed. That is, the preceding code represents a shorthand way of computing:

```
>> f=[1 1 1 1; 1 1 1 1; 1 1 1 1] % create numerator
f =
      1     1     1     1
      1     1     1     1
      1     1     1     1
>> f./g
ans =
      1.0000     0.5000     0.3333     0.2500
      0.2000     0.1667     0.1429     0.1250
      0.1111     0.1000     0.0909     0.0833
>> f./h
ans =
      1.0000     1.0000     1.0000     1.0000
      0.5000     0.5000     0.5000     0.5000
      0.3333     0.3333     0.3333     0.3333
```

This process of automatically expanding scalar values so that element-by-element arithmetic applies is called ***scalar expansion***. Scalar expansion is used extensively in MATLAB.

Division without the dot is the matrix division or matrix inversion operation, which is an entirely different operation, as the following code shows:

```
>> g/h
Warning: Rank deficient, rank = 1 tol = 5.3291e-015.
Ans =
          0         0      0.8333
          0         0      2.1667
          0         0      3.5000
>> h/g
Warning: Rank deficient, rank = 2, tol = 1.8757e-14.
ans =
    -0.1250         0      0.1250
    -0.2500         0      0.2500
    -0.3750         0      0.3750
```

Matrix division gives results that are not necessarily the same size as g and h. (Matrix operations are discussed in Chapter 16.)

Array exponentiation is defined in several ways. As used with multiplication and division, $\wedge$ is reserved for matrix exponentiation, and .$\wedge$ is used to denote element-by-element exponentiation. When the exponent is a scalar, the scalar is applied to each element of the array. For example,

```
>> g, h % recalls the arrays used earlier
g =
     1     2     3     4
     5     6     7     8
     9    10    11    12
h =
     1     1     1     1
     2     2     2     2
     3     3     3     3

>> g.^2
ans =
     1     4     9    16
    25    36    49    64
    81   100   121   144
```

squares the individual elements of g, whereas

```
>> g^2
??? Error using ==> mpower
Matrix must be a scalar and a square matrix.
```

is matrix exponentiation, which is defined only for square matrices—that is, matrices with equal row and column counts. The code

```
>> g.^-1
ans =
```

1.0000	0.5000	0.3333	0.2500
0.2000	0.1667	0.1429	0.1250
0.1111	0.1000	0.0909	0.0833

finds the reciprocal of each element in g. The code

```
>> 1./g
ans =
```

1.0000	0.5000	0.3333	0.2500
0.2000	0.1667	0.1429	0.1250
0.1111	0.1000	0.0909	0.0833

produces the same result as the scalar expansion approach seen earlier.
When the exponent is an array operating on a scalar, each element of the array is applied to the scalar. For example, the code

```
>> 2.^g
ans =
```

2	4	8	16
32	64	128	256
512	1024	2048	4096

raises 2 to the power of each element in the array g.
If both components are arrays of the same size, exponentiation is applied element by element. Thus,

```
>> g.^h
ans =
```

1	2	3	4
25	36	49	64
729	1000	1331	1728

raises the elements of g to the corresponding elements in h. In this case, the first row is unchanged, since the first row of h contains ones, the second row is squared, and the third row is cubed.

The following example shows that the scalar and array operations can be combined:

```
>> g.^(h-1)
ans =
```

1	1	1	1
5	6	7	8
81	100	121	144

The two forms of exponentiation that have scalar parts are further examples of scalar expansion. The results make intuitive sense if the scalars involved are first expanded to the size of the array and element-by-element exponentiation is then applied.

The following table summarizes basic array operations:

Element-by-Element Operation	**Representative Data** $A = [a_1\ a_2\ \ldots\ a_n]$, $B = [b_1\ b_2\ \ldots\ b_n]$, $c =$ <a *scalar*>
Scalar addition	$A+c = [a_1+c\ a_2+c\ \ldots\ a_n+c]$
Scalar subtraction	$A-c = [a_1-c\ a_2-c\ \ldots\ a_n-c]$
Scalar multiplication	$A*c = [a_1*c\ a_2*c\ \ldots\ a_n*c]$
Scalar division	$A/c = c \backslash A = [a_1/c\ a_2/c\ \ldots\ a_n/c]$
Array addition	$A+B = [a_1+b_1\ a_2+b_2\ \ldots\ a_n+b_n]$
Array multiplication	$A.*B = [a_1*b_1\ a_2*b_2\ \ldots\ a_n*b_n]$
Array right division	$A./B = [a_1/b_1\ a_2/b_2\ \ldots\ a_n/b_n]$
Array left division	$A.\backslash B = [a_1\backslash b_1\ a_2\backslash b_2\ \ldots\ a_n\backslash b_n]$
Array exponentiation	$A.\^{}c = [a_1\^{}c\ a_2\^{}c\ \ldots\ a_n\^{}c]$
	$c.\^{}A = [c\^{}a_1\ c\^{}a_2\ \ldots\ c\^{}a_n]$
	$A.\^{}B = [a_1\^{}b_1\ a_2\^{}b_2\ \ldots\ a_n\^{}b_n]$

5.7 STANDARD ARRAYS

Because of the general utility of standard arrays, MATLAB provides functions for creating a number of them. Standard arrays include those containing all ones or zeros or the special variables NaN or Inf. Others include identity matrices, arrays of random numbers, diagonal arrays, and arrays whose elements are a given constant. The following are examples:

```
>> ones(3)
ans =
     1     1     1
     1     1     1
     1     1     1
>> zeros(2,5)
ans =
     0     0     0     0     0
     0     0     0     0     0
>> size(g)
ans =
     3     4
>> nan(size(g))
ans =
    NaN    NaN    NaN    NaN
    NaN    NaN    NaN    NaN
    NaN    NaN    NaN    NaN
```

When called with a single input argument, for example ones(n) or zeros(n), MATLAB creates an n-by-n array containing ones or zeros, respectively. When called with two input arguments, ones(r,c), MATLAB creates an array having r rows and c columns. To create an array that is the same size as another array, use the size function (discussed later in this chapter) in the argument of the array-creation function as shown above.

In the code

```
>> eye(4)
ans =
     1     0     0     0
     0     1     0     0
```

```
        0     0     1     0
        0     0     0     1
>> eye(2,4)
ans =
        1     0     0     0
        0     1     0     0
>> eye(4,2)
ans =
        1     0
        0     1
        0     0
        0     0
```

the function eye produces identity matrices by using the same syntax style as that used to produce arrays of zeros and ones. An identity matrix or array is all zeros, except for the elements $A(i,i)$, where $i = 1:min(r,c)$, in which $min(r,c)$ is the minimum of the number of rows and columns in A. In the code

```
>> rand(3)
ans =
        0.8147     0.9134     0.2785
        0.9058     0.6324     0.5469
        0.1270     0.0975     0.9575
>> rand(1,5)
ans =
        0.9649     0.1576     0.9706     0.9572     0.4854


>> b = eye(3)
b =
        1     0     0
        0     1     0
        0     0     1
>> rand(size(b))
ans =
        0.8003     0.9157     0.6557
```

```
     0.1419      0.7922      0.0357
     0.4218      0.9595      0.8491
```

the function rand produces uniformly distributed random arrays whose elements lie between 0 and 1. On the other hand, the function randn produces arrays whose elements are samples from a *zero-mean*, *unit-variance* normal distribution.

```
>> randn(2)
ans =
     1.6302      1.0347
     0.4889      0.7269
>> randn(2,5)
ans =
    -0.3034     -0.7873     -1.1471     -0.8095      1.4384
     0.2939      0.8884     -1.0689     -2.9443      0.3252
```

The function diag creates diagonal arrays in which a vector can be placed at any location parallel to the main diagonal of an array:

```
>> a = 1:4 % start with a simple vector
a =
     1    2    3    4
>> diag(a) % place elements on the main diagonal
ans =
     1    0    0    0
     0    2    0    0
     0    0    3    0
     0    0    0    4
>> diag(a,1) % place elements 1 place up from diagonal
ans =
     0    1    0    0    0
     0    0    2    0    0
     0    0    0    3    0
     0    0    0    0    4
     0    0    0    0    0
```

```
>> diag(a,-2) % place elements 2 places down from diagonal
ans =
      0    0    0    0    0    0
      0    0    0    0    0    0
      1    0    0    0    0    0
      0    2    0    0    0    0
      0    0    3    0    0    0
      0    0    0    4    0    0
```

The function B = eps(A), where A is a numeric array, produces an array B the same size as A such that B(i,j) contains eps(abs(A(i,j))), that is, the smallest increment that can be added to the absolute value of A(i,j) to produce a number larger than abs(A(i,j)):

```
>> A = [0 1 2; -1 -2 0.5]
A =
           0      1.0000      2.0000
     -1.0000     -2.0000      0.5000
>> B = eps(A)
B =
     1.0e-15 *
      0.0000      0.2220      0.4441
      0.2220      0.4441      0.1110
```

With the preceding standard arrays, there are several ways to create an array whose elements all have the same value. Some of them are as follows:

```
>> d = pi; % choose pi for this example
>> d*ones(3,4) % slowest method (scalar-array multiplication)
ans =
       3.1416     3.1416     3.1416     3.1416
       3.1416     3.1416     3.1416     3.1416
       3.1416     3.1416     3.1416     3.1416
>> d+zeros(3,4) % slower method (scalar-array addition)
ans =
       3.1416     3.1416     3.1416     3.1416
       3.1416     3.1416     3.1416     3.1416
       3.1416     3.1416     3.1416     3.1416
```

```
>> d(ones(3,4)) % fast method (array addressing)
ans =
```

3.1416	3.1416	3.1416	3.1416
3.1416	3.1416	3.1416	3.1416
3.1416	3.1416	3.1416	3.1416

```
>> repmat(d,3,4) % fastest method (optimum array addressing)
ans =
```

3.1416	3.1416	3.1416	3.1416
3.1416	3.1416	3.1416	3.1416
3.1416	3.1416	3.1416	3.1416

For small arrays, all of the methods used in this example are fine. However, as the array grows in size, the multiplications required in the scalar multiplication approach slow the procedure down. Since addition is often faster than multiplication, the next best approach is to add the desired scalar to an array of zeros. Although they are not intuitive, the last two methods explained are the fastest for large arrays. They both involve array indexing, as described earlier.

The solution d(ones(r,c)) creates an r-by-c array of ones and then uses this array to index and duplicate the scalar d. Creating the temporary array of ones takes time and uses memory, thereby slowing down this approach, despite the fact that no floating-point mathematics are used. The solution repmat(d,r,c) calls the function repmat, which stands for replicate matrix. For scalar d, this function performs the following steps:

```
D(r*c) = d;        % a row vector whose (r*c)-th element is d
D(:) = d;          % scalar expansion to fill all elements of D with d
D = reshape(D,r,c); % reshape the vector into the desired r-by-c shape
```

This MATLAB code uses scalar expansion to create a vector having r*c elements all equal to d. The vector is then reshaped, by using the function reshape, into an r-by-c array. (The functions repmat and reshape are discussed in more detail later.)

5.8 ARRAY MANIPULATION

Because arrays are fundamental to MATLAB, there are many ways to manipulate them in MATLAB. Once arrays are formed, MATLAB provides powerful ways to insert, extract, and rearrange subsets of arrays by identifying subscripts of interest. Knowledge of these features is a key to using MATLAB efficiently. To illustrate the array-manipulation features of MATLAB, consider the following examples:

```
>> A = [1 2 3;4 5 6;7 8 9]
A =
     1    2    3
     4    5    6
     7    8    9
>> A(3,3) = 0 % set element in 3rd row, 3rd column to zero
A =
     1    2    3
     4    5    6
     7    8    0
```

The preceding code changes the element in the third row and third column to zero. The code

```
>> A(2,6) = 1 % set element in 2nd row, 6th column to one
A =
     1    2    3    0    0    0
     4    5    6    0    0    1
     7    8    0    0    0    0
```

places the number one in the second row, sixth column. Since A does not have six columns, the size of A is increased as necessary, and filled with zeros, so that the array remains rectangular.

The code

```
>> A(:,4) = 4
A =
     1    2    3    4    0    0
     4    5    6    4    0    1
     7    8    0    4    0    0
```

sets the fourth column of A equal to 4. Since 4 is a scalar, it is expanded to fill all of the elements specified. This is another example of scalar expansion. MATLAB performs scalar expansion to simplify statements that can be interpreted unambiguously.

For example, the preceding statement is equivalent to the following, more cumbersome, statement:

```
>> A(:,4) = [4;4;4]
A =
      1    2    3    4    0    0
      4    5    6    4    0    1
      7    8    0    4    0    0
```

Let's start over and look at other array manipulations. The code

```
>> A = [1 2 3;4 5 6;7 8 9]; % restore original data
>> B = A(3:-1:1,1:3)
B =
      7    8    9
      4    5    6
      1    2    3
>> B = A(end:-1:1,1:3) % same as above
B =
      7    8    9
      4    5    6
      1    2    3
```

creates an array B by taking the rows of A in reverse order. The word end automatically denotes the final or largest index for a given dimension. In this example, end signifies the largest row index 3. The code

```
>> B = A(3:-1:1,:)
B =
      7    8    9
      4    5    6
      1    2    3
```

does the same as the previous example. The final single colon means take all columns. That is, : is equivalent to 1:3 in this example, because A has three columns. The code

```
>> C = [A B(:,[1 3])]
```

```
C =

     1    2    3    7    9
     4    5    6    4    6
     7    8    9    1    3
```

creates C by appending, or concatenating, all rows in the first and third columns of B to the right of A. Similarly, the code

```
>> B = A(1:2,2:3)
B =

     2    3
     5    6
>> B = A(1:2,2:end)  % same as above
B =

     2    3
     5    6
```

creates B by extracting the first two rows and last two columns of A. Once again, colon notation is used to create index vectors that identify the array elements to extract. In the second instance just shown, end is used to denote the final or largest column index.

   The statement

```
>> C = [1 3]
C =

     1    3
>> B = A(C,C)
B =

     1    3
     7    9
```

uses the array C to index the array A, rather than specify them directly by using the colon notation start:increment:end or start:end. In this example, B is formed from the first and third rows and first and third columns of A. The code

```
>> B = A(:)
B =

     1

     4
```

```
7
2
5
8
3
6
9
```

builds B by stretching A into a column vector and taking its columns one at a time, in order, which is the simplest form of reshaping an array into an array that has different dimensions, but the same number of total elements.

In MATLAB, the statement

```
>> B = B.'
B =
    1   4   7   2   5   8   3   6   9
>> B = reshape(A,1,9) % reshape A into 1-by-9
B =
    1   4   7   2   5   8   3   6   9
>> B = reshape(A,[1 9])
B =
    1   4   7   2   5   8   3   6   9
```

illustrates the dot-transpose operation introduced earlier, as well as the function reshape. In this case, reshape works with the indices, supplied as separate function arguments or as a single-vector argument. The code

```
>> B = A % copy A into B
B =
    1   2   3
    4   5   6
    7   8   9
>> B(:,2) = []
B =
    1   3
    4   6
    7   9
```

redefines B by throwing away all rows in the second column of the original B. When you set something equal to the empty matrix or empty array [], it is deleted, causing the array to collapse into what remains. Note that you must delete whole rows or columns, so that the result remains rectangular.

The statement

```
>> C = B.'
C =
    1    4    7
    3    6    9
>> reshape(B,2,3) % reshape is not equivalent to transpose
ans =
    1    7    6
    4    3    9
```

illustrates the transpose of an array and demonstrates that reshape is not the same as transpose. The transpose operation converts the *i*th row to the *i*th column of the result, and so the original 3-by-2 array becomes a 2-by-3 array.

The code

```
>> C(2,:) = []
C =
    1    4    7
```

throws out the second row of C, leaving a row vector, while the code

```
>> A(2,:) = C
A =
    1    2    3
    1    4    7
    7    8    9
```

replaces the second row of A with C. The MATLAB code

```
>> B = A(:,[2 2 2 2]) % create new B array
B =
    2    2    2    2
    4    4    4    4
    8    8    8    8
```

```
>> B = A(:,2+zeros(1,4)) % [2 2 2 2]=2+zeros(1,4)
B =
        2     2     2     2
        4     4     4     4
        8     8     8     8
>> B = repmat(A(:,2),1,4) % replicate 2nd column into 4 columns
B =
        2     2     2     2
        4     4     4     4
        8     8     8     8
```

creates B in three ways by duplicating all rows in the second column of A four times.
The last approach is fastest for large arrays. The statement

```
>> A, C % show A and C again
A =
        1     2     3
        1     4     7
        7     8     9
C =
        1     4     7
>> A(2,2) = []
??? Subscripted assignment dimension mismatch.
```

shows that you can throw out only entire rows or columns. MATLAB simply does not
know how to collapse an array when partial rows or columns are thrown out. Finally,

```
>> C = A(4,:)
??? Attempted to access A(4,:); index out of bounds because
size(A)=[3,3].
```

produces an error since A does not have a fourth row. According to this result,
indices must adhere to the following guidelines:

If A(r,c) appears on the left-hand side of an equal sign and one or more elements
specified by (r,c) do not exist, then zeros are added to A as needed, so that A(r,c)
addresses known elements. However, on the right-hand side of an equal sign, all elements addressed by A(r,c) must exist or an error is returned.

Continuing on, consider the following:

```
>> C(1:2,:) = A
??? Subscripted assignment dimension mismatch.
```

This example shows that you can't squeeze one array into another array of a different size.

However, the statement

```
>> C(3:4,:) = A(2:3,:)
C =
     1    4    7
     0    0    0
     1    4    7
     7    8    9
```

reveals that you can place the second and third rows of A into the same-size area of C. Since the second through fourth rows of C did not exist, they are created as necessary. Moreover, the second row of C is unspecified, and so it is filled with zeros.

The code

```
>> A = [1 2 3;4 5 6;7 8 9] % fresh data
A =
     1    2    3
     4    5    6
     7    8    9
>> A(:,2:3)           % a peek at what's addressed next
ans =
     2    3
     5    6
     8    9
>> G(1:6) = A(:,2:3)
G =
     2    5    8    3    6    9
```

creates a row vector G by extracting all rows in the second and third columns of A. Note that the shapes of the matrices are different on both sides of the equal sign. The elements of A are inserted into the elements of G by going down the rows of the first column and then down the rows of the second column.

When (:) appears on the left-hand side of the equal sign, it signifies that elements will be taken from the right-hand side and placed into the array on the left-hand side without changing its shape. In the following example, this process extracts the second and third columns of A and inserts them into the column vector H:

```
>> H = ones(6,1); % create a column array
>> H(:) = A(:,2:3) % fill H without changing its shape
H =
     2
     5
     8
     3
     6
     9
```

Of course, for this code to work, both sides must address the same number of elements.

When the right-hand side of an assignment is a scalar and the left-hand side is an array, *scalar expansion* is used. For example, the code

```
>> A(2,:) = 0
A =
     1    2    3
     0    0    0
     7    8    9
```

replaces the second row of A with zeros. The single zero on the right-hand side is expanded to fill all the indices specified on the left. This example is equivalent to

```
>> A(2,:) = [0 0 0]
A =
     1    2    3
     0    0    0
     7    8    9
```

Scalar expansion occurs whenever a scalar is used in a location calling for an array. MATLAB automatically expands the scalar to fill all requested locations and then performs the operation dictated. The code

```
>> A(1,[1 3]) = pi
A =
    3.1416    2.0000    3.1416
         0         0         0
    7.0000    8.0000    9.0000
```

is yet another example in which the scalar π is expanded to fill two locations. Consider again what the function `reshape` does with scalar input. Let's create a 2-by-4 array containing the number 2:

```
>> D(2*4) = 2 % create array with 8 elements
D =
    0   0   0   0   0   0   0   2
>> D(:) = 2 % scalar expansion
D =
    2   2   2   2   2   2   2   2
>> D = reshape(D,2,4) % reshape
D =
    2   2   2   2
    2   2   2   2
```

The first line `D(2*4) = 2` causes a row vector of length 8 to be created and places 2 in the last column. Next, `D(:) = 2` uses scalar expansion to fill all elements of D with 2. Finally, the result is reshaped into the desired dimensions.

Sometimes, it is desirable to perform some mathematical operation between a vector and a two-dimensional (2-D) array. For example, consider these arrays:

```
>> A = reshape(1:12,3,4)'
A =
     1     2     3
     4     5     6
     7     8     9
    10    11    12
>> r = [3 2 1]
r =
     3     2     1
```

Suppose that we wish to subtract $r(i)$ from the ith column of A. One way of accomplishing it is as follows:

```
>> Ar = [A(:,1)-r(1) A(:,2)-r(2) A(:,3)-r(3)]
Ar =
      -2     0     2
       1     3     5
       4     6     8
       7     9    11
```

Alternatively, one can use indexing:

```
>> R = r([1 1 1 1],:) % duplicate r to have 4 rows
R =
       3     2     1
       3     2     1
       3     2     1
       3     2     1
>> Ar = A - R % now use element by element subtraction
Ar =
      -2     0     2
       1     3     5
       4     6     8
       7     9    11
```

The array R can also be computed faster and more generally by using the functions ones and size or by using the function repmat, all of which are discussed later. Consider the following example:

```
>> R = r(ones(size(A,1),1),:) % historically this is Tony's trick
R =
       3     2     1
       3     2     1
       3     2     1
       3     2     1
>> R = repmat(r,size(A,1),1) % often faster than Tony's trick
```

```
R =

     3      2      1
     3      2      1
     3      2      1
     3      2      1
```

In this example, size(A,1) returns the number of rows in A.

Sometimes it is more convenient to address array elements with a single *index*. **When a single index is used in MATLAB, the index counts down the rows, proceeding column by column starting with the first column.** Here is an example:

```
>> D = reshape(1:12,3,4) % new data
D =

     1      4      7     10
     2      5      8     11
     3      6      9     12

>> D(2)

ans =

     2

>> D(5)

ans =

     5

>> D(end)

ans =

    12

>> D(4:7)

ans =

     4      5      6      7
```

The MATLAB functions sub2ind and ind2sub perform the arithmetic to convert to and from a single index to row and column subscripts:

```
>> sub2ind(size(D),2,4) % find single index from row and column

ans =

    11

>> [r,c] = ind2sub(size(D),11) % find row and column from single index
```

r =

    2

c =

    4

The element in the second row, fourth column is the 11th indexed element. Note that these two functions need to know the size of the array to search (i.e., size(D), rather than the array D itself).

Subscripts refer to the row and column locations of elements of an array; for instance, A(2,3) refers to the element in the second row and third column of A. The *index* of an element in an array refers to its position relative to the element in the first row, first column, having index 1. Indices count down the rows first and then proceed across the columns of an array. That is, the indices count in the order in which the subscripts appear; for example, if D has three rows, D(8) is the second element in the third column of D.

In addition to addressing arrays on the basis of their subscripts, *logical arrays* that result from logical operations (to be discussed more thoroughly later) can also be used if the size of the array is equal to that of the array it is addressing. In this case, True (1) elements are retained, and False (0) elements are discarded. For example,

```
>> x = -3:3 % Create data
x =
    -3    -2    -1    0    1    2    3
>> abs(x)>1
ans =
    1    1    0    0    0    1    1
```

returns a logical array with ones (True), where the absolute value of x is greater than one and zeros (False) elsewhere. (Chapter 10 contains more detailed information.)
The statement

```
>> y = x(abs(x)>1)
y =
    -3    -2    2    3
```

creates y by taking those values of x where its absolute value is greater than 1.
Note, however, that even though abs(x)>1 produces the array [1 1 0 0 0 1 1], it is not equivalent to a numerical array containing those values; that is, the code

```
>> y = x([1 1 0 0 0 1 1])
??? Subscript indices must either be real positive integers or logicals.
```

gives an error, even though the abs(x)>1 and [1 1 0 0 0 1 1] appear to be the same vector. In the second case, [1 1 0 0 0 1 1] is a *numeric array* as opposed to a *logical array*. The difference between these two visually equal arrays can be identified by using the function class:

```
>> class(abs(x)>1) % logical result from logical comparison
ans =
logical
>> class([1 1 0 0 1 1]) % double precision array
ans =
double
```

Alternatively, the functions islogical and isnumeric return logical True (1) or False (0) to identify data types:

```
>> islogical(abs(x)>1)
ans =
     1
>> islogical([1 1 0 0 0 1 1])
ans =
     0
>> isnumeric(abs(x)>1)
ans =
     0
>> isnumeric([1 1 0 0 0 1 1])
ans =
     1
```

Because of the differences between logical and numeric arrays, MATLAB generates an error when executing x([1 1 0 0 0 1 1]) because MATLAB arrays do not have indices at zero.

Quite naturally, MATLAB provides the function logical for converting numeric arrays to logical arrays:

```
>> y = x(logical([1 1 0 0 0 1 1]))
y =
    -3    -2     2     3
```

Once again we have the desired result.
   To summarize,

> Specifying array subscripts with numerical arrays extracts the elements having the
> given numerical indices. On the other hand, specifying array subscripts with logical
> arrays, which are returned by logical expressions and the function `logical`, extracts
> elements that are logical True (1).

   Just as the functions `ones` and `zeros` are useful for creating numeric arrays,
the functions `true` and `false` are useful for creating logical arrays containing True
and False values, respectively:

```
>> true
ans =
     1
>> true(2,3)
ans =
     1     1     1
     1     1     1
>> false
ans =
     0
>> false(1,6)
ans =
     0     0     0     0     0     0
```

Although these results appear to be equal to results returned by equivalent statements that use `ones` and `zeros`, they are logical arrays, not numeric.
   Logical arrays work on 2-D arrays, as well as on vectors:

```
>> B = [5 -3;2 -4] % new data
B =
     5    -3
     2    -4
>> x = abs(B)>2 % logical result
x =
     1     1
     0     1
```

```
>> y = B(x) % grab True values
y =
          5
         -3
         -4
```

However, the final result is returned as a column vector, since there is no way to define a 2-D array having only three elements. No matter how many elements are extracted, MATLAB extracts all of the true elements by using single-index order and then forms or reshapes the result into a column vector.

The preceding array-addressing techniques are summarized in the following table:

Array Addressing	Description
A(r,c)	Addresses a subarray within A defined by the index vector of desired rows in r and an index vector of desired columns in c
A(r,:)	Addresses a subarray within A defined by the index vector of desired rows in r and all columns
A(:,c)	Addresses a subarray within A defined by all rows and the index vector of desired columns in c
A(:)	Addresses all elements of A as a column vector taken column by column. If A(:) appears on the left-hand side of the equal sign, it means to fill A with elements from the right hand-side of the equal sign without changing A's shape
A(k)	Addresses a subarray within A defined by the single-index vector k, as if A were the column vector A(:)
A(x)	Addresses a subarray within A defined by the logical array x. Note that x should be the same size as A. If x is shorter than A, the missing values in x are assumed to be False. If x is longer than A, all extra elements in x must be False.

5.9 ARRAY SORTING

Given a data vector, a common task required in numerous applications is sorting. In MATLAB, the function sort performs this task:

```
>> x = randperm(8) % new data
x =
        7    5    2    1    3    6    4    8
```

```
>> xs = sort(x) % sort ascending by default
xs =
        1    2    3    4    5    6    7    8
>> xs = sort(x,'ascend') % sort ascending
xs =
        1    2    3    4    5    6    7    8
>> [xs,idx] = sort(x) % return sort index as well
xs =
        1    2    3    4    5    6    7    8
idx =
        4    3    5    7    2    6    1    8
```

As shown, the sort function returns one or two outputs. The first is an ascending sort of the input argument, and the second is the sort index—for example, xs(k) = x(idx(k)).

Note that when a MATLAB function returns two or more variables, they are enclosed by square brackets on the left-hand side of the equal sign. This syntax is different from the array-manipulation syntax discussed previously, in which [a,b] on the right-hand side of the equal sign builds a new array, with b appended to the right of a.

In prior versions of MATLAB, sorting in the descending direction required turning around the sort function output by using array-indexing techniques. For example, consider the following code:

```
>> xsd = xs(end:-1:1)
xsd =
        8    7    6    5    4    3    2    1
>> idxd = idx(end:-1:1)
idxd =
        8    1    6    2    7    5    3    4
```

With MATLAB version 7 and above, the sort direction can be specified when calling the sort function:

```
>> xs = sort(x,'descend') % sort descending
xs =
        8    7    6    5    4    3    2    1
```

When presented with a 2-D array, `sort` acts differently. For example, in the code

```
>> A = [randperm(6);randperm(6);randperm(6);randperm(6)] % new data
A =
     1     2     5     6     4     3
     4     2     6     5     3     1
     2     3     6     1     4     5
     3     5     1     2     4     6
>> [As,idx] = sort(A)
As =
     1     2     1     1     3     1
     2     2     5     2     4     3
     3     3     6     5     4     5
     4     5     6     6     4     6
idx =
     1     1     4     3     2     2
     3     2     1     4     1     1
     4     3     2     2     3     3
     2     4     3     1     4     4
```

the `sort` function sorts, in ascending order, each column independently of the others, and the indices returned are those for each column. In many cases, we are more interested in sorting an array on the basis of the sort of a specific column. In MATLAB, this task is easy. For example, in the code

```
>> [tmp,idx] = sort(A(:,4)); % sort 4-th column only
>> As = A(idx,:) % rearrange rows in all columns using idx
As =
     2     3     6     1     4     5
     3     5     1     2     4     6
     4     2     6     5     3     1
     1     2     5     6     4     3
```

the rows of As are just the rearranged rows of A, in which the fourth column of As is sorted in ascending order.

MATLAB is a flexible program. Rather than sort each column, it is also possible to sort each row. By using a second argument to sort, you can specify in which direction to sort:

```
>> As = sort(A,2) % sort across 2-nd dimension
As =
     1    2    3    4    5    6
     1    2    3    4    5    6
     1    2    3    4    5    6
     1    2    3    4    5    6
>> As = sort(A,1) % same as sort(A)
As =
     1    2    1    1    3    1
     2    2    5    2    4    3
     3    3    6    5    4    5
     4    5    6    6    4    6
```

Since in A(r,c), the row dimension appears first, sort(A,1) means to sort down the rows. Because the column dimension appears second, sort(A,2) means to sort across the columns. Although not shown, 'descend' can be appended to the function calls (e.g., sort(A,1,'descend')), to return a descending sort. (Chapter 17 contains much more information about the sort function.)

5.10 SUBARRAY SEARCHING

Many times, it is desirable to know the indices or subscripts of the elements of an array that satisfy some relational expression. In MATLAB, this task is performed by the function find, which returns the subscripts when a relational expression is True as, for example, in the following code:

```
>> x = -3:3
x =
    -3   -2   -1    0    1    2    3
>> k = find(abs(x)>1) % finds those subscripts where abs(x)>1
k =
     1    2    6    7
```

```
>> y = x(k) % creates y using the indices in k.
y =
    -3    -2    2    3
>> y = x(abs(x)>1) % creates the same y vector by logical addressing
y =
    -3    -2    2    3
```

The find function also works for 2-D arrays:

```
>> A = [1 2 3;4 5 6;7 8 9] % new data
A =
    1    2    3
    4    5    6
    7    8    9
>> [i,j] = find(A>5) % i and j are not equal to sqrt(-1) here
i =
    3
    3
    2
    3
j =
    1
    2
    3
    3
```

Here, the indices stored in i and j are the associated row and column indices, respectively, where the relational expression is True. That is, A(i(1),j(1)) is the first element of A, where A>5, and so on.

Alternatively, find returns single indices for 2-D arrays:

```
>> k = find(A>5)
k =
    3
    6
    8
    9
```

Of the two index sets returned for 2-D arrays, this latter single-index form is often more useful:

```
>> A(k) % look at elements greater than 5
ans =
     7
     8
     6
     9

>> A(k) = 0 % set elements addressed by k to zero
A =
     1    2    3
     4    5    0
     0    0    0
>> A = [1 2 3;4 5 6;7 8 9] % restore data
A =
     1    2    3
     4    5    6
     7    8    9
>> A(i,j) % this is A([3 3 2 3],[1 2 3 3])
ans =
     7    8    9    9
     7    8    9    9
     4    5    6    6
     7    8    9    9
>> A(i,j) = 0 % this is A([3 3 2 3],[1 2 3 3]) also
A =
     1    2    3
     0    0    0
     0    0    0
```

These A(i,j) cases are not as easy to understand as the preceding single-index cases. To assume that A(k) is equivalent to A(i,j) is a common MATLAB indexing mistake. Rather, A(i,j) in the previous example is equivalent to

```
>> [A(3,1) A(3,2) A(3,3) A(3,3)
    A(3,1) A(3,2) A(3,3) A(3,3)
    A(2,1) A(2,2) A(2,3) A(2,3)
    A(3,1) A(3,2) A(3,3) A(3,3)]
ans =
      7    8    9    9
      7    8    9    9
      4    5    6    6
      7    8    9    9
```

where the first-row index in i is coupled with all of the column indices in j to form the first row in the result. Then, the second-row index in i is coupled with the column indices in j to form the second row in the result, and so on. Given the form in this example, the diagonal elements of A(i,j) are equal to those of A(k). Therefore, the two approaches are equal if the diagonal elements of A(i,j) are retained. In the code

```
>> diag(A(i,j))
ans =
      7
      8
      6
      9
```

while diag(A(i,j)) = A(k), simply entering A(k) is preferred, as it does not create an intermediate square array.

Similarly, A(i,j) = 0 is equivalent to

```
>> A(3,1)=0; A(3,2)=0; A(3,3)=0; A(3,3)=0; % i(1) with all j
>> A(3,1)=0; A(3,2)=0; A(3,3)=0; A(3,3)=0; % i(2) with all j
>> A(2,1)=0; A(2,2)=0; A(2,3)=0; A(2,3)=0; % i(3) with all j
>> A(3,1)=0; A(3,2)=0; A(3,3)=0; A(3,3)=0 % i(4) with all j
A =
      1    2    3
      0    0    0
      0    0    0
```

Based on the above equivalents, it is clear that A(i,j) is not generally as useful as A(k) for subarray searching using find.

When you only need to find a few indices, the `find` function offers an alternative syntax that avoids searching the entire array. For example, in the code

```
>> x = randperm(8) % new data
x =
      8    2    7    4    3    6    5    1
>> find(x>4) % find all values greater than 4
ans =
      1    3    6    7
>> find(x>4,1) % find first value greater than 4
ans =
      1
>> find(x>4,1,'first') % same as above
ans =
      1
>> find(x>4,2) % find first two values greater than 4
ans =
      1    3
>> find(x>4,2,'last') % find last two values greater than 4
ans =
      6    7
```

`find(expr,n)`, `find(expr,n,'first')`, and `find(expr,n,'last')`, where the expression `expr` is True, return up to n indices. That is, n defines the maximum number of indices returned. If fewer indices satisfying `expr` exist, fewer are returned.

The preceding concepts are summarized in the following table:

Array Searching	Description
`i=find(X)`	Returns single *indices* of the array X, where its elements are nonzero or True
`[r,c]=find(X)`	Returns row and column *subscripts* of the array X, where its elements are nonzero or True
`find(X,n)` or `find(X,n,'first')`	Starting at the beginning of the array X, returns up to n indices, where X is nonzero or True
`find(X,n,'last')`	Starting at the end of the array X, returns up to n indices, where X is nonzero or True

In addition to using the function find to identify specific values within an array, using the maximum and minimum values and their locations within an array is often helpful. MATLAB provides the functions max and min to accomplish these tasks:

```
>> v = rand(1,6) % new data
v =
      0.3046      0.1897      0.1934      0.6822      0.3028      0.5417

>> max(v) % return maximum value
ans =
      0.6822
>> [mx,i] = max(v) % maximum value and its index
mx =
      0.6822
i =
      4

>> min(v) % return minimum value
ans =
      0.1897

>> [mn,i] = min(v) % minimum value and its index
mn =
      0.1897
i =
      2
```

For 2-D arrays, min and max behave a little differently. For example, in the code

```
>> A = rand(4,6) % new data
A =
      0.1509      0.8537      0.8216      0.3420      0.7271      0.3704
      0.6979      0.5936      0.6449      0.2897      0.3093      0.7027
      0.3784      0.4966      0.8180      0.3412      0.8385      0.5466
      0.8600      0.8998      0.6602      0.5341      0.5681      0.4449
>> [mx,r] = max(A)
```

```
mx =

      0.8600      0.8998      0.8216      0.5341      0.8385      0.7027
r =

      4    4    1    4    3    2
>> [mn,r] = min(A)
mn =

      0.1509      0.4966      0.6449      0.2897      0.3093      0.3704
r =

      1    3    2    2    2    1
```

mx is a vector containing the maximum of each column of A, and r is the row index where the maximum appears. The same principle applies to mn and r, relative to the function min. To find the overall minimum or maximum of a 2-D array, one can take one of two approaches:

```
>> mmx = max(mx) % apply max again to prior result
mmx =

      0.8998
>> [mmx,i] = max(A(:)) % reshape A as a column vector first
mmx =

      0.8998
i =

      8
```

The first of these is essentially max(max(A)), which requires two function calls. The second is preferred in many situations, because it also returns the single index where the maximum occurs (i.e., mmx = A(i)). (The latter approach also works for multidimensional arrays, which are discussed in the next chapter.)

When an array has duplicate minima or maxima, the indices returned by min and max are the first ones encountered. To find all minima and maxima requires the use of the find function:

```
>> x = [1 4 6 3 2 1 6]
x =

      1    4    6    3    2    1    6
>> mx = max(x)
mx =

      6
```

```
>> i = find(x==mx) % indices of values equal to mx
i =
      3     7
```

(Chapter 17 contains more information about the functions min and max.)

5.11 ARRAY-MANIPULATION FUNCTIONS

In addition to the arbitrary array-addressing and manipulation capabilities described in the preceding sections, MATLAB provides several functions that implement common array manipulations. Many of these manipulations are easy to follow:

```
>> A = [1 2 3;4 5 6;7 8 9] % fresh data
A =
      1     2     3
      4     5     6
      7     8     9
>> flipud(A) % flip array in up-down direction
ans =
      7     8     9
      4     5     6
      1     2     3
>> fliplr(A) % flip array in the left-right direction
ans =
      3     2     1
      6     5     4
      9     8     7
>> rot90(A) % rotate array 90 degrees counterclockwise
ans =
      3     6     9
      2     5     8
      1     4     7
>> rot90(A,2) % rotate array 2*90 degrees counterclockwise
ans =
      9     8     7
      6     5     4
      3     2     1
```

```
>> A = [1 2 3;4 5 6;7 8 9] % recall data
A =
        1    2    3
        4    5    6
        7    8    9
>> circshift(A,1)  % circularly shift rows down by 1
ans =
        7    8    9
        1    2    3
        4    5    6
>> circshift(A,[0 1])  % circularly shift columns right by 1
ans =
        3    1    2
        6    4    5
        9    7    8
>> circshift(A,[-1 1])  % shift rows up by 1 and columns right by 1
ans =
        6    4    5
        9    7    8
        3    1    2
>> B = 1:12  % more data
B =
    1    2    3    4    5    6    7    8    9    10    11    12
>> reshape(B,2,6) % reshape to 2 rows, 6 columns, fill by columns
ans =
        1    3    5    7    9    11
        2    4    6    8    10   12
>> reshape(B,[2 6]) % equivalent to above
ans =
        1    3    5    7    9    11
        2    4    6    8    10   12
>> reshape(B,3,4) % reshape to 3 rows, 4 columns, fill by columns
```

ans =

```
   1    4    7    10
   2    5    8    11
   3    6    9    12
```

```
>> reshape(B,3,[]) % MATLAB figures out how many columns are needed
```

ans =

```
   1    4    7    10
   2    5    8    11
   3    6    9    12
```

```
>> reshape(B,[],6) % MATLAB figures out how many rows are needed
```

ans =

```
   1    3    5    7     9    11
   2    4    6    8    10    12
```

```
>> reshape(A,3,2) % A has more than 3*2 elements, OOPS!
??? Error using ==> reshape
To RESHAPE the number of elements must not change.
```

```
>> reshape(A,1,9) % stretch A into a row vector
```

ans =

```
   1    4    7    2    5    8    3    6    9
```

```
>> A(:)' % convert to column and transpose; same as the above
```

ans =

```
   1    4    7    2    5    8    3    6    9
```

```
>> reshape(A,[],3) % MATLAB figures out how many rows are needed
```

ans =

```
   1    2    3
   4    5    6
   7    8    9
```

The following functions extract parts of an array to create another array:

```
>> A % remember what A is
A =
   1    2    3
   4    5    6
   7    8    9
```

```
>> diag(A) % extract diagonal using diag
ans =
     1
     5
     9

>> diag(ans) % remember this? same function, different action
ans =
     1    0    0
     0    5    0
     0    0    9

>> triu(A) % extract upper triangular part
ans =
     1    2    3
     0    5    6
     0    0    9

>> tril(A) % extract lower triangular part
ans =
     1    0    0
     4    5    0
     7    8    9

>> tril(A) - diag(diag(A)) % lower triangular part with no diagonal
ans =
     0    0    0
     4    0    0
     7    8    0
```

The following functions create arrays from other arrays:

```
>> a = [1 2;3 4] % a smaller data array
a =
     1    2
     3    4
>> b = [0 1;-1 0] % another smaller data array
```

b =

```
    0    1
   -1    0
```

>> kron(a,b) % the Kronecker tensor product of a and b

ans =

```
    0    1    0    2
   -1    0   -2    0
    0    3    0    4
   -3    0   -4    0
```

The preceding kron(a,b) is equivalent to

>> [1*b 2*b
     3*b 4*b]

ans =

```
    0    1    0    2
   -1    0   -2    0
    0    3    0    4
   -3    0   -4    0
```

Now consider

>> kron(b,a) % the Kronecker tensor product of b and a

ans =

```
    0    0    1    2
    0    0    3    4
   -1   -2    0    0
   -3   -4    0    0
```

The preceding kron(b,a) is equivalent to

>> [0*a 1*a
    -1*a 0*a]

ans =

```
    0    0    1    2
    0    0    3    4
   -1   -2    0    0
   -3   -4    0    0
```

So, kron(a,b) takes each element of its first argument, multiplies it by the second argument, and creates a block array.

One of the most useful array-manipulation function is repmat, which was introduced earlier:

```
>> a % recall data
a =
     1    2
     3    4

>> repmat(a,1,3) % replicate a once down, 3 across
ans =
     1    2    1    2    1    2
     3    4    3    4    3    4

>> repmat(a,[1 3]) % equivalent to above
ans =
     1    2    1    2    1    2
     3    4    3    4    3    4

>> [a a a] % equivalent to above
ans =
     1    2    1    2    1    2
     3    4    3    4    3    4

>> repmat(a,2,2) % replicate a twice down, twice across
ans =
     1    2    1    2
     3    4    3    4
     1    2    1    2
     3    4    3    4

>> repmat(a,2) % same as repmat(a,2,2) and repmat(a,[2 2])
ans =
     1    2    1    2
     3    4    3    4
     1    2    1    2
     3    4    3    4
```

```
>> [a a; a a] % equivalent to above
ans =
     1    2    1    2
     3    4    3    4
     1    2    1    2
     3    4    3    4
```

As illustrated, the functions repmat and reshape accept indexing arguments in two ways. The indexing arguments can be passed as separate input arguments, or they can be passed as individual elements in a single-vector argument. In addition, for repmat, a single second-input argument repmat(A,n) is interpreted as repmat(A,[n n]).

Finally, to replicate a scalar to create an array that is the same size as another array, one can simply use repmat(d,size(A)):

```
>> A = reshape(1:12,[3 4]) % new data
A =
     1    4    7    10
     2    5    8    11
     3    6    9    12
>> repmat(pi,size(A)) % pi replicated to be the size of A
ans =
    3.1416    3.1416    3.1416    3.1416
    3.1416    3.1416    3.1416    3.1416
    3.1416    3.1416    3.1416    3.1416
```

(Note that the function size is discussed in the next section.)

5.12 ARRAY SIZE

In cases where the size of an array or vector is unknown and is needed for some mathematical manipulation, MATLAB provides the utility functions size, length, and numel:

```
>> A = [1 2 3 4;5 6 7 8]
A =
     1    2    3    4
     5    6    7    8
```

```
>> s = size(A)
s =
     2    4
```

With one output argument, the size function returns a row vector whose first element is the number of rows, and whose second element is the number of columns:

```
>> [r,c] = size(A)
r =
     2
c =
     4
```

With two output arguments, size returns the number of rows in the first variable and the number of columns in the second variable:

```
>> r = size(A,1) % number of rows
r =
     2
>> c = size(A,2) % number of columns
c =
     4
```

Called with two input arguments, size returns either the number of rows or columns. The correspondence between the second argument to size and the number returned follows the order in which array elements are indexed. That is, A(r,c) has its row index r specified first and hence size(A,1) returns the number of rows. Likewise, A(r,c) has its column index c specified second and so size(A,2) returns the number of columns.

The function numel returns the total number of elements in an array:

```
>> numel(A)
ans =
     8
```

The function length returns the number of elements along the largest dimension. For example, the code

```
>> length(A)
ans =
     4
```

returns the number of rows or the number of columns, whichever is larger. For vectors, length returns the vector length:

```
>> B = -3:3
B =
     -3    -2    -1    0    1    2    3
>> length(B) % length of a row vector
ans =
     7
>> length(B') % length of a column vector
ans =
     7
```

The functions size and length also work for an array of zero dimension:

```
>> c = [] % you can create an empty variable!
C =
     []
>> size(c)
ans =
     0    0
>> d = zeros(3,0) % an array with one dimension nonzero!
d =
     Empty matrix: 3-by-0
>> size(d)
ans =
     3    0
>> length(d)
ans =
     0
>> max(size(d)) % maximum of elements of size(d)
ans =
     3
```

As shown, MATLAB allows arrays to have one zero and one nonzero dimension. For these arrays, length and maximum dimension are not the same.

These array size concepts are summarized in the following table:

Array Size	Description
s=size(A)	Returns a row vector s whose first element is the number of rows in A, and whose second element is the number of columns in A
[r,c]=size(A)	Returns two scalars, r and c, containing the number of rows and columns, respectively
r=size(A,1)	Returns the number of rows in A
c=size(A,2)	Returns the number of columns in A
n=length(A)	Returns max(size(A)) for nonempty A, 0 when A has either zero rows or zero columns and the length of A if A is a vector
n=max(size(A))	Returns length(A) for nonempty A, and for empty A returns the length of any nonzero dimension of A
n=numel(A)	Returns the total number of elements in A

5.13 ARRAYS AND MEMORY UTILIZATION

In most modern computers, data transfer to and from memory is more time consuming than floating-point arithmetic, which has been fully integrated into most processors. Memory speed just hasn't kept pace with processor speed, and this has forced computer manufacturers to incorporate multiple levels of memory cache into their products in an attempt to keep processors supplied with data. In addition, computer users work with increasingly larger data sets (variables), which can easily exceed cache capacity. Consequently, efficient memory utilization is critical to effective computing.

MATLAB itself does not perform any explicit memory management. Memory allocation and deallocation within MATLAB use calls to standard C functions (malloc, calloc, free). Therefore, MATLAB relies on the compiler's implementation of these library functions to take appropriate, efficient, system-specific steps for memory allocation and deallocation. Given that there is an inherent trade-off between memory use and execution speed, MATLAB purposely chooses to use more memory when doing so increases execution speed. Since MATLAB uses memory to gain performance, it is beneficial to consider how memory is allocated in the program and what can be done to minimize memory overuse and fragmentation.

When a variable is created with an assignment statement such as

```
>> P = zeros(100);
```

MATLAB requests a contiguous chunk of memory to store the variable, letting the compiler and the operating system determine where that chunk is allocated. If a variable is reassigned, as in

```
>> P = rand(5,6);
```

the original memory is deallocated, and a new allocation request is made for the new variable in its new size. Once again, it is up to the compiler and the operating system to figure out how to implement these tasks. Clearly, when the reassigned variable is larger than the original, a different chunk of contiguous memory is required to store the new data.

In a case where the variable reassignment just so happens to use exactly the same amount of memory, MATLAB still goes through the memory allocation and deallocation process:

```
>> P = zeros(5,6); % same size as earlier
>> P = ones(6,5);  % same number of elements as earlier
```

As a result, reusing variable names does not eliminate memory allocation or deallocation overhead. However, it does clear the memory used by prior data that are no longer needed. The exception to this reallocation is with scalars. For example, a = 1 followed by a = 2 simply copies the new value into the old memory location.

On the other hand, if an assignment statement addresses indices that already exist in a variable, the associated memory locations are updated, and no allocation or deallocation overhead is incurred. For example,

```
>> P(3,3) = 1;
```

does not force any memory allocation calls because P(3,3) already exists. The variable size remains the same, and hence, there's no need to find a different contiguous chunk of memory to store it. However, if an assignment statement increases the size of a variable, as in the code

```
>> P(8,1) = 1;
>> size(P)
ans =
     8     5
```

which makes P grow from 6-by-5 or 30 elements to 8-by-5 or 40 elements, MATLAB requests memory for the revised variable, copies the old variable into the new memory, and then deallocates the old memory chunk, thereby incurring allocation or deallocation overhead.

When all the indices of a variable are addressed in an assignment statement, such as

```
>> P(:) = ones(1,40);
```

the number of elements on the left- and right-hand sides must be equal. When this is true, MATLAB simply copies the data on the right-hand side into the memory that already exists on the left-hand side. No allocation or deallocation overhead is incurred. Furthermore, the dimensions of P remain unchanged:

```
>> size(P)
ans =
     8     5
```

In MATLAB, variables can be declared global in scope using the global command. (This is discussed elsewhere in the text.) From a memory-management point of view, global variables do not behave any differently than ordinary variables. So there is neither benefit from, nor penalty for, using them.

One interesting thing that MATLAB does to improve performance is a feature called ***delayed copy***. For example, in the code

```
>> A = zeros(10);
>> B = zeros(10);
>> C = B;
```

the variables A and B each have memory for 100 elements allocated to them. However, the variable C does not have any memory allocated to it. It shares the memory allocated to B. Copying of the data in B to a chunk of memory allocated to C does not take place until either B or C is modified. That is, a simple assignment, such as that shown, does *not* immediately create a copy of the right-hand-side array in the left-hand-side variable. When an array is large, it is advantageous to delay the copy. That is, future references to C simply access the associated contents of B, and so to the user it appears that C is equal to B. Time is taken to copy the array B into the variable C only if the contents of B are about to change or if the contents of C are about to be assigned new values by some MATLAB statement. While the time saved by this delayed-copy feature is insignificant for smaller arrays, it can lead to significant performance improvements for very large arrays.

The delayed-copy scheme applies to functions as well. When a function is called, for example, myfunc(a,b,c), the arrays a, b, and c are not copied into the workspace of the function, unless the function modifies them in some way. By implementing a delayed copy, memory-allocation overhead is avoided, unless the function modifies the variable. As a result, if that function only reads data from the array, there is no performance penalty for passing a large array to a function. (This feature is covered in Chapter 12 as well.)

When you call a function, it is common to pass the result of a computation or a function output directly to another function. For example, in the code

```
>> prod(size(A))
ans =
    100
```

the results from the function `size(A)` are passed directly to the function `prod`, without explicitly storing the result in a named variable. The memory used to execute this statement is identical to

```
>> tmp = size(A);
>> prod(tmp)
ans =
     100
```

In other words, even though the first statement did not explicitly create a variable to store the results from `size(A)`, MATLAB created an implicit variable and then passed it to the function `prod`. The only benefit gained by the first approach is that the implicit variable is automatically cleared after the function call.

In many applications, it is convenient to increase the size of an array as part of a computational procedure. Usually, this procedure is part of some looping structure, but for now, let's consider a simpler example:

```
>> A = 1:5
A =
     1    2    3    4    5
>> B = 6:10;

>> A = [A;B]
A =
     1    2    3    4    5
     6    7    8    9   10
>> C = 11:15;

>> A = [A;C]
A =
     1    2    3    4    5
     6    7    8    9   10
    11   12   13   14   15
```

Every time the size of the variable A is increased, new memory is allocated, old data is copied, and old memory is deallocated. If the arrays involved are large or the reassignment occurs numerous times, then memory overhead can significantly reduce algorithm speed. To alleviate the problem, you should ***preallocate*** all the memory needed and then fill it as required. Doing so for the preceding example gives

```
>> A = zeros(3,5)  % grab all required memory up front
A =
       0    0    0    0    0
       0    0    0    0    0
       0    0    0    0    0
>> A(1,:) = 1:5;   % no memory allocation here

>> B = 6:10;
>> A(2,:) = B;     % no memory allocation here
>> C = 11:15;
>> A(3,:) = C      % no memory allocation here
A =
       1    2    3    4    5
       6    7    8    9   10
      11   12   13   14   15
```

While this example is somewhat silly, it serves to illustrate that the memory-allocation process for the variable A is performed only once, rather than once at every reassignment of A.

If the size of a variable is decreased, no additional memory allocation is performed. The deleted elements are removed and the memory for the unused elements is deallocated, that is, the memory is *compacted*.

```
>> A = A(:,1:3)    % no memory allocation; deallocation does occur
A =
       1    2    3
       6    7    8
      11   12   13
```

The following table summarizes the facts discussed in this section:

Syntax	Description
P=zeros(100); P=rand(5,6);	Reassignment of a variable incurs memory allocation and deallocation overhead.
P(3,3)=1;	If the indices on the left exist, no memory allocation and deallocation is performed.

`P(8,1)=1;`	If the indices on the left do not exist, memory allocation and deallocation occurs.
`P=zeros(5,6);` `P=ones(5,6);`	Reassignment of a variable incurs memory allocation and deallocation overhead, even if the reassignment does not change the number of elements involved.
`P(:)=rand(1,30)`	If P exists, contents from the right-hand side are copied into the memory allocated on the left. No memory allocation and deallocation is performed.
`B=zeros(10); C=B;`	Delayed copy. C and B share the memory allocated to B until either B or C is modified.

`prod(size(A))`
`tmp=size(A);`
`prod(tmp)`
	Implicit and explicit variables require the same memory allocation. However, the implicit variable is automatically cleared.

`myfunc(a,b,c)`	Delayed copy. The variables a, b, and c are not copied into the function workspace unless the function modifies them. No memory allocation and deallocation overhead is incurred if the variables are not modified within the function.

`A=zeros(3,5);`
`A(1,:)=1:5;`
`A(2,:)=6:10;`
`A(3,:)=11:15;`
	Preallocate all memory for a variable that grows as an algorithm progresses. Memory allocation and deallocation overhead is incurred only once, rather than once per iteration or reassignment.

`A=A(:,1:3);`	When the size of a variable is decreased, the variable uses the same memory location so no new memory is allocated, but the unused portion is deallocated.

6

Multidimensional Arrays

In the previous chapter, 1- and 2-D arrays and their manipulation were illustrated. Since MATLAB version 5 appeared several years ago, MATLAB has added support for arrays of arbitrary dimensions. For the most part, MATLAB supports multidimensional arrays (i.e., n-D arrays) by using the same functions and addressing techniques that apply to 1- and 2-D arrays. In general, the third dimension is numbered by *pages*, while higher dimensions have no generic name. Thus, a 3-D array has rows, columns, and pages. Each page contains a 2-D array of rows and columns. In addition, just as all of the columns of a 2-D array must have the same number of rows and vice versa, all of the pages of a 3-D array must have the same number of rows and columns. One way to visualize 3-D arrays is to think of the residential listings (white pages) in a phone book. Each page has the same number of columns and the same number of names (rows) in each column. The stack of all of the pages forms a 3-D array of names and phone numbers.

Even though there is no limit to the number of dimensions, 3-D arrays are predominately used as examples in this chapter, because they are easily visualized and displayed.

6.1 ARRAY CONSTRUCTION

Multidimensional arrays can be created in several ways:

```
>> A = zeros(4,3,2)
A(:,:,1) =
   0   0   0
   0   0   0
```

```
      0     0     0
      0     0     0
A(:,:,2) =
      0     0     0
      0     0     0
      0     0     0
      0     0     0
```

This is an array of zeros, having four rows, three columns, and two pages. The first page is displayed first, followed by the second page. The other common array-generation functions ones, rand, and randn work the same way, simply by adding dimensions to the input arguments.

Direct indexing also works, as in the following example:

```
>> A = zeros(2,3) % start with a 2-D array
A =
      0     0     0
      0     0     0
>> A(:,:,2) = ones(2,3) % add a second page to go 3-D!
A(:,:,1) =
      0     0     0
      0     0     0
A(:,:,2) =
      1     1     1
      1     1     1
>> A(:,:,3) = 4 % add a third page by scalar expansion
A(:,:,1) =
      0     0     0
      0     0     0
A(:,:,2) =
      1     1     1
      1     1     1
A(:,:,3) =
      4     4     4
      4     4     4
```

This approach starts with a 2-D array, which is the first page of a 3-D array. Then, additional pages are added by straightforward array addressing.

The functions `reshape` and `repmat` can also be used to create *n*-D arrays. For example, in the code

```
>> B = reshape(A,2,9) % 2-D data, stack pages side-by-side
B =
     0    0    0    1    1    1    4    4    4
     0    0    0    1    1    1    4    4    4
>> B = [A(:,:,1) A(:,:,2) A(:,:,3)] % equivalent to above
B =
     0    0    0    1    1    1    4    4    4
     0    0    0    1    1    1    4    4    4
>> reshape(B,2,3,3) % recreate A
ans(:,:,1) =
     0    0    0
     0    0    0
ans(:,:,2) =
     1    1    1
     1    1    1
ans(:,:,3) =
     4    4    4
     4    4    4

>> reshape(B,[2 3 3]) % alternative to reshape(B,2,3,3)
ans(:,:,1) =
     0    0    0
     0    0    0
ans(:,:,2) =
     1    1    1
     1    1    1
ans(:,:,3) =
     4    4    4
     4    4    4
```

`reshape` can change any dimensional array into any other dimensional array.

The code

```
>> C = ones(2,3) % new data
C =
     1    1    1
     1    1    1
```

```
>> repmat(C,1,1,3) % this form not allowed above 2-D!
??? Error using ==> repmat
Too many input arguments.
```

```
>> repmat(C,[1 1 3]) % this is how to do it
ans(:,:,1) =
     1    1    1
     1    1    1
ans(:,:,2) =
     1    1    1
     1    1    1
ans(:,:,3) =
     1    1    1
     1    1    1
```

replicates C once in the *row* dimension, once in the *column* dimension, and three times in the *page* dimension.

The cat function creates *n*-D arrays from lower dimensional arrays:

```
>> a = zeros(2); % new data
>> b = ones(2);
>> c = repmat(2,2,2);
>> D = cat(3,a,b,c) % conCATenate a,b,c along the 3rd dimension
D(:,:,1) =
     0    0
     0    0
D(:,:,2) =
     1    1
     1    1
```

```
D(:,:,3) =
      2    2
      2    2
>> D = cat(4,a,b,c) % try the 4th dimension!
D(:,:,1,1) =
      0    0
      0    0
D(:,:,1,2) =
      1    1
      1    1
D(:,:,1,3) =
      2    2
      2    2
>> D(:,1,:,:) % look at elements in column 1
ans(:,:,1,1) =
      0
      0
ans(:,:,1,2) =
      1
      1
ans(:,:,1,3) =
      2
      2
>> size(D)
ans =
      2    2    1    3
```

Note that D has two rows, two columns, one page, and three fourth-dimension parts.

6.2 ARRAY MATHEMATICS AND MANIPULATION

As additional dimensions are created, array mathematics and manipulation become more cumbersome. Scalar–array arithmetic remains straightforward, but array–array arithmetic requires that the two arrays have the same size in all dimensions. Since scalar–array and array–array arithmetic remains unchanged from the 2-D case presented in the previous chapter, further illustrations are not presented here.

MATLAB provides several functions for the manipulation of n-D arrays. The function squeeze eliminates *singleton dimensions*; that is, it eliminates dimensions of size 1. For example, in the code

```
>> E = squeeze(D) % squeeze dimension 4 down to dimension 3
E(:,:,1) =
      0    0
      0    0
E(:,:,2) =
      1    1
      1    1
E(:,:,3) =
      2    2
      2    2
>> size(E)
ans =
      2    2    3
```

E contains the same data as D, but has two rows, two columns, and three pages.
How about a 3-D vector? This case is shown as follows:

```
>> v(1,1,:) = 1:6 % a vector along the page dimension
v(:,:,1) =
      1
v(:,:,2) =
      2
v(:,:,3) =
      3
v(:,:,4) =
      4
v(:,:,5) =
      5
v(:,:,6) =
      6
>> squeeze(v) % squeeze it into a column vector
```

```
ans =
     1
     2
     3
     4
     5
     6
>> v(:) % this always creates a column vector
ans =
     1
     2
     3
     4
     5
     6
```

The function reshape allows you to change the row, column, page, and higher order dimensions, without changing the total number of elements:

```
>> F = cat(3,2+zeros(2,4),ones(2,4),zeros(2,4)) % new 3-D array
F(:,:,1) =
     2    2    2    2
     2    2    2    2
F(:,:,2) =
     1    1    1    1
     1    1    1    1
F(:,:,3) =
     0    0    0    0
     0    0    0    0
>> G = reshape(F,[3 2 4]) % change it to 3 rows, 2 columns, 4 pages
G(:,:,1) =
     2    2
     2    2
     2    2
```

```
G(:,:,2) =
        2    1
        2    1
        1    1
G(:,:,3) =
        1    1
        1    0
        1    0
G(:,:,4) =
        0    0
        0    0
        0    0

>> H = reshape(F,[4 3 2]) % or 4 rows, 3 columns, 2 pages
H(:,:,1) =
        2    2    1
        2    2    1
        2    2    1
        2    2    1
H(:,:,2) =
        1    0    0
        1    0    0
        1    0    0
        1    0    0

>> K = reshape(F,2,12) % 2 rows, 12 columns, 1 page
K =
    2   2   2   2   1   1   1   1   0   0   0   0
    2   2   2   2   1   1   1   1   0   0   0   0
```

Reshaping can be confusing until you become comfortable with visualizing arrays in *n*-D space. In addition, some reshaping requests make more practical sense than others. For example, G in the previously displayed code has little practical value, whereas K is much more practical, because it stacks the pages of F side by side as additional columns.

The reshaping process follows the same pattern as that for 2-D arrays. Data are gathered first by rows, followed by columns, then by pages, and so on into higher dimensions. That is, all the rows in the first column are gathered, then all the rows in the second column, and so on. Thus, when the first page has been gathered, MATLAB moves on to the second page and starts over with all the rows in the first column.

The order in which array elements are gathered is the order in which the functions sub2ind and ind2sub consider single-index addressing:

```
>> sub2ind(size(F),1,1,1) % 1st row, 1st column, 1st page is element 1
ans =
     1
>> sub2ind(size(F),1,2,1) % 1st row, 2nd column, 1st page is element 3
ans =
     3

>> sub2ind(size(F),1,2,3) % 1st row, 2nd column, 3rd page is element 19
ans =
     19
>> [r,c,p] = ind2sub(size(F),19) % inverse of above
r =
     1
c =
     2
p =
     3
```

The *n*-D equivalent to flipud and fliplr is flipdim:

```
>> M = reshape(1:18,2,3,3) % new data
M(:,:,1) =
     1    3    5
     2    4    6
M(:,:,2) =
     7    9   11
     8   10   12
M(:,:,3) =
    13   15   17
    14   16   18

>> flipdim(M,1) % flip row order
```

```
ans(:,:,1)
       2       4       6
       1       3       5
ans(:,:,2) =
       8      10      12
       7       9      11
ans(:,:,3) =
      14      16      18
      13      15      17
>> flipdim(M,2) % flip column order
ans(:,:,1) =
       5       3       1
       6       4       2
ans(:,:,2) =
      11       9       7
      12      10       8
ans(:,:,3) =
      17      15      13
      18      16      14

>> flipdim(M,3) % flip page order
ans(:,:,1) =
      13      15      17
      14      16      18
ans(:,:,2) =
       7       9      11
       8      10      12
ans(:,:,3) =
       1       3       5
       2       4       6
```

The function shiftdim shifts the dimensions of an array. That is, if an array has r rows, c columns, and p pages, a shift by one dimension creates an array with c rows, p columns, and r pages, as shown in the following example:

```
>> M % recall data
M(:,:,1) =
        1      3      5
        2      4      6
M(:,:,2) =
        7      9     11
        8     10     12
M(:,:,3) =
       13     15     17
       14     16     18
>> shiftdim(M,1) % shift one dimension
ans(:,:,1) =
        1      7     13
        3      9     15
        5     11     17
ans(:,:,2) =
        2      8     14
        4     10     16
        6     12     18
```

Shifting dimensions by 1 causes the first row on page 1 to become the first column on page 1, the second row on page 1 to become the first column on page 2, and so on.

In the code

```
>> shiftdim(M,2) % shift two dimensions
ans(:,:,1) =
        1      2
        7      8
       13     14
ans(:,:,2) =
        3      4
        9     10
       15     16
```

```
ans(:,:,3) =
        5        6
       11       12
       17       18
```

the first column on page 1 of M becomes the first row on page 1, the first column on page 2 becomes the second row on page 1, and so on. If you are like the authors of this text, shifting dimensions is not immediately intuitive. For the 3-D case, it helps if you visualize M forming a rectangular box with page 1 on the front, followed by page 2 behind page 1, and then page 3 forming the back of the box. Hence, shifting dimensions is equivalent to rotating the box so that a different side faces you.

The function shiftdim also accepts negative shifts. In this case, the array is pushed into higher dimensions, leaving singleton dimensions behind, as in this example:

```
>> M % recall data
M(:,:,1) =
        1        3        5
        2        4        6
M(:,:,2) =
        7        9       11
        8       10       12
M(:,:,3) =
       13       15       17
       14       16       18
>> size(M) % M has 2 rows, 3 columns, and 3 pages
ans =
        2        3        3

>> shiftdim(M,-1) % shift dimensions out by 1
ans(:,:,1,1) =
        1        2
ans(:,:,2,1) =
        3        4
ans(:,:,3,1) =
        5        6
ans(:,:,1,2) =
        7        8
```

```
ans(:,:,2,2) =
     9    10
ans(:,:,3,2) =
    11    12
ans(:,:,1,3) =
    13    14
ans(:,:,2,3) =
    15    16
ans(:,:,3,3) =
    17    18
>> size(ans)
ans =
    1    2    3    3
```

The result now has four dimensions. The reader is left to figure out the correspondence between the original data and the shifted result.

In 2-D arrays, the transpose operator swapped rows and columns, converting an r-by-c array into a c-by-r array. The functions permute and ipermute are the n-D equivalents of the transpose operator. By itself, permute is a generalization of the function shiftdim. In the code

```
>> M % recall data
M(:,:,1) =
     1    3    5
     2    4    6
M(:,:,2) =
     7    9   11
     8   10   12
M(:,:,3) =
    13   15   17
    14   16   18
>> permute(M,[2 3 1]) % same as shiftdim(M,1)
ans(:,:,1) =
     1    7   13
     3    9   15
     5   11   17
```

```
ans(:,:,2) =
        2       8      14
        4      10      16
        6      12      18

>> shiftdim(M,1)
ans(:,:,1) =
        1       7      13
        3       9      15
        5      11      17
ans(:,:,2) =
        2       8      14
        4      10      16
        6      12      18
```

[2 3 1] instructs the function to make the second dimension the first, the third dimension the second, and the first dimension the third.

    Now, in the simpler example

```
>> permute(M,[2 1 3])
ans(:,:,1) =
        1       2
        3       4
        5       6
ans(:,:,2) =
        7       8
        9      10
       11      12
ans(:,:,3) =
       13      14
       15      16
       17      18
```

[2 1 3] instructs permute to transpose the rows and columns, but to leave the third dimension alone. As a result, each page in the result is a conventional transpose of the original data.

    The second argument to permute, called ORDER, must be a permutation of the dimensions of the array, passed as the first argument; otherwise, the requested permutation doesn't make sense, as shown in the following example:

```
>> permute(M,[2 1 1])
??? Error using ==> permute
ORDER cannot contain repeated permutation indices.
>> permute(M,[2 1 4])
??? Error using ==> permute
ORDER contains an invalid permutation index.
```

The function permute can also be used to push an array into higher dimensions. For example, shiftdim(M,-1), shown earlier, is equivalent to

```
>> permute(M,[4 1 2 3])
ans(:,:,1,1) =
     1    2
ans(:,:,2,1) =
     3    4
ans(:,:,3,1) =
     5    6
ans(:,:,1,2) =
     7    8
ans(:,:,2,2) =
     9   10
ans(:,:,3,2) =
    11   12
ans(:,:,1,3) =
    13   14
ans(:,:,2,3) =
    15   16
ans(:,:,3,3) =
    17   18
```

An array always has a unit dimension beyond its size; for example, a 2-D array has one page. That is, all dimensions past the nonunity size of an array are singletons. As a result, in the preceding example, the singleton fourth dimension in M is made the first dimension of the result shown.

For 2-D arrays, issuing the transpose operator a second time returns the array to its original form. Because of the added generality of *n*-D arrays, the function ipermute is used to undo the actions performed by permute:

```
>> M % recall data
M(:,:,1) =
     1      3      5
     2      4      6
M(:,:,2) =
     7      9     11
     8     10     12
M(:,:,3) =
    13     15     17
    14     16     18
>> permute(M,[3 2 1]) % sample permutation
ans(:,:,1) =
     1      3      5
     7      9     11
    13     15     17
ans(:,:,2) =
     2      4      6
     8     10     12
    14     16     18
>> ipermute(ans,[3 2 1]) % back to original data
ans(:,:,1) =
     1      3      5
     2      4      6
ans(:,:,2) =
     7      9     11
     8     10     12
ans(:,:,3) =
    13     15     17
    14     16     18
```

6.3 ARRAY SIZE

As demonstrated in the prior chapter and earlier in this chapter, the function size returns the size of an array along each of its dimensions. The functionality of size is unchanged from the features demonstrated in the last chapter. In addition, the function numel remains unchanged. Consider the following example:

```
>> size(M) % return array of dimensions
ans =
      2    3    3
>> numel(M) % number of elements
ans =
     18
>> [r,c,p] = size(M) % return individual variables
r =
     2
c =
     3
p =
     3
>> r = size(M,1) % return just rows
ans =
     2
>> c = size(M,2) % return just columns
c =
     3
>> p = size(M,3) % return just pages
p =
     3
>> v = size(M,4) % default for all higher dimensions
v =
     1
```

When the number of dimensions is unknown or variable, the function `ndims` is useful:

```
>> ndims(M)
ans =
     3
>> ndims(M(:,:,1)) % just the 2-D first page of M
ans =
     2
```

In this last example, M(:,:,1) is a 2-D array, because it has only one page. That is, it has a singleton third dimension. The function ndims is equivalent to the following simple code fragment:

```
>> length(size(M))
ans =
    3
```

The following table summarizes the functions illustrated in this chapter:

n-D Function	Description
ones(r,c, ...) zeros(r,c, ...) rand(r,c, ...) randn(r,c, ...)	Basic *n*-D array creation
reshape(B,2,3,3) reshape(B,[2 3 3])	Reshapes array into arbitrary dimensions
repmat(C,[1 1 3])	Replicates array into arbitrary dimensions
cat(3,a,b,c)	Concatenates array along a specified dimension.
squeeze(D)	Eliminates dimensions of size equal to 1 (i.e., singleton dimensions)
sub2ind(size(F),1,1,1) [r,c,p]=ind2sub(size(F),19)	Subscript to single-index conversion, and single-index to subscript conversion
flipdim(M,1)	Flips order along a given dimension; *n*-D equivalent to flipud and fliplr
shiftdim(M,2)	Shifts dimensions; circular shift for positive second argument; push-out for negative second argument
permute(M,[2 1 3]) ipermute(M,[2 1 3])	Arbitrary permutation and inverse of dimensions; generalization of transpose operator to *n*-D arrays
size(M) [r,c,p]=size(M) r=size(M,1) c=size(M,2) p=size(M,3)	Size of *n*-D array along its dimensions
ndims(M)	Number of dimensions in an array
numel(M)	Number of elements in an array

7

Numeric Data Types

In the preceding chapters, numeric variables were real or complex arrays containing values stored in double precision. Historically, all data in MATLAB was stored in double-precision format. Character strings and logical data were stored as double-precision, 8 byte, real arrays. Needless to say, this resulted in very inefficient memory usage. Character strings need, at most, two bytes per character, and logical arrays require only one bit per element to distinguish between True and False.

Over time, these storage inefficiencies were eliminated. First, character strings became a separate *data type* or *variable class* and were changed to a two-bytes-per-character representation. Then, logical arrays became a separate data type and were changed to a one-byte-per-value representation. Most recently, single-precision data types and a variety of signed and unsigned integer data types were introduced.

Prior to MATLAB 7, arithmetic operations on single-precision and integer data were undefined. However, sorting, searching, logical comparisons, and array manipulation were supported. To perform arithmetic operations on these data types, it was necessary to convert the data to double precision before performing the desired operation. Then, if desired, the result could be converted back to the original data type. With the release of MATLAB 7, most operations on these data types are performed natively, without explicit conversions.

7.1 INTEGER DATA TYPES

MATLAB supports signed and unsigned integer data types having 8-, 16-, 32-, and 64-bit lengths. These data types are summarized in the following table:

Data Type	Description
uint8	Unsigned 8-bit integer in the range 0 to 255 (or 0 to 2^8)
int8	Signed 8-bit integer in the range -128 to 127 (or -2^7 to 2^7-1)
uint16	Unsigned 16-bit integer in the range 0 to 65,535 (or 0 to 2^{16})
int16	Signed 16-bit integer in the range $-32,768$ to $32,767$ (or -2^{15} to $2^{15}-1$)
uint32	Unsigned 32-bit integer in the range 0 to 4,294,967,295 (or 0 to 2^{32})
int32	Signed 32-bit integer in the range $-2,147,483,648$ to $2,147,483,647$ (or -2^{31} to $2^{31}-1$)
uint64	Unsigned 64-bit integer in the range 0 to 18,446,744,073,709,551,615 (or 0 to 2^{64})
int64	Signed 64-bit integer in the range $-9,223,372,036,854,775,808$ to $9,223,372,036,854,775,807$ (or -2^{63} to $2^{63}-1$)

With the exception of the range of definition, each of these integer data types has the same properties. The upper and lower limits of their ranges are given by the intmax and intmin functions, as shown in the following example:

```
>> intmax('int8')
ans =
      127
>> intmin('uint32')
ans =
      0
```

Variables containing integer data can be created in a number of ways. When an array of zeros or ones is desired, the functions zeros and ones can be used:

```
>> m = zeros(1,6,'uint32') % specify data type as last argument
m =
Columns 1 through 5
      0    0    0    0    0
Column 6
      0
>> class(m) % confirm class of result
ans =
      uint32
>> n = ones(4,'int8') % again specify data type as last argument
```

```
n =
     1     1     1     1
     1     1     1     1
     1     1     1     1
     1     1     1     1
```

```
>> class(n) % confirm class of result
ans =
int8
```

For other values, one must convert or cast results into the desired data type:

```
>> k = 1:7 % create default double precision
k =
     1     2     3     4     5     6     7
>> class(k)
ans =
double
```

```
>> kk = uint8(k) % convert using uint8 function
kk =
     1     2     3     4     5     6     7
>> class(kk)
ans =
uint8
```

```
>> kkk = cast(k,'uint8') % use more general cast function
kkk =
     1     2     3     4     5     6     7
>> class(kkk)
ans =
uint8
```

Once a variable of a given data type exists, it retains that data type even if other data are inserted into it:

```
>> kkk(3:5) = zeros(1,3)
kkk =
     1     2     0     0     0     6     7
```

```
>> class(kkk) % class remains unchanged
ans =
uint8

>> kkk(5:7) = ones(1,3,'uint8')
kkk =
     1     2     0     0     1     1     1

>> class(kkk) % class remains unchanged
ans =
uint8

>> kkk(1:2:end) = exp(1)
kkk =
     3     2     3     0     3     1     3
>> class(kkk)
ans =
uint8
```

Note that when a noninteger is inserted, it is first rounded to the nearest integer and then inserted into the array.

Mathematical operations on integer data types of the same kind are defined, as in the following code:

```
>> k = int8(1:7) % create new data
k =
     1     2     3     4     5     6     7
>> m = int8(randperm(7)) % more new data
m =
     7     2     3     6     4     1     5

>> k+m % addition
ans =
     8     4     6    10     9     7    12
>> k-m % subtraction
ans =
    -6     0     0    -2     1     5     2
```

```
>> k.*m % element by element multiplication
ans =
      7     4     9    24    20     6    35

>> k./m % element by element division
ans =
      0     1     1     1     1     6     1

>> k % recall data
k =
      1     2     3     4     5     6     7

>> k/k(2)
ans =
      1     1     2     2     3     3     4
```

Addition, subtraction, and multiplication are straightforward. However, in many cases, integer division does not produce an integer result. As shown, MATLAB performs the integer division as if the arrays were double precision and then rounds the result to the nearest integer.

When dividing integers, MATLAB effectively performs the operation in double precision, rounds the result to the nearest integer, and converts the result back to the integer data type involved.

Mathematical operations between variables of different integer data types are not defined. However, mathematical operations between a double-precision scalar and an integer data type implicitly convert the double-precision scalar to the corresponding integer type and then performs the requested operation, as the following code illustrates:

```
>> m % recall data and data type
m =
      0    0    0    0    0    0
>> class(m)
ans =
int8
```

```
>> n = cast(k,'uint16') % new data of type uint16
n =
      1    2    3    4    5    6    7

>> m+n % try mixed type addition
??? Error using ==> plus
Integers can only be combined with integers of the same class, or scalar doubles.
>> n+3 % try adding default double precision constant 3
ans =
      4    5    6    7    8    9    10
>> class(ans)
ans =
uint16
>> n-(1:7) % try nonscalar double precision subtraction
??? Error using ==> minus
Integers can only be combined with integers of the same class, or scalar doubles.
```

> MATLAB supports mixed mathematical operations between a **scalar** double-precision value and an integer data type, but it does not support operations between an array of double-precision values and an integer array.

Given the limited range of each integer data type, mathematical operations may produce results that exceed the data type's range. In this case, MATLAB implements saturation. That is, when the result of an operation exceeds that specified by intmin and intmax, the result becomes either intmin or intmax, depending on which limit is exceeded:

```
>> k = cast('hellothere','uint8') % convert a string to uint8
k =
     104   101   108   108   111   116   104   101   114   101
>> double(k)+150 % perform addition in double precision
ans =
     254   251   258   258   261   266   254   251   264   251
>> k+150 % perform addition in uint8, saturate at intmax('uint8')=255
ans =
     254   251   255   255   255   255   254   251   255   251
```

```
>> k-110 % perform subtraction in uint8, saturation at intmin('uint8')=0
ans =
     0    0    0    0    1    6    0    0    4    0
```

In sum, MATLAB supports a variety of integer data types. Except for the 64-bit types, these data types are more storage efficient than double-precision data. Mathematical operations on identical integer data types produce results in the same data type. Operations between mixed data types are defined only between a double-precision scalar and the integer data types. Although not illustrated previously, the double-precision values inf and NaN map to intmax and zero respectively for all integer data types.

7.2 FLOATING-POINT DATA TYPES

The default data type in MATLAB is double precision, or simply double. This floating-point data type conforms to the IEEE standard for double-precision arithmetic. MATLAB supports arrays containing single-precision data as a storage-saving alternative. Mathematical operations on single-precision data are defined and perform similarly to the integer data types illustrated in the preceding section. As shown in the following example, the values of realmin, realmax, and eps reflect the reduced range and precision of single-precision data:

```
>> realmin('single')
ans =
    1.1754944e-38
>> realmax('single')
ans =
    3.4028235e+38
>> eps('single')
ans =
    1.1920929e-07

>> realmin('double') % compare to corresponding double values
ans =
    2.225073858507201e-308
>> realmax('double')
ans =
    1.797693134862316e+308
```

```
>> eps % same as eps(1) and eps('double')
ans =
      2.220446049250313e-16
```

The creation of single-precision data follows the approach used for integer data types:

```
>> a = ones(1,5,'single') % specify data type as last argument
a =
     1    1    1    1    1
>> b = *eye(3,'single') % specify data type as last argument
b =
     2    0    0
     0    2    0
     0    0    2
>> c = single(11:17) % convert default double precision to single
c =
    11   12   13   14   15   16   17
>> d = cast(3:-1:-3,'single') % use more general cast function
d =
     3    2    1    0   -1   -2   -3
```

Mathematical operations between single-precision data and between single- and double-precision data produce single-precision results:

```
>> c.^d % element by element exponentiation of singles
ans =
     1.0e+003 *
    Columns 1 through 4
    1.3310000    0.1440000    0.0130000    0.0010000
    Columns 5 through 7
    0.0000667    0.0000039    0.0000002
>> c*pi % multiplication by a scalar double
ans =
    Columns 1 through 4
    34.5575180  37.6991119  40.8407059  43.9822960
```

```
Columns 5 through 7
47.1238899  50.2654839  53.4070740
>> d.*rand(size(d)) % element by element multiplication by a double array
ans =
    4.8017    0.7094    1.6870    2.7472    1.5844    0.9595    0
>> class(ans)

ans =

single
```

> MATLAB supports mathematical operations between double- and single-precision
> arrays, returning single-precision results.

Single-precision data share the special floating-point values of inf and NaN that are
well known in double precision:

```
c = % recall data
    11    12    13    14    15    16    17
>> c(1:2:end) = 0 % inserting double precision does not change data type
c =
    0    12    0    14    0    16    0
>> c./c % create 0/0 values
Warning: Divide by zero.
ans =
    NaN    1    NaN    1    NaN    1    NaN
>> 1./c % create 1/0 values
Warning: Divide by zero.
ans =
    Columns 1 through 4
       Inf   0.0833333       Inf   0.0714286
    Columns 5 through 7
       Inf   0.0625000       Inf
```

7.3 SUMMARY

The following table identifies functions that pertain to the numeric data types supported in MATLAB:

Function	Description
double	Double-precision data type creation and conversion
single	Single-precision data type creation and conversion
int8, int16, int32, int64	Signed integer data type creation and conversion
uint8, uint16, uint32, uint64	Unsigned integer data type creation and conversion
isnumeric	True for integer or floating-point data types
isinteger	True for integer data types
isfloat	True for single or double data types
isa(x,'type')	True if x has class 'type', including 'numeric', 'integer', and 'float'
cast(x,'type')	Casts x to class 'type'
intmax('type')	Maximum integer value for class 'type'
intmin('type')	Minimum integer value for class 'type'
realmax('type')	Maximum floating-point real value for class 'type'
realmin('type')	Minimum floating-point real value for class 'type'
eps('type')	eps value for floating-point value for class 'type'
eps(x)	Distance between x and next larger representable value of same type as x
zeros(...,'type')	Creates array containing zeros of class 'type'
ones(...,'type')	Creates array containing ones of class 'type'
eye(...,'type')	Creates identity array of class 'type'
rand(...,'type')	Creates an array of pseudorandom numbers of class 'type' from the standard uniform distribution
randn(...,'type')	Creates an array of pseudorandom numbers of class 'type' from the standard normal distribution
randi(...,'type')	Creates an array of pseudorandom integers of class 'type' from a uniform discrete distribution

8

Cell Arrays and Structures

MATLAB version 5 introduced two container data types called ***cell arrays*** and ***structures***. These data types allow the grouping of dissimilar, but related, arrays into a single variable. Data management then becomes easier, since groups of related data can be organized and accessed through a cell array or structure. Because cell arrays and structures are *containers* for other data types, mathematical operations on them are not defined. The *contents* of a cell or structure must be addressed to perform mathematical operations.

One way to visualize cell arrays is to imagine a collection of post office boxes covering a wall at the post office. The collection of boxes is the cell array, with each box being one cell in the array. The contents of each post office box are different, just as the contents of each cell in a cell array are different types or sizes of MATLAB data, such as character strings or numerical arrays of varying dimensions. Just as each post office box is identified by a number, each cell in a cell array is indexed by a number. When you send mail to a post office box, you identify the box number you want it put in. When you put data in a particular cell, you identify the cell number you want them put in. The number is also used to identify which box or cell to take data out of.

Structures are almost identical to cell arrays, except that the individual post office boxes (or data-storage locations) are not identified by number. Instead, they are identified by name. Continuing the post office box analogy, the collection of post office boxes is the structure, and each box is identified by its owner's name. To send mail to a post office box, you identify the name of the box you want it put in. To place data in a particular structure element, you identify the name (i.e., field) of the structure element to put the data in.

8.1 CELL ARRAY CREATION

Cell arrays are MATLAB arrays whose elements are cells. Each cell in a cell array can hold any MATLAB data type, including numerical arrays, character strings, symbolic objects, other cell arrays, and structures. For example, one cell of a cell array might contain a numerical array; another, an array of character strings; and another, a vector of complex values. Cell arrays can be created with any number of dimensions, just as numerical arrays can. However, in most cases, cell arrays are created as a simple vector of cells.

Cell arrays can be created by using assignment statements or by preallocating the array with the `cell` function and then assigning data to the cells. If you have trouble with these examples, it is very likely that you have another variable of the same name in the workspace. If any of the examples that follow give unexpected results, clear the array from the workspace and try again.

If you assign a cell to an existing variable that is not a cell, MATLAB will stop and report an error.

Like other kinds of arrays, cell arrays can be built by assigning data to individual cells, one at a time. There are two different ways to access cells. If you use standard array syntax to index the array, you must enclose the cell contents in curly braces, { }. These curly braces construct a cell array. They act like a call to the `cell` function:

```
>> clear A % make sure A is not being used
>> A(1,1) = { [3 2 1; 12 4 86; -23 2 8] };
>> A(1,2) = { 3-3i };                    % semicolons suppress display
>> A(2,1) = { 'Some character string' };
>> A(2,2) = { 8:-2:0 } % no semicolon, so display requested
A =

        [3x3 double]    [3.0000- 3.0000i]

        [1x21 char ]          [1x5 double]
```

Note that MATLAB shows that A is a 2-by-2 cell array, but it does not show the contents of all cells. Basically, MATLAB shows the cell contents if they do not take up much space, but only describes the cell contents if they do take up significant space. The curly braces on the right side of the equal sign indicate that the right-hand-side expression returns a cell, rather than numerical values. This is called *cell indexing*. Alternatively, the following statements create the same cell array:

```
>> A{1,1} = [3 2 1; 12 4 86; -23 2 8];
>> A{1,2} = 3-3i; % semicolons suppress display
```

```
>> A{2,1} = 'Some character string';
>> A{2,2} = 8:-2:0 % no semicolon, so display requested
A =

        [3x3 double]    [3.0000+ 3.0000i]
        [1x21 char ]              [1x5 double]
```

Here, the curly braces appear on the left-hand side of the equal sign. The term A{1,1} indicates that A is a cell array and that the contents of the first row and first column of the cell array are on the right-hand side of the equal sign. This is called *content addressing*. Both methods can be used interchangeably.

As the example demonstrates, curly braces { } are used to access or specify the contents of cells, whereas parentheses ( ) are used to identify cells, but not their contents. To apply the post office box analogy, curly braces are used to look at the contents of post office boxes, whereas parentheses are used to identify post office boxes without looking inside at their contents.

The two commands A(i,j) = {x}; and A{i,j} = x; both tell MATLAB to store the content of variable x in the (i,j) element of the cell array A. The notation A(i,j) is called **cell indexing**, and the notation A{i,j} is called **content addressing**. That is, curly braces { } access the contents of cells, whereas parentheses ( ) identify cells without looking at their contents.

The celldisp function forces MATLAB to display the contents of cells in the usual manner:

```
>> celldisp(A)
A{1,1} =
    1    2    3
    4    5    6
    7    8    9
A{2,1} =
    A character string
A{1,2} =
    2.0000+   3.0000i
A{2,2} =
    12    10    8    6    4    2    0
```

If a cell array has many elements, celldisp can produce a lot of output to the *Command* window. As an alternative, requesting the contents of the cell by using content addressing displays the contents of a single cell. This is different from cell indexing, which identifies the cell, but not its contents:

```
>> A{2,2} % content addressing
ans =
     8    6    4    2    0
>> A(2,2) % cell indexing
ans =
     [1x5 double]
```

```
>> A{1,:} % address contents of the first row
ans =
          3     2     1
         12     4    86
        -23     2     9
ans =
     3.0000 - 3.0000i
>> A(1,:)
ans =
     [3x3 double]    [3.0000- 3.0000i]
```

Note that the contents of all of the cells shown are generically named ans, because the cells store data that do not have associated variable names.

Square brackets were used in previous chapters to create numerical arrays. Curly braces work the same way for cells. In the following example, commas separate columns, and semicolons separate rows:

```
>> B = { [1 2], 'John Smith'; 2+3i, 5 }
B =
          [1x2 double]    'John Smith'
       [2.0000+ 3.0000i]    [          5]
```

When dealing with numerical arrays, it is common to preallocate an array with zeros and then fill the array as needed. The same can be done with cell arrays. The cell function creates a cell array and fills it with empty numerical matrices, []:

```
>> C = cell(2,3)
C =
     []    []    []
     []    []    []
```

Once the cell array has been defined, both cell indexing and content addressing can be used to populate the cells. In the code

```
>> C(1,1) = 'This doesn''t work'
??? Conversion to cell from char is not possible.
```

the left-hand side uses cell indexing. As a result, the right-hand side must be a cell, but it is not, because it lacks curly braces surrounding its contents. This problem can be resolved by addressing the cell correctly:

```
>> C(1,1) = { 'This does work' }
C =
    'This does work'    []      []
                        []      []      []
>> C{2,3} = 'This works too'
C =
    'This does work'    []                      []
                        []      []      'This works too'
```

Because curly braces appear on the left-hand side of the last statement, MATLAB makes the character string the contents of the addressed cell. Once again, this is an example of content addressing, whereas the prior statement is an example of cell indexing.

8.2 CELL ARRAY MANIPULATION

The numerical array-manipulation techniques presented in preceding chapters apply to cell arrays as well. In a sense, cell array manipulation is just a natural extension of these techniques to a different type of array. If you assign data to a cell outside of the dimensions of the current cell array, MATLAB automatically expands the array and fills the intervening cells with the empty numerical array [].

Square brackets are used to combine cell arrays into larger cell arrays, just as they are used to construct larger numerical arrays:

```
>> A % recall prior cell arrays
A =
    [3x3 double]    [3.0000- 3.0000i]
    [1x21 char]         [1x5 double]
>> B
B =
    [1x2 double]    'John Smith'
    [2.0000+ 3.0000i]    [5]
```

```
>> C = [A;B]
C =
            [3x3 double]     [3.0000- 3.0000i]
            [1x21 char]          [1x5 double]
            [1x2 double]     'John Smith'
        [2.0000+ 3.0000i]    [               5]
```

A subset of cells can be extracted to create a new cell array by conventional array-addressing techniques:

```
>> D = C([1 3],:) % first and third rows
D =
    [3x3 double]      [2.0000+ 3.0000i]
    [1x2 double]      'John Smith'
```

An entire row or column of a cell array can be deleted by using the empty array:

```
>> C(3,:) = []
C =
            [3x3 double]      [2.0000+ 3.0000i]
      'A character string'         [1x7 double]
      [      2.0000+ 3.0000i]    [               5]
```

Note that curly braces do not appear in either of the preceding expressions, because we are working with the cell array itself, not the contents of the cells. Once again, curly braces are used to address the contents of cells, whereas parentheses are used to identify the cells without regard for their content.

The reshape function can be used to change the configuration of a cell array, but cannot be used to add or remove cells. For example, in the code

```
>> X = cell(3,4);
>> size(X) % size of the cell array, not the contents
ans =
    3    4

>> Y = reshape(X,6,2);
>> size(Y)
ans =
    6    2
```

the function `reshape` naturally reshapes any array, without regard to its type. Similarly, the `size` function returns the size of any array type.

The function `repmat` also works, even though its name implies that it replicates matrices. The function `repmat` was created when matrices were MATLAB's only data type. If MATLAB had been written after the introduction of cell arrays and other array types, perhaps this function would have been called `reparray`. The following example is illustrative:

```
>> Y % recall data
Y =
    []    []
    []    []
    []    []
    []    []
    []    []
    []    []
>> Z = repmat(Y,1,3)
Z =
    []    []    []    []    []    []
    []    []    []    []    []    []
    []    []    []    []    []    []
    []    []    []    []    []    []
    []    []    []    []    []    []
    []    []    []    []    []    []
```

8.3 RETRIEVING CELL ARRAY CONTENT

To retrieve the contents of a cell in a cell array, you must apply content addressing, which involves the use of curly braces. For instance, in the code

```
>> A % recall cell array
A =
    [3x3 double]     [3.0000- 3.0000i]
    [1x21 char ]        [1x5 double]
>> x = A{2,2} % content addressing uses { }
x =
    8    6    4    2    0
```

the variable x now contains the numerical value 5, making x a numerical array (a scalar, in this case). The class function can be used to confirm that this is true, as in the following example:

```
>> class(x) % return argument's data type
ans =
     double
```

Because numerical arrays are double precision, class returns a character string that identifies x as being double. If cell indexing had been used mistakenly, the result would be different:

```
>> y = A(2,2)
y =
    [1x5 double]
>> y = A(4)      % same as above using single index
y =
    [1x5 double]

>> class(y)      % y is not a double, but a cell!
ans =
cell
>> class(y{1})   % but the contents of y is a double!
ans =
double
```

By now, you are either bored with this distinction between cell indexing and content addressing, or your head is spinning from the confusion.

There are also other functions for testing variable types. The following examples return logical results, where True = 1 and False = 0:

```
>> iscell(y)        % yes, y is a cell
ans =
     1
>> iscell(y{1})     % contents of y is NOT a cell
ans =
     0
```

```
>> isa(y,'cell')   % yes, y is a cell
ans =
     1
>> isdouble(y{1})  % this function doesn't exist
??? Undefined function or method 'isdouble' for input arguments of
type 'double'.
>> isnumeric(y{1}) % contents of y is numerical
ans =
     1
> isfloat(y{1}) % contents of y is floating point
ans =
     1
>> isa(y{1},'double')   % contents of y is a double
ans =
     1
>> isa(y{1},'numeric') % contents of y is also numeric
ans =
     1
>> isa(y{1},'cell')     % contents of y is NOT a cell
ans =
     0
```

While you can display the contents of more than one cell array at a time, it is not possible to assign more than one at a time to a single variable by using a standard assignment statement. Consider the following code:

```
>> B{:,2}
ans =
John Smith
ans =
     5
>> d = B{:,2}
d =
John Smith
>> class(d)
```

```
ans =

char
```

If you think about it, this behavior makes sense. How can two pieces of data be assigned to a single variable? In the preceding example, d is assigned the contents of the first cell referenced by B{:,2}, that is, B{1,2}. The cell content referenced by B{2,2} is superfluous and is silently dropped. Extracting data from multiple cells at a time simply cannot be done so casually.

When addressing the contents of a single cell, it is possible to further address a subset of the contents by simply appending the desired subscript range:

```
>> A % recall prior data
A =
   [3x3 double]     [3.0000- 3.0000i]
   [1x21 char  ]            [1x5 double]
>> celldisp(A) % display contents
A{1,1} =
          3    2    1
         12    4   86
        -23    2    8
A{2,1} =
Some character string
A{1,2} =
     3.0000 - 3.0000i
A{2,2} =
     8   6   4   2   0
>> A{1,1}(3,:)  % third row of 3-by-3 array
ans =
       -23   2   8
>> A{4}(2:5)    % second through fifth elements of A{2,2}
ans =
     6   4   2   0
>> A{1,2}(2)    % second element doesn't exist
??? Index exceeds matrix dimensions.
>> A{2,1}(6:14)
ans =
character
```

8.4 COMMA-SEPARATED LISTS

To extract the contents of more than one cell at a time, MATLAB provides comma–separated list syntax. This syntax applies in any place where variables or constants appear in a list separated by commas. For example, comma-separated lists appear in array construction:

```
>> a = zeros(2,3);
>> b = ones(2,1);
>> c = (23:24)';
>> d = [a,b,c] % same as [a b c]
d =
 1   1   1   0   23
 1   1   1   0   24
```

They also appear in function input and output argument lists:

```
>> d = cat(2,a,b,c)
d =
    1   1   1   0   23
    1   1   1   0   24
>> [m,n] = size(d')
m =
 5
n =
 2
```

Comma–separated list syntax is implemented as follows: Placing a content-addressed cell array in locations where comma-separated lists appear causes MATLAB to extract the contents of the addressed cell arrays and place them sequentially, separated by commas. In the code

```
>> F = {c b a} % create a cell array
F =
    [2x1 double]    [2x1 double]    [2x3 double]
>> d = cat(2,F{:}) % same as cat(2,a,b,c)
d =
    23   1   0   0   0
    24   1   0   0   0
```

```
>> d = cat(2,F(:)) % not content addressing
d =
    [2x1 double]
    [2x1 double]
    [2x3 double]
```

`cat(2,F{:})` is interpreted by MATLAB as `cat(2,F{1},F{2},F{3})`, which is equal to `cat(2,c,b,a)`. That is, comma–separated list syntax dictates that `F{:}` is interpreted as a listing of all addressed parts of `F{:}`, separated by commas. Note that content addressing must be used. Cell indexing as shown in the second case in the preceding example, acts on the cells, not on their contents. As a result, there is no comma-separated list in the second case, `cat(2,F(:))`. Consider this example:

```
>> d = [F{:}]
d =
    23   1   0   0   0
    24   1   0   0   0
>> d = [F{1},F{2},F{3}] % what is implied by the above
d =
    23   1   0   0   0
    24   1   0   0   0
>> e = [F{2:3}] % can also content address any subset
e =
    1   0   0   0
    1   0   0   0
>> e = [F{2},F{3}] % what is implied by the above
e =
    1   0   0   0
    1   0   0   0
```

At first, comma–separated list syntax may seem strange. However, once you become familiar with it, you will recognize its power. (For more information, see the online help for `lists`, or search for the phrase `comma separated list`.)

Given comma–separated list syntax, MATLAB provides a way to extract the contents of numerous cells into separate variables. In the code

```
>> celldisp(F) % recall data
F{1} =
```

```
            23

            24

F{2} =

            1

            1

F{3} =

      0    0    0

      0    0    0

>> [r,s,t] = F{:} % extract the contents of F into separate
variables

r =

            23

            24

s =

            1

            1

t =

      0    0    0

      0    0    0
```

the variables r, s, and t are numerical variables, with r = F{1}, s = F{2}, t = F{3}.

Comma-separated lists can be used to assign the contents of multiple unrelated variables in one statement. If you have ever used playing cards, you know that "to deal" means to pass out the cards in some organized way. The function deal does just that; it passes out the contents of a list of variables to another set of variables. In the code

```
>> a,b,c % recall data

a =

      0    0    0

      0    0    0

b =

            1

            1

c =

            23

            24
```

```
>> [r,s,t] = deal(a,b,c) % deal out contents of a, b, and c
r =
    0   0   0
    0   0   0
s =
    1
    1
t =
    23
    24
```

the variables r, s, and t are numerical variables, with r = a, s = b, t = c. So, the function deal is, in reality, pretty simple. It just assigns the contents of the first input argument to the first output argument, the second to the second, and so on. Despite its simplicity, deal is a useful tool for extracting data from multiple variables with one statement when content addressing is not easily accomplished.

Because the output of deal is also a comma-separated list, deal can be used to assign the contents of multiple types of variables in one statement:

```
>> d = eye(2);
>> e = 'This is a character string.';
>> f = {[1 2 3]};

>> [x,y,z] = deal(d,e,f)
x =
    1   0
    0   1
y =
This is a character string.
z =
    [1x3 double]

>> class(x)
ans =
double
>> class(y)
```

```
ans =
char
>> class(z)
ans =
cell
```

8.5 CELL FUNCTIONS

Besides the functions celldisp, cell, iscell, deal, and isa, there are several other functions that are useful when dealing with cell arrays. The function cellfun provides a way to apply certain functions to all cells in a cell array, thereby eliminating the need to apply them to each cell individually:

```
>> A % recall data
A =
    [3x3 double]    [3.0000- 3.0000i]
    [1x21 char ]        [1x5 double]

>> cellfun('isreal',A) % True=1 where not complex
ans =
    1    0
    1    1
>> cellfun('length',A) % length of contents
ans =
    3    1
   21    5
>> cellfun('prodofsize',A) % number of elements in each cell
ans =
    9    1
   21    5
>> cellfun('isclass',A,'char') % True for character strings
ans =
    0    0
    1    0
```

The function cellfun offers other options as well. For further information, see online help.

Another function that is sometimes useful is num2cell. This function takes an array of any type (not just numbers as the function name suggests) and fills a cell array with its components:

```
>> a = rand(3,5) % new numerical data
a =
    0.7922    0.0357    0.6787    0.3922    0.7060
    0.9595    0.8491    0.7577    0.6555    0.0318
    0.6557    0.9340    0.7431    0.1712    0.2769
>> c = num2cell(a) % c{i,j}=a(i,j)
c =
    [0.7922]    [0.0357]    [0.6787]    [0.3922]    [0.7060]
    [0.9595]    [0.8491]    [0.7577]    [0.6555]    [0.0318]
    [0.6557]    [0.9340]    [0.7431]    [0.1712]    [0.2769]
>> d = num2cell(a,1) % d{i}=a(:,i)
d =
    [3x1 double]    [3x1 double]    [3x1 double]    [3x1 double]
>> e = num2cell(a,2) % e{i}=a(i,:)
e =
    [1x5 double]
    [1x5 double]
    [1x5 double]
```

With numerical data as input, num2cell(a) isn't useful in many applications, but packing larger pieces of an array into cells is often more useful, as illustrated in the last two cases.

A similar function is mat2cell. This function breaks up a two-dimensional matrix of any type into a cell array of adjacent matrices. The arguments are an input matrix, a vector of the distribution of rows, and a vector of the distribution of columns. The distribution vectors must sum to the size of the input matrix. Consider the following code:

```
>> M=[1 2 3 4 5; 6 7 8 9 10; 11 12 13 14 15; 16 17 18 19 20; 21 22 23 24 25];
>> size(M)
```

```
ans =

     5     5


>> C = mat2cell(M,[1 2 2],[2 3]) % Break up matrix into six pieces
C =

     [1x2 double]  [1x3 double]
     [2x2 double]  [2x3 double]
     [2x2 double]  [2x3 double]
>> celldisp(C)
C{1,1} =

     1     2
C{2,1} =

     6     7
    11    12
C{3,1} =

    16    17
    21    22
C{1,2} =

     3     4     5
C{2,2} =

     8     9    10
    13    14    15
C{3,2} =

    18    19    20
    23    24    25
```

The corresponding function is cell2mat:

```
>> N = cell2mat(C)
N =

     1     2     3     4     5
     6     7     8     9    10
    11    12    13    14    15
    16    17    18    19    20
    21    22    23    24    25
```

8.6 CELL ARRAYS OF STRINGS

Although a rigorous discussion of character strings doesn't occur until the next chapter, the use of character strings and their storage in cell arrays are described in this section. If the concept of a character string is foreign to you, read the next chapter and then come back to this material.

In MATLAB, character strings are formed by enclosing characters within quotes. For example, in the code

```
>> s = 'Rarely is the question asked: is our children learning?'
s =
Rarely is the question asked: is our children learning?
```

the variable s contains a string of characters. In many applications, groups of character strings are associated with each other. When this occurs, it is convenient to group them into a cell array rather than store them in individual variables. In the code

```
>> cs = {'My answer is bring them on.'
s
'I understand small business growth. I was one.'}
cs =
    'My answer is bring them on.'
    'Rarely is the question asked: is our children learning?'
    'I understand small business growth. I was one.'

>> size(cs) % a column cell array
ans =
    3    1
>> iscell(cs) % yes, it is a cell array
ans =
    1
```

the cell array cs has three cells, each containing a character string. In MATLAB, this is simply called a ***cell array of strings***. Because cell arrays are commonly used to store sets of character strings, MATLAB provides several functions to support them, like the one in the following example:

```
>> iscellstr(cs)
ans =
    1
```

The function `iscellstr` returns logical True (1) if all the cells in its cell array argument contain character strings; otherwise, it returns logical False (0).

Before cell arrays were introduced in MATLAB 5, groups of character strings were stored in character-string arrays—2-D arrays just like the numerical arrays discussed earlier—with each string occupying a separate row in the array. Each character in the *string array* occupies its own location and is indexed just like a numerical array.

MATLAB provides functions to convert a cell array of strings to a string array and vice versa:

```
>> cs % recall previous cell array of strings
cs =
    'My answer is bring them on.'
    'Rarely is the question asked: is our children learning?'
    'I understand small business growth. I was one.'
>> sa = char(cs) % convert to a string array
sa =
My answer is bring them on.
Rarely is the question asked: is our children learning?
I understand small business growth. I was one.
>> ischar(sa) % True for string array
ans =
    1
>> iscell(sa) % True for cell array
ans =
    0
>> size(sa) % size of string array
ans =
    3    34
>> size(cs) % size of cell array
ans =
    3    1
>> cst = cellstr(sa) % convert back to cell array
cst =
    'My answer is bring them on.'
    'Rarely is the question asked: is our children learning?'
    'I understand small business growth. I was one.'
```

```
>> iscell(cst)          % True for cell array
ans =
    1
>> isequal(cs,cst)      % True for equal variables
ans =
    1
>> isequal(cs,sa)       % cell array not equal to string array
ans =
    0
```

So, the MATLAB functions char and cellstr are inverses of each other. In addition, since a string array must have the same number of columns in each row, blank spaces are added to rows, as necessary, to make the string array rectangular.

8.7 STRUCTURE CREATION

Structures are like cell arrays in that they allow you to group collections of dissimilar data into a single variable. However, instead of addressing elements by number, structure elements are addressed by names called *fields*. Like cell arrays, structures can have any number of dimensions, but a simple scalar or vector array is the most common type.

Whereas cell arrays use curly braces to access data, structures use dot notation to access data in fields. Creating a structure can be as simple as assigning data to individual fields. For example, in the code

```
>> circle.radius = 3;         % semicolon, no display
>> circle.center = [-1 2];
>> circle.linestyle = '==';
>> circle.color = 'blue'      % no semicolon, so display
circle =
        radius: 3
        center: [-1 2]
     linestyle: '=='
         color: 'blue'
```

the data are stored in a structure variable called circle. The case-sensitive fields are entitled radius, center, linestyle, and color. Structure field names have the same restrictions as variable names: They can contain up to 63 characters, and they must begin with a letter. For example, in the statement

```
>> size(circle)
ans =
     1    1
>> whos
```

Name	Size	Bytes	Class Attributes
ans	1x2	16	double array
circle	1x1	532	struct array

circle is a scalar structure, since size says that it is a 1-by-1 array. The whos command shows its size, the fact that it is a structure array, and the fact that it uses 532 bytes of memory. If there is more than one circle, it can be stored as a second element in the circle variable. For example, in the code

```
>> circle(2).radius = exp(1);
>> circle(2).color = 'black';
>> circle(2).linestyle = '--';
>> circle(2).center = [pi 2]
circle =
1x2 struct array with fields:
    radius
    center
    linestyle
    color
```

circle is a structure array having two elements. The (2) appears immediately after the variable name, because it is the variable that is having an element added to it. The .fieldname suffix identifies the field where data is to be placed. Note that the structure fields are filled in a different order this time and that the size of the data differs between the two elements. That is, the color fields are 'blue' and 'black'. There are no restrictions on what can be placed in fields from one array element to the next. For instance, in the MATLAB statement

```
>> circle(2).radius = 'pi'
circle =
1x2 struct array with fields:
    radius
    center
    linestyle
    color
```

```
>> circle.radius % display radius contents
ans =

            3

ans =

pi
```

circle(1).radius holds numerical data and circle(2).radius holds a character string.

If the value of structures is not apparent to you, consider how the data would be stored without structures:

```
>> Cradius = [pi exp(1)]; % ignore changes above
>> Ccenter = [1 -1 ; 3.14 -3.14];
>> Clinestyle = {'= =' '--'}; % cell array of strings
>> Ccolor = {'blue' 'black'};
```

Now, rather than having a single variable storing the information for two circles, there are four variables that must be indexed properly to extract data for each circle. It's easy to add another circle to the structure, but more cumbersome to add them to the four variables. Now, consider the following example:

```
>> circle(3).radius = 2.4;
>> circle(3).center = [1 2];
>> circle(3).linestyle = ':';
>> circle(3).color = 'red'  % third circle added
circle =
1x3 struct array with fields:
    radius
    center
    linestyle
    color
>> Cradius(3) = 3.4
Cradius =
            3.1416        2.7183        3.4000
>> Ccenter(3,:) = [1 2]
Ccenter =
            1.0000       -1.0000
            3.1400       -3.1400
            1.0000        2.0000
```

```
>> Cradius(3) = 2.4
Cradius =
            3.1416        2.7183        2.4000

>> Clinestyle{3} = '->'
Clinestyle =
    '=='     '--'    '->'
>> Ccolor(3) = {'orange'}
Ccolor =
  'blue'    'black'    'orange'
```

The clarity provided by structures should be readily apparent. In addition, consider passing the circle data to a function. For example, as a structure, the data are passed simply as

```
    myfunc(circle)
```

whereas the other approach requires

```
    myfunc(Cradius,Ccenter,Clinestyle,Ccolor)
```

Suppose that, at some later date, you wanted to add another field to circle:

```
>> circle(1).filled = 'no'
circle =
1x3 struct array with fields:
    radius
    center
    linestyle
    color
    filled
>> circle.filled % display all .filled fields
ans =
no
ans =
    []
ans =
    []
```

Now, all of the elements of circle have the field filled. Those not assigned by default contain the empty array []. The other filled fields are easily assigned:

```
>> circle(2).filled = 'yes';
>> circle(3).filled = 'yes';
>> circle.filled

ans =

no

ans =

yes

ans =

yes
```

When structure creation by direct assignment isn't possible, MATLAB provides the function struct. For example, to re-create the previous structure, the following code is used:

```
>> values1 = {3.2 'pi' 2.27}; % cell arrays with field data
>> values2 = {[1 1] [2 -2] [-1 exp(1)]};
>> values3 = {'<-' '--' '=='};
>> values4 = {'black' 'red' 'green'};
>> values5 = {'no' 'yes' 'yes'};
>> CIRCLE = struct('radius',values1,'center',values2,...
            'linestyle',values3,'color',values4,'filled',values5)

CIRCLE =

1x3 struct array with fields:

    radius

    center

    linestyle

    color

    filled

>> isequal(circle,CIRCLE) % False since we did not use the same data

ans =

    0
```

8.8 STRUCTURE MANIPULATION

Structures are arrays and therefore can be combined and indexed like numerical arrays and cell arrays. When combining structure arrays, the only restriction is that the arrays combined must share the same fields. Thus, in the code

```
>> square.width = 5; % a new structure
>> square.height = 14;
>> square.center = zeros(1,2);
>> square.rotation = pi/4
square =
        width: 5
       height: 14
       center: [0 0]
     rotation: 0.7854
>> A = [circle CIRCLE]
A =
1x6 struct array with fields:
    radius
    center
    linestyle
    color
    filled
>> B = [circle square]
??? Error using ==> horzcat
CAT arguments are not consistent in structure field number.
```

the structures circle and CIRCLE are both 1 by 3 and share the same fields; hence, concatenating them produces a 1-by-6 structure array. However, the structures square and circle do not have the same fields and therefore cannot be concatenated.

It is also possible to address a subarray of a structure. For example, in the code

```
>> C = [circle(1:2) CIRCLE(3)]
C =
1x3 struct array with fields:
    radius
    center
    linestyle
```

```
    color
    filled
>> isequal(C,circle) % True since equal
ans =

     1
```

C and `circle` are equal because `CIRCLE(3)` = `circle(3)`. Once again, basic array addressing and concatenation apply to structure arrays as well. For concatenation, field names must match exactly.

Although they are not as useful in this case, structures can also be manipulated with the functions `reshape` and `repmat`:

```
>> Aa = reshape(A,3,2)
Aa =
3x2 struct array with fields:
    radius
    center
    linestyle
    color
    filled
>> Aaa = reshape(A,1,2,3)
Aaa =
1x2x3 struct array with fields:
    radius
    center
    linestyle
    color
    filled
```

Since there is seldom a practical reason to have anything other than a structure array with vector orientation, `reshape` is seldom needed or used with structures. However, it is often convenient to create a structure array with default data in all fields of all array elements. The function `repmat` performs this task with ease:

```
>> S = repmat(square,3,1)
S =
3x1 struct array with fields:
    width
    height
```

```
    center
    rotation
>> S.width % look at all width fields
ans =
     5
ans =
     5
ans =
     5
```

All three elements of the structure S contain the data originally assigned to the structure square. At this point, the structure fields can be modified, as needed, to describe different squares.

8.9 RETRIEVING STRUCTURE CONTENT

When you know the names of the fields associated with a structure array, retrieving the data in a particular structure element and field simply requires identifying them. For example, in the code

```
>> rad2 = circle(2).radius
rad2 =
pi

>> circle(1).radius
ans =
          3

>> area1 = pi*circle(1).radius^2
area1 =
          28.2743
```

circle(1).radius identifies the value 3, which is used to compute the area of the first circle.

When the contents of a field are an array, it is also possible to retrieve a subset of that field by appending an array index to the structure request:

```
>> circle(1).filled          % the entire field
ans =
no
```

```
>> circle(1).filled(1)          % first element of field
ans =
n
>> circle(1).filled(2:end)      % rest of field
ans =
0
```

As is true for cell arrays, retrieving the contents of more than one structure array element and field cannot be accomplished by direct addressing. For example, the command

```
>> col = circle.color
col =
blue
```

attempted to extract three pieces of data and store them in one variable, but only stored one. However, MATLAB solves this problem the same way it does for cell arrays by using comma-separated lists.

When the structure field to be accessed is stored in a character string, MATLAB provides *dynamic addressing* to gain access to the data:

```
>> fldstr = 'color';   % store desired field in variable

>> circle.(fldstr)     % get all color fields
ans =
blue
ans =
black
ans =
red

>> fldstr = 'radius';
>> area=pi*circle(1).(fldstr)^2 % compute area of first circle
area =
       28.2743
```

Dynamic addressing is similar to using subscripts to identify specific elements in an array. Both methods use parentheses to hold the chosen elements. For example, A(r,c) uses the information in variables r and c to identify the desired rows and columns of array A, whereas `circle.(fldstr)` uses the information in variable `fldstr` to identify the desired field in the structure `circle`.

8.10 COMMA-SEPARATED LISTS (AGAIN)

To extract the contents of more than one structure array element at a time, MATLAB provides comma–separated list syntax. This syntax applies in any place where variables or constants appear in a list separated by commas. For example, it appears in array construction:

```
>> a = ones(2,3);
>> b = zeros(2,1);
>> c = (3:4)';
>> d = [a,b,c] % same as [a b c]
d =
   1   1   1   0   3
   1   1   1   0   4
```

It also appears in function input and output argument lists:

```
>> d = cat(2,a,b,c)
d =
     1    1    1    0    3
     1    1    1    0    4
>> [m,n] = size(d)
m =
  2
n =
  5
```

Building on this idea, comma–separated list syntax is implemented as follows: Placing a structure array with appended field name in locations where comma-separated lists appear causes MATLAB to extract the contents of the addressed fields and place them sequentially, separated by commas. In the code

```
>> cent = cat(1,circle.center) % comma-separated list syntax
cent =
          -1.0000        2.0000
           3.1416        2.0000
           1.0000        2.0000
```

```
>> cent = cat(1,circle(1).center,circle(2).center,circle(3).center)
cent =
            -1.0000        2.0000
             3.1416        2.0000
             1.0000        2.0000
>> some = cat(1,circle(2:end).center)
some =
             3.1416        2.0000
             1.0000        2.0000
```

the cat function concatenates the circle centers into rows of the numerical array cent. According to comma–separated list syntax, the first two statements in the preceding example are identical. The third statement shows that you can index and extract a subarray as well.

Since the color fields of the structure circle are character strings of different lengths, they cannot be extracted into a string array, but they can be extracted into a cell array, as in the following example:

```
>> circle.color
ans =
blue
ans =
black
ans =
red

>> col = cat(1,circle.color) % elements have different lengths
??? Error using ==> cat
CAT arguments dimensions are not consistent.
>> col = [circle.color] % no error but not much use!
col =
blueblackred
>> col = {circle.color} % cell array of strings
col =
    'blue' 'black' 'red'

>> col = char(col) % if needed, convert to string array
col =
```

blue

black

red

MATLAB does not provide tools for extracting all the fields of a single structure array element. For instance, there is no way to retrieve the radius, center, linestyle, color, and filled fields of the first circle in one statement. In a sense, there is no need to have such a function, because each of these fields can be directly addressed and used in computations, such as area1 = pi*circle(1). radius^2.

Given comma–separated list syntax, you can extract the contents of numerous structure elements into separate variables, as in the following example:

```
>> [c1,c2,c3] = circle.color % get all colors
c1 =
blue
c2 =
black
c3 =
red
>> [rad1,rad3] = circle ([1 3]).radius % 1st and 3rd radius
rad1 =
            3
rad3 =
         2.4000
```

Because the output of deal is also a comma-separated list, deal can be used to assign the contents of multiple structure array elements, with a single field in one statement. For example, consider the following code:

```
>> [circle.radius] = deal(5,14,83)
circle =
1x3 struct array with fields:
    radius
    center
    linestyle
    color
    filled
```

```
>> circle.radius % confirm assignments
ans =
     5
ans =
    14
ans =
    83
>> [triangle(:).type] = deal('right','isosceles','unknown')
??? Error using ==> deal at 38
The number of outputs should match the number of inputs.
>> [triangle(1:3).type] = deal('right','isosceles','unknown')
triangle =
1x3 struct array with fields:
    type
>> triangle.type
ans =
right
ans =
isosceles
ans =
unknown
```

In the first statement, the structure circle already existed and has three elements. Therefore, the output argument was expanded into three elements. In the second statement containing deal, the structure triangle didn't exist. Because it wasn't possible to determine how many elements to create in a comma-separated list, an error was returned. However, in the last statement containing deal, the number of elements was given explicitly, and the type property of the newly created structure triangle was populated with the given data.

8.11 STRUCTURE FUNCTIONS

In the *Command* window, it is easy to identify the field names of a given structure by simply entering the structure name at the MATLAB prompt:

```
>> CIRCLE
CIRCLE =
```

```
1x3 struct array with fields:
    radius
    center
    linestyle
    color
    filled
>> square
square =
        width: 5
       height: 14
       center: [0 0]
     rotation: 0.7854
```

When a structure is passed to a function—for example, myfunc(circle)—the function internally must know or have some way to obtain the field names of the structure. (Writing functions in MATLAB is covered in Chapter 12.) In MATLAB, the function fieldnames provides this information:

```
>> fieldnames(CIRCLE)
ans =
    'radius'
    'center'
    'linestyle'
    'color'
    'filled'
```

The output of fieldnames is a cell array of strings identifying the fields that are associated with the input structure.

It is also possible to guess the field names and ask whether they exist by using the logical function isfield:

```
>> isfield(CIRCLE,'height')
ans =
     0
>> isfield(CIRCLE,'filled')
ans =
     1
```

If you don't know whether a variable is a structure or not, the functions class and isstruct are helpful, as the following example shows:

```
>> class(square)          % ask for the class of variable square
ans =
struct
>> isstruct(CIRCLE)     % True for structures
ans =
     1
>> d = pi;
>> isstruct(d) % False for doubles
ans =
     0
```

When the field names of a structure are known, the function rmfield allows you to remove one or more fields from the structure. For example, the code

```
>> fnames = fieldnames(CIRCLE)
fnames =
        'radius'
        'center'
        'linestyle'
        'color'
        'filled'
```

stores in fnames the field names of circle in a cell array. Arbitrarily choosing the last field name, we can remove the filled field from the structure by calling rmfield. This code:

```
>> circle2 = rmfield(circle,fnames{5})
circle2 =
1x3 struct array with fields:
    radius
    center
    linestyle
    color
```

removes the field filled from circle and assigns the result to a new structure circle2. Similarly,

```
>> circle3 = rmfield(circle,fnames(1:3))
circle3 =
1x3 struct array with fields:
    'color'
    'filled'
```

removes the fields radius, center, and linestyle from circle, and assigns the result to a new structure circle3.

In some cases, it is beneficial to rearrange the order in which the field names of a structure are presented. The function orderfields performs this task and places the fields in ASCII order:

```
>> circleA = orderfields(circle)
circleA =
1x3 struct array with fields:
    center
    color
    filled
    linestyle
    radius
```

Other orderings are possible as well, as in the following code:

```
>> circleB = orderfields(circleA,CIRCLE) % match fields of structure CIRCLE
circleB =
1x3 struct array with fields:
    radius
    center
    linestyle
    color
    filled
>> circleC = orderfields(circle,[2 5 1 4 3]) % provide permutation vector
circleC =
1x3 struct array with fields:
    center
    filled
    radius
```

```
    color
    linestyle
>> circleD = orderfields(circle,fnames(end:-1:1)) % reverse original order
circleD =
1x3 struct array with fields:
    filled
    color
    linestyle
    center
    radius
```

Finally, given the similarity between cell arrays and structures and applying the post office box analogy, it's not hard to believe that MATLAB provides the functions `cell2struct` and `struct2cell` to convert cell arrays to structures and back in an organized fashion. (More detailed explanations of these functions can be found in the online documentation.)

8.12 SUMMARY

MATLAB supports arrays with an unlimited number of dimensions. Cell arrays and structures can store any array type, including cell arrays and structures. So, it is possible to have a cell in a cell array that contains a structure with a field that contains another structure that has a field containing a cell array, of which one cell contains another cell array. Needless to say, there comes a point where the power of cell arrays and structures becomes indecipherable and of little practical use. Thus, to end this chapter, try to decipher the following legal MATLAB statements:

```
>> one(2).three(4).five = {circle}
one =
1x2 struct array with fields:
    three
>> test = {{{circle}}}
test =
    {1x1 cell}
```

How many structures are involved in the first statement? What is the total number of structure elements, including empty arrays, created by the statement? Can the structure `circle` be extracted from the second statement with a single MATLAB statement?

9

Character Strings

MATLAB's true power is in its ability to crunch numbers. However, there are times when it is desirable to manipulate text, such as when putting labels and titles on plots. In MATLAB, text is referred to as character strings, or simply strings. Character strings represent another variable class or data type in MATLAB.

9.1 STRING CONSTRUCTION

Character strings in MATLAB are special numerical arrays of ASCII values that are displayed as their character-string representation, as in the following code:

```
>> t = 'This is a character string'
t =
This is a character string
>> size(t)
ans =
     1    26
>> whos
  Name          Size          Bytes     Class Attributes

  ans           1x2              16     double
  t             1x26             52     char
```

A character string is simply text surrounded by single quotes. Each character in a string is one element in an array that requires 2 bytes per character for storage, which is different from the 8 bytes per element that is required for numerical or double arrays.

To see the underlying ASCII representation of a character string, you need only perform some arithmetic operation on the string or use the dedicated function `double`, as in the following code:

```
>> u = double(t)
u =
  Columns 1 through 14
      84   104   105   115    32   105   115    32    97    32    99   104    97   114
  Columns 15 through 26
      97    99   116   101   114    32   115   116   114   105   110   103

>> abs(t)
ans =
  Columns 1 through 14
      82   104   105   115    32   115    32    97    32    32    99   104    97   114
  Columns 15 through 26
      97    99   116   101   114    32   115   116   114   105   110   103
```

The function `char` performs the inverse transformation:

```
>> char(u)
ans =
This is a character string
```

Numerical values less than 0 produce a warning message when converted to character; values greater than 255 simply address characters in the font beyond `char(255)`:

```
>> a = double('a')
a =
    97
```

```
>> char(a)
ans =
a
>> char(a+256) % adding 256 does change the result
ans =
š
>> char(a-256) % negative value produces a blank character
Warning: Out of range or non-integer values truncated during conversion
to character.
ans =
```

Since strings are arrays, they can be manipulated with all the array-manipulation tools available in MATLAB. For example, in the code

```
>> u = t(16:24)
u =
cter stri
```

strings are addressed just as arrays are. Here, elements 16 through 24 contain the word `character`.

```
>> u = t(24:-1:16)
u =
irts retc
```

The result here is the word `character` spelled backwards.

Using the transpose operator changes the word `character` to a column:

```
>> u = t(16:24)'
u =
c
t
e
r
s
t
r
i
```

Single quotes within a character string are symbolized by two consecutive quotes:

```
>> v = 'I can''t find the manual!'
v =
I can't find the manual!
```

Also, string concatenation follows directly from array concatenation:

```
>> u = 'If a woodchuck could chuck wood,';
>> v = 'how much wood could a woodchuck chuck?';


>> w = [u v]
w =
If a woodchuck could chuck wood, how much wood could a woodchuck chuck?
```

The function disp allows you to display a string without printing its variable name:

```
>> disp(u)
If a woodchuck could chuck wood,
```

Note that the u = statement is suppressed. This feature is useful for displaying help text within a script file.

Like other arrays, character strings can have multiple rows, but each row must have an equal number of columns. Therefore, blanks are explicitly required to make all rows the same length, as in the following code:

```
>> v = ['Character strings having more than'
         'one row must have the same number '
         'of columns just like arrays!      ']
v =
Character strings having more than
one row must have the same number
of columns just like arrays!
```

The functions char and strvcat create multiple-row string arrays from individual strings of varying lengths:

```
>> legends = char('John','Tom','Smith','Marlon','Morgan','Jim')
legends =
John
Tom
Smith
Marlon
Morgan
Jim
>> legends = strvcat('John','Tom','Smith','Marlon','Morgan','Jim')
legends =
John
Tom
Smith
Marlon
Morgan
Jim
>> size(legends)
ans =
     6      6
```

The only difference between char and strvcat is that strvcat ignores empty string inputs, whereas char inserts blank rows for empty strings, as in the following example:

```
>> char('one','','two','three')
ans =
one

two
three

>> strvcat('one','','two','three')
ans =
one
two
three
```

Horizontal concatenation of string arrays that have the same number of rows is accomplished by the function strcat. Padded blanks are ignored. The following code is illustrative:

```
>> a = char('apples','bananas')
a =
apples
bananas
>> b = char('oranges','grapefruit')
b =
oranges
grapefruit
>> strcat(a,b)
ans =
applesoranges
bananasgrapefruit
```

Once a string array is created with padded blanks, the function deblank is useful for eliminating the extra blanks from individual rows extracted from the array:

```
>> c = legends(4,:)
c =
Marlon

>> size(c)
ans =
     1     6

>> c = deblank(legends(4,:))
c =
Marlon
>> size(c)
ans =
     1     6
```

9.2 NUMBERS TO STRINGS TO NUMBERS

There are numerous contexts in which it is desirable to convert numerical results to character strings and to extract numerical data from character strings. MATLAB provides the functions int2str, num2str, mat2str, sprintf, and fprintf for converting numerical results to character strings. Examples of the first three functions are in the following code:

```
>> int2str(eye(3)) % convert integer arrays
ans =
1  0  0
0  1  0
0  0  1
>> size(ans) % it's a character array, not a numerical matrix
ans =
     3       7

>> num2str(rand(2,4)) % convert noninteger arrays
ans =
0.95013     0.60684     0.8913     0.45647
0.23114     0.48598     0.7621     0.01850
>> size(ans) % again it is a character array
ans =

     2      43

>> mat2str(pi*eye(2)) % convert to MATLAB input syntax form!
ans =
[3.14159265358979 0;0 3.14159265358979]
>> size(ans)
ans =
     1      39
```

The following code illustrates the two functions sprintf and fprintf:

```
>> fprintf('%.4g\n',sqrt(2)) % display in Command window
1.414
```

```
>> sprintf('%.4g',sqrt(2)) % create character string
ans =
1.414

>> size(ans)
ans =
    1     5
```

sprintf and fprintf are general-purpose conversion functions that closely resemble their ANSI C language counterparts. As a result, these two functions offer the most flexibility. Normally fprintf is used to convert numerical results to ASCII format and append the converted results to a data file. However, if no file identifier is provided as the first argument to fprintf or if a file identifier of one is used, the resulting output is displayed in the *Command* window. The functions sprintf and fprintf are identical, except that sprintf simply creates a character array that can be displayed, passed to a function, or modified like any other character array. Because sprintf and fprintf are nearly identical, consider the usage of sprintf in the following example:

```
>> radius = sqrt(2);
>> area = pi * radius^2;

>>s = sprintf('A circle of radius %.5g has an area of %.5g.',radius,area)
s =
A circle of radius 1.4142 has an area of 6.2832.
```

Here, %.5g, the format specification for the variable radius, indicates that five significant digits in general-conversion format are desired. The most common usage of sprintf is to create a character string for the purposes of annotating a graph, displaying numerical values in a graphical user interface, or creating a sequence of data file names. A rudimentary example of this last usage is as follows:

```
>> i = 3;
>> fname = sprintf('mydata%.0f.dat',i)
fname =
mydata3.dat
```

In the past, the functions int2str and num2str were nothing more than a simple call to sprintf, with %.0f and %.4g as format specifiers, respectively. In MATLAB 5, int2str and num2str were enhanced to work with numerical arrays. As a result of these functions' former simplicity, in prior versions of MATLAB it was common to use the following code:

```
s = ['A circle of radius' num2str(radius)' has an area of' ...
num2str(area) '.']
s =
A circle of radius 1.4142 has an area of 6.2832.
```

Even though we obtain the same result when using int2str and num2str as we do when using sprintf, these functions require more computational effort, are more prone to typographical errors (such as missing spaces or single quotes), and require more effort to read. As a result, it is suggested that usage of int2str and num2str be limited to the conversion of arrays, as illustrated earlier in this section. In almost all other cases, it is more productive to use sprintf directly.

The help text for sprintf concisely describes its use:

```
>> help sprintf
 SPRINTF Write formatted data to string.
    STR = SPRINTF(FORMAT, A,...) applies the FORMAT to all elements of
    array A and any additional array arguments in column order, and returns
    the results to string STR.

    [STR, ERRMSG] = SPRINTF(FORMAT, A,...) returns an error message when
    the operation is unsuccessful. Otherwise, ERRMSG is empty.

    SPRINTF is the same as FPRINTF except that it returns the data in a
    MATLAB string rather than writing to a file.

    FORMAT is a string that describes the format of the output fields, and
    can include combinations of the following:

    * Conversion specifications, which include a % character, a
      conversion character (such as d, i, o, u, x, f, e, g, c, or s),
      and optional flags, width, and precision fields. For more
      details, type "doc sprintf" at the command prompt.
    * Literal text to print.
    * Escape characters, including:
            \b     Backspace            ''  Single quotation mark
            \f     Form feed            %%  Percent character
            \n     New line             \\  Backslash
            \r     Carriage return      \xN Hexadecimal number N
            \t     Horizontal tab       \N  Octal number N%
```

where \n is a line termination character on all platforms.

Notes:

If you apply an integer or string conversion to a numeric value that contains a fraction, MATLAB overrides the specified conversion, and uses %e.

Numeric conversions print only the real component of complex numbers.

Examples

```
sprintf('%0.5g',(1+sqrt(5))/2)          % 1.618
sprintf('%0.5g',1/eps)                  % 4.5036e+15
sprintf('%15.5f',1/eps)                 % 4503599627370496.00000
sprintf('%d',round(pi))                 % 3
sprintf('%s','hello')                   % hello
sprintf('The array is %dx%d.',2,3)      % The array is 2x3.
```

See also fprintf, sscanf, num2str, int2str, char.

Reference page in Help browser

doc sprintf

The following table shows how pi is displayed under a variety of conversion specifications:

Command	Result
sprintf('%.0e',pi)	3e+00
sprintf('%.1e',pi)	3.1e+00
sprintf('%.3e',pi)	3.142e+00
sprintf('%.5e',pi)	3.14159e+00
sprintf('%.10e',pi)	3.1415926536e+00
sprintf('%.0f',pi)	3
sprintf('%.1f',pi)	3.1
sprintf('%.3f',pi)	3.142
sprintf('%.5f',pi)	3.14159
sprintf('%.10f',pi)	3.1415926536
sprintf('%.0g',pi)	3

`sprintf('%.1g',pi)`	3
`sprintf('%.3g',pi)`	3.14
`sprintf('%.5g',pi)`	3.1416
`sprintf('%.10g',pi)`	3.141592654
`sprintf('%8.0g',pi)`	3
`sprintf('%8.1g',pi)`	3
`sprintf('%8.3g',pi)`	3.14
`sprintf('%8.5g',pi)`	3.1416
`sprintf('%8.10g',pi)`	3.141592654

In the preceding table, the format specifier e signifies exponential notation, f signifies fixed-point notation, and g signifies the use of e or f, whichever is shorter. Note that for the e and f formats, the number to the right of the decimal point indicates how many digits to the right of the decimal point are to be displayed. On the other hand, in the g format, the number to the right of the decimal specifies the total number of digits that are to be displayed. In addition, note that in the last five entries, a width of eight characters is specified for the result, and the result is right justified. In the very last case, the 8 is ignored because more than eight digits were specified.

Although it is not as common, sometimes it is necessary to convert or extract a numerical value from a character string. The MATLAB functions str2num, sscanf, and str2double provide this capability:

```
>> s = num2str(pi*eye(2)) % create string data
s =
3.1416          0
     0     3.1416

>> ischar(s) % True for string
ans =
     1

>> m = str2num(s) % convert string to number
m =
        3.1416              0
             0         3.1416
```

```
>> isdouble(m) % Oops, this function doesn't exist
??? Undefined function or method 'isdouble' for input arguments of
type 'double'.

>> isnumeric(m) % True for numbers
ans =
    1
>> isfloat(m) % True for floating point numbers
ans =
    1
>> pi*eye(2) - m % accuracy is lost
ans =
-7.3464e-06          0
         0   -7.3464e-06
```

The function `str2num` can contain expressions, but not variables, in the workspace:

```
>> x = pi; % create a variable

>> ss = '[sqrt(2) j; exp(1) 2*pi-x]' % string with variable x
ss =
[sqrt(2) j; exp(1) 2*pi-x]

>> str2num(ss) % conversion fails because of x
ans =
    []

>> ss = '[sqrt(2) j; exp(1) 2*pi-6]' % replace x with 6
ss =
[sqrt(2) j; exp(1) 2*pi-6]

>> str2num(ss) % now it works
ans =
  1.4142                    0 +           1i
  2.7183              0.28319
```

```
>> class(ans) % yes, its a double
ans =
double
```

The function sscanf is the counterpart to sprintf. The function sscanf reads data from a string under format control:

```
>> v = version       % get MATLAB version as a string
v =
7.11.0.584 (R2010b)

>> sscanf(v,'%f')    % get floating point numbers
ans =
             7.11
                0
           0.584

>> sscanf(v,'%f',1) % get just one floating point number
ans =
       7.11

>> sscanf(v,'%d')    % get an integer
ans =
       7

>> sscanf(v,'%s')    % get a string
ans =
7.11.0.584(R2010b)
```

You can specify the format under which sscanf operates; this capability makes sscanf a very powerful and flexible function. (See the online help text for sscanf for more thorough information about the capabilities of this function.)

When the conversion to a single double-precision value is required, the function str2double is useful. While the function str2num performs this task as well, str2double is generally quicker because of its more limited scope:

```
>> str2double('Inf') % It does convert infinity
ans =
Inf
```

```
>> class(ans)
ans =
double

>> str2double('34.6 - 23.2j') % complex numbers work
ans =
      34.6 -          23.2i

>> str2double('pi') % variables and expressions don't work
ans =
 NaN
```

9.3 STRING EVALUATION

There are some applications in which it is convenient to evaluate a character string as if it were a MATLAB expression. This is a generalization of what str2num and str2double do. These functions are limited to extracting numerical values from strings. The MATLAB functions eval and evalc bring in the entire MATLAB interpreter to evaluate any string that conforms to MATLAB syntax. For example, in the code

```
>> funs = char('floor','fix','round','ceil')
funs =
floor
fix
round
ceil
>> [deblank(funs(1,:)) '(pi)'] % display string to evaluate
ans =
floor(pi)
>> f = eval([deblank(funs(1,:)) '(pi)'])
f =
      3
>> class(f) % data type of output is numeric double
ans =
double
>> fc = evalc([deblank(funs(1,:)) '(pi)']) % try evalc
```

```
fc =
ans =
     3
>> class(fc) % output of evalc is a character string
ans =
char
>> size(fc)
ans =
     1    16
```

the function `eval` uses the MATLAB command interpreter to evaluate a character-string input and returns the results into its output argument. The function `evalc` also evaluates an input string, but its output is the character-string representation of the output. In other words, it returns the results seen in the *Command* window.

　　If at all possible, for the most part, the use of `eval` and `evalc` should be avoided. Since these functions bring in the entire MATLAB interpreter to evaluate a string expression, they incur significant overhead. In addition, the functions cannot be compiled by the MATLAB compiler, which is a MATLAB add-on product that converts MATLAB code into executable C code.

9.4 STRING FUNCTIONS

MATLAB provides a variety of string-related functions, some of which have been discussed already. The following table briefly describes many of the string functions in MATLAB:

Function	Description
char(S1,S2,...)	Creates character array from strings or cell arrays
double(S)	Converts string to ASCII representation
cellstr(S)	Creates cell array of strings from character array
blanks(n)	Creates string of n blanks
deblank(S)	Removes trailing blanks
eval(S), evalc(S)	Evaluates string expression with MATLAB interpreter
ischar(S)	True for string array
iscellstr(C)	True for cell array of strings

Function	Description
`isletter(S)`	True for letters of the alphabet
`isspace(S)`	True for white-space characters
`isstrprop(S,'property')`	True for elements that have specified property
`strcat(S1,S2,...)`	Concatenates strings horizontally
`strvcat(S1,S2,...)`	Concatenates strings vertically, ignoring blanks
`strcmp(S1,S2)`	True if strings are equal
`strncmp(S1,S2,n)`	True if n characters of strings are equal
`strcmpi(S1,S2)`	True if strings are equal, ignoring case
`strncmpi(S1,S2,n)`	True if n characters of strings are equal, ignoring case
`strtrim(S1)`	Trims leading and trailing white space
`findstr(S1,S2)`	Finds shorter string within longer string (depreciated; use `strfind`)
`strfind(S1,S2)`	Finds S2 string in S1 string
`strjust(S1,type)`	Justifies string array left, right, or center
`strmatch(S1,S2)`	Finds indices of matching strings
`strrep(S1,S2,S3)`	Replaces occurrences of S2 in S1 with S3
`strtok(S1,D)`	Finds tokens in string given delimiters
`upper(S)`	Converts to uppercase
`lower(S)`	Converts to lowercase
`num2str(x)`	Converts number to string
`int2str(k)`	Converts integer to string
`mat2str(X)`	Converts matrix to string for `eval`
`str2double(S)`	Converts string to double-precision value
`str2num(S)`	Converts string array to numerical array
`sprintf(S)`	Creates string under format control
`sscanf(S)`	Reads string under format control
`textscan(S)`	Reads formatted data from a string

Now consider some examples of the usage of some of the functions that were defined in the preceding table. For instance, the function `strfind` returns the starting indices of one string within another:

```
>> b = 'She sells sea shells on the sea shore';
>> findstr(b,' ')    % find indices of spaces
ans =
     4    10    14    21    24    28    32
>> findstr(b,'s')    % find the letter s
ans =
     5     9    11    15    20    29    33

>> find(b=='s')      % for single character searches find works too
ans =
     5     9    11    15    20    29    33
>> findstr(b,'ocean')% 'ocean' does not exist
ans =
     []
>> findstr(b,'sells') % find the string sells
ans =
     5
```

Note that `strfind` is case sensitive and returns the empty matrix when no match is found. `strfind` does not work on string arrays with multiple rows.

Tests on character strings include the following:

```
>> c = 'a2 : b_c'
c =
a2 : b_c
>> ischar(c) % it is a character string
ans =
     1
>> isletter(c) % where are the letters?
ans =
     1    0    0    0    0    1    0    1
```

```
>> isspace(c) % where are the spaces?
ans =
     0    0    1    0    1    0    0    0

>> isstrprop(c,'wspace') % same as above
ans =
     0    0    1    0    1    0    0    0

>> isstrprop(c,'digit') % True for digits
ans =
     0    1    0    0    0    0    0    0
```

To illustrate string comparison, consider the situation where a user types (perhaps into an editable text uicontrol) a string that must match, at least in part, one of a list of strings. The function strmatch provides this capability:

```
>> S = char('apple','banana','peach','mango','pineapple')
S =
apple
banana
peach
mango
pineapple
>> strmatch('pe',S) % pe is in 3rd row
ans =
     3
>> strmatch('p',S) % p is in 3rd and 5th rows
ans =
     3
     5
>> strmatch('banana',S) % banana is in 2nd row
ans =
     2
>> strmatch('Banana',S) % but Banana is nowhere
ans =
     Empty matrix: 0-by-1
```

```
>> strmatch(lower('Banana'),S) % changing B to b finds banana
ans =

     2
```

9.5 CELL ARRAYS OF STRINGS

The fact that all rows in string arrays must have the same number of columns is sometimes cumbersome, especially when the nonblank portions vary significantly from row to row. This issue of cumbersomeness is eliminated by using cell arrays. All data forms can be placed in cell arrays, but their most frequent use is with character strings. A cell array is simply a data type that allows you to name a group of data of various sizes and types, as in the following example:

```
>> C = {'This';'is';'a cell';'array of strings'}
C =

    'This'

    'is'

    'a cell'

    'array of strings'

>> size(C)
ans =

     4     1
```

Note that curly brackets { } are used to create cell arrays and that the quotes around each string are displayed. In this example, the cell array C has four rows and one column. However, each element of the cell array contains a character string of different length.

Cell arrays are addressed just as other arrays are:

```
>> C(2:3)
ans =

    'is'

    'a cell'
>> C([4 3 2 1])
ans =

    'array of strings'

    'a cell'
```

```
    'is'
    'This'
>> C(1)
ans =
    'This'
```

Here, the results are still cell arrays. That is, C(indices) addresses given cells, but not the contents of those cells. To retrieve the contents of a particular cell, use curly brackets, as in the following example:

```
>> s = C{4}
s =
array of strings
>> size(s)
ans =
      1      16
```

To extract more than one cell, the function deal is useful:

```
>> [a,b,c,d] = deal(C{:})
a =
This
b =
is
c =
a cell
d =
array of strings
```

Here, C{:} denotes all the cells as a list. That is, it's the same as

```
>> [a,b,c,d] = deal(C{1},C{2},C{3},C{4})
a =
This
b =
is
c =
a cell
```

```
d =
array of strings
>> [a,b,c,d] = C{:} % since C{:} is a list, deal is optional
a =
This
b =
is
c =
a cell
d =
array of strings
```

Partial cell array contents can also be dealt, as in the following code:

```
>> [a,b] = deal(C{2:2:4}) % get 2nd and 4th cell contents
a =
is
b =
array of strings
```

```
>> [a,b] = C{2:2:4} % the right hand side is a list, so no deal
a =
is
b =
array of strings
```

A subset of the contents of a particular cell array can be addressed as well:

```
>> C{4}(1:15) % 4th cell, elements 1 through 15
ans =
array of string
```

The multipurpose function char converts the contents of a cell array to a conventional string array, as in the following example:

```
>> s = char(C)
s =
```

```
This
is
a cell
array of strings
>> size(s) % result is a standard string array with blanks
ans =
      4    16
>> ss = char(C(1:2)) % naturally you can convert subsets
ss =
This
is
>> size(ss) % result is a standard string array with blanks
ans =
      2    4
```

The inverse conversion is performed by the function cellstr, which deblanks the strings as well:

```
>> cellstr(s)
ans =
     'This'
     'is'
     'a cell'
     'array of strings'
```

You can test whether a particular variable is a cell array of strings by using the function iscellstr:

```
>> iscellstr(C)    % True for cell arrays of strings
ans =
     1
>> ischar(C)       % True for string arrays not cell arrays of strings
ans =
     0
>> ischar(C{3})    % Contents of 3rd cell is a string array
ans =
     1
```

```
>> iscellstr(C(3))      % but 3rd cell itself is a cell
ans =
     1
>> ischar(C(3))         % and not a string array
ans =
     0

>> class(C)             % get data type or class string
ans =
cell
>> class(s)             % get data type or class string
ans =
char
```

Most of the string functions in MATLAB work with either string arrays or cell arrays of strings. In particular, examples include the functions `deblank`, `strcat`, `strcmp`, `strncmp`, `strcmpi`, `strncmpi`, `strmatch`, and `strrep`. (Further information regarding cell arrays in general can be found in Chapter 8.)

9.6 SEARCHING USING REGULAR EXPRESSIONS

The standard string functions, like `strfind`, are useful to search for specific sequences of characters within strings and to replace specific character sequences. More generally, you may want to find such *patterns* in strings as repeated characters, all uppercase letters, any capitalized words, or all dollar amounts (i.e., strings of digits preceded by a dollar sign and containing a decimal point). MATLAB supports **regular expressions**, which is a powerful tool to search for character strings within character strings. A regular expression is a formula for matching strings that follow some pattern. The Unix world has used regular expressions in many of its tools (such as `grep`, `awk`, `sed`, and `vi`) for years. Programmers who use Perl and other languages make extensive use of regular expressions. MATLAB's implementation of regular expressions is very comprehensive and has many features including some called *extended regular expressions*. And, if you are familiar with regular expressions, MATLAB's implementation will seem to you to follow naturally. If you are new to regular expressions, you can make use of the simpler features immediately and the more complex features later on.

Here are a few simple rules to get you started. A formula string, or *expression*, that describes the criteria for identifying matching portions of a character string is created. The expression consists of characters and optional modifiers that define the criteria to be used for the substring match. The simplest expression is a string of literal characters, as in the following code:

```
>> str = 'Peter Piper picked a peck of pickled peppers.'; % create a string

>> regexp(str,'pe')       % return the indices of substrings matching 'pe'
ans =
     9    22    38    41
```

Here, we found the letter combination 'pe' in the words Piper and peck and twice in the word peppers.

      Character classes are used to match a specific type of character—such as a letter, a number, or a white-space character—or to match a specific set of characters. The most useful character class is a period (.), which represents *any single character*. Another useful character class is a list of characters, or range of characters, within square brackets ([]) that matches *any member of the list or range of characters in the set*. For example, consider searching for a sequence of three characters consisting of a p, followed by any single character, followed by the letter c or the letter r:

```
>> regexpi(str,'p.[cr]')
ans =
     9    13    22    30    41
>> regexp(str,'p.[cr]','match')           % list the substring matches
ans =
     'per'    'pic'    'pec'    'pic'    'per'
```

You could also use a range of characters to find all uppercase letters in the string:

```
>> regexp(str,'[A-Z]')                    % match any uppercase letter
ans =
     1     7
```

    The character expressions in the following table match a single character with the indicated characteristics:

Character Expression	Description and Usage
.	Any single character (including white space)
[abcd35]	Any single character in the set enclosed by square brackets []
[a-zA-Z]	Any single character in the range a-z or A-Z; the dash - between characters defines an inclusive range of characters
[^aeiou]	Any single character NOT in the set of lowercase vowels; the ^ used as the first character in the set negates the sense of the set

\s	Any white-space character (space, tab, form feed, new line, or carriage return), equivalent to the set [\t\f\n\r]
\S	Any nonwhite-space character: [^ \t\f\n\r]
\w	Any "word character" (an upper- or lowercase letter, a digit, or an underscore): [a-zA-Z_0-9]
\W	Any character that is not a "word character": [^a-zA-Z_0-9]
\d	Any numeric digit: [0-9]
\D	Any nondigit character: [^0-9]
\xN or \x{N}	The character with hexadecimal value N
\oN or \o{N}	The character with octal value N
\a	The alarm, bell, or beep character: \o007 or \x07
\b	The backspace character: \o010 or \x08
\t	The horizontal tab character: \o011 or \x09
\n	The line-feed or newline character: \o012 or \x0A
\v	The vertical tab character: \o013 or \x0B
\f	The form-feed character: \o014 or \x0C
\r	The carriage-return character: \o015 or \x0D
\e	The escape character: \o033 or \x1B
\	Use a backslash to force a literal match on the next character. This enables matching on characters that have special meaning in regular expressions; for example, the characters \.*\?\\ represent a literal period, asterisk, question mark, and backslash, respectively.

The character expressions in the preceding table can be modified by using regular-expression *modifiers* or *quantifiers*. Here is an example of using the \w class, along with a modifier, to find all of the words in the string:

```
>> regexp(str,'\w+','match')          % find all individual words
ans =
        'Peter'   'Piper'   'picked'   'a'   'peck'   'of'
        'pickled' 'peppers'
```

This example shows how the + modifier changes the sense of the expression. Without the modifier, the \w expression matches any single "word character." Adding the modifier changes the expression to match all words (groups of one or more "word characters") in the string.

Character modifiers are listed in the following table:

Modifier	Description and Usage
?	Matches the preceding element zero or one time
*	Matches the preceding element zero or more times
+	Matches the preceding element one or more times
{n}	Matches the preceding element exactly n times
{n,}	Matches the preceding element at least n times
{n,m}	Matches the preceding element at least n times, but no more than m times

Character modifiers are considered *greedy* (i.e., they match the longest string they can). For example, in the code

```
>> regexp(str,'p.*p','match')

ans =

        'per picked a peck of pickled pepp'
```

the modifier matches a lowercase p, followed by zero or more of any character, followed by a lowercase p. This behavior can be changed by using the quantifier expressions listed in the following table:

Quantifier	Description and Usage
q	Matches as much of the string as possible (a *greedy quantifier*). This is the default. Here, q represents one of the modifiers in the previous table (e.g., ?, *, +, {n,m}).
q+	Matches as many components of the string as possible, without rescanning any portions of the string should the initial match fail (a *possessive quantifier*). Again, q represents one of the modifiers in the previous table (e.g., ?, *, +, {n,m}).
q?	Matches as little of the string as possible, while scanning the string (a *lazy quantifier*). Again, q represents one of the modifiers in the previous table (e.g., ?, *, +, {n,m}).

The following code matches shorter strings bracketed by lowercase p characters:

```
>> regexp(str,'p.*?p','match')

ans =

        'per p'  'peck of p'  'pep'
```

Notice that the string `'picked a p'` was not returned. Lazy quantifiers cause the string to be scanned from beginning to end. If a match is found, scanning continues, beginning with the next character in the string; previous portions are not rescanned.

Parentheses can be used to group patterns. For example, the expression in the following code matches a p, followed by one or more characters that are not the letter i:

```
>> regexp(str,'p[^i]+','match')
ans =
       'per p' 'peck of p' 'peppers'
```

The modifier applies to the preceding character. The sense of the expression changes slightly when parentheses are added. The expression in the following example matches one or more sequences of a p followed by a character that is not the letter i:

```
>> regexp(str,'(p[^i])+','match')
ans =
       'pe' 'pe' 'pepp'
```

The modifier now applies to the two-character sequence within the parentheses.

Matches can be made conditional to the context of a regular-expression match. MATLAB supports both *logical* operators and *lookaround* operators. Lookaround operators can be used to condition a match only if the string is preceded or followed by the presence or absence of another string match. The next example matches all words preceded by words ending in d:

```
>> regexp(str,'(?<=d\s)\w+','match')
ans =
       'a' 'peppers'
```

The expression matches groups of word characters only if preceded by the character d, followed by a white-space character.

Lookaround operators are listed in the following table:

Lookaround Operator	Description and Usage
p(?=q)	Lookahead. Matches pattern p only if followed by a match for pattern q
p(?!q)	Negative lookahead. Matches pattern p only if not followed by a match for pattern q
(?<=q)p	Lookbehind. Matches pattern p only if preceded by a match for pattern q
(?<!q)p	Negative lookbehind. Matches pattern p only if not preceded by a match for pattern q

Logical operators are also available. The following example matches words containing the string `'ip'` or the string `'ck'`:

```
>> regexp(str,'\w*(ip|ck)\w*','match')
ans =
    'Piper' 'picked' 'peck' 'pickled'
```

Logical operators are listed in the following table:

Logical Operator	Description and Usage
p|q	Matches the pattern p or the pattern q
^p	Matches the pattern p only if it occurs at the beginning of the string
p$	Matches the pattern p only if it occurs at the end of the string
\<p	Matches the pattern p only if it occurs at the beginning of a word
p\>	Matches the pattern p only if it occurs at the end of a word
\<p\>	Matches the pattern p only if it exactly matches a word

Repeated sequences can be matched by using tokens. When expressions contain parentheses, the string matching the expression within the parentheses (a *token*) is stored and can be reused in the expression. Up to 255 tokens can be used in an expression. Tokens can be referenced by using the syntax \d, where d is the index of the desired token. For example, \1 represents the first token, \2 represents the second token, and so on. The following example matches any doubled word character:

```
>> regexp(str,'(\w)\1','match')
ans =
    'pp'
```

The next example matches any character followed by zero or more word characters, followed by a white-space character, followed by the same initial character, followed by zero or more word characters, followed by the same white-space character:

```
>> regexp(str,'(.)\w*(\s)\1\w*\2','match')
ans =
    'Peter Piper'
```

Notice that the string `'pickled  peppers'` did not match, since there is no space after the string `'peppers'`.

The following table lists the token expressions available in MATLAB:

Expression	Description and Usage
(p)	Captures all characters matched by expression p in a token
(?:p)	Groups all characters matched by expression p, but do not save in a token
(?>p)	Groups atomically, but do not save in a token
(?#A Comment)	Inserts a comment into an expression (The comment is ignored.)
\N	Matches the Nth token in this expression (\1 is the first token, \2 is the second token, etc.)
$N	Inserts a match for the Nth token in a replacement string (regexprep function only)
(?<name>p)	Captures all characters matched by the pattern p in a token and assigns a name to the token
\k<name>	Matches the token referred to by name
S<name>	Inserts a match for the named token in a replacement string (regexprep function only)
(?(T)p)	If token T is generated (i.e., the match for token T was successful), then matches pattern p. This is an IF/THEN construction. The token can be a named token or a positional token.
(?(T)p|q)	If token T is generated, then matches pattern p; otherwise, matches pattern q. This is an IF/THEN/ELSE construction. The token can be a named token or a positional token.

There are four different regular-expression functions in MATLAB. The preceding examples used regexp. The regexpi function ignores case when matching, whereas the regexprep function replaces strings by using regular expressions. The regexptranslate function can help to generate regular expressions from character strings.

The following example searches for words beginning with the string 'pi', reverses the words, and returns the resulting string:

```
>> regexprep(str,'(pi\w*)(.*)(pi\w*)','$3$2$1')
ans =
    'Peter Piper pickled a peck of picked peppers'
```

This example generates three tokens: Token 1 is the string 'picked', token 2 is 'a peck of', and token 3 is 'pickled'. The replacement string '$3$2$1' instructs regexprep to remove the part of the original string matched by the first argument, and replace it with a string consisting of the values of the tokens, in the reverse order. The rest of the original string is returned unchanged. Numbered tokens are referenced in the search pattern by using the \N notation. The $N notation is used to reference the tokens in the replacement string.

The regular-expression functions are listed in the following table:

Function	Description and Usage
regexp	Searches for substrings by using regular expressions
regexpi	Searches for substrings by using regular expressions, ignoring case
regexprep	Searches and replaces substrings by using regular expressions
regexptranslate	Translates a wildcard string into a regular expression

There are many options available for applying and modifying the regular-expression functions. One option is the 'match' argument illustrated earlier. The regexprep function is normally case sensitive and replaces all matches it finds. Options are available to change that default behavior.

All four functions operate on cell arrays of strings, as well as on individual strings. One pattern can be used to search many strings, or different patterns can be used for different strings. Any or all of the input parameters can be a cell array of strings.

Dynamic regular expressions are available for the more adventurous or those more familiar with the usage of regular expressions. Dynamic regular expressions use independent regular expression matching or MATLAB command output to modify a regular expression dynamically. They allow you to make the pattern you want to match dependent on the string you are searching.

Dynamic operators available in MATLAB are listed in the following table:

Dynamic Operator	Description and Usage
(??p)	Parses p as a separate regular expression and includes the resulting string in the match expression
(??@cmd)	Executes the MATLAB command cmd and includes the string returned in the match expression
(?@cmd)	Executes the MATLAB command cmd but throws away any string returned by the command (often used to debug regular expressions. e.g., disp())
${cmd}	Executes the MATLAB command cmd and includes the string returned in the replacement expression

For additional examples and more thorough information about these extremely powerful and complex capabilities, see the MATLAB help documentation.

10

Relational and Logical Operations

In addition to traditional mathematical operations, MATLAB supports relational and logical operations. You may be familiar with these if you've had some experience with other programming languages. The purpose of these operators and functions is to provide answers to True/False questions. One important use of this capability is to control the flow or order of execution of a series of MATLAB commands (usually in a M-file) based on the results of True/False questions.

As inputs to all relational and logical expressions, nonzero values are considered True, and zero values are considered False. The output of all relational and logical expressions produces logical arrays with True (1) and False (0).

Logical arrays are a special type of numerical array that can be used for logical array addressing (discussed in Chapter 5), as well as in any numerical expression. Internally, logical arrays are a separate variable class that uses one byte of storage per value.

10.1 RELATIONAL OPERATORS

MATLAB relational operators include all common comparisons, as shown in the following table:

Relational Operator	Description
<	Less than
<=	Less than or equal to
>	Greater than
>=	Greater than or equal to
==	Equal to (not to be confused with =)
~=	Not equal to

MATLAB relational operators can be used to compare two arrays of the same size or to compare an array with a scalar. In the latter case, scalar expansion is used to compare the scalar with each array element and the result has the same size as the array. For example, the code

```
>> A = 8:-1:0, B = 9-A
A =
      8     7     6     5     4     3     2     1     0
B =
      1     2     3     4     5     6     7     8     9
>> tf = A>4
tf =
      1     1     1     1     0     0     0     0     0
```

finds elements of A that are greater than 4. Zeros appear in the result where A<=4, and ones appear where A>4.

The code

```
>> tf = (A==B-1)
tf =
      0     0     0     0     1     0     0     0     0
```

finds elements of A that have a value of one less than those in B.

Note that = and == mean two different things: == compares two variables and returns ones (True) where they are equal and zeros (False) where they are not; =, on the other hand, is used to assign the output of an operation to a variable.

Note that testing for equality sometimes produces confusing results for floating-point numbers:

```
>> tf = (-0.04 + 0.25 -0.21)==(0.25 - 0.21 - 0.04) % equal?
tf =
        0
>> tf = (-0.04 + 0.25 -0.21)~=(0.25 - 0.21 - 0.04) % equal?
tf =
        1
>> (-0.04 + 0.25 -0.21)-(0.25 - 0.21 - 0.04) % find the difference
ans =
  -6.9389e-018
```

This is the example we used in Chapter 2 to illustrate that arithmetic is not *exactly* commutative when using finite precision. We would expect the result to be True = 1; but it is not, because the two expressions differ by a number smaller than eps. On the MATLAB newsgroup, this fundamental fact of finite-precision arithmetic is regularly posed as a *bug* in MATLAB.

It is possible to combine relational expressions with mathematical expressions. For example, the code

```
>> tf = B - (A>2)
tf =
        0    1    2    3    4    5    7    8    9
```

finds where A>2 and subtracts the resulting vector from B. The code

```
>> A = A + (A==0)*eps
A =
        8    7    6    5    4    3    2    1  2.2204e-016
```

is a demonstration of how to replace zero elements in an array with the special MATLAB number eps, which is approximately 2.2e–16 for double-precision values. This particular expression is sometimes useful to avoid dividing by zero, as in the code

```
>> x = (-3:3)/3
x =
  -1.0000   -0.6667   -0.3333        0    0.3333    0.6667    1.0000
```

```
>> sin(x)./x
ans =
    0.8415    0.9276    0.9816    NaN    0.9816    0.9276    0.8415
```

Since `sin(0)/0` is undefined, MATLAB returns NaN (meaning Not-a-Number) at that location in the result. This can be avoided by replacing the zero with eps, as the following code shows:

```
>> x = x + (x==0)*eps;
>> sin(x)./x
ans =
Columns 1 through 6
    0.8415    0.9276    0.9816    1.0000    0.9816    0.9276
Column 7
    0.8415
```

Now `sin(x)/x` for x=0 gives the correct limiting answer. Alternatively, you can avoid the computation at x=0, for example, by using the code

```
>> x = (3:9)/3 % create new x
x =
    1.0000    1.3333    1.6667    2.0000    2.3333    2.6667    3.0000
>> y = zeros(size(x)) % create default output
y =
    0    0    0    0    0    0    0
>> tf = x~=0 % find nonzero locations
tf =
    1    1    1    1    1    1    1
>> y(tf) = sin(x(tf))./x(tf) % operate only on nonzeros
y =
Columns 1 through 4
    0.84147    0.72895    0.59724    0.45465
Columns 5 through 7
    0.30989    0.17148    0.04704
```

While this approach may seem cumbersome compared with adding eps to x, the concept of avoiding computations with selected components of an array is used often in efficient MATLAB programming.

10.2 LOGICAL OPERATORS

Logical operators provide a way to combine or negate relational expressions. MATLAB logical operators include those shown in the following table:

Logical Operator	**Description**
&	Element-by-element AND for arrays
|	Element-by-element OR for arrays
~	NOT
&&	Scalar AND with short circuiting
||	Scalar OR with short circuiting

The next four examples illustrate the use of logical operators. The code

```
>> A = 8:-1:0;  B = 9-A; % recall data
>> tf = A>4
tf =
    1   1   1   1   0   0   0   0   0
```

finds where A is greater than 4.
The code

```
>> tf = ~(A>4)
tf =
    0   0   0   0   1   1   1   1   1
```

negates the preceding result; that is, it swaps the positions of the logical ones and zeros.
The code

```
>> tf = (A>2) & (A<6)
tf =
    0   0   0   1   1   1   0   0   0
```

returns True where A is greater than 2 AND less than 6.
The code

```
>> tf = A<2 | A>7
```

```
tf =

     1     0     0     0     0     0     0     1     1
```

returns True where A is less than 2 OR greater than 7.

The scalar short-circuiting logical operators permit an early exit from logical comparisons where the operator arguments are scalars. That is, when a logical result is known before performing all relational tests, unneeded comparisons are skipped. The following code is illustrative:

```
>> a = 0; b = pi; % new data
>> a==0 || b~=1 % a=0 so this is True and b~=1 is not evaluated
ans =
     1
>> b==1 && a==0 % b~=1 so this is False and a==0 is not evaluated
ans =
     0
>> a==0 || (1/a)<1 % a==0 so 1/a is not computed
ans =
     1
```

This last example demonstrates the utility of short circuiting. Since a is equal to zero, the second expression is never evaluated because it does not change the True result. Therefore, 1/a is not computed. If it was, a divide-by-zero error would have been returned.

10.3 OPERATOR PRECEDENCE

When evaluating an expression, MATLAB follows a set of rules governing operator precedence. Operators having higher precedence are evaluated before operators of lower precedence. Operators of equal precedence are evaluated left to right. The following table illustrates the operator precedence rules used by MATLAB:

Operator	Precedence Level
Parentheses ( )	Highest
Transpose (.'), conjugate transpose ('), power (.^), matrix power (^)	
Unary plus (+), unary minus (−), negation (~)	

Multiplication (.*), matrix multiplication (*), right division (./), left division (.\\), matrix right division (/), matrix left division (\\)	
Addition (+), subtraction (−), logical negation (~)	
Colon operator (:)	
Less than (<), less than or equal to (<=), greater than (>), greater than or equal to (>=), equal to (==), not equal to (~=)	
Element-wise logical AND (&)	
Element-wise logical OR (|)	
Short-circuiting logical AND (&&)	
Short-circuiting logical OR (||)	Lowest

This order of precedence is similar, or exactly equal, to the order used by most computer programming languages. As a result, you are probably already comfortable with writing expressions that conform to these rules. As in other programming languages, parentheses can and should be used to control the order in which an expression is evaluated. Within each level of parentheses, the rules hold; for example, the code

```
>> 1|0&0
ans =
     1
>> 1|(0&0)
ans =
     1
>> (1|0)&0
ans =
     0
```

points out that the order of precedence table is different from the tables included in MATLAB versions prior to 6. In version 5 and prior versions of MATLAB, logical AND and logical OR shared the same order of precedence, whereas logical AND now has higher precedence than logical OR.

10.4 RELATIONAL AND LOGICAL FUNCTIONS

In addition to basic relational and logical operators, MATLAB provides a number of other relational and logical functions, including the following:

Function	Description
and(x,y)	Functional equivalent of the & (AND) operator: x & y
or(x,y)	Functional equivalent of the | (OR) operator: x | y
not(x,y)	Functional equivalent of the ~ (NOT) operator: ~x
xor(x,y)	Exclusive OR operation. Returns True for each element where either x or y is nonzero. Returns False where both x and y are zero, or both are nonzero.
any(x)	Returns True if any element in a vector x is nonzero. Returns True for each column in an array x that has any nonzero elements.
all(x)	Returns True if all elements in a vector x are nonzero. Returns True for each column in an array x that has all nonzero elements.

In addition to these functions, MATLAB provides numerous functions that test for the existence of specific values or conditions and return logical results as in the following table:

Function	Description
ispc	True for the PC (Windows) version of MATLAB
ismac	True for the Macintosh OS X version of MATLAB
isunix	True for the UNIX version of MATLAB
isstudent	True for MATLAB student edition
ismember	True for set member
isglobal	True for global variables
mislocked	True if an M-file cannot be cleared
isdir	True if input is a directory or folder
isempty	True for an empty matrix
isequal	True if arrays are numerically equal
isequalwithequalnans	True if arrays are equal with NaNs considered equal
isevent	True if input is object event
isfinite	True for finite elements
isfloat	True for floating-point numbers
isscalar	True for a scalar

`isinf`	True for infinite elements
`islogical`	True for a logical array
`isnan`	True for Not-a-Number
`isnumeric`	True for a numeric array
`isinteger`	True for an integer array
`isreal`	True for a real array
`isprime`	True for prime numbers
`issorted`	True if an array is sorted
`automesh`	True if the inputs should be automatically meshgridded
`inpolygon`	True for points inside a polygonal region
`isvarname`	True for a valid variable name
`iskeyword`	True for keywords or reserved words
`issparse`	True for a sparse matrix
`isvector`	True for a vector
`isappdata`	True if application-defined data exists
`ishandle`	True for graphics handles
`ishold`	True if the graphics hold state is On
`figflag`	True if a figure is currently displayed on screen
`iscellstr`	True for a cell array of strings
`ischar`	True for a character-string array
`isletter`	True for letters of the alphabet
`isspace`	True for white-space characters
`isa`	True if an object is a given class
`isstrprop`	True if a string is of a given category
`iscell`	True for a cell array
`isfield`	True if a field is in structure array
`isjava`	True for Java object arrays
`ismethod`	True for object methods
`isobject`	True for objects
`isprop`	True for object properties

Function	Description
isstruct	True for structures
isvalid	True for valid timer, handle, serial port, or instrument or device group objects
iscom	True for Component Object Model (COM) objects
isinterface	True for Component Object Model (COM) interfaces

10.5 NANS AND EMPTY ARRAYS

NaNs (Not-a-Number) and empty arrays ([]) require special treatment in MATLAB, especially when used in logical or relational expressions. According to IEEE mathematical standards, almost all operations on NaNs result in NaNs. Consider the following example:

```
>> a = [1 2 nan inf nan] % note that NaN is not case-sensitive
a =
     1     2    NaN    Inf    NaN
>> b = 2*a
b =
     2     4    NaN    Inf    NaN
>> c = sqrt(a)
c =
    1.0000    1.4142    NaN    Inf    NaN
>> d = (a==nan)
d =
     0     0     0     0     0
>> f = (a~=nan)
f =
     1     1     1     1     1
```

The first two computations give NaN results for NaN inputs. However, the final two relational computations produce somewhat surprising results: (a==nan) produces all zeros or False results even when NaN is compared with NaN; at the same time, (a~=nan) produces all ones or True results. **Thus, individual NaNs are not equal to each other.** Because of this property of NaNs, MATLAB has a built-in logical function for finding NaNs, called isnan:

```
>> g = isnan(a)
g =
     0    0    1    0    1
```

Moreover, the isnan function makes it possible to find the indices of NaNs with the find command:

```
>> i = find(isnan(a))  % find indices of NaNs
i =
     3    5
>> a(i) = zeros(size(i))  % changes NaNs in a to zeros
a =
     1    2    0    Inf    0
```

Whereas NaNs are well-defined mathematically by IEEE standards, empty arrays are defined by the creators of MATLAB and have their own interesting properties. Empty arrays are just that: They are MATLAB variables having zero length in one or more dimensions, as in the following code:

```
>> size([]) % simplest empty array
ans =
     0    0
>> c = zeros(0,5) % how about an empty array with multiple columns!
c =
   Empty matrix: 0-by-5

>> size(c)
ans =
     0    5
>> d = ones(4,0) % an empty array with multiple rows!
d =
   Empty matrix: 4-by-0
>> size(d)
ans =
     4    0
>> length(d) % its length is zero even though it has 4 rows
```

```
ans =
     0
```

This may seem strange, but allowing an empty array to have zero length in any dimension is sometimes useful. [] is just the simplest empty array.

In MATLAB, many functions return empty arrays when no other result is appropriate. Perhaps the most common example is the find function:

```
>> x = -2:2 % new data
x =
    -2    -1    0    1    2
>> y = find(x>2)
y =
Empty matrix: 1-by-0
```

In this example, x contains no values greater than 2, and so, there are no indices to return. To test for empty results, MATLAB provides the logical function isempty:

```
>> isempty(y)
ans =
     1
```

When performing relational tests where empty arrays may appear, it is important to use isempty, as the following code shows:

```
>> c == [] % comparing 0-by-5 to 0-by-0 arrays produces an error
??? Error using ==> eq
Matrix dimensions must agree.

>> isempty(c) % isempty returns the desired result.
ans =
     1
>> a = []; % create an empty variable

>> a == [] % comparing equal size empties gives empty results
ans =
     []
```

```
>> b = 1; % create nonempty variable
>> b == [] % comparing nonempty to empty produces an empty result.
ans =
     []

>> b ~= [] % even not equal comparison produces and empty result.
ans =
     []
```

The general rule is that relational operations on empty arrays produce either an error or an empty array result. Therefore, it is important to use isempty to consider the empty array case whenever it may appear.

11

Control Flow

Computer programming languages and programmable calculators offer features that allow you to control the flow of command execution using decision-making structures. If you have used these features before, this section will be very familiar to you. On the other hand, if control flow is new to you, the material may seem complicated the first time through.

Control flow is extremely powerful, since it lets past computations influence future operations. MATLAB offers five decision-making or control-flow structures: (1) For Loops, (2) While Loops, (3) If-Else-End constructions, (4) Switch-Case constructions, and (5) Try-Catch blocks. Because these constructions often encompass numerous MATLAB commands, they frequently appear in M-files, rather than having to be typed directly at the MATLAB prompt.

11.1 FOR LOOPS

For Loops allow a group of commands to be repeated for a fixed, predetermined number of times. The general form of a For Loop is

```
for x = array
    (commands)
end
```

The *(commands)* between the for and end statements are executed once for every **column** in array. At each iteration, x is assigned to the next column of array; that is, during the nth time through the loop, x = array(:,n). For example, in the code

```
>> for n = 1:10
      x(n) = cos(n*pi/10);
   end
>> x
x =
  Columns 1 through 8
    0.9511    0.8090    0.5878    0.3090    0.0000   -0.3090   -0.5878   -0.8090
  Columns 9 through 10
   -0.9511       -1.0000
```

the first statement says, "For n equals 1 to 10, evaluate all statements until the next end statement." The first time through the For Loop, n=1; the second time, n=2; and so on through the n=10 case. After the n=10 case, the For Loop ends, and any commands after the end statement are evaluated, which in this case results in the computed elements of x.

Since the loop variable is assigned to successive columns of the array on the right-hand side of the equal sign, arbitrary indexing, or inadvertent errors, can occur. For example, in the code

```
>> for n = 10:-1:1 % decrementing loop
      x(n) = cos(n*pi/10);
   end
>> x
x =
  Columns 1 through 8
    0.9511    0.8090    0.5878    0.3090    0.0000   -0.3090   -0.5878   -0.8090
  Columns 9 through 10
   -0.9511       -1.0000
```

the loop variable n counts down from 10 to 1. The expression 10:-1:1 is a standard array-creation statement that creates a row vector with multiple columns. Any numerical array can be used. In the code

```
>> i = 0; % count loop iterations
>> for n = (1:10)'
      i = i+1;
      x(n) = cos(n*pi/10);
   end
```

```
>> i % Only one time through the loop!
i =
    1
>> x
x =
  Columns 1 through 8
    0.9511    0.8090    0.5878    0.3090    0.0000   -0.3090   -0.5878   -0.8090
  Columns 9 through 10
    -0.9511      -1.0000
```

the For Loop executes only one pass! The expression (1:10)' is a column vector, and so n is set equal to the entire array (1:10)' on its first pass. Since there are no additional columns in the right-hand-side array, the loop terminates. In the code

```
>> array = randperm(10)
array =
    8    2    10    7    4    3    6    9    5    1
>> for n = array
     x(n) = sin(n*pi/10);
   end
>> x
x =
  Columns 1 through 6
    0.30902    0.58779    0.80902    0.95106    1    0.95106
  Columns 7 through 10
    0.80902    0.58779    0.30902    1.2246e-016
```

In this example the loop variable n takes on the numbers 1–10 in the random order given by array.

A For Loop cannot be terminated by reassigning the loop variable n within the loop:

```
>> for n = 1:10
     x(n) = sin(n*pi/10);
     n = 10;
   end

>> x
x =
```

```
Columns 1 through 6
   0.30902       0.58779       0.80902       0.95106           1       0.95106
Columns 7 through 10
   0.80902       0.58779       0.30902       1.2246e-016
```

To repeat, the right-hand-side array in the For Loop statement can be any valid array-creation statement:

```
>> i = 1;
>> for x = rand(4,5)
      y(i) = sum(x);
      i = i+1;
   end
>> y
y =
   2.4385     1.5159     2.5453     1.8911     1.5018
```

Here, the loop variable x is assigned to the successive 4-by-1 columns of a random array. Since the For Loop has no natural loop index, i was added.

Naturally, For Loops can be nested as desired:

```
>> for n = 1:5
      for m = 5:-1:1
          A(n,m) = n^2 + m^2;
      end
      disp(n)
   end
   1
   2
   3
   4
   5
>> A
A =
    2     5    10    17    26
    5     8    13    20    29
```

10	13	18	25	34
17	20	25	32	41
26	29	34	41	50

Just because the preceding examples were used to illustrate For Loop usage, it doesn't mean that they are examples of efficient MATLAB programming. Historically, For Loops represented poor programming practice whenever an equivalent array approach existed. The equivalent array approach, called a *vectorized* solution, often is orders of magnitude faster than the scalar approaches just shown. For example, in the code

```
>> n = 1:10;
>> x = cos(n*pi/10)
x =
  Columns 1 through 8
    0.9511   0.8090   0.5878   0.3090   0.0000   -0.3090   -0.5878   -0.8090
  Columns 9 through 10
   -0.9511   -1.0000
```

the two statements duplicate the repeated example of computing the sine function at 10 angles. In addition to being orders of magnitude faster, the preceding vectorized solution is more intuitive, is easier to read, and requires less typing.

The earlier nested For Loop is equivalent to the following vectorized code:

```
>> n = 1:5;
>> m = 1:5;
>> [nn,mm] = meshgrid(n,m);

>> A = nn.^2 + mm.^2
A =
     2     5    10    17    26
     5     8    13    20    29
    10    13    18    25    34
    17    20    25    32    41
    26    29    34    41    50
```

As discussed in Chapter 5, arrays should be preallocated before a For Loop (or While Loop) is executed. Doing so minimizes the amount of memory allocation required. For example, in the first case considered in this section, every time the commands within the For Loop are executed, the size of the variable x is increased by 1, which forces MATLAB to take the time to allocate more memory for x every time it goes through the loop. To eliminate this step, the For Loop example should be rewritten as follows:

```
>> x = zeros(1,10); % preallocated memory for x
>> for n = 1:10
       x(n) = sin(n*pi/10);
   end
```

In this case, only the values of x(n) need to be changed each time through the loop. Memory allocation occurs once outside the loop, so no memory allocation overhead bogs down the operations within the loop.

Starting with MATLAB 6.5, improvements were made to the MATLAB interpreter to minimize the processing overhead involved in executing loops. These improvements are collectively known as the ***JIT-accelerator***. As stated in the documentation, such improvements will appear over a series of MATLAB releases. When the JIT-accelerator was introduced, MATLAB code containing loops would benefit from JIT-acceleration if the code had the following features and properties:

1. The loop structure is a For Loop.
2. The loop contains only logical, character-string, double-precision, and less than 64-bit integer data types.
3. The loop uses arrays that are three dimensional or less.
4. All variables within a loop are defined prior to loop execution.
5. Memory for all variables within the loop are preallocated and maintain constant size and data type for all loop iterations.
6. Loop indices are scalar quantities, such as the index i in for i=1:N.
7. Only built-in MATLAB functions are called within the loop.
8. Conditional statements with *if-then-else* or *switch-case* constructions (introduced later in this chapter) involve scalar comparisons.
9. All lines within the block contain no more than one assignment statement.
10. The number of iterations is significant (more than a few).

JIT-acceleration provides the greatest benefit when the arrays that are addressed within the loop are relatively small and the number of iterations is large. As array sizes increase, computational time increases and the percentage of time spent on processing overhead decreases, thereby leading to less dramatic improvements in overall execution time. The JIT-accelerator has continued to evolve and now handles much more of the MATLAB language (e.g., objects), but still incurs costs the first time (or few times) through a loop as it analyzes and compiles the code. At this time, the JIT-accelerator offers significant improvements for a large subset of MATLAB syntax, data types, and array sizes.

The following code demonstrates the capabilities of the JIT-accelerator:

```
N = 1e6;
% generate sin(x) at 1e6 points by using array mathematics
% this is often called a 'vectorized' solution.

x = linspace(0,2*pi,N);
y = sin(x);          % vectorized solution requires two lines

% redo code using JIT-acceleration.
i = 0;
y = zeros(1,N);     % initialize all variables within loop
x = zeros(1,N);     % and allocate all memory

for i=1:N                    % scalar loop variable
    x(i) = 2*pi*(i-1)/N;   % only built-in function calls
    y(i) = sin(x(i));
end
```

With JIT-acceleration, both approaches take approximately the same time to execute. However, prior to the existence of the JIT-accelerator, the For Loop approach would have been orders of magnitude slower than the vectorized solution. For this particular case, the array mathematics approach is much shorter and much easier to read, so the JIT-acceleration approach has little value for solving this problem. JIT-acceleration proves its value in more substantial problems where it is difficult or impossible to compose a vectorized solution by using array mathematics.

11.2 WHILE LOOPS

As opposed to a For Loop that evaluates a group of commands a fixed number of times, a While Loop evaluates a group of statements an indefinite number of times.
The general form of a While Loop is

```
while expression
    (commands)
end
```

The *(commands)* between the while and end statements are executed as long as *all* elements in *expression* are True. Usually, evaluation of *expression* gives a scalar result, but array results are also valid. In the array case, *all* elements of the resulting array must be True.

One way of computing the double-precision value eps, which is the smallest number that can be added to 1 such that the result is greater than 1, using finite precision is

```
>> num = 0; EPS = 2;
>> while (1+EPS)>2
       EPS = EPS/2;
       num = num+1;
   end

>> num
num =
    1

>> EPS = 2*EPS
EPS =
    2
```

Here, we used uppercase EPS so that the MATLAB value eps is not overwritten. In this example, EPS starts at 1. As long as (1+EPS)>2 is True (nonzero), the commands inside the While Loop are evaluated. Since EPS is continually divided in two, it eventually gets so small that adding EPS to 1 is no longer greater than 2. (Recall that this happens because a computer uses a fixed number of digits to represent numbers. Double precision specifies approximately 16 digits, so we would expect eps to be near 10^{-16}.) At this point, (1+EPS)>2 is False (zero) and the While Loop terminates. Finally, EPS is multiplied by 2, because the last division by 2 made it too small by a factor of 2.

For array expressions, the While Loop continues only when *all* elements in *expression* are True. If you want the While Loop to continue when *any* element is True, simply use the function any. For instance, while any(*expression*) returns a scalar logical True whenever any of its contents are True.

11.3 IF-ELSE-END CONSTRUCTIONS

Many times, sequences of commands must be conditionally evaluated on the basis of a relational test. In programming languages, this logic is provided by some variation of an If-Else-End construction. The simplest If-Else-End construction is

```
if expression
    (commands)
end
```

The *(commands)* between the if and end statements are evaluated if *all* elements in *expression* are True (nonzero).

In cases where *expression* involves several logical subexpressions, only the minimum number required to determine the final logical state are evaluated. For example, if *expression* is *(expression1 | expression2)*, then *expression2* is evaluated only if *expression1* is False. Similarly, if *expression* is *(expression1 & expression2)*, then *expression2* is not evaluated if *expression1* is False. Note that this short circuiting occurs in If-Else-End constructions even if the specific short-circuiting operators || or && are not used.

The following example is illustrative:

```
>> apples = 10;            % number of apples
>> cost = apples*25        % cost of apples
cost =
    250

>> if apples>5             % give 20% discount for larger purchases
       cost = (1-20/100)*cost;
end
>> cost
cost =
200
```

In cases where there are two alternatives, the If-Else-End construction is

```
if expression
    (commands evaluated if True)
else
    (commands evaluated if False)
end
```

Here, the first set of commands is evaluated if *expression* is True; the second set is evaluated if *expression* is False.

When there are three or more alternatives, the If-Else-End construction takes the form

```
if expression1
    (commands evaluated if expression1 is True)
```

```
elseif expression2

    (commands evaluated if expression2 is True)
elseif expression3

    (commands evaluated if expression3 is True)
elseif expression4

    (commands evaluated if expression4 is True)
elseif expression5

    .

    .

    .

else

    (commands evaluated if no other expression is True)
end
```

In this last form, only the commands associated with the first True expression encountered are evaluated; ensuing relational expressions are not tested; and the rest of the If-Else-End construction is skipped. Furthermore, the final `else` command may or may not appear.

Now that we know how to make decisions with If-Else-End constructions, it is possible to show a legal way to break out of For Loops and While Loops:

```
>> EPS = 2;
>> for num = 1:1000
    EPS = EPS/2;
    if (1+EPS)<=2
        EPS = EPS*2
        break
    end
end
EPS =
    2

>> num
num =
    1
```

This example demonstrates another way of estimating the double-precision value eps. In this case, the For Loop is instructed to run some sufficiently large number of times. The If-Else-End structure tests to see if EPS has gotten small enough. If it has, EPS is multiplied by 2, and the `break` command forces the For Loop to end prematurely, which, in this case is at `num=53`.

Furthermore, when the `break` statement is executed, MATLAB jumps to the next statement outside of the loop in which the `break` statement appears. Therefore, it returns to the MATLAB prompt and displays EPS. If a `break` statement appears in a nested For Loop or While Loop structure, MATLAB only jumps out of the immediate loop in which the `break` statement appears. It does not jump all the way out of the entire nested structure.

MATLAB version 6 introduced the command `continue` for use in For Loops and While Loops. When MATLAB encounters a `continue` statement inside of a For Loop or While Loop, it immediately jumps to the `end` statement of the loop, bypassing all of the commands between the `continue` command and the `end` statement. In doing so, the `continue` command moves immediately to the expression test for the next pass through the loop:

```
>> EPS = 2;
>> for num = 1:1000
       EPS = EPS/2;
       if (1+EPS)>2
           continue
       end
       EPS = EPS*2
       break
   end
EPS =
    2
```

Here, the previous example is rewritten to use the `continue` command. Note that the `continue` command has no effect on the If-End construction.

11.4 SWITCH-CASE CONSTRUCTIONS

When sequences of commands must be conditionally evaluated on the basis of repeated use of an equality test with one common argument, a Switch-Case construction is often easier. Switch-Case constructions have the form

```
switch expression
   case test_expression1
      (commands1)
```

```
    case {test_expression2, test_expression3, test_expression4}
        (commands2)
    otherwise
        (commands3)
end
```

where *expression* must be either a scalar or a character string. If *expression* is a scalar, *expression==test_expressionN* is tested by each case statement. If *expression* is a character string, strcmp(*expression, test_expression*) is tested. In this example, *expression* is compared with *test_expression1* at the first case statement. If they are equal, (*commands1*) are evaluated, and the rest of the statements before the end statement are skipped. If the first comparison is not true, the second is considered. In this example, *expression* is compared with *test_expression2, test_expression3*, and *test_expression4*, which are contained in a cell array. If any of these are equal to *expression*, (*commands2*) are evaluated, and the rest of the statements before end are skipped. If all case comparisons are false, (*commands3*) following the optional otherwise statement are executed. *Note that this implementation of the Switch-Case construction allows at most one of the command groups to be executed.*

A simple example demonstrating the Switch-Case construction is

```
x = 2.7;
units = 'm';
switch units % convert x to centimeters
    case {'inch','in'}
        y = x*2.54;
    case {'feet','ft'}
        y = x*2.54*12;
    case {'meter','m'}
        y = x/100;
    case {'millimeter','mm'}
        y = x*10;
    case {'centimeter','cm'}
        y = x;
    otherwise
        disp(['Unknown Units: ' units])
        y = nan;
end
```

Executing this code gives a final value of y = 0.027.

11.5 TRY-CATCH BLOCKS

A Try-Catch block provides user-controlled error-trapping capabilities. That is, with a Try-Catch block, errors found by MATLAB are captured, giving the user the ability to control the way MATLAB responds to errors. Try-Catch blocks have the form

```
try
    (commands1)
catch
    (commands2)
end
```

Here, all MATLAB expressions in (commands1) are executed. If no MATLAB errors are generated, control is passed to the end statement. However, if a MATLAB error appears while executing (commands1), control is immediately passed to the catch statement and subsequent expressions in (commands2). The code within (commands2) can make use of the functions lasterr and lasterror to access information about the error and act accordingly.

Consider the following example, implemented in a script M-file for convenience:

```
x = ones(4,2);
y = 4*eye(2);
try
    z = x*y;
catch
    z = nan;
    disp('X and Y are not conformable.')
end
z
```

With the preceding data for x and y, this code segment produces the following output:

```
z =
     4     4
     4     4
     4     4
     4     4
```

In this case, only the code in the Try block was executed. Changing the variable y creates an error:

```
x = ones(4,2);
y = 4*eye(3); % now wrong size
try
    z = x*y;
catch
    z = nan;
    disp('X and Y are not conformable.')
end
z
```

Executing the code this time generates the following output in the *Command* window:

```
X and Y are not conformable.
z =
    NaN
```

In addition, the function `lasterr` describes the error found:

```
>> lasterr
ans =
Error using ==> mtimes
Inner matrix dimensions must agree.
```

More detailed information is returned by the structure output of the function `lasterror`:

```
>> lasterror
ans =
    message: 'Error using ==> mtimes
Inner matrix dimensions must agree.'
   identifier: 'MATLAB:innerdim'
        stack: [0x1 struct]
```

Here, the content of the `message` field is the same as the output from `lasterr`. The `identifier` field describes the message identifier, which in this case classifies the error type as being generated by an inner-dimension error in MATLAB.

The Catch block also may perform tasks and then reissue the original error by using the `rethrow` function. For example, revising and rerunning the code from the previous example produces

```
x = ones(4,2);
y = 4*eye(3); % now wrong size
try
    z = x*y;
catch ERR        % capture the error in an MException object (ERR)
    z = nan;
    disp('X and Y are not conformable.')
    rethrow(ERR) % process error as if Try-Catch did not happen.
end
z
```

In this case, the *Command* window displays

```
X and Y are not conformable.
??? Error using ==> mtimes
Inner matrix dimensions must agree.
```

The `rethrow` function reissues the error, terminates execution, and therefore does not display the contents of z as requested by the last code line.

Functions

When you use MATLAB functions such as inv, abs, angle, and sqrt, MATLAB takes the variables that you pass to it, computes the required results using your input, and then passes these results back to you. The commands evaluated by the function, as well as any intermediate variables created by these commands, are hidden. All you see is what goes in and what comes out. In other words, a function is a black box.

This property makes functions very powerful tools for evaluating commands that encapsulate useful mathematical functions or sequences of commands that often appear when you are solving some larger problem. Because of the usefulness of this power, MATLAB provides several structures that enable you to create functions of your own. These structures include **M-file** functions, **anonymous** functions, and **inline** functions. Of these, M-file functions are the most common. M-file functions are text files that contain MATLAB code and a function header. The function mmempty is a good example of an M-file function:

```
function d = mmempty(a,b)
%MMEMPTY Substitute Value if Empty.
% MMEMPTY(A,B) returns A if A is not empty,
% otherwise B is returned.
%
% Example: The empty array problem in logical statements
% let a = []; then use MMEMPTY to set default logical state
        % (a==1) is [], but MMEMPTY(a,1)==1 is true
        % (a==0) is [], but MMEMPTY(a,0)==0 is true
```

```
    o:
    sum(a) is 0, but sum(MMEMPTY(a,b)) = sum(b)
% prod(a) is 1, but prod(MMEMPTY(a,b)) = prod(b)
%
% See also ISEMPTY, SUM, PROD, FIND
if isempty(a)
        d = b;
else
        d = a;
end
```

A function M-file is similar to a script M-file in that it is a text file having a .m extension. Like script M-files, function M-files are not entered in the *Command* window, but rather are external text files created with a text editor (probably the Editor/Debugger that comes with MATLAB). A function M-file is different from a script file in that a function communicates with the MATLAB workspace only through the variables passed to it and through the output variables it creates. Intermediate variables within the function do not appear in, or interact with, the MATLAB workspace. As can be seen in the example, the first line defines the M-file as a function and specifies its name, which is the same as its filename without the .m extension. The first line also defines the M-file's input and output variables. The next continuous sequence of comment lines comprises the text displayed in response to the help command `help mmempty` or `helpwin mmempty`. The first help line, called the H1 line, is the line searched by the `lookfor` command. Finally, the remainder of the M-file contains MATLAB commands that create the output variables. Note that there is no `return` command in mmempty; the function simply terminates after it executes the last command. However, you can use the `return` command to terminate execution before reaching the end of the M-file.

12.1 M-FILE FUNCTION CONSTRUCTION RULES

Function M-files must satisfy a number of criteria and should have a number of desirable features:

1. The function M-file name and the function name (e.g., mmempty) that appear in the first line of the file should be identical. In reality, MATLAB ignores the function name in the first line and executes functions on the basis of the file name stored on disk.

2. Function M-file names can have up to 63 characters. This maximum may be limited by the operating system, in which case the lower limit applies. MATLAB ignores characters beyond the 63rd or the operating system limit, and so longer names can be used, provided the legal characters point to a unique file name.

3. Function M-file names are case sensitive on UNIX platforms, and as of MATLAB 7, they are now case sensitive on Windows platforms as well. To avoid platform dependencies across MATLAB versions, it is beneficial to use only lowercase letters in M-file names.

4. Function names must begin with a letter. Any combination of letters, numbers, and underscores can appear after the first character. Function names cannot contain spaces or punctuation characters. This naming rule is identical to that for variables.

5. The first line of a function M-file is called the ***function-declaration line*** and must contain the word `function` followed by the calling syntax for the function in its most general form. The input and output variables identified in the first line are variables local to the function. The input variables contain data passed to the function, and the output variables contain data passed back from the function. It is not possible to pass data back through the input variables.

6. The first set of contiguous comment lines after the function-declaration line are the help text for the function. The first comment line is called the H1 line and is the line searched by the `lookfor` command. The H1 line typically contains the function name in uppercase characters and a concise description of the function's purpose. Comment lines after the first describe possible calling syntaxes, algorithms used, and simple examples, if appropriate.

7. Function names appearing in the help text of a function are normally capitalized only to give them visual distinction. Functions are called by matching the exact case of the letters making up their filenames.

8. All statements following the first set of contiguous comment lines compose the body of the function. The body of a function contains MATLAB statements that operate on the input arguments and produce results in the output arguments.

9. A function M-file terminates after the last line in the file is executed or whenever a `return` statement is encountered. If an M-file contains nested functions, each function in the M-file requires a terminating `end` statement.

10. A function can abort operation and return control to the *Command* window by calling the function `error`. This function is useful for flagging improper function usage, as shown in the following code fragment:

```
if length(val) > 1
    error('VAL must be a scalar.')
end
```

When the function `error` in the preceding code is executed, the string `'VAL must be a scalar.'` is displayed in the *Command* window, after a line identifying the file that produced the error message. Passing an empty string to `error` (e.g., error('')) causes no action to be taken. After being issued, the error character string is passed to the functions lasterror and lasterr for later recall. It is also possible to pass numerical data to the displayed error string by using the function error as one would use the function sprintf. For instance, the previous example could be revised as follows:

```
val = zeros(1,3);
if length(val) > 1
    error('VAL has %d elements but must be a scalar.',length(val))
end
??? VAL has 3 elements but must be a scalar.
```

The %d specification indicates that an integer format should be used to insert the value of length(val) into the given place in the error string. When MATLAB identifies an error, it also creates an error message identifier string that is returned by the function lasterror, as in the following code:

```
>> eig(eye(2,4))
??? Error using ==> eig
Matrix must be square.
>> lasterror
ans =
        message: 'Error using ==> eig
Matrix must be square.'
    identifier: 'MATLAB:square'
         stack: [0x1 struct]
```

Here, the identifier field identifies the error source as coming from MATLAB and having to do with square matrices. To include this information in an M-file function, simply add it as a first argument to the error function, as illustrated in the following code:

```
>> val = zeros(1,3);
>> msg = 'VAL has %d elements but must be a scalar.';
if length(val) > 1
    error('MyToolbox:scalar',msg,length(val))
end
??? VAL has 3 elements but must be a scalar.
>> lasterror
```

```
ans =

    message: 'VAL has 3 elements but must be a scalar.'
  identifier: 'MyToolbox:scalar'
       stack: [0x1 struct]
```

The `identifier` field now shows that this error was flagged by a function in MyToolbox, rather than by MATLAB itself, and that the error was related to a scalar.

11. A function can report a warning and then continue operation by calling the function `warning`. This function has the same calling syntax as the function `error`. Warnings can contain simple character strings, strings containing formatted data, and an optional initial message identifier string. The difference between the `warning` and `error` functions is that warnings can be turned Off globally, or have warnings associated with specific message identifiers turned Off. Warning states can also be queried. (See MATLAB documentation for more thorough information about the features of the function `warning`.)

12. Function M-files can contain calls to script files. When a script file is encountered, it is evaluated in the function's workspace, not in the MATLAB workspace.

13. Multiple functions can appear in a single function M-file. Additional functions, called subfunctions or local functions, are simply appended to the end of the primary function. Subfunctions begin with a standard function statement line and follow all function construction rules.

14. Subfunctions can be called by the primary function in the M-file, as well as by other subfunctions in the same M-file, but subfunctions cannot be called directly from outside the M-file. Like all functions, subfunctions have their own individual workspaces.

15. Subfunctions can appear in any order after the primary function in an M-file. Help text for subfunctions can be displayed by entering >> `helpwin` *func>subfunc*, where *func* is the main function name and *subfunc* is the subfunction name.

16. It is suggested that subfunction names begin with the word `local`, for example, `local_myfun`. This practice improves the readability of the primary function, because calls to local functions are clearly identifiable. All local function names can have up to 63 characters. The following function `mmclass` demonstrates the use of subfunctions:

```
function c = mmclass(arg)
%MMCLASS MATLAB Object Class Existence.
% MMCLASS returns a cell array of strings containing the
% names of MATLAB object classes available with this license.
%
```

```
% MMCLASS('ClassName') returns logical True (1) if the class
% having the name 'ClassName' exists with this license.
% Otherwise logical False (0) is returned.
%
% MMCLASS searches the MATLABPATH for class directories.
% Classes not on the MATLABPATH are ignored.
%
% See also CLASS, ISA, METHODS, ISOBJECT
persistent clist    % save data for future calls

if isempty(clist)   % clist contains no data, so create it
    clist = local_getclasslist;
end
if nargin==0
    c = clist;
elseif ischar(arg)
    c = ~isempty(strmatch(arg,clist));
else
    error('Character String Argument Expected.')
end

function clist = local_getclasslist
%LOCAL_GETCLASSLIST Get list of MATLAB classes
%
% LOCAL-GETCLASSLIST returns a list of all MATLAB classes
% in a cell array of strings.
clist = cell(0);
cstar = [filesep '@*'];
dlist = [pathsep matlabpath];
sidx = findstr(pathsep,dlist)+1;      % path segment starting indices
eidx = [sidx(2:end)-2 length(dlist)]; % path segment ending indices

for i = 1:length(sidx)-1  % look at each path segment
cdir = dir([dlist(sidx(i):eidx(i)) cstar]);  % dir @* on segment
```

```
clist = [clist {cdir.name}];                    % add results to list
end
cstr = char(clist);        % convert to string array
cstr(:,1) = [];            % eliminate initial '@'
cstr = unique(cstr,'rows');% alphabetize and make unique
clist = cellstr(cstr);     % back to a cell array
% end of subfunction
```

17. In addition to subfunctions, M-file functions can contain *nested* functions. Nested functions are defined completely within another function in an M-file. (Refer to Section 12.8 for more information on nested functions.)

18. In addition to subfunctions and nested functions, M-files can call private M-files, which are standard function M-files that reside in a subdirectory of the calling function entitled `private`. Only functions in the immediate parent directory of private M-files have access to private M-files. Private subdirectories are meant to contain utility functions useful to several functions in the parent directory. Private function M-file names need not be unique, because of their higher precedence and limited scope.

19. It is suggested that private M-file names begin with the word `private`, for example, `private_myfun`. This practice improves the readability of the primary function, because calls to private functions are clearly identifiable. Like other function names, the names of all private M-files can have up to 63 characters.

12.2 INPUT AND OUTPUT ARGUMENTS

MATLAB functions can have any number of input and output arguments. The features of, and criteria for, these arguments are as follows:

1. Function M-files can have zero input arguments and zero output arguments.

2. Functions can be called with fewer input and output arguments than are specified in the function-definition line in the M-file. Functions cannot be called with more input or output arguments than the M-file specifies.

3. The number of input and output arguments used in a function call can be determined by calls to the functions `nargin` and `nargout`, respectively. Since `nargin` and `nargout` are functions, not variables, one cannot reassign them with statements such as `nargin = nargin - 1`. The function `mmdigit` illustrates the use of `nargin`:

```
function y = mmdigit(x,n,b,t)
%MMDIGIT Round Values to Given Significant Digits.
% MMDIGIT(X,N,B) rounds array X to N significant places in base B.
% If B is not given, B = 10 is assumed.
% If X is complex the real and imaginary parts are rounded separately.
% MMDIGIT(X,N,B,'fix') uses FIX instead of ROUND.
% MMDIGIT(X,N,B,'ceil') uses CEIL instead of ROUND.
% MMDIGI T(X,N,B,'floor') uses FLOOR instead of ROUND.

if nargin<2
    error('Not enough input arguments.')
elseif nargin==2
    b = 10;
    t = 'round';
elseif nargin==3
    t = 'round';
end
n = round(abs(n(1)));
if isempty(b),     b = 10;
else               b = round(abs(b(1)));
end
if isreal(x)
    y = abs(x)+(x==0);
    e = floor(log(y)./log(b)+1);
    p = repmat(b,size(x)).^(n-e);
    if strncmpi(t,'round',1)
        y = round(p.*x)./p;
    elseif strncmpi(t,'fix',2)
        y = fix(p.*x)./p;
    elseif strncmpi(t,'ceil',1)
        y = ceil(p.*x)./p;
    elseif strncmpi(t,'floor',2)
        y = floor(p.*x)./p;
```

```
    else
        error('Unknown rounding requested')
    end
else % complex input
    y = complex(mmdigit(real(x),n,b,t),mmdigit(imag(x),n,b,t));
end
```

In `mmdigit`, `nargin` is used to assign default values to input arguments not supplied by the user.

4. When a function is called, the input variables are not copied into the function's workspace, but their values are made ***readable*** within the function. However, if any values in the input variables are changed, the array is then copied into the function's workspace. Thus, to conserve memory and increase speed, it is better to extract elements from large arrays and then modify them, rather than to force the entire array to be copied into the function's workspace. Note that using the same variable name for both an input and an output argument causes an immediate copying of the contents of the variable into the function's workspace. For example, `function y = myfunction(x,y,z)` causes the variable y to be immediately copied into the workspace of `myfunction`.

5. If a function declares one or more output arguments, but no output is desired, simply do not assign the output variable (or variables) any values. Alternatively, the function `clear` can be used to delete the output variables before terminating the function.

6. Functions can accept a variable and an unlimited number of input arguments by specifying `varargin` as the last input argument in the function-declaration line. The argument `varargin` is a predefined cell array whose ith cell is the ith argument, starting from where `varargin` appears. For example, consider a function having the following function-declaration line:

```
function a = myfunction(varargin)
```

If this function is called as `myfunction(x,y,z)`, then `varargin{1}` contains the array x, `varargin{2}` contains the array y, and `varargin{3}` contains the array z. Likewise, if the function is called as `myfunction(x)`, then `varargin` has length 1 and `varargin{1}` = x. Every time `myfunction` is called, it can be called with a different number of arguments.

   In cases where one or more input arguments are fixed, `varargin` must appear as the last argument:

```
function a = myfunction(x,y,varargin)
```

If this function is called as myfunction(x,y,z), then, inside of the function, x and y are available, and varargin{1} contains z. In any case, the function nargin returns the actual number of input arguments used. (For further information on cell arrays, see Chapter 8.)

7. Functions can accept a variable, unlimited number of output arguments by specifying varargout as the last output argument in the function-declaration line. The argument varargout is a predefined cell array whose ith cell is the ith argument, starting from where varargout appears. For example, consider a function having the following function-declaration line:

```
function varargout = myfunction(x)
```

If this function is called as [a,b] = myfunction(x), then, inside of the function, the contents of varargout{1} must be assigned to the data that become the variable a, and the contents of varargout{2} must be assigned to the data that become the variable b. As with varargin, discussed previously, the length of varargout is equal to the number of output arguments used and nargout returns this length. In cases where one or more output arguments are fixed, varargout must appear as the last argument in the function-declaration line, that is, function [a,b,varargout] = myfunction(x) (see Chapter 8).

8. The functions nargchk and nargoutchk provide simple error checking for the number of valid input and output arguments, respectively. Since functions automatically return an error if called with more input or output arguments that appear in their function definitions, these functions have limited value. They may be useful, however, when a function definition declares an arbitrary number of input or output arguments.

12.3 FUNCTION WORKSPACES

As stated earlier, functions are black boxes. They accept inputs, act on these inputs, and create outputs. Any and all variables created within the function are hidden from the MATLAB (or base) workspace. Each function has its own temporary workspace that is created with each function call and deleted when the function completes execution. MATLAB functions can be called recursively, and each call has a separate workspace. In addition to furnishing input and output arguments, MATLAB provides several techniques for communicating among function workspaces and the MATLAB (or base) workspace:

1. Functions can share variables with other functions, the MATLAB workspace, and recursive calls to themselves if the variables are declared global. To gain access to a global variable within a function or the MATLAB workspace, the variable must be declared global—for example, global myvariable.

As a matter of programming practice, the use of global variables is discouraged whenever possible. However, if they are used, it is suggested that global variable names be long, contain all capital letters, and optionally start with the name of the M-file where they appear—for example, MYFUN_ALPHA. If followed, these suggestions will minimize unintended conflicts among global variables.

2. In addition to sharing data through global variables, function M-files can have restricted access to variables for repeated or recursive calls to themselves by declaring a variable persistent, such as persistent myvariable. Persistent variables act like global variables whose scope is limited to the function where they are declared. Persistent variables exist as long as an M-file remains in memory in MATLAB. The function mmclass illustrates the use of persistent variables:

```
function c = mmclass(arg)
%MMCLASS MATLAB Object Class Existence.
% MMCLASS returns a cell array of strings containing the
% names of MATLAB object classes available with this license.
%
% MMCLASS('ClassName') returns logical True (1) if the class
% having the name 'ClassName' exists with this license.
% Otherwise logical False (0) is returned.
%
% MMCLASS searches the MATLABPATH for class directories.
% Classes not on the MATLABPATH are ignored.
%
% See also CLASS, ISA, METHODS, ISOBJECT
persistent clist % save data for future calls

if isempty(clist) % clist contains no data, so create it
    clist = cell(0);
    cstar = [filesep '@*'];
    dlist = [pathsep matlabpath];
    sidx = findstr(pathsep,dlist)+1;       % path segment starting indices
    eidx = [sidx(2:end)-2 length(dlist)]; % path segment ending indices
```

```
    for i = 1:length(sidx)-1      % look at each path segment
        cdir = dir([dlist(sidx(i):eidx(i)) cstar]); % dir @* on segment
        clist = [clist {cdir.name}];              % add results to list
    end
    cstr = char(clist);          % convert to string array
    cstr(:,1) = [];              % eliminate initial '@'
    cstr = unique(cstr,'rows');  % alphabetize and make unique
    clist = cellstr(cstr);       % back to a cell array
end
if nargin==0
    c = clist;
elseif ischar(arg)
    c = ~isempty(strmatch(arg,clist));
else
    error('Character String Argument Expected.')
end
```

In `mmclass`, the variable `clist` is declared persistent. The first time `mmclass` is called during a MATLAB session, `clist` is created as an empty array. When the function finds it empty, it fills it with data in the first If-End construction in the function. In future calls to `mmclass`, `clist` exists because of its persistence, and re-creating `clist` in the future is unnecessary. For these subsequent function calls, the data previously stored in `clist` is simply reused.

3. MATLAB provides the function `evalin` to allow you to reach into another workspace, evaluate an expression, and return the result to the current workspace. The function `evalin` is similar to `eval`, except that the string is evaluated in either the **caller** or the **base** workspace. The caller workspace is the workspace where the current function was called from. The base workspace is the MATLAB workspace in the *Command* window. For example, A=evalin('caller', 'expression') evaluates 'expression' in the caller workspace and returns the results to the variable A in the current workspace. Alternatively, A=evalin ('base','expression') evaluates 'expression' in the MATLAB workspace and returns the results to the variable A in the current workspace. The function evalin also provides error trapping with the syntax evalin('workspace', 'try','catch'), where 'workspace' is either 'caller' or 'base'; 'try' is the first expression evaluated; and 'catch' is an expression that is evaluated in the **current** workspace if the evaluation of 'try' produces an error.

4. Since you can evaluate an expression in another workspace, it makes sense that you can also assign the results of some expression in the current workspace to a variable in another workspace. The function `assignin` provides this capability. For example, `assignin('`*`workspace`*`','`*`vname`*`',X)`, where `'`*`workspace`*`'` is either `'caller'` or `'base'`, assigns the contents of the variable X in the current workspace to a variable named `'`*`vname`*`'` in the `'caller'` or `base'` workspace.

5. The function `inputname` provides a way to determine the variable names used when a function is called. For example, suppose a function is called as

```
>> y = myfunction(xdot,time,sqrt(2))
```

Issuing `inputname(1)` inside of `myfunction` returns the character string `'xdot'`, `inputname(2)` returns `'time'`, and `inputname(3)` returns an empty array, because `sqrt(2)` is not a variable, but rather an expression that produces an unnamed temporary result.

The function `mmswap` illustrates the use of `evalin`, `assignin`, and `inputname`:

```
function mmswap(x,y)
%MMSWAP Swap Two Variables.
% MMSWAP(X,Y) or MMSWAP X Y swaps the contents of the
% variable X and Y in the workspace where it is called.
% X and Y must be variables not literals or expressions.
%
% For example: Rat = ones(3); Tar=pi; MMSWAP(Rat,Tar) or MMSWAP Rat Tar
% swaps the contents of the variables named Rat and Tar in the
% workspace where MMSWAP is called giving Rat = pi and Tar = ones(3).
if nargin~=2
    error('Two Input Arguments Required.')
end
if ischar(x) & ischar(y)   % MMSWAP X Y 'string input arguments'

    % check existence of arguments in caller
    estr = sprintf('[exist(''%s'',''var'') exist(''%s'',''var'')]',x,y);
    t = evalin('caller',estr);
    if all(t)                          % both x and y are valid
        xx = evalin('caller',x);       % get contents of x
```

```
      yy = evalin('caller',y);           % get contents of y
      assignin('caller',y,xx)            % assign contents of x to y
      assignin('caller',x,yy)            % assign contents of y to x

   elseif isequal(t,[0 1])               % x is not valid
      error(['Undefined Variable: ''' x '''])

   elseif isequal(t,[1 0])               % y is not valid
      error(['Undefined Variable: ''' y ''''])

   else                                  % neither is valid
      error(['Undefined Variables: ''' x ''' and ''' y ''''])
   end
else                         % MMSWAP(X,Y) 'numerical input arguments'
   xname = inputname(1);     % get x argument name if it exists
   yname = inputname(2);     % get y argument name if it exists

   if ~isempty(xname) & ~isempty(yname)  % both x and y are valid
      assignin('caller',xname,y)         % assign contents of y to x
      assignin('caller',yname,x)         % assign contents of x yo y

   else
      error('Arguments Must be Valid Variables.')
   end
end
```

6. The name of the M-file being executed is available within a function in the variable mfilename. For example, when the M-file myfunction.m is being executed, the workspace of the function contains the variable mfilename, which contains the character string 'myfunction'. This variable also exists within script files, in which case it contains the name of the script file being executed.

12.4 FUNCTIONS AND THE MATLAB SEARCH PATH

Function M-files and their powerful features are among the fundamental strengths of MATLAB. They allow you to encapsulate sequences of useful commands and apply them over and over. Since M-files exist as text files on disk, it is important that

MATLAB maximize the speed at which the files are found, opened, and executed. The following are the techniques that MATLAB uses to maximize speed:

1. The first time MATLAB executes a function M-file, it opens the corresponding text file and **_compiles_** the commands into an internal pseudocode representation in memory that speeds execution for all later calls to the function. If the function contains references to other M-file functions and script M-files, they too are compiled into memory.

2. The function `inmem` returns a cell array of strings containing a list of functions and script files that are currently compiled into memory.

3. Issuing the command `mlock` within an M-file locks the compiled function so that it cannot be cleared from memory. For example, `clear functions` does not clear a locked function from memory. By locking an M-file, persistent variables declared in a function are guaranteed to exist from one call to the next. Issuing the command `munlock` within an M-file unlocks the compiled function. The function call `munlock('FUN')` unlocks the function FUN, so that it can be cleared from memory. The function `mislocked('FUN')` returns True if the function FUN is currently locked in memory. By default, function M-files are unlocked.

4. It is possible to store the pseudocode, or P-code, version of a function M-file to disk by using the `pcode` command. When this is done, MATLAB loads the P-file, rather than the M-file, into memory. For most functions, this step does not significantly shorten the amount of time required to execute a function for the first time. However, it can speed up large M-files associated with complex GUI functions. P-code files are created by issuing the command

```
>> pcode myfunction
```

where _myfunction_ is the M-file name to be compiled. P-code files are encrypted, platform-independent binary files that have the same name as the original M-file but end in .p rather than in .m. P-code files provide a level of security, since they are visually indecipherable and can be executed without the corresponding M-file. Furthermore, it is not possible to convert a P-code file into an M-file. Because of their binary nature, P-code files are not necessarily backward compatible across MATLAB versions. That is, a P-code file created by using MATLAB 7.5 will not run on MATLAB 6 or even 7.4. However, they are generally forward compatible. That is, a P-code file created using MATLAB 7.4 will usually run on MATLAB 7.5 or even 7.10.

5. When MATLAB encounters a name it doesn't recognize, it follows a set of precedence rules to determine what to do. For example, when you enter `cow` at the MATLAB prompt or if MATLAB encounters a reference to `cow` in a script or function M-file,

   a. MATLAB checks to see if `cow` is a **_variable_** in the current workspace; if not,
   b. it checks to see if `cow` is a **_nested_** function in the function in which `cow` appears, if not,

c. it checks to see if cow is a ***subfunction*** in the file in which cow appears, if not,

d. it checks to see if cow is a ***private*** function to the file in which cow appears, if not,

e. it checks to see if cow exists in the ***current directory***, if not,

f. it checks to see if cow exists in each directory specified on the ***MATLAB search path***, by searching in the order in which the search path is specified.

MATLAB uses the first match it finds. In addition, in steps d, e, and f, it prioritizes by type considering MEX-files first, followed by P-code files, and then by M-files. So if cow.*mex*, cow.p, and cow.m exist, MATLAB uses cow.*mex*, where *mex* is replaced by the platform-dependent MEX-file extension. If cow.p and cow.m exist, MATLAB uses cow.p. In step f, if cow.m exists on the search path and cow is a *built-in* function, the built-in function is executed. If cow.m exists and is not a built-in function, the M-file is executed.

6. When MATLAB is started, it ***caches*** the name and location of all M-files stored within the toolbox subdirectory and within all subdirectories of the toolbox directory. This allows MATLAB to find and execute function M-files much faster.

   M-file functions that are cached are considered read-only. If they are executed and then later altered, MATLAB simply executes the function that was previously compiled into memory, ignoring the changed M-files. Moreover, if new M-files are added within the toolbox directory after MATLAB is already running, their presence will not be noted in the cache, and thus they will be unavailable for use.

   As a result, as you develop M-file functions, it is best to store them outside of the toolbox directory, perhaps in the MATLAB directory, until you consider them to be complete. When they are complete, move them to a subdirectory inside of the read-only toolbox directory. Finally, make sure the MATLAB search path cache is changed to recognize their existence.

7. When new M-files are added to a ***cached*** location, MATLAB finds them only if the cache is refreshed by the command rehash toolbox. On the other hand, when cached M-files are modified, MATLAB recognizes the changes only if a previously compiled version is dumped from memory by issuing the clear command; for example, >> clear *myfun* clears the M-file function *myfun* from memory. Alternatively, >> clear functions clears all unlocked, compiled functions from memory.

8. MATLAB keeps track of the modification date of M-files outside of the toolbox directory. As a result, when an M-file function that was previously compiled into memory is encountered, MATLAB compares the modification date of the compiled M-file with that of the M-file on disk. If the two dates are the same, MATLAB executes the compiled M-file. On the other hand, if the M-file on disk is newer, MATLAB dumps the previously compiled M-file and compiles the newer, revised M-file for execution.

9. It is possible to check all of the file dependencies for a function M-file by using the function `depfun`. This function rigorously parses an M-file for all calls to other M-file functions, built-in functions, and function calls in `eval` strings and callbacks, and identifies variable and Java classes used. This function is helpful for identifying function dependencies in M-files that are being shared with others who may not have the same *toolboxes* installed. The function `depdir` uses `depfun` to return a listing of the dependent directories of an M-file.

12.5 CREATING YOUR OWN TOOLBOX

It is common to organize a group of M-files into a subdirectory on the MATLAB search path. If the M-files are considered complete, the subdirectory should be placed in the `toolbox` directory so that the M-file names are cached, as described earlier. When a `toolbox` subdirectory is created, it is beneficial to include two additional script M-files containing only MATLAB comments (i.e., lines that begin with a percent sign %). These M-files, named `Readme.m` and `Contents.m`, have the following properties:

1. The script file `Readme.m` typically contains comment lines that describe late breaking changes or descriptions of undocumented features. Issuing the command `>> whatsnew MyToolbox` (where `MyToolbox` is the name of the directory containing the group of M-files) displays this file in the *Help* window. If the toolbox is posted to *MATLAB Central* at *The Mathworks* website, the `Readme.m` file should include a disclaimer, such as the following, to avoid legal problems:

```
% These M-files are User Contributed Routines that are being redistributed
% by The Mathworks, upon request, on an "as is" basis. A User Contributed
% Routine is not a product of The Mathworks, Inc. and The Mathworks assumes
% no responsibility for any errors that may exist in these routines.
```

2. The script file `Contents.m` contains comment lines that list all the M-files in the toolbox. Issuing the command `>> doc MyToolbox` (where `MyToolbox` is the name of the directory containing the group of M-files) displays the file listing in the *Help* window. The first line in the `Contents.m` file should specify the name of the toolbox, and the second line should state the toolbox version and date, as follows:

```
% Toolbox Description
% Version xxx dd-mmm-yyyy
```

This information is used by the `ver` command, which lists installed toolbox information.

3. When writing a collection of M-files to form a toolbox, it is sometimes convenient to allow the user to maintain a set of preferences for toolbox use or for one or more functions in the toolbox. While it is always possible to store this information in a MAT-file and retrieve it in later MATLAB sessions, doing so requires choosing a directory location for the preferences file and guaranteeing that the file isn't moved or deleted between sessions. To eliminate these weaknesses of the data file approach, MATLAB provides the functions `getpref`, `setpref`, `addpref`, and `rmpref`. These functions allow you to get, set, add, and remove preferences, respectively. Preferences are organized in groups so that preferences for multiple activities are supported. Within each group, individual preferences are named with character strings, and the values stored can be any MATLAB variable. The function `ispref` is provided to verify the existence of a specific group or preference. When these functions are used, the handling of preference files is hidden from the user. Where they are stored is system dependent, and they remain persistent from one MATLAB session to the next.

12.6 COMMAND–FUNCTION DUALITY

In addition to creating function M-files, it is also possible to create MATLAB *commands*. Examples of MATLAB commands include `clear`, `who`, `dir`, `ver`, `help`, and `whatsnew`. MATLAB commands are very similar to functions. In fact, there are only two differences between commands and functions:

1. Commands do not have output arguments.
2. Input arguments to commands are not enclosed in parentheses.

For example, `clear functions` is a command that accepts the input argument `functions` without parentheses, performs the action of clearing all compiled functions from memory, and produces no output. A function, on the other hand, usually places data in one or more output arguments and must have its input arguments separated by commas and enclosed in parentheses (e.g., `a=atan2(x,y)`).

In reality, MATLAB commands are function calls that obey the two differences. For example, the command `whatsnew` is a function M-file. When called from the MATLAB prompt as

```
>> whatsnew MyToolbox
```

MATLAB interprets the command as a call to the function `whatsnew` with the following syntax:

```
>> whatsnew('MyToolbox')
```

In other words, as long as there are no output arguments requested, MATLAB interprets command arguments as character strings, places them in parentheses, and then calls the requested function. This interpretation applies to all MATLAB commands.

Both command and function forms can be entered at the MATLAB prompt, although the command form generally requires less typing. A function M-file can also be interpreted as a function, if it obeys the rules for calling functions. For example, the command

```
>> which fname
```

displays the directory path string to the M-file *fname*, and the function call

```
>> s = which('fname')
```

returns the directory path string in the variable s. At the same time, the code

```
>> s = which fname
??? s = which fname
            |
Error: Unexpected MATLAB expression.
```

causes an error because it mixes function and command syntaxes. ***Whenever MATLAB encounters an equal sign, it interprets the rest of the statement as a function, which requires a comma-separated list of arguments enclosed in parentheses.***

To summarize, both commands and functions call functions. Commands are translated into function calls by interpreting command arguments as character strings, placing them in parentheses, and then calling the requested function. Any function call can be made in the form of a command if it produces no output and if it requires only character-string input.

12.7 FUNCTION HANDLES AND ANONYMOUS FUNCTIONS

There are a number of occasions when the identity of a function must be passed to another function for evaluation. For example, many of the numerical analysis functions in MATLAB evaluate a function provided by the user as part of the function's input arguments. For example, Chapter 23 discusses the function quad, which, when called as quad(Fun,low,high), computes the area under the function Fun over the range from low to high. Historically, the function argument Fun was specified by the character-string name of the function to be evaluated (e.g., *sin*(x) was denoted as 'sin'). This method works for both built-in functions and M-file functions. Alternatively, MATLAB 5 introduced *inline functions*, which create functions from character-string expressions. In MATLAB 7.0 and above, the use of strings identifying function names and the use of inline functions remain supported, but are discouraged in favor of using ***anonymous functions*** and their corresponding ***function handles***.

Anonymous functions are created as shown in the following example:

```
>> af_humps = @(x) 1./((x-.3).^2 +.01) +1./((x-.9).^2 +.04) - 6;
```

Here, the @ symbol identifies that the left-hand side is to be a *function handle*. The (x) defines the list of function arguments, and the remainder of the line describes the function expression. Evaluation of this function uses the function handle name itself to perform function evaluation:

```
>> z = af_humps([-1 0 1])
z =
    -5.1378       5.1765          16
```

The definition of an anonymous function can access any MATLAB function as well as the present content of variables that exist in the workspace where the anonymous function is created. For example, in the code

```
>> a = -.3; b = -.9;
>> af_humpsab = @(x) 1 ./ ((x+a).^2 + .01) + 1 ./ ((x+b).^2 + .04) - 6;

>> af_humpsab([-1 0 1])
ans =
    -5.1378       5.1765          16
```

the values of the previously defined values of a and b become part of the anonymous function definition. If the values of a or b change, the anonymous function does not change. The function handle af_humpsab captures and holds a snapshot of the function at the time it is created:

```
>> a = 0; % changing the value of a does not change the function
>> af_humpsab([-1 0 1]) % evaluate again, get the same results
ans =
    -5.1378       5.1765          16
```

The concept of creating function handles applies to built-in and M-file functions as well. For example, in the code

```
>> fh_Mfile = @humps % function handle for M-file function
fh_Mfile =
    @humps
```

```
>> fh_Mfile(1) % evaluate humps(1)
ans =
      16
>> fh_builtin = @cos % function handle for built-in function
fh_builtin =
      @cos

>> fh_builtin(pi) % evaluate cos(pi)
ans =
      -1
```

a function handle for an M-file function or built-in function is created by using the @ symbol, followed immediately by the name of the function to be converted. Function handles can also be placed in cell arrays and evaluated by using the same approach:

```
>> fhan = {@humps @cos}
fhan =
      [@humps] [@cos]

>> fhan{1}(1) % evaluate humps(1)
ans =
      16
>> fhan{2}(pi) % evaluate cos(pi)
ans =
      -1
```

Here, the cell array fhan contains a function handle to the M-file function humps.m and a function handle to the built-in function cos. Evaluation of the two functions follows the approach outlined earlier, with fhan{1} and fhan{2} addressing the first and second function handles, respectively. In MATLAB 6, function handles were standard arrays, so the preceding handles could be created using standard brackets, as in fhan = [@humps @cos]. This standard-array format for function handles is no longer supported.

To support function handles, MATLAB offers a number of useful functions:

```
>> functions(fh_Mfile)
ans =
      function: 'humps'
          type: 'simple'
          file: '/usr/local/matlab/toolbox/matlab/demos/humps.m'
```

```
>> functions(fh_builtin)
ans =
    function: 'cos'
        type: 'simple'
        file:''

>> functions(af_humps)
ans =
    function: '@(x)1./((x-.3)^2+.01)+1./((x-.9).^2+.04)-6'
        type: 'anonymous'
        file: ''
    workspace: {[1x1 struct]}
```

The function functions returns information about the function handle. (For the most part, this information is used for debugging purposes only. MATLAB warns that the content of the structure returned by functions is subject to change.)

When the name of a function is stored in a character-string variable, the function str2func provides a means for creating a function handle:

```
>> myfunc = 'humps' % place name of humps.m in a string variable
myfunc =
humps
>> fh2 = @myfunc % this doesn't work!
??? Error: "myfunc" was previously used as a variable,
conflicting with its use here as the name of a function or command.
See MATLAB Programming, "How MATLAB Recognizes Function Calls That Use
Command Syntax" for details.
>> fh2 = str2func(myfunc) % this works
fh2 =
    @humps

>> isequal(fh2,fh_Mfile) % these are equal function handles
ans =
    1
```

The inverse operation of str2func is func2str. As shown next, it simply extracts the name of the function or the string identifying the anonymous function:

```
>> func2str(fh2) % M-file function
ans =
humps
>> func2str(af_humps) % anonymous function
ans =
@(x)1./((x-.3).^2+.01)+1./((x-.9).^2+.04)-6
>> class(ans) % output is a character string
ans =
char


>> isa(fh2,'function_handle') % True for function handles
ans =
     1
```

Function handles are an extremely powerful and beneficial feature in MATLAB. First, they capture all of the information needed to evaluate the function at the time of function handle creation. As a result, when a function defined by a function handle is evaluated, MATLAB does not need to search for it on the MATLAB path. It immediately evaluates it. For example, as shown before, the file location for humps.m is stored as part of the function handle fh_Mfile. The overhead time required to find the function is eliminated, which improves performance, especially when a function is evaluated repeatedly.

Another powerful and beneficial feature of function handles is that they can point to subfunctions, private functions, and nested functions (to be discussed next) that normally are not visible from outside of the functions where they appear or are referenced. That is, if the output of a function contains a function handle to a subfunction, private function, or nested function that is visible or within the scope of the function returning the function handle, the returned function handle can be evaluated. For example, the following M-file function skeleton demonstrates returning a function handle to a subfunction:

```
function out = myfunction(select)
%MYFUNCTION Return function handle to a subfunction.
% Example function demonstrating function handles to subfunctions.

switch select
```

```
case 'case1'
    out = @local_subfun1;
case 'case2'
    out = @local_subfun2;
otherwise
    out = [];
    error('Unknown Input.')
end

function a = local_subfun1(b,c)
%LOCAL_SUBFUN Some function operation.

% code that operates on the input arguments b and c
% and returns content in the variable a

% end of local_subfun1

function d = local_subfun2(e,f)
%LOCAL_SUBFUN Some function operation.

% code that operates on the input arguments e and f
% and returns content in the variable d

% end of local_subfun2
```

On the basis of this function skeleton, the following sample code creates a function handle to a subfunction in myfunction and then evaluates the subfunction outside of the context of the original function myfunction:

```
>> h_subfun = myfunction('case2'); % handle to local_subfun2
>> dout = h_subfun(x,y); % execute local_subfun2(e,f)
```

This works because all of the information required to execute either subfunction is captured by the function handle when it is created. Therefore, the subfunction local_subfun2 can be evaluated, because h_subfun has complete knowledge of the content and location of local_subfun2.

12.8 NESTED FUNCTIONS

Nested functions were introduced in MATLAB 7.0. If you are unfamiliar with the concept of nested functions, they can appear to be strange and confusing, and they can promote poor programming practice. Indeed, they can be all of these things if used improperly. However, they can also be immensely helpful in some situations.

Fundamentally, nested functions provide a way to pass information to and from a function without using global variables, and without passing information through the input and output arguments. The following is the basic form of a simple nested function:

```
function out = primary_function(...)
%PRIMARYFUNCTION primary function.

% code in primary function
% this code can call nested functions

    function nout1 = nested_function1(...)
    % Code in nested_function1.
    % In addition to variables passed through the input arguments
    % this nested function has access to all variables in existence
    % in the primary function at the point where this function
    % definition appears. This access permits reading and writing
    % to all variables available.

    end % required to mark the end of nested_function1
% other code in primary_function, including other
% nested functions terminated with end statements

end % end statement required here for primary_function
```

As shown in the preceding code, nested functions are functions that are fully contained within the definition of another function, with end statements marking the end of each nested function and the primary function as well. As opposed to subfunctions, which are functions appended to the primary function, nested functions have access not only to data passed through the nested function input arguments, but also to variables in the parent function. A nested function has its own workspace that includes read and write access to the primary function workspace. In addition, the primary function has read and write access to variables defined in functions nested within it.

In general, a primary function can have any number of nested functions. In addition, nested functions can have nested functions within them. Arbitrary nesting is permitted, but usually is not that useful. In addition, a nested function has access to variables passed to it, and to the workspaces of all functions in which it is nested. If this sounds confusing to you, you are not alone. Nested functions add a level of complexity that can make M-file debugging difficult. Moreover, it may not be clear what beneficial purpose this apparent complexity provides.

The example that follows illustrates a prime use for nested functions. In this rational polynomial example, the primary function returns a function handle to a nested function, which obtains data directly from the primary function workspace:

```
function fhandle = nestexample(num,den)
%NESTEXAMPLE Example Nested Function.
%  NESTEXAMPLE(Num,Den) returns a function handle to a function that can
%  be used to evaluate a rational polynomial. Num and Den are vectors
%  containing the numerator and denominator polynomial coefficients.
%
%  For example, ratpoly = nestexample([1 2],[1 2 3]) returns a function
%  handle that facilitates evaluation of the rational polynomial
%
%     x + 2
%  ------------
%  x^2 + 2x + 3
if ~isnumeric(num) || ~isnumeric(den)
    error('Num and Den Must be Numeric Vectors.')
end
num = reshape(num,1,[]); % make num into a row vector
den = reshape(den,1,[]); % make den into a row vector
fhandle = @nested_ratpoly; % create function handle for return
    function out = nested_ratpoly(x)
    % Nested function that evaluates a rational polynomial, where the
    % numerator and denominator coefficients are obtained from the primary
    % function workspace. Only the evaluation points x appear as an input
    % argument.

    out = polyval(num,x)./polyval(den,x);
```

```
end % nested function terminated with an end statement
end % primary function terminated with an end statement too!
```

Using this function, function handles to specific rational polynomials can be created, as demonstrated by the following code:

```
>> ratpoly1 = nestexample([1 2],[1 2 3]) % (x + 2)/(x^2 + 2x + 3);

>> ratpoly2 = nestexample([2 1],[3 2 1]) % (2x +1)/(3x^2 + 2x +1);

>> x = linspace(-10,10); % independent variable data
>> y1 = ratpoly1(x);      % evaluate first rational polynomial
>> y2 = ratpoly2(x);      % evaluate second rational polynomial

>> plot(x,y1,x,y2) % plot created data
>> xlabel('X')
>> ylabel('Y')
>> title('Figure 12.1: Rational Polynomial Evaluation')
```

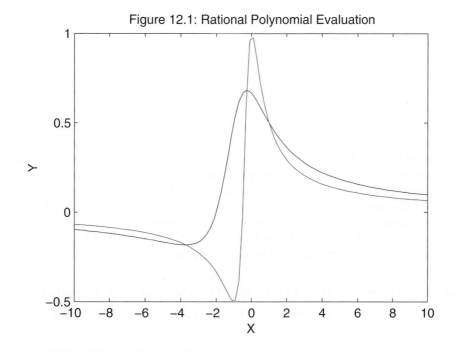

Figure 12.1: Rational Polynomial Evaluation

The two calls to `nestexample` in the preceding code return function handles to two different rational polynomials. Because each rational polynomial obtained its coefficients from the input arguments to `nestexample`, and these coefficients were available to the nested function through a shared workspace, the coefficients are not needed for evaluating the rational polynomials. The coefficients are contained in their respective function handles.

Prior to the existence of nested functions, rational polynomial evaluation using function handles could have been implemented with the following function:

```
function out = ratpolyval(num,den,x)
%RATPOLYVAL Evaluate Rational Polynomial.
% RATPOLYVAL(Num,Den,X) evaluates the rational polynomial, whose
% numerator and denominator coefficients are given by the vectors
% Num and Den respectively, at the points given in X.
if ~isnumeric(num) || ~isnumeric(den)
    error('Num and Den Must be Numeric Vectors.')
end
num = reshape(num,1,[]); % make num into a row vector
den = reshape(den,1,[]); % make den into a row vector
out = polyval(num,x)./polyval(den,x);
```

Evaluation of the two rational polynomials used earlier proceeds as follows:

```
>> yy1 = ratpolyval([1 2],[1 2 3],x); % same as y1 above

>> yy2 = ratpolyval([2 1],[3 2 1],x); % same as y2 above
```

In this case, the numerator and denominator polynomials are required every time a rational polynomial is evaluated. While this is straightforward and common in MATLAB, the nested function approach allows you to encapsulate all information required for a particular rational polynomial into a function handle that is easily evaluated.

Because rational polynomial evaluation can be written as a single statement, this example can also be implemented by using anonymous functions, while retaining the features of the nested function approach. That is, the two function handles `ratpoly1` and `ratpoly2` could be created by the following statements:

```
>> num = [1 2]; den = [1 2 3]; % specify numerator and denominator
>> ratpoly1 = @(x) polyval(num,x)./polyval(den,x); % create anonymous fun
```

```
>> num = [2 1]; den = [3 2 1]; % redefine numerator and denominator
>> ratpoly2 = @(x) polyval(num,x)./polyval(den,x); % create anonymous fun

>> y1 = ratpoly1(x); % evaluate first rational polynomial
>> y2 = ratpoly2(x); % evaluate second rational polynomial
```

This process of creating handles to nested functions and using function handles to evaluate functions is extremely powerful. It allows you to create multiple instances of a function, each with a shared purpose (such as rational polynomial evaluation), while permitting distinct properties as well (such as different numerator and denominator coefficients). Equally important is the fact that each instance has its own workspace, and therefore, each operates independently of all others.

12.9 DEBUGGING M-FILES

In the process of developing function M-files, it is inevitable that errors (i.e., **bugs**) appear. MATLAB provides a number of approaches and functions to assist in **debugging** M-files.

Two types of errors can appear in MATLAB expressions: syntax errors and run-time errors. Syntax errors (such as misspelled variables or function names or missing quotes or parentheses) are found when MATLAB evaluates an expression or when a function is compiled into memory. MATLAB flags these errors immediately and provides feedback on the type of error encountered and the line number in the M-file where the error occurred. Given this feedback, these errors are usually easy to spot. An exception to this situation occurs in syntax errors within GUI callback strings. These errors are not detected until the strings themselves are evaluated during the operation of the GUI.

Run-time errors, on the other hand, are generally more difficult to find, even though MATLAB flags them also. When a run-time error is found, MATLAB returns control to the *Command* window and the MATLAB workspace. Access to the function workspace where the error occurred is lost, and so you cannot interrogate the contents of the function workspace in an effort to isolate the problem.

According to the authors' experience, the most common run-time errors occur when the result of some operation leads to empty arrays or to NaNs. All operations on NaNs return NaNs, so if NaNs are a possible result, it is good to use the logical function isnan to perform some default action when NaNs occur. Addressing arrays that are empty always leads to an error, since empty arrays have a zero dimension. The find function represents a common situation where an empty array may result. If the empty array output of the find function is used to index some other array, the result returned will also be empty. That is, empty matrices tend to propagate empty matrices, as, for example, in the following code:

```
>> x = pi*(1:4)  % example data
x =
   3.1416    6.2832    9.4248    12.5664

>> i = find(x>20)  % use find function
i =
  Empty matrix: 1-by-0

>> y = 2*x(i)  % propagate the empty matrix
y =
  Empty matrix: 1-by-0
```

Clearly, when y is expected to have a finite dimension and values, a run-time error is likely to occur. When performing operations or using functions that can return empty results, it helps to use the logical function isempty to define a default result for the empty matrix case, thereby avoiding a run-time error.

There are several approaches to debugging function M-files. For simple problems, it is straightforward to use a combination of the following:

1. Remove semicolons from selected lines within the function so that intermediate results are displayed in the *Command* window.
2. Add statements that display variables of interest within the function.
3. Place the keyboard command at selected places in the M-file to give temporary control to the keyboard. By doing so, the function workspace can be interrogated and values changed as necessary. Resume function execution by issuing a return command, K>> return, at the keyboard prompt.
4. Change the function M-file into a script M-file by placing a % before the function definition statement at the beginning of the M-file. When executed as a script file, the workspace is the MATLAB workspace, and thus it can be interrogated after the error occurs.

When the M-file is large, recursive, or highly nested (i.e., it calls other M-file functions that call still other functions, etc.) it is more convenient to use the MATLAB graphical debugging functions found on the **Debug** and **Breakpoints** menus of the Editor/Debugger. *Command* window equivalents of these functions exist, but are more cumbersome to use. If you insist on using these functions rather than the graphical debugger, see the online help text for debug—that is,

```
>> doc debug
```

The graphical debugging tools in MATLAB allow you to stop at user-set breakpoints, at MATLAB warnings and errors, and at expressions that create NaNs and Infs. You can also set conditional breakpoints on the basis of some condition tested in the breakpoint itself. When a breakpoint is reached, MATLAB stops execution before the affected line completes execution and returns results. While the program is stopped, the *Command* window keyboard prompt K>> appears, allowing you to interrogate the function workspace, change the values of variables in the workspace, and so on, as required to track down the bug. In addition, the *Editor/Debugger* window shows the line where execution stopped and provides the means for stepping into the workspace of other functions that were called by the M-file being debugged. When you are ready to move beyond a breakpoint, the Editor/Debugger provides menu items for single-stepping, continuing until the next breakpoint, continuing until the cursor position is reached, or terminating the debugging activity.

Describing the use of the graphical debugging tools is difficult because of their graphical nature and because the process of debugging is unique to each debugging session. However, the authors have found the debugging tools to be intuitive and easy to use. Once the basic steps are mastered, the graphical debugging features of MATLAB are extremely powerful and productive in assisting the process of creating good MATLAB code.

12.10 SYNTAX CHECKING AND FILE DEPENDENCIES

During the process of creating and debugging an M-file, it is advantageous to check the code for syntax errors and run-time errors. In the past, this was accomplished by executing the M-file and watching for warnings and errors reported in the *Command* window. Alternatively, it was done by using the pcode command to create a P-code file from the M-file. Now the function mlint (introduced in MATLAB 7) parses M-files for syntax errors and other possible problems and inefficiencies. For example, mlint will point out variables that are defined but never used, input arguments that are not used, output arguments that are not assigned, obsolete usages, statements that are unreachable, and so on. In addition, mlint will make suggestions to improve execution time. This function is commonly called as a command—for example,

```
>> mlint myfunction
```

where *myfunction* is the name of an M-file anywhere on the MATLAB search path. This function can also be called using function syntax, in which case the output can be captured in a number of different ways. For example, the function

```
>> out = mlint('myfunction','-struct');
```

returns the output of mlint in the structure variable out. This feature is also available as the Code Analyzer from the Tools menu of the MATLAB Editor. In addition to checking a single function by using mlint, the function mlintrpt accepts a directory name as an input argument, applies `mlint` to all M-files in the directory, and provides a report for each function in a separate HTML-based window. This feature is also available as a choice from the *Current Folder* window in the MATLAB desktop.

Given the ease with which M-files can be transferred electronically, it is not uncommon to run an M-file only to find out that it fails to run because it calls one or more M-file functions not found on your MATLAB path. MATLAB provides the function `depfun`, which parses an M-file for file dependencies. This function recursively searches for all function dependencies, including the dependencies within functions called by the function in question as well as any dependencies found in the callbacks of Handle Graphics objects. Because of the large amount of output produced by this function, it is difficult to demonstrate here and is best explored firsthand.

12.11 PROFILING M-FILES

Even when a function M-file works correctly, there may be ways to fine-tune the code to avoid unnecessary calculations or function calls. Performance improvements can be obtained by simply storing the result of a calculation to avoid a complex recalculation, by preallocating arrays before accessing them within loops, by storing and accessing data in columns, by using vectorization techniques to avoid an iterative procedure such as a For Loop, or by taking advantage of MATLAB JIT-acceleration features. When writing a function, it is difficult to guess where most execution time is spent. In today's high-speed processors, with integrated floating-point units, it may be faster to calculate a result more than once than to store it in a variable and recall it again later. There is an inherent trade-off between memory usage and the number of computations performed. Depending on the data being manipulated, it may be faster to use more memory to store intermediate results, or it may be faster to perform more computations. Furthermore, the trade-off between memory and computations is almost always dependent on the size of the data set being considered. If a function operates on large data sets, the optimum implementation may be much different from that of a small data set. Complicating things even further is the fact that the best computer implementation of a given algorithm is often much different from how one writes the algorithm on paper.

MATLAB provides ***profiling*** tools to optimize the execution of M-file functions. These tools monitor the execution of M-files and identify which lines consume the greatest amount of time relative to the rest of the code. For example, if one line (or function call) consumes 50 percent of the time in a given M-file, the attention paid to this line (or function call) will have the greatest impact on overall execution speed. Sometimes, you can rewrite the code to eliminate the offending line or lines. Other times, you can minimize the amount of data manipulated in the line, thereby speeding it up. And still other times, there may be nothing you can do to increase speed. In any case, a great deal of insight is gained by profiling the operation of M-file functions.

In addition to using the profiler through the *Profiler* window on the MATLAB desktop, MATLAB uses the `profile` command to determine which lines of code in an M-file take the most time to execute. Using `profile` is straightforward. For example, the execution profile of `myfunction` is found by executing the following commands:

```
>> profile on

>> for i = 1:100
       out = myfunction(in);
end
>> profile viewer
```

First, the profiler is turned On; then, `myfunction` is executed some sufficient number of times to gather sufficient data; finally, a profile report is generated. The profile report is an HTML file, displayed in the *Profiler* window in the MATLAB desktop, which in turn is displayed by choosing it from the **Desktop** menu or by issuing the last command in the preceding example. The profile report generates a variety of data. Clicking on various links in the window supplies additional information.

13

File and Directory Management

MATLAB opens and saves data files in a variety of file formats. Some are formats custom to MATLAB, others are industry standards, and still others are file formats custom to other applications. The techniques used to open and save data files include GUIs, as well as *Command* window functions.

Like most modern applications, MATLAB uses the current directory as the default location for data files and M-files. Implementing directory management tools and changing the current directory are accomplished through GUIs, as well as through the *Command* window functions.

This chapter covers file and directory management features in MATLAB.

13.1 NATIVE DATA FILES

Variables in the MATLAB workspace can be saved in a format native to MATLAB by using the `save` command. For example,

```
>> save
```

stores all variables from the MATLAB workspace, in MATLAB binary format, to the file `matlab.mat` in the current directory. These native binary MAT-files maintain full double precision, as well as the names of the variables saved. MAT-files are not platform-independent, but are completely cross-platform compatible. Variables saved on one platform can be opened on other MATLAB platforms without any special treatment.

The `save` command can be used to store specific variables as well. For example, the code

```
>> save var1 var2 var3
```

saves just the variables var1, var2, and var3 to matlab.mat. The file name can be specified as a first argument to save, as in the code

```
>> save filename var1 var2 var3
```

which saves var1, var2, and var3 to the file named filename.mat.

Using command–function duality, the preceding command form can also be written in function form as

```
>> save('filename','var1','var2,','var3')
```

This particular format is useful if the file name is stored in a MATLAB character string:

```
>> fname = 'myfile';
>> save(fname,'var1','var2,','var3')
```

Here, the named variables are stored in a file named myfile.mat.

In addition to these simple forms, the `save` command supports options for saving in uncompressed format and in ASCII text formats, and the command can be used to append data to a file that already exists. (For help with these features, refer to the online help.)

The complement to save is the `load` command. This command opens data files that were created by the `save` command or that are compatible with the `save` command. For example,

```
>> load
```

loads all variables found in matlab.mat, wherever it is first found in the current directory or on the MATLAB search path. The variable names originally stored in matlab.mat are restored in the workspace, and they overwrite any like-named variables that may exist there.

To load specific variables from a MAT-file, you must include the file name and a variable list:

```
>> load filename var1 var2 var3
>> load('filename','var1','var2','var3')
```

Here, filename.mat is opened, and variables var1, var2, and var3 are loaded into the workspace. The second statement demonstrates the functional form of the load command, which allows the data file to be specified as a character string. Although not shown, the filename string can include a complete or a partial directory path, thereby restricting load to a search in a specific directory for the data file.

The latter example provides a way to open a sequence of enumerated data files, such as mydata1.mat and mydata2.mat, as in the following example:

```
for i = 1:N
    fname = sprintf('mydata%d',i);
    load(fname)
end
```

This code segment uses sprintf to create file-name strings inside of a For Loop, so that a sequence of data files is loaded into the workspace.

When you do not wish to overwrite workspace variables, you can write the load command in function form and give it an output argument. For example,

```
>> vnew = load('filename','var1','var2');
```

opens the file filename.mat and loads the variables var1 and var2 into a **structure** variable named vnew that has fields var1 and var2—that is, vnew.var1 = var1 and vnew.var2 = var2.

The load command can also open ASCII text files. In particular, if the data file consists of MATLAB comment lines and rows of space-separated values, the syntax

```
>> load filename.ext
```

opens the file filename.ext and loads the data into a single double-precision data array named filename. (For further information regarding the load command, see the online help.)

To find out whether a data file exists and what variables it holds, the MATLAB commands exist and whos are valuable. For example, the command

```
>> exist('matlab.mat','file')
```

returns 0 if the file doesn't exist and 2 if it does, and the command

```
>> whos -file matlab.mat
```

returns the standard whos *Command* window display for the variables contained in the file matlab.mat. Alternatively, the code

```
>> w = whos('-file','matlab.mat')
w =
3x1 struct array with fields:
```

```
name

size

bytes

class

global

sparse

complex

nesting

persistent
```

returns a structure array with fields named for the columns of the whos display along with potential attributes. Used in this way, the variable names, array sizes, storage requirements, class, and any attributes are stored in variables.

Last, but not least, data files can be deleted by using the *Command* window command delete. For example,

```
>> delete filename.ext
```

deletes the file named filename.ext.

In MATLAB, data file management functions can be accessed from the *Current Folder* browser, as well as from the *Import* wizard. The *Current Folder* browser can be viewed by choosing **Current Folder** from the **Desktop** menu on the *MATLAB* desktop. The *Import* wizard, which appears by selecting **Import Data...** from the **File** menu or by typing uiimport in the *Command* window, is a general-purpose GUI that facilitates loading data in a variety of formats, not just MATLAB's native MAT-file format.

13.2 DATA IMPORT AND EXPORT

In addition to supporting MATLAB's native MAT-file format and conventional ASCII text format, MATLAB supports a variety of industry standard formats and other custom file formats. Some formats are restricted to reading, others to writing. Some formats are restricted to images, others to multimedia or spreadsheets. These data import and export functions and capabilities make it possible for MATLAB to exchange data with other programs.

Figure window images can be saved in a native MATLAB FIG-file format or exported to a variety of standard graphics file formats by selecting **Save** or **Save As...** from the **File** menu of a *Figure* window. The *Command* window function saveas provides an alternative to this GUI-based approach. (For assistance in using saveas, see the online documentation.)

Data-specific import and export functions available in MATLAB include those listed in the following table:

Function	Description
load	Loads data from MAT file
save	Saves data to MAT file
dlmread	Reads delimited text file
dlmwrite	Writes delimited text file
textread	Reads formatted text from file (depreciated; use textscan instead)
textscan	Reads formatted text from file after opening with fopen or from a string
xlsread	Reads spreadsheet file
xlswrite	Writes spreadsheet file
importdata	Reads data file
VideoReader	Reads movie file (replaces mmreader)
VideoWriter	Writes movie file
imread	Reads image file
imwrite	Writes image file
auread	Reads Sun sound file
auwrite	Writes Sun sound file
wavread	Reads Microsoft sound file
wavwrite	Writes Microsoft sound file
hdfread	Reads data from an HDF4 or HDF-EOS file
hdf5read	Reads data from an HDF5 file
hdf5write	Writes data to an HDF5 file
xmlread	Parses an XML document and returns a Document Object Model node
xmlwrite	Writes a DOM node to an XML file
cdfread	Reads from a CDF file
cdfwrite	Writes to a CDF file
fitsread	Reads data from a FITS file
netcdf.*	Low-level access to netCDF files

(The help text for each of these functions provides information on their use.)
The functions imread and imwrite in particular support multiple formats, including JPEG, TIFF, BMP, PNG, HDF, PCX, and XWD. The help text for fileformats provides a more complete listing of the file formats supported by MATLAB, such as the following:

```
>> help fileformats
   Supported file formats.
```

NOTE: You can import any of these file formats with the Import Wizard or the IMPORTDATA function, except netCDF, H5, Motion JPEG 2000, and platform-specific video. The IMPORTDATA function cannot read HDF files.

NOTE: '.' indicates that no existing high-level functions export the given data format.

Format	Import	Export
------	--------	--------
MAT - MATLAB workspace	load	save
DAQ - Data Acquisition Toolbox	daqread	.
Text formats		
any - White-space delimited numbers	load	save -ascii
any - Delimited numbers	dlmread	dlmwrite
any - Any above text format, or a mix of strings and numbers	textscan	.
XML - Extended Markup Language	xmlread	xmlwrite
Spreadsheet formats		
XLS - Excel worksheet XLSX, XLSB, XLSM require Excel 2007 for Windows	xlsread	xlswrite
Scientific data formats		
CDF - Common Data Format	cdfread	cdfwrite
	cdflib	cdflib
FITS - Flexible Image Transport System	fitsread	.

HDF	– Hierarchical Data Format v.4	hdfread	.
H5	– Hierarchical Data Format v.5	hdf5read	hdf5write
NC	– network Common Data Form v.3	netcdf	netcdf

Video formats (All Platforms)

AVI	– Audio Video Interleave	VideoReader	VideoWriter
MJ2	– Motion JPEG 2000	VideoReader	VideoWriter

Video formats (Windows and Mac)

MPEG	– Motion Picture Experts Group, phases 1 and 2 (Includes MPG)	VideoReader	.

Video formats (Windows Only)

WMV	– Windows Media Video	VideoReader	.
ASF	– Windows Media Video	VideoReader	.
ASX	– Windows Media Video	VideoReader	.
any	– formats supported by DirectShow	VideoReader	.

Video formats (Mac Only)

MOV	– QuickTime Movie	VideoReader	.
MP4	– MPEG-4 Video (Includes M4V)	VideoReader	.
3GP	– 3GPP Mobile Video	VideoReader	.
3G2	– 3GPP2 Mobile Video	VideoReader	.
DV	– Digital Video Stream	VideoReader	.
any	– formats supported by QuickTime	VideoReader	.

Video formats (Linux Only)

any	– formats supported by GStreamer plug-ins on your system	VideoReader	.

Image formats

BMP	– Windows Bitmap	imread	imwrite
CUR	– Windows Cursor resources	imread	.
FITS	– Flexible Image Transport System Includes FTS	imread	.
GIF	– Graphics Interchange Format	imread	imwrite

HDF	- Hierarchical Data Format	imread	imwrite
ICO	- Icon image	imread	.
JPEG	- Joint Photographic Experts Group	imread	imwrite
	Includes JPG		
JP2	- JPEG 2000	imread	imwrite
	Includes JPF, JPX, J2C, J2K		
PBM	- Portable Bitmap	imread	imwrite
PCX	- Paintbrush	imread	imwrite
PGM	- Portable Graymap	imread	imwrite
PNG	- Portable Network Graphics	imread	imwrite
PNM	- Portable Any Map	imread	imwrite
PPM	- Portable Pixmap	imread	imwrite
RAS	- Sun Raster	imread	imwrite
TIFF	- Tagged Image File Format	imread	imwrite
	Includes TIF		
XWD	- X Window Dump	imread	imwrite

Audio formats

AU	- NeXT/Sun sound	auread	auwrite
SND	- NeXT/Sun sound	auread	auwrite
WAV	- Microsoft Wave sound	wavread	wavwrite

See also uiimport, fscanf, fread, fprintf, fwrite, hdf, hdf5, Tiff, imformats.

Reference page in Help browser
    doc fileformats

13.3 LOW-LEVEL FILE I/O

Because an infinite variety of file types exists, MATLAB provides low-level file I/O functions for reading or writing any binary or formatted ASCII file imaginable. These functions closely resemble their ANSI C programming language counterparts, but do not necessarily exactly match their characteristics. In fact, many of the special-purpose file I/O commands described previously use these commands internally. The low-level file I/O functions in MATLAB are given in the following table:

Category	Function	Description/Syntax Example
File opening and closing	fopen	Opens file fid = fopen('$filenename$','$permission$')
	fclose	Closes file status = fclose(fid)
Binary I/O	fread	Reads part or all of a binary file A = fread(fid,num,$precision$)
	fwrite	Writes array to a binary file count = fwrite(fid,$array$,$precision$)
Formatted I/O	fscanf	Reads formatted data from file A = fscanf(fid,$format$,num)
	fprintf	Writes formatted data to file count = fprintf(fid,$format$,A)
	fgetl	Reads line from file; discards newline character line = fgetl(fid)
	fgets	Reads line from file; keeps newline character line = fgets(fid)
String conversion	sprintf	Writes formatted data to string S = sprintf($format$,A)
	sscanf	Reads string under format control A = sscanf($string$,$format$,num)
File positioning	ferror	Inquires about file I/O status message = ferror(fid)
	feof	Tests for end of file TF = feof(fid)
	fseek	Sets file position indicator status = fseek(fid,$offset$,$origin$)
	ftell	Gets file position indicator position = ftell(fid)
	frewind	Rewinds file frewind(fid)

In this table, fid is a file identifier number and $permission$ is a character string identifying the permissions requested. Possible strings include 'r' for reading only, 'w' for writing only, 'a' for appending only, and 'r+' for both reading and writing. Since the PC distinguishes between text and binary files, a 't' must often be appended when you are working with text files—for example, 'rt' and 'wt' cause the carriage return character preceding a newline character to be deleted on input and added before the newline character on output, respectively. In the preceding table, *format* is a character string defining the desired formatting. Note that *format* follows ANSI standard C very closely. (More information regarding the use of these functions can be found in the online documentation for each function.)

13.4 DIRECTORY MANAGEMENT

With all the windows in MATLAB, it makes sense to have management of the current directory and its files available in a GUI. The *Current Folder* window, displayed by choosing **Current Folder** from the **Desktop** menu in the *MATLAB* desktop window, performs these tasks. In addition to traversing the directory tree, this GUI allows you to preview the files in the current directory, see their modification dates, search for text in M-files, create new directories and new M-files, and so on. Because of the fundamental utility of knowing the current directory, the current directory is also displayed in a pop-up menu in the *MATLAB* desktop window. Therefore, the current directory always can be known simply by looking at the toolbar on the desktop.

Prior to MATLAB version 6, directory management was conducted through the use of *Command* window functions. Although these functions are not as important now, they still serve a valuable purpose. In particular, most functions have the capability of returning directory and file information in MATLAB variables, thereby allowing complex manipulation of files and directories to be completed within function M-files. The *Command* window directory management functions available in MATLAB are summarized in the following table:

Function	Description
cd, pwd S = cd;	Shows present working directory Returns present working directory as a string in S
cd *dirname*	Changes present working directory to *dirname*
copyfile(*oldname, dirname*) copyfile(*oldname, newname*)	Copies file *oldname* to directory *dirname* Copies file *oldname* to *newname*
delete *filename.ext*	Deletes file *filename.ext*
dir, ls S = dir;	Displays files in current directory Returns directory information in structure S
fileattrib	Gets or sets file attributes
mkdir *dirname*	Makes directory *dirname* in current directory
movefile(*source, destination*)	Moves source file or directory to a new destination
rmdir *dirname*	Removes directory *dirname*

Function	Description
what	Displays an organized listing of all MATLAB files in the current directory
S = what;	Returns listing information in structure S
which *filename*	Displays directory path to *filename*
S = which('*filename*');	Returns directory path to *filename* as a string in S
who	Displays variables in workspace
who -file *filename*	Displays variables in MAT-file *filename.mat*
S = who('-file','*filename*');	Returns variables names in *filename.mat* in a cell array S
whos	Displays variables, size, and class in workspace
whos -file *filename*	Displays variables, size, and class in MAT-file *filename.mat*
S = whos('-file', '*filename*')	Returns variables, size, and class in *filename.mat* in structure S
help *filename*	Displays help text for *filename* in *Command* window
S = help('*filename*');	Returns help text for *filename* in a character string S
type *filename*	Displays M-file *filename* in *Command* window

Most of these functions require only partial path information to locate a particular file. That is, *filename* may or may not include part of its directory path. If no directory path is included in *filename*, then the MATLAB search path is used to find the requested file. If some part of the directory path is provided, MATLAB traverses the MATLAB search path to find the subdirectory and file specified. For example, if *filename* = 'mystuff/myfile', MATLAB restricts its search to a subdirectory named mystuff on the MATLAB path.

To illustrate the usefulness of these functions, consider the function mmbytes:

```
function y = mmbytes(arg)
%MMBYTES Variable Memory Usage.
% MMBYTES and MMBYTES('base') returns the total memory in bytes
% currently used in the base workspace.
% MMBYTES('caller') returns the total memory in bytes currently
% used in the workspace where MMBYTES is called from.
% MMBYTES('global') returns the total memory in bytes currently
```

```
% used in the global workspace.
if nargin==0
    arg = 'base';
end
if strcmp(arg,'global')
    x = evalin('base','whos(''global'')');
else
    x = evalin(arg,'whos');
end
y = sum(cat(1,x.bytes));
```

This function uses the whos function to gather information about the variables that exist within MATLAB. The output of the whos function generates a structure array x. The bytes field of this structure contains the memory allocated to all variables. The final statement in the function concatenates all of the memory-allocated numbers into a vector, which is summed using the sum function.

Because of its varied uses, the function exist was not listed in the preceding table. This function tests for the existence of variables, files, directories, and so on. The help text for this function describes its many uses:

```
>> help exist
EXIST Check if variables or functions are defined.
    EXIST('A') returns:
        0 if A does not exist
        1 if A is a variable in the workspace
        2 if A is an M-file on MATLAB's search path. It also returns 2 when
            A is the full pathname to a file or when A is the name of an
            ordinary file on MATLAB's search path
        3 if A is a MEX-file on MATLAB's search path
        4 if A is a MDL-file on MATLAB's search path
        5 if A is a built-in MATLAB function
        6 if A is a P-file on MATLAB's search path
        7 if A is a directory
        8 if A is a class (EXIST returns 0 for Java classes if you
            start MATLAB with the -nojvm option.)
```

EXIST('A') or EXIST('A.EXT') returns 2 if a file named 'A' or 'A.EXT' and the extension isn't a P or MEX function extension.

EXIST('A','var') checks only for variables.
EXIST('A','builtin') checks only for built-in functions.
EXIST('A','file') checks for files or directories.
EXIST('A','dir') checks only for directories.
EXIST('A','class') checks only for classes.

If A specifies a filename, MATLAB attempts to locate the file, examines the filename extension, and determines the value to return based on the extension alone. MATLAB does not examine the contents or internal structure of the file.

When searching for a directory, MATLAB finds directories that are part of MATLAB's search path. They can be specified by a partial path. It also finds the current working directory specified by a partial path, and subdirectories of the current working directory specified by a relative path.

EXIST returns 0 if the specified instance isn't found.

See also dir, what, isempty, partialpath.

Overloaded methods:
    inline/exist

Reference page in Help browser
    doc exist

To facilitate the manipulation of character strings containing directory paths and filenames, MATLAB provides several useful functions, summarized in the following table:

Function	Description
addpath('*dirname*')	Prepends directory *dirname* to the MATLAB search path
[path,name,ext] = ... fileparts(*filename*)	Returns path, name, and extension for file *filename*

`filesep`	Returns file separator character for this computer platform. The file separator is the character used to separate directories and the file name. For example, on the PC the file separator is '\'.
`fullfile(d1, d2, ..., filename)`	Returns full path and file specification for `filename` by using directory tree strings, `d1, d2, ....`
`matlabroot`	Returns a string containing the path to the root directory of MATLAB
`mexext`	Returns MEX-file extension for this computer platform
`pathsep`	Returns path separator for this platform. The path separator is the character used to separate entries on the MATLAB search path string.
`prefdir`	Returns MATLAB preferences directory for this platform
`rmpath('dirname')`	Removes directory `dirname` from the MATLAB search path
`tempdir`	Returns name of a temporary directory for this platform
`tempname`	Returns name of a temporary file for this platform

These functions, along with those in the preceding table, facilitate the creation of file and directory management functions as M-files. (More assistance for each function can be found by consulting the online help.)

13.5 FILE ARCHIVES AND COMPRESSION

MATLAB supports file archives and file compression in a number of common formats using graphical tools and *Command* window functions. The *Current Folder* browser treats ZIP file archives as folders. The contents are visible simply by opening the "folder," for example, `myfiles.zip`, in the file and folder list. *Command* window functions support ZIP, GZIP, and TAR formats.

The `zip` function creates a compressed ZIP archive given a list of files and directories to be included in the archive as a string or cell array of strings. Directories recursively add all of their contents to the archive. The `unzip` function extracts the files and folders from a zip archive. All paths are relative in the archive. The `unzip` function also accepts URL input. The zip archive given in the URL is downloaded to a temporary directory, the files and directories are extracted, and the zip file is deleted. Password-protected and encrypted zip archives are not supported.

The `gzip` function, given a list of files and directories as a string or cell array of strings, creates compressed versions of the files using GNU GZIP compression. Compressed files are given the `.gz` extension. Directories recursively compress all of their content. The `gunzip` function uncompresses gziped files. Folders are recursively searched for gzipped files to uncompress. The `gunzip` function also accepts URL input. The gzipped file given in the URL is downloaded and uncompressed.

The `tar` function creates a TAR archive given a list of files and directories to be included in the archive as a string or cell array of strings. Directories recursively add all of their contents to the archive. The archive that is created is given the `.tar` extension if an extension is omitted in the archive filename. If the filename ends in `.tgz` or `.gz`, the archive is automatically compressed using `gzip`. All paths are relative in the archive. The `untar` function extracts the files and folders from a tar archive into the current directory and sets the file attributes. If the archive filename ends in `.tgz` or `.gz`, the tar file is uncompressed into a temporary directory, the contents are extracted into the current directory, and the temporary file is deleted. The `untar` function also accepts URL input. Consider these examples:

```
>> zip('mfiles.zip','*.m'); % zip up all M-files in the current directory
>> unzip('mfiles.zip','backup'); % extract into the directory 'backup'

>> gzip('*.m','backup'); % create compressed M-files in 'backup' directory
>> gunzip('backup'); % ungzip the .gz files in the 'backup' directory

>> tar('backup/mfiles.tgz','*.m'); % create gzipped tar archive in 'backup'
```

MATLAB file archive and compression utility functions are summarized in the following table.

Function	Description
zip	Compresses files and directories into ZIP archive
unzip	Extracts contents of ZIP file
gzip	Compresses files into GZIP files
gunzip	Uncompresses GZIP files
tar	Archive files and directories into TAR archive with optional compression
untar	Extracts contents of TAR archive

13.6 INTERNET FILE OPERATIONS

MATLAB includes several built-in Internet file access capabilities. From within MATLAB, the function `ftp` creates an ***FTP object*** that acts like the file identifier used in low-level file I/O. For example,

```
>> ftp_object = ftp('ftp.somewhere.org','username','password')
```

establishes an FTP connection to the site *ftp.somewhere.org* and returns an identifier to this connection in the variable `ftp_object`. If the FTP server supports anonymous connections, the username and password arguments can be omitted. Once this FTP session is established, common FTP commands can be used to change directories and perform file operations. Many of these operations use the same functions used for local file and directory manipulation. For instance,

```
>> dir(ftp_object)
```

returns a directory listing of the default directory. Once FTP operations are completed, the function `close` closes the FTP session. For example,

```
>> close(ftp_object)
```

terminates the FTP session at *ftp.somewhere.org*.

The following table identifies a list of FTP functions available in MATLAB:

Function	Description
`ascii`	Sets FTP transfer type to ASCII
`binary`	Sets FTP transfer type to binary
`cd`	Changes current directory
`close`	Closes connection
`delete`	Deletes file
`dir`	Lists directory contents
`ftp`	Connects to FTP server
`mget`	Downloads file
`mkdir`	Creates new directory
`mput`	Uploads file or directory
`rename`	Renames file
`rmdir`	Removes directory

Web page content is also accessible from within MATLAB. The `urlread` and `urlwrite` functions access the contents of a URL and store the result in a string variable or file, respectively. The URL string must include the protocol being used, that is, `'http://'`, `'ftp://'`, or `'file://'`. Some examples are shown below.

```
>> G = urlread('http://www.google.com');
>> urlwrite('ftp://ftp.mathworks.com/README','/tmp/README.txt');
>> S = urlread('file:///tmp/README.txt');
```

Information may be passed to the web server using the POST or GET methods as well. For example, to access data from a company intranet server that requires a query string such as '*http://qserver.intra.net/query.php?user=john&pass =secret&page=3*', the following code

```
>> myurl = 'http://qserver.intra.net/query.php';
>> mymethod = 'get';
>> myparams = {'user','john','pass','secret','page','3'};
>> myfile = 'page3.html';
>> urlwrite(myurl, myfile, mymethod, myparams);
```

downloads the requested page and saves the result to the file page3.html.

You can even send email from within MATLAB. The sendmail function sends the email including attachments, if desired, from the *Command* window or even from within an M-file. Two preferences have to be set before sendmail is invoked or an error will occur. In the following example, an email with two attachments is sent to the *Mastering MATLAB* authors.

```
>> setpref('Internet','SMTP_Server','mail.example.edu');
>> setpref('Internet','E_mail','jsmith@example.edu');
>> mail_to = {'MasteringMatlab@gmail.com','jsmith@example.edu'};
>> mail_subj = 'Mastering MATLAB example';
>> mail_body = {'I may have found a slightly faster algorithm for one of the ',
               'example M-files. Both versions are attached to this email.',
               'Please check it out.',
               'Sincerely, John Smith'};
>> mail_attach = {'mm4001.m','js4001.m'};
>> sendmail(mail_to, mail_subj, mail_body, mail_attach);
```

14

Set, Bit, and Base Functions

14.1 SET FUNCTIONS

Since arrays are ordered collections of values, they can be thought of as sets. With that understanding, MATLAB provides several functions for testing and comparing sets. The simplest test is for equality, as in the following code:

```
>> a = rand(2,5); % random array
>> b = randn(2,5); % a different random array
>> isequal(a,b) % a and b are not equal
ans =
     0
>> isequal(a,a) % but a is certainly equal to a
ans =
     1

>> isequal(a,a(:)) % a with a as a column
ans =
     0
```

For two arrays to be equal, they must have the same dimensions and the same contents. This function applies to all MATLAB data types, not just numerical arrays:

```
>> a = 'a string';
>> b = 'a String';
>> isequal(a,b) % character string equality
ans =
      0
>> a = {'four' 'five' 'six'};
>> b = {'four' 'two' 'three'};
>> isequal(a,b) % cell array equality
ans =
      0
>> isequal(a,a)
ans =
      1
>> c.one = 'two';
>> c.two = 4;
>> c.three = pi;
>> d.two = 4;
>> d.one = 'two';
>> d.three = pi;
>> isequal(c,d) % structure equality
ans =
      1
>> isequal(c,c)
ans =
      1
```

MATLAB variables are equal if they have the same size and exactly the same content.

The function unique removes duplicate items from a given set:

```
>> a = [2:2:10;4:2:12] % new data
a =
     2    4    6    8   10
     4    6    8   10   12
```

```
>> unique(a) % unique elements sorted into a column
ans =
     2
     4
     6
     8
    10
    12
```

The `unique` function returns a sorted column vector because removal of duplicate values makes it impossible to maintain the array dimensions. The function `unique` also applies to cell arrays of strings, as in the following example:

```
>> c = {'Tom' 'Bob' 'Tom' 'Shaun' 'Frida' 'Shaun' 'Bob'};
>> unique(c)
ans =
     'Bob'    'Frida'    'Shaun'    'Tom'
```

Set membership is determined with the function `ismember`, as in the following code:

```
>> a = 2:10
a =
     2    3    4    5    6    7    8    9    10
>> b = 2:2:10
b =
     2    4    6    8    10
>> ismember(a,b) % which elements in a are in b
ans =
     1    0    1    0    1    0    1    0    1
>> ismember(b,a) % which elements in b are in a
ans =
     1    1    1    1    1
```

For numeric arrays, `ismember` returns a logical array the same size as its first argument, with ones appearing at the indices where the two vectors share common values.

```
>> A = [1 2 3; 4 5 6; 7 8 9]  % new data
```

```
A =
     1    2    3
     4    5    6
     7    8    9

>> B = [4 5 6; 3 2 1] % more new data
B =
     4    5    6
     3    2    1

>> ismember(A,B) % which elements in A are in the set B
ans =
     1    1    1
     1    1    1
     0    0    0

>> ismember(B,A) % which elements in B are in the set A
ans =
     1    1    1
     1    1    1
```

If both arrays contain the same number of columns, the following example returns a logical column vector containing ones where the corresponding row in A is also a row in B:

```
>> ismember(A,B,'rows')  % which rows of A are also rows of B
ans =
     0
     1
     0
```

The function ismember also applies to cell arrays of strings:

```
>> c                      % recall prior data
c =
     'Tom'  'Bob'  'Tom'  'Shaun'  'Frida'  'Shaun'  'Bob'

>> ismember(c,'Tom')
ans =
     1    0    1    0    0    0    0
```

Set arithmetic is accomplished with the functions union, intersect, setdiff, and setxor. Examples of the use of these functions include the following:

```
>> a,b                    %  recall prior data
a =
    2   3   4   5   6   7   8   9   10
b =
    2   4   6   8   10

>> union(a,b) % union of a and b
ans =
    2   3   4   5   6   7   8   9   10

>> intersect(a,b) % intersection of a and b
ans =
    2   4   6   8   10

>> setxor(a,b) % set exclusive or of a and b
ans =
    3   5   7   9

>> setdiff(a,b) % values in a that are not in b
ans =
    3   5   7   9

>> setdiff(b,a) % values in b that are not in a
ans =
  Empty matrix: 1-by-0

>> union(A,B,'rows') % matrix inputs give rows; no repetitions
ans =
    1   2   3
    3   2   1
    4   5   6
    7   8   9
```

Like prior functions discussed in this chapter, these set functions also apply to cell arrays of strings.

14.2 BIT FUNCTIONS

In addition to the logical operators discussed in Chapter 10, MATLAB provides functions that allow logical operations on individual bits of any unsigned integer data. For backward compatibility, these bit functions also work on floating-point integers, namely integers stored in double-precision floating-point variables. The MATLAB bitwise functions bitand, bitcmp, bitor, bitxor, bitset, bitget, and bitshift work on integers. Examples of bit operations include:

```
>> format hex
>> intmax('uint16') % largest unsigned 16-bit number
ans =
     ffff
>> a = uint16(2^10 -1) % first data value
a =
     03ff
>> b = uint16(567) % second data value
b =
     0237
>> bitand(a,b) % (a & b)
ans =
     0237
>> bitor(a,b) % (a | b)
ans =
     03ff
>> bitcmp(a) % complement a
ans =
     fc00
>> bitxor(a,b) % xor(a,b)
ans =
     01c8
>> bitget(b,7) % get 7th bit of b
ans =
     0000
```

```
>> bitset(b,7) % set 7th bit of b to 1
ans =
     0277
>> b, swapbytes(b) % swap byte ordering (little-endian<=>big-endian)
b =
     0237
ans =
     3702
>> format short g % reset display format
```

14.3 BASE CONVERSIONS

MATLAB provides a number of utility functions for converting decimal numbers to other bases *in the form of character strings*. Conversions between decimals and binary numbers are performed by the functions dec2bin and bin2dec, as in the following example:

```
>> a = dec2bin(17) % find binary representation of 17
a =
10001
>> class(a) % result is a character string
ans =
char
>> bin2dec(a) % convert a back to decimal
ans =
     17
>> class(ans) % result is a double precision decimal
ans =
double
```

Conversions between decimals and hexadecimals are performed by dec2hex and hex2dec, as in the following example:

```
>> a = dec2hex(2047) % hex representation of 2047
a =
7FF
>> class(a) % result is a character string
```

```
ans =
char
>> hex2dec(a) % convert a back to decimal
ans =
          2047
>> class(ans) % result is a double precision decimal
ans =
double
```

Conversions between decimals and any base between 2 and 36 are performed by dec2base and base2dec:

```
>> a = dec2base(26,3)
a =
222
>> class(a)
ans =
char
>> base2dec(a,3)
ans =
        26
```

Base 36 is the maximum usable base, because it uses the numbers 0 through 9 and the letters A through Z to represent the 36 distinct digits of a base 36 number.

15

Time Computations

MATLAB offers a number of functions to manipulate time. You can do arithmetic with dates and times, print calendars, and find specific days. MATLAB does this by storing the date and time as a double-precision number representing the number of days since the beginning of year zero. For example, midnight, January 1, 2000, is represented as 730486, and the same day at noon is 730486.5. This format may make calculations easier for a computer, but it is difficult to interpret visually. That's why MATLAB supplies a number of functions to convert between date numbers and character strings, and to manipulate dates and times.

15.1 CURRENT DATE AND TIME

The function clock returns the current date and time in an array:

```
>> T = clock
T =
1.0e+003 *
      2.0110      0.0090      0.0010      0.0190      0.0470      0.0232
```

This was the time when this part of the text was written. The preceding data are organized as

  T = [year month day hour minute seconds], so the time shown is the year 2011, the 9th month, 1st day, 19th hour, 47th minute, and 23.2 seconds.

  The function now returns the current date and time as a double-precision date number, or simply a date number:

```
>> format long g

>> t = now
t =
      734747.825187164

>> format short g
```

Both T and t represent essentially the same information.

The function `date` returns the current date as a character string in the dd-mmm-yyyy format:

```
>> date
ans =
01-Sep-2011
>> class(t)
ans =
double
```

15.2 DATE FORMAT CONVERSIONS

In general, mathematics with time involves converting time to date number format, performing standard mathematical operations on the date numbers, and then converting the result back to a format that makes human sense. As a result, converting time among different formats is very important. MATLAB supports three formats for dates: (1) double-precision date number, (2) date (character) strings in a variety of styles, and (3) numerical date vector, where each element contains a different date component, that is, [year, month, day, hour, minute, seconds].

The function `datestr` converts the date number to a date string. The syntax for using `datestr` is `datestr(date, dateform)`, where `dateform` is described by the help text for `datestr`:

```
>> help datestr
  DATESTR String representation of date.
    S = DATESTR(V) converts one or more date vectors V to date strings S.
    Input V must be an M-by-6 matrix containing M full (six-element) date
    vectors. Each element of V must be a positive double-precision number.
    DATESTR returns a column vector of M date strings, where M is the total
    number of date vectors in V.
```

S = DATESTR(N) converts one or more serial date numbers N to date strings S. Input argument N can be a scalar, vector, or multidimensional array of positive double-precision numbers. DATESTR returns a column vector of M date strings, where M is the total number of date numbers in N.

S = DATESTR(D, F) converts one or more date vectors, serial date numbers, or date strings D into the same number of date strings S. Input argument F is a format number or string that determines the format of the date string output. Valid values for F are given in Table 1, below. Input F may also contain a free-form date format string consisting of format tokens as shown in Table 2, below.

Date strings with 2-character years are interpreted to be within the 100 years centered around the current year.

S = DATESTR(S1, F, P) converts date string S1 to date string S, applying format F to the output string, and using pivot year P as the starting year of the 100-year range in which a two-character year resides. The default pivot year is the current year minus 50 years. F = -1 uses the default format.

S = DATESTR(...,'local') returns the string in a localized format. The default (which can be called with 'en_US') is US English. This argument must come last in the argument sequence.

Note: The vectorized calling syntax can offer significant performance improvement for large arrays.

TABLE 1: Standard MATLAB date format definitions

Number	String	Example
0	'dd-mmm-yyyy HH:MM:SS'	01-Mar-2000 15:45:17
1	'dd-mmm-yyyy'	01-Mar-2000
2	'mm/dd/yy'	03/01/00
3	'mmm'	Mar
4	'm'	M

5	'mm'	03
6	'mm/dd'	03/01
7	'dd'	01
8	'ddd'	Wed
9	'd'	W
10	'yyyy'	2000
11	'yy'	00
12	'mmmyy'	Mar00
13	'HH:MM:SS'	15:45:17
14	'HH:MM:SS PM'	3:45:17 PM
15	'HH:MM'	15:45
16	'HH:MM PM'	3:45 PM
17	'QQ-YY'	Q1-96
18	'QQ'	Q1
19	'dd/mm'	01/03
20	'dd/mm/yy'	01/03/00
21	'mmm.dd,yyyy HH:MM:SS'	Mar.01,2000 15:45:17
22	'mmm.dd,yyyy'	Mar.01,2000
23	'mm/dd/yyyy'	03/01/2000
24	'dd/mm/yyyy'	01/03/2000
25	'yy/mm/dd'	00/03/01
26	'yyyy/mm/dd'	2000/03/01
27	'QQ-YYYY'	Q1-1996
28	'mmmyyyy'	Mar2000
29 (ISO 8601)	'yyyy-mm-dd'	2000-03-01
30 (ISO 8601)	'yyyymmddTHHMMSS'	20000301T154517
31	'yyyy-mm-dd HH:MM:SS'	2000-03-01 15:45:17

TABLE 2: Free-form date format symbols

```
Symbol   Interpretation of format symbol
===================================================================
yyyy     full year, e.g. 1990, 2000, 2002
yy       partial year, e.g. 90, 00, 02
mmmm     full name of the month, according to the calendar locale, e.g.
         "March", "April" in the UK and USA English locales.
mmm      first three letters of the month, according to the calendar
         locale, e.g. "Mar", "Apr" in the UK and USA English locales.
```

mm   numeric month of year, padded with leading zeros, e.g ... /03/.
          or ../12/.

m   capitalized first letter of the month, according to the
          calendar locale; for backwards compatibility.

dddd   full name of the weekday, according to the calendar locale, e.g.
          "Monday", "Tuesday", for the UK and USA calendar locales.

ddd   first three letters of the weekday, according to the calendar
          locale, e.g. "Mon", "Tue", for the UK and USA calendar locales.

dd   numeric day of the month, padded with leading zeros, e.g.
          05/../.. or 20/../.

d   capitalized first letter of the weekday; for backwards
          compatibility

HH   hour of the day, according to the time format. In case the time
          format AM | PM is set, HH does not pad with leading zeros. In
          case AM | PM is not set, display the hour of the day, padded
          with leading zeros. e.g 10:20 PM, which is equivalent to 22:20;
          9:00 AM, which is equivalent to 09:00.

MM   minutes of the hour, padded with leading zeros, e.g. 10:15,
          10:05, 10:05 AM.

SS   second of the minute, padded with leading zeros, e.g. 10:15:30,
          10:05:30, 10:05:30 AM.

FFF   milliseconds field, padded with leading zeros, e.g.
          10:15:30.015.

PM   set the time format as time of morning or time of afternoon. AM
          or PM is appended to the date string, as appropriate.

Examples:
DATESTR(now) returns '24-Jan-2003 11:58:15' for that particular date,
on an US English locale DATESTR(now,2) returns 01/24/03, the same as
for DATESTR(now,'mm/dd/yy') DATESTR(now,'dd.mm.yyyy') returns
24.01.2003 To convert a non-standard date form into a standard MATLAB
dateform, first convert the non-standard date form to a date number,
using DATENUM, for example,
DATESTR(DATENUM('24.01.2003','dd.mm.yyyy'),2) returns 01/24/03.

See also date, datenum, datevec, datetick.

Reference page in Help browser
doc datestr

Some examples of datestr usage include the following:

```
>> t = now
t =
        734747.842392396
>> datestr(t)
ans =
01-Sep-2011 20:13:02

>> datestr(t,12)
ans =
Sep11
>> datestr(t,23)
ans =
09/01/2011
>> datestr(t,25)
ans =
11/09/01
>> datestr(t,13)
ans =
20:13:02
>> datestr(t,29)
ans =
2011-09-01
```

The function datenum is the inverse of datestr. That is, datenum converts a date string to a date number using the form datenum(str). Alternatively, it converts individual date specifications using the form `datenum(year,month,day)` or `datenum(year,month,day,hour,minute,second)`, as in the following example:

```
>> format long
>> t = now
t =
     7.347478436599305e+005
>> ts = datestr(t)
ts =
01-Sep-2011 20:14:52
>> datenum(ts)
ans =
     7.347478436574074e+005
>> datenum(2011,9,1,20,15,07)
ans =
     7.347478438310185e+005
>> datenum(2011,9,1)
ans =
     734747
```

The `datevec` function converts a date string to a numerical vector containing the date components using a format string consisting of format symbols from Table 2 in `datestr` help or formats 0, 1, 2, 6, 13, 14, 15, 16, or 23 from Table 1. Note that supplying the format string, if known, is significantly faster than omitting it. Alternatively, it converts a date number to a numerical vector of date components, as in the following code:

```
>> c = datevec('12/24/1984','mm/dd/yy')
c =
    1984    12    24    0    0    0

>> [yr,mo,day,hr,min,sec] = datevec('24-Dec-1984 08:22')
yr =
    1984
mo =
    12
day =
    24
```

```
hr =

    8

min =

    22

sec =

    0


>> [yr,mo,day,hr,min,sec] = datevec(t)

yr =

    2010

mo =

    11

day =

    2

hr =

    15

min =

    38

sec =

    29.848297119140625
```

15.3 DATE FUNCTIONS

The numerical day of the week can be found from a date string or a date number by using the function weekday (*MATLAB uses the convention that Sunday is day 1 and Saturday is day 7*):

```
>> [d,w] = weekday(734699)

d =

    6

w =

    Fri


>> [d,w] = weekday('21-Dec-1994')

d =

    4
```

```
w =

    Wed
```

The last day of any month can be found by using the function eomday. Because of leap year, both the year and the month are required, as in the following example:

```
>> eomday(2008,2) % divisible by 4 is a leap year
ans =
    29


>> eomday(1800,2) % divisible by 100 not a leap year
ans =
    28


>> eomday(1600,2) % divisible by 400 is a leap year
ans =
    29
```

MATLAB can generate a calendar for any month you request and display it in the *Command* window or place it in a 6-by-7 matrix by using the function calendar:

```
>> calendar(date)
                    Sep 2011
      S     M    Tu     W    Th     F     S
      0     0     0     0     1     2     3
      4     5     6     7     8     9    10
     11    12    13    14    15    16    17
     18    19    20    21    22    23    24
     25    26    27    28    29    30     0
      0     0     0     0     0     0     0
>> calendar(1986,1)
                    Jan 1986
      S     M    Tu     W    Th     F     S
      0     0     0     1     2     3     4
      5     6     7     8     9    10    11
     12    13    14    15    16    17    18
     19    20    21    22    23    24    25
```

26	27	28	29	30	31	0
0	0	0	0	0	0	0

```
>> x = calendar(2011,7)
x =
```

0	0	0	0	0	1	2
3	4	5	6	7	8	9
10	11	12	13	14	15	16
17	18	19	20	21	22	23
24	25	26	27	28	29	30
31	0	0	0	0	0	0

```
>> class(x)
ans =
double
```

Arithmetic can be performed on dates using the addtodate function. There are some limitations, however. The addtodate function requires a date number as input and returns a date number as output. The number to be added must be an integer value and can affect only one field at a time. The field specification must be one of the strings 'year', 'month', 'day', 'hour', 'minute', 'second', or 'millisecond'. Date format conversion functions are used to convert other date forms as needed:

```
>> datestr(addtodate(now,20,'month')) % 20 months from now
ans =
03-Jul-2012 17:53:28

>> datestr(addtodate(addtodate(now,6,'month'),29,'day')) % 6 months 29 days
ans =
01-Jun-2011 17:53:50

>> datestr(addtodate(datenum('22-Jan-2012','dd-mmm-yyyy'),-6,'day'))
ans =
16-Jan-2012
```

15.4 TIMING FUNCTIONS

The functions tic and toc are used to time a sequence of MATLAB operations. The function tic starts a stopwatch, while toc stops the stopwatch and displays the elapsed time:

```
>> tic; plot(rand(50,5)); toc
Elapsed time is 0.999414 seconds.
```

```
>> tic; plot(rand(50,5)); toc
Elapsed time is 0.151470 seconds.
```

Note the difference in elapsed times for identical plot commands. The second plot was significantly faster, because MATLAB had already created the *Figure* window and compiled the functions it needed into memory.

The function cputime returns the amount of central processing unit (CPU) time, in seconds, that MATLAB has used since the current session was started. The function etime calculates the elapsed time between two time vectors, in six-element row vector form such as that returned by the functions clock and datevec. Both cputime and etime are based on the system clock and can be used to compute the amount of system time it takes for an operation to be completed. Usage of cputime and etime are demonstrated by the following examples, in which myoperation is a script file containing a number of MATLAB commands:

```
>> t0 = cputime; myoperation; cputime - t0 % CPU time
ans =
      0.130000000000109
```

```
>> t = clock; myoperation; etime(clock, t) % system clock time
ans =
      0.108533999999999
```

```
>> t1 = tic; myoperation; toc(t1) % elapsed time (most accurate)
Elapsed time is 0.107682 seconds.
```

15.5 PLOT LABELS

Sometimes it is useful to plot data and use dates or time strings for one or more of the axis labels. The datetick function automates this task. *Use of this function requires that the axis to be marked be plotted with a vector of date numbers, such as the output of the datenum function.* The following code is illustrative:

```
>> t = (1920:10:2010)';
>> p = [ 75.995;  91.972; 105.711; 123.203; 131.669;
        150.697; 179.323; 203.212; 226.505; 249.633];
>> plot(datenum(t,1,1),p)
```

```
>> datetick('x','yyyy') % use 4-digit year on the x-axis
>> title('Figure 15.1: Population by Year')
```

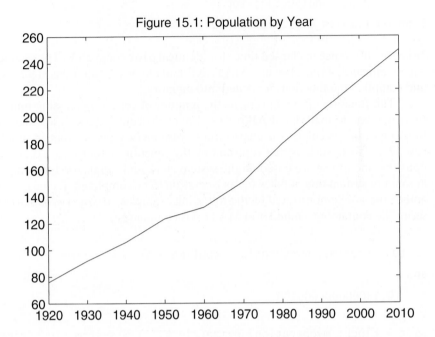

Figure 15.1: Population by Year

Next, we create a bar chart of company sales from November 2009 to December 2010:

```
>> y = [2009 2009 2010*ones(1,12)]';
>> m = [11 12 (1:12)]';
>> s = [1.1 1.3 1.2 1.4 1.6 1.5 1.7 1.6 1.8 1.3 1.9 1.7 1.6 1.95]';
>> bar(datenum(y,m,1),s)

>> datetick('x','mmmyy')
>> ylabel('$ Million')
>> title('Figure 15.2: Monthly Sales')
```

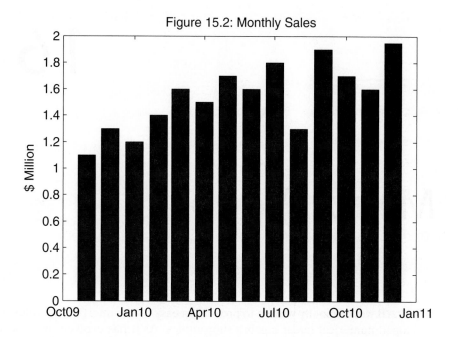

Figure 15.2: Monthly Sales

Time and date functions in MATLAB are summarized in the following table.

Date and Time Function	Description
clock	Current date and time as a date vector
now	Current date and time as a date number
date	Current date as a date string
datestr	String representation of a date
datenum	Numeric representation of a date
datevec	Vector of date components
addtodate	Adds an integer to a date field
weekday	Day of the week for a specified date
eomday	Last day number for a specified month and year
calendar	Generates a calendar for the current or a specified month
tic	Starts the stopwatch timer
toc	Reads the stopwatch timer
cputime	Elapsed CPU time in seconds
etime	Elapsed time between date vectors
datetick	Date formatted tick labels

16

Matrix Algebra

MATLAB was originally written to provide an easy-to-use interface to professionally developed numerical linear algebra subroutines. As it has evolved over the years, other features, such as graphics and graphical user interfaces, have made the numerical linear algebra routines less prominent. Nevertheless, MATLAB offers a wide range of valuable matrix algebra functions.

> It is important to note that while MATLAB supports n-dimensional arrays, matrix algebra is defined only for 2-D arrays—that is, vectors and matrices.

16.1 SETS OF LINEAR EQUATIONS

One of the most common linear algebra problems is finding the solution of a linear set of equations. For example, consider the set of equations

$$\begin{bmatrix} 4 & 5 & 6 \\ 23 & 2 & 84 \\ -3 & -1 & 1 \end{bmatrix} \cdot \begin{bmatrix} x_1 \\ x_2 \\ x_3 \end{bmatrix} = \begin{bmatrix} 232 \\ 401 \\ 198 \end{bmatrix}$$

$$A \cdot x = y$$

where the mathematical multiplication symbol $(\cdot)$ is now defined in the matrix sense, as opposed to the array sense discussed earlier. In MATLAB, this matrix multiplication is denoted with the asterisk notation *. The above equations define the product of the matrix A and the vector x as being equal to the vector y. The very existence of solutions of this equation is a fundamental issue in linear algebra. Moreover, when a solution does exist, there are numerous approaches to finding that solution, such as Gaussian elimination, LU factorization, or the direct use of A^{-1}. Clearly, it is beyond the scope of this text to discuss the many analytical and numerical issues of matrix algebra. We wish only to demonstrate how MATLAB can be used to solve problems like the one in the preceding example.

To solve the problem at hand, it is necessary to enter A and y:

```
>> A = [4:6;23 2 84;-3:2:2
A =

     4     5     6
    23     2    84
    -3    -1     1
>> y = [232;401;198]
y =

   232

   401

   198
```

As discussed earlier, the entry of the matrix A shows the two ways that MATLAB distinguishes between rows. The semicolon between the 3 and 4 signifies the start of a new row, as does the new line between the 6 and 7. The vector y is a column, because each semicolon signifies the start of a new row.

If you have a background in linear algebra, you will find it easy to see that this problem has a unique solution if the rank of A and the rank of the augmented matrix [A y] are both equal to 3. Alternatively, you can check the condition number of A. If the condition number is not excessively large, then A has an inverse with good numerical properties. Testing this problem produces the following result:

```
>> rank(A)
ans =

     3
>> rank([A y])
ans =

     3

>> cond(A) % close to one is best
ans =

    39.619
```

Since the rank and condition number tests hold, MATLAB can find the solution of $A\cdot x = y$ in two ways, one of which is preferred. The less favorable method is to take $x = A^{-1}\cdot y$ literally, as in the following code:

```
>> x = inv(A)*y % Avoid this approach if possible
x =
          -85.018
           83.126
           26.073
```

Here, `inv(A)` is a MATLAB function that computes A^{-1}, and the multiplication operator * denotes matrix multiplication. The preferable solution is found by using the matrix left-division operator, or backward slash, as in this example:

```
>> x = A\y % Recommended approach to solving sets of equations
x =
          -85.018
           83.126
           26.073
```

This equation uses an LU factorization approach and expresses the answer as the left division of A into y. The left-division operator \ has no preceding dot, as this is a matrix operation, not an element-by-element array operation. There are many reasons to prefer this second solution. Of these, the simplest is that the latter method requires fewer floating-point operations and, as a result, is significantly faster. In addition, this solution is generally more accurate, especially for larger problems. In either case, if MATLAB cannot find a solution or cannot find one accurately, it displays a warning message.

If the transpose of the above set of linear equations is taken (i.e., $(A\cdot x)' = y'$), then the previous set of linear equations can be written as $x'\cdot A' = y'$, where x' and y' are now row vectors. As a result, it is equally valid to express a set of linear equations in terms of the product of a row vector and a matrix being equal to another row vector—for example, $u\cdot B = v$. In MATLAB, this case is solved by the same internal algorithms by using the matrix right-division operator, or forward slash, as in u = v/B.

It is important to note that when MATLAB encounters a forward slash (/) or backward slash (\), it checks the structure of the coefficient matrix to determine what internal algorithm to use to find the solution. In particular, if the matrix is upper or lower triangular or a permutation of an upper or lower triangular matrix, MATLAB does not refactor the matrix, but rather just performs the forward or backward substitution steps required to find the solution. As a result, MATLAB makes use of the properties of the coefficient matrix to compute the solution as quickly as possible.

When the structure of the coefficient matrix is known ahead of time, the function linsolve solves the set of equations by using matrix structure information provided by the user. By using this function, MATLAB does not spend time analyzing the structure of the coefficient matrix and thereby minimizes the time required to solve the set of equations. The calling syntax for this function is linsolve(A,y,options), where options is a structure containing fields that can be set to True or False to identify the structure of the A matrix.

If you've studied linear algebra rigorously, you know that when the number of equations and number of unknowns differ, a single unique solution usually does not exist. However, with further constraints, a practical solution can usually be found. In MATLAB, when rank(A) = min(r,c), where r and c are the number of rows and columns in A, respectively, and there are more equations than unknowns ($r > c$) (i.e., the **_over-determined_** case), a division operator / or \ automatically finds the solution that minimizes the norm of the squared residual error $e = A \cdot x - y$. This solution is of great practical value and is called the **_least-squares solution_**. The following example is illustrative:

```
>> A = [1 2 3;4 5 6;7 8 0;2 5 8]    % 4 equations in 3 unknowns
A =
     1     2     3
     4     5     6
     7     8     0
     2     5     8
>> y = [366 804 351 514]'    % a new r.h.s. vector
y =
   366
   804
   351
   514
>> x = A\y      % least squares solution
x =
      247.98
     -173.11
      114.93
>> e = A*x-y    % this residual has the smallest norm.
e =
  -119.45
    11.945
     2.2737e-013
    35.836
```

```
>> norm(e)
ans =
   125.28
```

In addition to the least-squares solution that is computed with the left- and right-division operators, MATLAB offers the functions lscov and lsqnonneg. The function lscov solves the weighted least-squares problem when the covariance matrix (or weighting matrix) of the data is known, and lsqnonneg finds the nonnegative least-squares solution where all solution components are constrained to be positive.

When there are fewer equations than unknowns ($r < c$) (i.e., the **_under-determined_** case), an infinite number of solutions exist. Of these solutions, MATLAB computes two in a straightforward way. Use of the division operator gives a solution that has a maximum number of zeros in the elements of *x*. Alternatively, computing x = pinv(A)*y gives a solution where the length or norm of x is smaller than all other possible solutions. This solution, based on the *pseudoinverse*, also has great practical value and is called the **_minimum norm solution_**:

```
>> A = A'      % create 3 equations in 4 unknowns
A =
       1      4      7      2
       2      5      8      5
       3      6      0      8
>> y = y(1:3)    % new r.h.s. vector
y =
   366
   804
   351
>> x = A\y      % solution with maximum zero elements
x =
           0
   -165.9000
     99.0000
    168.3000
>> xn = pinv(A)*y    % minimum norm solution
xn =
     30.8182
   -168.9818
     99.0000
    159.0545
```

```
>> norm(x) % norm of solution with zero elements
ans =
   256.2200
>> norm(xn) % minimum norm solution has smaller norm!
ans =
   254.1731
```

16.2 MATRIX FUNCTIONS

In addition to the solution of linear sets of equations, MATLAB offers numerous matrix functions that are useful for solving numerical linear algebra problems. A thorough discussion of these functions is beyond the scope of this text. In general, MATLAB provides functions for all common, and some uncommon, numerical linear algebra problems. A brief description of many of the matrix functions is given in the following table:

Function	Description
/ and \	Solves Ax = y (much better than using inv(A)*y)
accumarray(ind,val)	Constructs array with accumulation
A^n	Exponentiation, for example, A^3 = A*A*A
balance(A)	Scale to improve eigenvalue accuracy
[V,D] = cdf2rdf(V,D)	Complex diagonal form to real block diagonal form
chol(A)	Cholesky factorization
cholupdate(R,X)	Rank 1 update to Cholesky factorization
cond(A)	Matrix condition number using singular value decomposition
condest(A)	1-norm condition number estimate
[V,D,s] = condeig(A)	Condition number with respect to eigenvalues
det(A)	Determinant
dmperm(A)	Dulmage–Mendelsohn permutation
eig(A)	Vector of eignenvalues
[V,D] = eig(A)	Matrix of eignenvectors, and diagonal matrix containing eigenvalues
expm(A)	Matrix exponential
funm(A,@*fun*)	General matrix function

Function	Description
gsvd(A,B)	Generalized singular values
[U,V,X,C,S] = gsvd(A)	Generalized singular value decomposition
hess(A)	Hessenburg form of a matrix
inv(A)	Matrix inverse (use only when / or \ won't do)
linsolve(A,y,options)	Solves Ax = y quickly when structure of A is given by options
logm(A)	Matrix logarithm
lscov(A,y,V)	Weighted least squares with covariance matrix
lsqnonneg(A,y)	Nonnegative least-squares solution
[L,U,P]=lu(A)	LU decomposition
minres(A,y)	Minimum residual method
norm(A,type)	Matrix and vector norms
null(A)	Null space
ordeig(T)	Eigenvalues of quasitriangular Schur matrix
ordeig(A,B)	Generalized eigenvalues of quasitriangular matrices
ordschur(U,T,select)	Reorders eigenvalues in Schur factorization
ordqz(A,B,Q,Z,select)	Reorders eigenvalues in QZ factorization
orth(A)	Orthogonal range space using singular value decomposition
pinv(A)	Pseudoinverse using singular value decomposition
planerot(X)	Givens plane rotation
poly(A)	Characteristic polynomial
polyeig(A0,A1,...)	Polynomial eigenvalue solution
polyvalm(A)	Evaluates matrix polynomial
qr(A)	Orthogonal-triangular decomposition
qrdelete(Q,R,J)	Deletes column or row from QR factorization
qrinsert(Q,R,J,X)	Inserts column or row into QR factorization
qrupdate(Q,R,U,V)	Rank 1 update to QR factorization
qz(A,B)	Generalized eigenvalues
rank(A)	Matrix rank using singular value decomposition

Function	Description
rcond(A)	LAPACK reciprocal condition estimator
rref(A)	Reduced row echelon form
rsf2csf(A)	Real Schur form to complex Schur form
schur(A)	Schur decomposition
sqrtm(A)	Matrix square root
subspace(A,B)	Angle between two subspaces
svd(A)	Singular values
[U,S,V] = svd(A)	Singular value decomposition
trace(A)	Sum of matrix diagonal elements

16.3 SPECIAL MATRICES

MATLAB offers a number of special matrices; some of them are general utilities, while others are matrices of interest to specialized disciplines. These and other special matrices include those given in the following table (use the online help to learn more about these matrices):

Matrix	Description
[ ]	Empty matrix
blkdiag(A0,A1, . . .)	Block diagonal concatenation of input arguments
compan(P)	Companion matrix of a polynomial
eye(r,c)	Identity matrix
gallery	More than 60 test matrices
hadamard(n)	Hadamard matrix of order n
hankel(C)	Hankel matrix
hilb(n)	Hilbert matrix of order n
invhilb(n)	Inverse Hilbert matrix of order n
magic(n)	Magic matrix of order n
ones(r,c)	Matrix containing all ones
pascal(n)	Pascal matrix of order n.
rand(r,c)	Uniformly distributed random matrix with elements between 0 and 1

Matrix	Description
`randi(n,r,c)`	Uniformly distributed random integers over the range `1:n`
`randn(r,c)`	Normally distributed random matrix with elements having zero mean and unit variance
`rosser`	Classic symmetric eigenvalue test problem
`toeplitz(C,R)`	Toeplitz matrix
`vander(C)`	Vandermonde matrix
`wilkinson(n)`	Wilkinson's eigenvalue test matrix of order `n`
`zeros(r,c)`	Matrix containing all zeros

16.4 SPARSE MATRICES

In many practical applications, matrices are generated that contain only a few nonzero elements. As a result, these matrices are said to be ***sparse***. For example, circuit simulation and finite element analysis programs routinely deal with matrices containing less than 1 percent nonzero elements. If a matrix is large—for example, `max(size(A)) > 100`—and has a high percentage of zero elements, it is wasteful, both of computer storage to store the zero elements and of computational power to perform arithmetic operations by using the zero elements. To eliminate the storage of zero elements, it is common to store only the nonzero elements of a sparse matrix and two sets of indices identifying the row and column positions of these elements. Similarly, to eliminate arithmetic operations on the zero elements, special algorithms have been developed to solve typical matrix problems (such as solving a set of linear equations in which operations involving zeros are minimized, and intermediate matrices have minimum nonzero elements).

The techniques used to optimize sparse matrix computations are complex in implementation, as well as in theory. Fortunately, MATLAB hides this complexity. In MATLAB, sparse matrices are stored in variables, just as regular, full matrices are. Moreover, most computations with sparse matrices use the same syntax as that used for full matrices. In particular, all of the array-manipulation capabilities of MATLAB work equally well on sparse matrices. For example, `s(i,j)` = *value* assigns `value` to the `i`th row and `j`th column of the sparse matrix `s`.

In this text, only the creation of sparse matrices and the conversion to and from sparse matrices are illustrated. In general, operations on full matrices produce full matrices and operations on sparse matrices produce sparse matrices. In addition, operations on a mixture of full and sparse matrices generally produce sparse matrices, unless the operation makes the result too densely populated with nonzeros to make sparse storage efficient.

Sparse matrices are created using the MATLAB function sparse. For example, the code

```
>> As = sparse(1:9,1:9,ones(1,9))
As =
    (1,1)        1
    (2,2)        1
    (3,3)        1
    (4,4)        1
    (5,5)        1
    (6,6)        1
    (7,7)        1
    (8,8)        1
    (9,9)        1
```

creates a 9-by-9 identity matrix. In this usage, sparse(i,j,s) creates a sparse matrix whose kth nonzero element is s(k), which appears in the row i(k) and column j(k). Note the difference in how sparse matrices are displayed. Nonzero elements and their row and column positions are displayed. The preceding sparse matrix can also be created by conversion. For example, the code

```
>> As = sparse(eye(9))
As =
    (1,1)        1
    (2,2)        1
    (3,3)        1
    (4,4)        1
    (5,5)        1
    (6,6)        1
    (7,7)        1
    (8,8)        1
    (9,9)        1
```

creates the 9-by-9 identity matrix again, this time by converting the full matrix eye(9) to sparse format. While this method of creating a sparse matrix works, it is seldom used in practice, because the initial full matrix wastes a great deal of memory.

Given a sparse matrix, the function full generates the conventional full matrix equivalent. The following example converts the sparse matrix As back to its full form:

```
>> A = full(As)
A =
```

1	0	0	0	0	0	0	0	0
0	1	0	0	0	0	0	0	0
0	0	1	0	0	0	0	0	0
0	0	0	1	0	0	0	0	0
0	0	0	0	1	0	0	0	0
0	0	0	0	0	1	0	0	0
0	0	0	0	0	0	1	0	0
0	0	0	0	0	0	0	1	0
0	0	0	0	0	0	0	0	1

To compare sparse matrix storage to full matrix storage, consider the following example:

```
>> B = eye(200);
>> Bs = sparse(B);
>> whos
  Name    Size      Bytes    Class      Attributes

  B       200x200   320000   double
  Bs      200x200   3204     double     sparse
```

Here, the sparse matrix Bs contains only 0.5 percent nonzero elements and requires 3204 bytes of storage. On the other hand, B, the same matrix in full matrix form requires two orders of magnitude more bytes of storage!

16.5 SPARSE MATRIX FUNCTIONS

MATLAB provides numerous sparse matrix functions. Many involve different aspects of, and techniques for, the solution of sparse simultaneous equations. A discussion of these functions is beyond the scope of this text. The functions available are listed in the following table:

Sparse Matrix Function	Description
bicg	Biconjugate gradient iterative linear equation solution
bicgstab	Biconjugate gradient stabilized iterative linear equation solution
bicgstabl	Biconjugate gradients stabilized(l) method
cgs	Conjugate gradients squared iterative linear equation solution
cholinc	Incomplete Cholesky factorization
colamd	Column approximate minimum degree reordering method
colperm	Column permutation
condest	1-norm condition number estimate
dmperm	Dulmage–Mendelsohn reordering method
eigs	A few eigenvalues using ARPACK
etree	Elimination tree
etreeplot	Plots elimination tree
find	Finds indices of nonzero elements
full	Converts sparse matrix to full matrix
gmres	Generalized minimum residual iterative linear equation solution
gplot	Constructs graph theory plot
ilu	Incomplete LU factorization
issparse	True for sparse matrix
lsqnonneg	Solves nonnegative least-squares constraints problem
lsqr	LSQR implementation of conjugate gradients on normal equations
luinc	Incomplete LU factorization
minres	Minimum residual iterative linear equation solution
nnz	Number of nonzero matrix elements
nonzeros	Nonzero matrix elements
normest	Estimate of matrix 2-norm
nzmax	Storage allocated for nonzero elements

Sparse Matrix Function	Description
pcg	Preconditioned conjugate gradients iterative linear equation solution
qmr	Quasi-minimal residual iterative linear equation solution
randperm	Random permutation
spalloc	Allocates space for sparse matrix
sparse	Creates sparse matrix
spaugment	Forms least-squares augmented system
spconvert	Import from sparse matrix external format
spdiags	Sparse matrix formed from diagonals
speye	Sparse identity matrix
spfun	Applies function to nonzero elements
spones	Replaces nonzeros with ones
spparms	Sets parameters for sparse matrix routines
sprand	Sparse uniformly distributed matrix
sprandn	Sparse normally distributed matrix
sprandsym	Sparse random symmetric matrix
sprank	Structural rank
spy	Visualizes sparsity pattern
svds	A few singular values
symamd	Symmetric approximate minimum degree reordering method
symbfact	Symbolic factorization analysis
symmlq	Symmetric LQ iterative linear equation solution
symrcm	Symmetric reverse Cuthill–Mckee reordering method
tfqmr	Transpose-free quasi-minimal residual method
treelayout	Lays out tree or forest
treeplot	Plots picture of tree

17

Data Analysis

Because of its array orientation, MATLAB readily performs statistical analyses on data sets. While MATLAB, by default, considers data sets stored in column-oriented arrays, data analysis can be conducted along any specified dimension.

> Unless specified otherwise, each column of an array represents a different measured variable, and each row represents individual samples or observations.

17.1 BASIC STATISTICAL ANALYSIS

For example, let's assume that the daily high temperature (in Celsius) of three cities over a 31-day month was recorded and assigned to the variable `temps` in a script M-file. Running the M-file puts the variable `temps` in the MATLAB workspace. When this work is done, the variable `temps` contains the following data:

```
>> temps
temps =
      12      8     18
      15      9     22
      12      5     19
      14      8     23
      12      6     22
      11      9     19
```

15	9	15
8	10	20
19	7	18
12	7	18
14	10	19
11	8	17
9	7	23
8	8	19
15	8	18
8	9	20
10	7	17
12	7	22
9	8	19
12	8	21
12	8	20
10	9	17
13	12	18
9	10	20
10	6	22
14	7	21
12	5	22
13	7	18
15	10	23
13	11	24
12	12	22

Each row contains the high temperatures for a given day, and each column contains the high temperatures for a different city. To visualize the data, plot it with the following code:

```
>> d = 1:31; % number the days of the month
>> plot(d,temps)

>> xlabel('Day of Month'), ylabel('Celsius')
>> title('Figure 17.1: Daily High Temperatures in Three Cities')
```

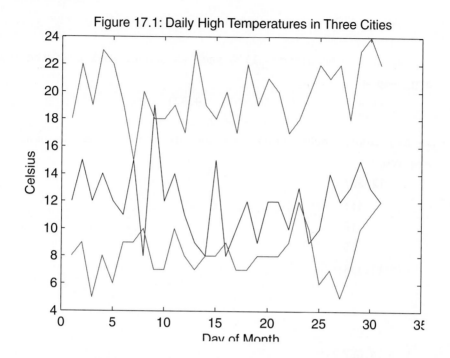

Figure 17.1: Daily High Temperatures in Three Cities

The `plot` command in this example illustrates yet another form of `plot` command usage. The variable d is a vector of length 31, whereas `temps` is a 31-by-3 matrix. Given this data, the `plot` command plots each column of `temps` versus d.

To illustrate some of the data-analysis capabilities of MATLAB, consider the following commands based on the temperature data stored in `temps`:

```
>> avg_temp = mean(temps)
avg_temp =
    11.968    8.2258    19.871
```

This result shows that the third city has the highest average temperature. Here, MATLAB found the average of each column individually. Taking the average again gives

```
>> avg_avg = mean(avg_temp)
avg_avg =
    13.355
```

which returns the overall average temperature of the three cities.

When the input to a data-analysis function is a row or column vector, MATLAB simply performs the operation on the vector, returning a scalar result.

Alternatively, you can specify the dimension to work on:

```
>> avg_temp = mean(temps,1) % same as above, work down the rows
avg_temp =
        11.968         8.2258         19.871

>> avg_tempr = mean(temps,2) % compute means across columns
avg_tempr =
        12.667
        15.333
           12
           15
        13.333
           13
           13
        12.667
        14.667
        12.333
        14.333
           12
           13
        11.667
        13.667
        12.333
        11.333
        13.667
           12
        13.667
        13.333
           12
        14.333
           13
        12.667
           14
           13
        12.667
```

```
             16
             16
          15.333
```

This output lists the three-city average temperature on each day. The scalar second argument to mean dictates the dimension to be analyzed; for example, 1 is the row dimension number, so MATLAB performs the analysis accumulating information down the rows; 2 is the column dimension number, so MATLAB performs the analysis accumulating information across the columns.

If temps were *n*-dimensional, the mean and other statistical functions could be used along any dimension, as in the following example:

```
>> temps2 = temps+round(2*rand(size(temps))-1);
>> temps3 = cat(3,temps,temps2);
>> size(temps3)
ans =
    31     3     2
>> mean(temps3)    % same as mean(temps3,1)
ans(:,:,1) =
        11.968          8.2258          19.871
ans(:,:,2) =
        11.935          8.1935          19.935

>> squeeze(mean(temps3))'    % squeeze to two dimensions
ans =
        11.968          8.2258          19.871
        11.935          8.1935          19.935

>> reshape(mean(temps3),3,2)'    % alternate squeeze
ans =
        11.968          8.2258          19.871
        11.935          8.1935          19.935
>> mean(temps3,3)
ans =
            12               8              18
          14.5             8.5              22
          12.5             5.5              19
          14.5             8.5            23.5
```

12	6	22.5
11	8.5	19.5
14.5	8.5	15
8	10	20
19	7	18
12	6.5	17.5
14.5	10	18.5
11	8	17.5
8.5	7	22.5
7.5	8	18.5
15	8	18
7.5	9.5	19.5
10	7.5	17
12.5	7	22
9	8	19.5
12	7.5	21.5
12	8	20.5
9.5	9	17
12.5	11.5	18
9	10	20
10.5	6	22
13.5	7	21
12	5	22
13.5	6.5	18
15	10	23
13	11.5	24
12.5	12.5	22

Here, temps2 contains randomly generated data for the temperatures in the three cities for a second month. The variable temps3 contains the first month's temperatures on page 1 and the second month's temperatures on page 2. The function call mean(temps3) computes the means down each column on each page, giving a result that has one row, three columns, and two pages. This data can be squeezed into two rows and three columns by using squeeze or reshape. The function call mean(temps3,3) computes the mean along the page dimension, which is the month-to-month mean of the temperatures on a given day, in a given city. The result is an array having 31 rows, three columns, and one page (i.e., a 2-D array).

Going back to the 2-D case, consider the problem of finding the daily deviation from the mean of each city. That is, avg_temp(i) must be subtracted from column i of temps. You cannot simply issue the statement

```
>> avg_temp
avg_temp =
        11.968        8.2258        19.871
```

```
>> temps - avg_temp
??? Error using ==> minus
Matrix dimensions must agree.
```

because the operation is not a defined array operation (temps is 31 by 3, and avg_temp is 1 by 3). Perhaps the most straightforward approach is to use a For Loop:

```
>> for c = 1:3
       tdev(:,c) = temps(:,c) - avg_temp(c);
   end
```

While this approach works, it was much slower before the introduction of JIT-acceleration in MATLAB. An alternative is to use the array-manipulation features of MATLAB. In this case, you duplicate avg_temp to make it the size of temps and then do the subtraction:

```
>> tdev = temps - repmat(avg_temp,31,1)
tdev =
        0.032258      -0.22581       -1.871
        3.0323         0.77419        2.129
        0.032258      -3.2258        -0.87097
        2.0323        -0.22581        3.129
        0.032258      -2.2258         2.129
       -0.96774        0.77419       -0.87097
        3.0323         0.77419       -4.871
       -3.9677         1.7742         0.12903
        7.0323        -1.2258        -1.871
        0.032258      -1.2258        -1.871
        2.0323         1.7742        -0.87097
       -0.96774       -0.22581       -2.871
```

-2.9677	-1.2258	3.129
-3.9677	-0.22581	-0.87097
3.0323	-0.22581	-1.871
-3.9677	0.77419	0.12903
-1.9677	-1.2258	-2.871
0.032258	-1.2258	2.129
-2.9677	-0.22581	-0.87097
0.032258	-0.22581	1.129
0.032258	-0.22581	0.12903
-1.9677	0.77419	-2.871
1.0323	3.7742	-1.871
-2.9677	1.7742	0.12903
-1.9677	-2.2258	2.129
2.0323	-1.2258	1.129
0.032258	-3.2258	2.129
1.0323	-1.2258	-1.871
3.0323	1.7742	3.129
1.0323	2.7742	4.129
0.032258	3.7742	2.129

Here, repmat(avg_temp,31,1) replicates the row vector avg_temp, 31 times in the row dimension and once in the column dimension, creating a 31-by-3 matrix the same size as temps. The first column of repmat(avg_temp,31,1) contains avg_tmp(1), the second column contains avg_temp(2), and the third column contains avg_temp(3).

A better solution is to use the function bsxfun introduced in MATLAB version 7.4. This function applies the element-by-element binary operation specified by a function handle to two arrays. The code

```
>> tdev = bsxfun(@minus, temps, mean(temps))
tdev =
      0.032258    -0.22581      -1.871
      3.0323       0.77419       2.129
      0.032258    -3.2258      -0.87097
      2.0323      -0.22581       3.129
      0.032258    -2.2258       2.129
```

-0.96774	0.77419	-0.87097
3.0323	0.77419	-4.871
-3.9677	1.7742	0.12903
7.0323	-1.2258	-1.871
0.032258	-1.2258	-1.871
2.0323	1.7742	-0.87097
-0.96774	-0.22581	-2.871
-2.9677	-1.2258	3.129
-3.9677	-0.22581	-0.87097
3.0323	-0.22581	-1.871
-3.9677	0.77419	0.12903
-1.9677	-1.2258	-2.871
0.032258	-1.2258	2.129
-2.9677	-0.22581	-0.87097
0.032258	-0.22581	1.129
0.032258	-0.22581	0.12903
-1.9677	0.77419	-2.871
1.0323	3.7742	-1.871
-2.9677	1.7742	0.12903
-1.9677	-2.2258	2.129
2.0323	-1.2258	1.129
0.032258	-3.2258	2.129
1.0323	-1.2258	-1.871
3.0323	1.7742	3.129
1.0323	2.7742	4.129
0.032258	3.7742	2.129

produces the same result without explicitly duplicating arrays. This is possible because bsxfun performs singleton expansion internally when needed.

MATLAB can also find minima and maxima. For example, the code

```
>> max_temp = max(temps)
max_temp =
    19    12    24
```

finds the maximum high temperature of each city over the month, while

```
>> [max_temp,maxday] = max(temps)
max_temp =
      19      12      24
maxday =
       9      23      30
```

finds the maximum high temperature of each city and the row index maxday where the maximum appears. In this example, maxday identifies the day of the month when the highest temperature occurred.

The code

```
>> min_temp = min(temps)
min_temp =
       8       5      15
```

finds the minimum high temperature of each city, and the code

```
>> [min_temp,minday] = min(temps)
min_temp =
       8       5      15
minday =
       8       3       7
```

finds the minimum high temperature of each city and the row index minday where the minimum appears. For this example, minday identifies the day of the month when the lowest high temperature occurred.

Other standard statistical measures are also provided in MATLAB, including the following:

```
>> s_dev = std(temps) % standard deviation in each city
s_dev =
     2.5098    1.7646    2.2322
>> median(temps) % median temperature in each city
ans =
      12       8      20
>> cov(temps) % covariance
```

```
ans =

       6.2989           0.04086         -0.13763
       0.04086          3.114            0.063441
      -0.13763          0.063441         4.9828
>> corrcoef(temps) % correlation coefficients
ans =

            1         0.0092259        -0.024567
       0.0092259            1            0.016106
      -0.024567       0.016106             1
```

You can also compute differences from day to day by using the function diff:

```
>> daily_change = diff(temps)
daily_change =
        3      1      4
       -3     -4     -3
        2      3      4
       -2     -2     -1
       -1      3     -3
        4      0     -4
       -7      1      5
       11     -3     -2
       -7      0      0
        2      3      1
       -3     -2     -2
       -2     -1      6
       -1      1     -4
        7      0     -1
       -7      1      2
        2     -2     -3
        2      0      5
       -3      1     -3
        3      0      2
        0      0     -1
       -2      1     -3
        3      3      1
```

-4	-2	2
1	-4	2
4	1	-1
-2	-2	1
1	2	-4
2	3	5
-2	1	1
-1	1	-2

This function computes the difference between daily high temperatures and describes how much the daily high temperature varied from day to day. For example, between the first and second days of the month, the first row of daily_change is the amount that the daily high changed. As with other functions, the difference can be computed along other dimensions as well:

```
>> city_change = diff(temps,1,2)
city_change =
```

-4	10
-6	13
-7	14
-6	15
-6	16
-2	10
-6	6
2	10
-12	11
-5	11
-4	9
-3	9
-2	16
0	11
-7	10
1	11
-3	10
-5	15
-1	11
-4	13

−4	12
−1	8
−1	6
1	10
−4	16
−7	14
−7	17
−6	11
−5	13
−2	13
0	10

Here, the function `diff(temps,1,2)` produces a first-order difference along dimension 2. Therefore, the result is the city difference. The first column is the difference between city 2 and city 1; the second column is the difference between city 3 and city 2.

It is common to use the value NaN to signify missing data. When this is done, most statistical functions require special treatment, because operations on NaNs generally produce NaNs, as in the following example:

```
>> temps4 = temps; % copy data
>> temps4(5,1) = nan; % insert some NaNs
>> temps4(29,2) = nan;
>> temps4(13:14,3) = nan
temps4 =
```

12	8	18
15	9	22
12	5	19
14	8	23
NaN	6	22
11	9	19
15	9	15
8	10	20
19	7	18
12	7	18
14	10	19
11	8	17

9	7	NaN
8	8	NaN
15	8	18
8	9	20
10	7	17
12	7	22
9	8	19
12	8	21
12	8	20
10	9	17
13	12	18
9	10	20
10	6	22
14	7	21
12	5	22
13	7	18
15	NaN	23
13	11	24
12	12	22

```
>> max(temps4) % max and min ignore NaNs
ans =
    19     12     24
>> mean(temps4) % other statistical functions propagate NaNs
ans =
    NaN    NaN    NaN
>> std(temps4)
ans =
    NaN    NaN    NaN
```

One solution to this NaN problem is to write your own functions that exclude NaN elements:

```
>> m = zeros(1,3); % preallocate memory for faster results

>> for j = 1:3 % find mean column by column
       idx = ~isnan(temps4(:,j));
```

```
     m(j) = mean(temps4(idx,j));
  end
>> m

m =

     11.967      8.1667      19.793
```

Here, the function isnan is used to locate elements containing NaN. Then, the mean down each column is found by indexing only non-NaN elements. It is not possible to exclude NaN elements over all columns simultaneously, because not all columns have the same number of NaN elements.

An alternative to the preceding approach is to replace the NaNs with zeros, and compute the mean accordingly, as in the following example:

```
>> lnan = isnan(temps4);  % logical array identifying NaNs
>> temps4(lnan) = 0;      % change all NaNs to zero
>> n = sum(~lnan);        % number of nonNaN elements per column

>> m = sum(temps4)./n     % find mean for all columns
m =
     11.967      8.1667      19.793
```

In this example, a logical array lnan containing True where the elements of temps4 are NaN was found. The NaN values were then set to zero. Then the mean was computed by dividing the column sums by the number of non-NaN elements per column.

17.2 BASIC DATA ANALYSIS

In addition to statistical data analysis, MATLAB offers a variety of general-purpose data-analysis functions. For example, the temperature data temps can be filtered by using the function filter(b,a,data):

```
>> filter(ones(1,4),4,temps)
ans =
           3          2         4.5
        6.75       4.25          10
        9.75        5.5       14.75
       13.25        7.5        20.5
       13.25          7        21.5
       12.25          7       20.75
```

13	8	19.75
11.5	8.5	19
13.25	8.75	18
13.5	8.25	17.75
13.25	8.5	18.75
14	8	18
11.5	8	19.25
10.5	8.25	19.5
10.75	7.75	19.25
10	8	20
10.25	8	18.5
11.25	7.75	19.25
9.75	7.75	19.5
10.75	7.5	19.75
11.25	7.75	20.5
10.75	8.25	19.25
11.75	9.25	19
11	9.75	18.75
10.5	9.25	19.25
11.5	8.75	20.25
11.25	7	21.25
12.25	6.25	20.75
13.5	7.25	21
13.25	8.25	21.75
13.25	10	21.75

Here, the filter implemented is $4y_n = x_n + x_{n-1} + x_{n-2} + x_{n-3}$, or, equivalently, $y_n = (x_n + x_{n-1} + x_{n-2} + x_{n-3})/4$. In other words, each column of temps is passed through a ***moving-average filter*** of length 4. Any realizable filter structure can be applied by specifying different coefficients for the input and output coefficient vectors.

The function y = filter(b,a,x) implements the following general tapped delay-line algorithm:

$$\sum_{k=0}^{N} a_{k+1} y_{n-k} = \sum_{k=0}^{M} b_{k+1} x_{n-k}$$

Here, the vector a is the tap weight vector a_{k+1} on the output, and the vector b is the tap weight vector b_{k+1} on the input. For $N = 2$ and $M = 3$, the preceding equation is equivalent to

$$a_1 y_n + a_2 y_{n-1} + a_3 y_{n-2} = b_1 x_n + b_2 x_{n-1} + b_3 x_{n-2} + b_4 x_{n-3}$$

The `filter` function uses a difference equation description of a filter. When a state space description of a filter is known, the built-in function `ltitr` is useful. The help text of this function contains the following:

```
>> help ltitr
LTITR              Linear time-invariant time response kernel.

X = LTITR(A,B,U) calculates the time response of the system:
                  x[n+1] = Ax[n] + Bu[n]
to input sequence U. The matrix U must have as many columns as
there are inputs u. Each row of U corresponds to a new time
point. LTITR returns a matrix X with as many columns as the
number of states x, and with as many rows as in U.

LTITR(A,B,U,X0) can be used if initial conditions exist.
Here is what it implements, in high speed:

   for i = 1:n
         x(:,i) = x0;
         x0 = a * x0 + b * u(i,:).';
   end
   x = x.';
```

In this case, if the state space output equation is $y[n] = Cx[n] + Du[n]$, where C and D are matrices of appropriate dimensions, $x = $ `ltitr(A,B,u)` can be used to compute the state response $x[n]$, followed by

```
y = x*C.' + u*D.';
```

to obtain the filter output. Here, the array y will have as many columns as there are outputs and as many rows as there are time points.

In MATLAB, data can also be sorted:

```
>> data = rand(10,1) % create some data
data =
      0.61543
      0.79194
```

```
         0.92181
         0.73821
         0.17627
         0.40571
         0.93547
          0.9169
         0.41027
         0.89365
>> [sdata,sidx] = sort(data) % sort in ascending order
sdata =
         0.17627
         0.40571
         0.41027
         0.61543
         0.73821
         0.79194
         0.89365
          0.9169
         0.92181
         0.93547
sidx =
          5
          6
          9
          1
          4
          2
         10
          8
          3
          7
```

The second output of the sort function is the sorted index order. That is, the fifth element in data has the lowest value, and the seventh element in data has the highest value.

Sometimes, it is important to know the ***rank*** of the data. For example, what is the rank or position of the ith data point in the unsorted array with respect to the sorted array? With MATLAB array indexing, the rank is found by the single statement

```
>> ridx(sidx) = 1:10 % ridx is rank
ridx =
     4    6    9    5    1    2   10    8    3    7
```

That is, the first element of the unsorted data appears fourth in the sorted data, and the last element is seventh. If a descending sort is required, it is simply a matter of calling the sort function as [sdata,sidx] = sort(data,'descend').

When the array to be sorted is a matrix, as temps in the preceding example, each column is sorted, and each column produces a column in the optional index matrix. As with the other data-analysis functions, the dimension to analyze can be specified as a second input argument before the optional final 'ascend' or 'descend' argument.

Very often, it is desirable to use the results of sorting one column of an array by applying that sort order to all remaining columns, as in the following example:

```
>> newdata = randn(10,4) % new data for sorting
newdata =
    -0.43256    -0.18671     0.29441    -0.39989
    -1.6656      0.72579    -1.3362      0.69
     0.12533    -0.58832     0.71432     0.81562
     0.28768     2.1832      1.6236      0.71191
    -1.1465     -0.1364     -0.69178     1.2902
     1.1909      0.11393     0.858       0.6686
     1.1892      1.0668      1.254       1.1908
    -0.037633    0.059281   -1.5937     -1.2025
     0.32729    -0.095648   -1.441      -0.01979
     0.17464    -0.83235     0.57115    -0.15672

>> [tmp,idx] = sort(newdata(:,2)); % sort second column

>> newdatas = newdata(idx,:) % shuffle rows using idx from 2nd column
newdatas =
     0.17464    -0.83235     0.57115    -0.15672
     0.12533    -0.58832     0.71432     0.81562
    -0.43256    -0.18671     0.29441    -0.39989
    -1.1465     -0.1364     -0.69178     1.2902
```

0.32729	-0.095648	-1.441	-0.01979
-0.037633	0.059281	-1.5937	-1.2025
1.1909	0.11393	0.858	0.6686
-1.6656	0.72579	-1.3362	0.69
1.1892	1.0668	1.254	1.1908
0.28768	2.1832	1.6236	0.71191

Here, the second column of the random array is sorted in increasing order. Then, the sort index is used to shuffle the rows in all columns. For example, the last row of newdata is now the first row in newdatas, because the last element in the second column is the smallest element in the second column.

A vector is strictly monotonic if its elements either always increase or always decrease as one proceeds down the array. The function diff is useful for determining monotonicity:

```
>> A = diff(data) % check random data
A =
       0.17651
       0.12987
      -0.1836
      -0.56194
       0.22944
       0.52976
      -0.01857
      -0.50663
       0.48338
>> mono = all(A>0) | all(A<0) % as expected, not monotonic
mono =
     0
>> B = diff(sdata) % check random data after sorting
B =
       0.22944
       0.00456
       0.20516
       0.12278
       0.05373
       0.10171
```

```
      0.02325
      0.00491
      0.01366
>> mono = all(B>0) | all(B<0) % as expected, monotonic
mono =
      1
```

Furthermore, a monotonic vector is equally spaced if the second difference is zero.

```
>> all(diff(diff(sdata))==0) % random data is not equally spaced
ans =
    0
>> all(diff(diff(1:25))==0) % but numbers from 1 to 25 are equally spaced
ans =
    1
```

17.3 DATA ANALYSIS AND STATISTICAL FUNCTIONS

By default, data analysis in MATLAB is performed on column-oriented matrices. Different variables are stored in individual columns, and each row represents a different observation of each variable. Many data-analysis functions work along any dimension, provided that the dimension is specified as an input argument. The data analysis and statistical functions in MATLAB are listed in the following table:

Function	Description
corrcoef(A)	Correlation coefficients
conv(A,B)	Convolution and polynomial multiplication
conv2(A,B)	Two-dimensional convolution
convn(A,B)	N-dimensional convolution
cov(A)	Covariance matrix
cplxpair(v)	Sorts vector into complex conjugate pairs
cumprod(A)	Cumulative product of elements
cumsum(A)	Cumulative sum of elements
cumtrapz(A)	Cumulative trapezoidal integration

Function	Description
deconv(B,A)	Deconvolution and polynomial division
del2(A)	Discrete Laplacian (surface curvature)
detrend(A)	Linear trend removal
diff(A)	Differences between elements
filter(B,A,X)	One-dimensional digital filter
filter2(B,X)	Two-dimensional digital filter
gradient(Z,dx,dy)	Approximate surface gradient
hist(X,M)	Bins the elements of X into M equally spaced containers
histc(X,edges)	Histogram count and bin locations using bins marked by edges
issorted(A)	True if A is sorted
max(A)	Maximum values
mean(A)	Mean values
median(A)	Median values
min(A)	Minimum values
mode(A)	Mode or most frequent sample
prod(A)	Product of elements
sort(A)	Sorts in ascending or descending order
sortrows(A)	Sorts rows in ascending order, that is, dictionary sort
std(A)	Standard deviation
subspace(A,B)	Finds the angle between two subspaces specified by columns of A and B
sum(A)	Sum of elements
trapz(A)	Trapezoidal integration
var(A)	Variance, that is, square of standard deviation

17.4 TIME SERIES ANALYSIS

An important subset of data analysis is concerned with *time series* data. Data sampled over time at regular intervals are analyzed to identify potential patterns or internal structure. These patterns can be used to model processes and possibly forecast future values. Time series data inherently include some random variation that may be reduced or eliminated by smoothing techniques to help reveal the underlying structure. MATLAB contains structures and tools (objects and methods) specifically designed for analysis of time series data.

MATLAB `timeseries` objects are structures optimized for storage and analysis of time series observations. They contain values for one or more variables associated with each specific sample time along with a number of properties associated with the `timeseries` object itself. Events may be created and associated with specific time values. Events are used to define intervals that may be analyzed or manipulated independently within a `timeseries` object. Time series functions are summarized in the following table.

Time Series Function	**Description**
addevent	Adds events
addsample	Adds sample(s) to a time series object
ctranspose	Transpose of time series data
delevent	Removes events
delsample	Deletes sample(s) from a time series object
detrend	Removes mean or best-fit line and all NaNs from time series data
fieldnames	Cell array of time series property names
filter	Shapes time series data
get	Query time series property values
getabstime	Extracts a date string time vector into a cell array
getdatasamplesize	Size of time series data
getinterpmethod	Interpolation method name for a time series object
getqualitydesc	Quality description of the time series data
getsampleusingtime	Extracts data in the specified time range to a new object
gettsafteratevent	Extracts samples occurring at or after a specified event
gettsafterevent	Extracts samples occurring after a specified event
gettsatevent	Extracts samples occurring at a specified event
gettsbeforeatevent	Extracts samples occurring at or before a specified event
gettsbeforeevent	Extracts samples occurring before a specified event
gettsbetweenevents	Extracts samples occurring between two specified events
idealfilter	Applies an ideal (noncausal) filter to time series data
iqr	Interquartile range of the time series data

Time Series Function	Description
isempty	True for empty time series object
ldivide (.\)	Left array divide time series
length	Length of the time vector
max	Max of the time series data
mean	Mean of the time series data
median	Median of the time series data
min	Min of the time series data
minus (-)	Subtracts time series data
mldivide (\)	Left matrix division of time series
mrdivide (/)	Right matrix division of time series
mtimes (*)	Matrix multiplication of time series
plot	Plots time series data
plus (+)	Adds time series data
rdivide (./)	Right array divide time series
resample	Resamples time series data
set	Sets time series property values
setabstime	Sets time using date strings
setinterpmethod	Sets default interpolation method in a time series
size	Size of the time series object
std	Standard deviation of the time series data
sum	Sum of the time series data
synchronize	Synchronizes two time series objects onto a common time vector
times (.*)	Multiplies time series data
timeseries	Creates a time series object
tsdata.event	Constructs an event object for a time series
tsprops	Time series object properties
var	Variance of the time series data
vertcat	Vertical concatenation of time series objects

Related time series objects with a common time vector may be combined into `tscollection` objects (time series collections) that may be analyzed or manipulated as a collection. Time series collection functions are summarized in the following table.

Time Series Collection Function	Description
addsampletocollection	Adds sample(s) to a collection
addts	Adds data vector or time series object to a collection
delsamplefromcollection	Removes sample(s) from a collection
fieldnames	Cell array of time series collection property names
get	Query time series collection property values
getabstime	Extracts a date string time vector into a cell array
getsampleusingtime	Extracts samples from a collection between specified time values
gettimeseriesnames	Cell array of names of time series in `tscollection`
horzcat	Horizontal concatenation of `tscollection` objects
isempty	True for empty `tscollection` objects
length	Length of the time vector
removets	Removes time series object(s) from a collection
resample	Resamples time series members of a collection
set	Sets time series collection property values
setabstime	Sets time of a collection using date strings
settimeseriesnames	Changes the name of a time series member of a collection
size	Size of a `tscollection` object
tscollection	Creates a time series collection object
vertcat	Vertical concatenation of `tscollection` objects

Time series objects and methods may be accessed using the ***Time Series Tools*** window invoked from the **Start** menu as **Start/MATLAB/Time Series Tools** or using the `tstool` command in the *Command* window.

18

Data Interpolation

Interpolation is a way of estimating values of a function between those given by some set of data points. In particular, interpolation serves as a valuable tool when you cannot quickly evaluate the function at the desired intermediate points—for example, when the data points are the result of some experimental measurements or lengthy computational procedure. MATLAB provides tools for interpolating in any number of dimensions by using multidimensional arrays. To illustrate interpolation, only 1- and 2-D interpolations are considered in depth here. However, the functions used for higher dimensions are briefly discussed.

18.1 ONE-DIMENSIONAL INTERPOLATION

Perhaps the simplest example of interpolation is MATLAB plots. By default, MATLAB draws straight lines connecting the data points used to make a plot. This linear interpolation guesses that intermediate values fall on a straight line between the entered points. Certainly, as the number of data points increases and the distance between them decreases, linear interpolation becomes more accurate:

```
>> x1 = linspace(0,2*pi,60);
>> x2 = linspace(0,2*pi,6);

>> plot(x1,sin(x1),x2,sin(x2),'--')
>> xlabel('x'),ylabel('sin(x)')
>> title('Figure 18.1: Linear Interpolation')
```

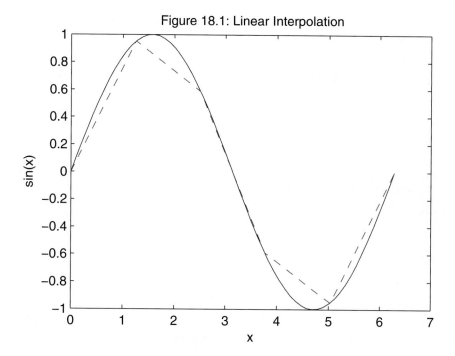

Figure 18.1: Linear Interpolation

Of the two plots of the sine function shown, the one using 60 points is much more accurate between the data points than the one using only 6 points.

To understand 1-D interpolation, consider the following illustration: The threshold of audibility (i.e., the lowest perceptible sound level) of the human ear varies with frequency. Typical data are as follows:

```
>> Hz = [20:10:100 200:100:1000 1500 2000:1000:10000]; % frequencies in Hertz
>> spl = [76 66 59 54  49  46  43 40 38 22 ... % sound pressure level in dB
       14  9  6 3.5 2.5 1.4 0.7  0 -1 -3 ...
       -8 -7 -2  2  7   9  11 12];
```

The sound pressure levels are normalized, so that 0 dB appears at 1000 Hz. Since the frequencies span such a large range, plot the data using a logarithmic x-axis:

```
>> semilogx(Hz,spl,'-o')

>> xlabel('Frequency, Hz')
>> ylabel('Relative Sound Pressure Level, dB')
>> title('Figure 18.2: Threshold of Human Hearing')
>> grid on
```

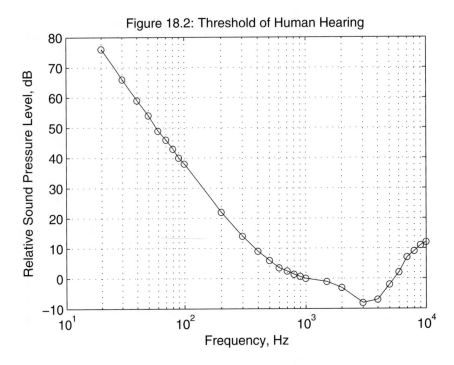

Figure 18.2: Threshold of Human Hearing

According to this plot, the human ear is most sensitive to tones around 3 kHz. Given these data, let's use the function `interp1` to estimate the sound pressure level in several different ways at a frequency of 2.5 kHz:

```
>> s = interp1(Hz,spl,2.5e3)              % linear interpolation
s =

          -5.5
>> s = interp1(Hz,spl,2.5e3,'linear')     % linear interpolation again
s =

          -5.5
>> s = interp1(Hz,spl,2.5e3,'cubic')      % cubic interpolation
s =

        -6.0488
>> s = interp1(Hz,spl,2.5e3,'spline')     % cubic spline interpolation
s =

        -5.869
>> s = interp1(Hz,spl,2.5e3,'nearest')    % nearest neighbor interpolation
s =

          -8
```

Note the differences in these results. The first two results return exactly what is shown in the figure at 2.5 kHz, since MATLAB linearly interpolates between data points on plots. Cubic and spline interpolation fit cubic, that is, third-order, polynomials to each data interval by using different constraints. Cubic interpolation maintains data monotonicity, whereas spline interpolation exhibits the greatest smoothness. The crudest interpolation in this case is the nearest-neighbor method, which returns the input data point nearest the given value.

So how do you choose an interpolation method for a given problem? In many cases, linear interpolation is sufficient. In fact, that's why it is the default method. While the nearest-neighbor method produced poor results here, it is often used when speed is important or when the data set is large. The most time-consuming method is 'spline', but it frequently produces the most desirable results.

While the preceding case considered only a single interpolation point, interp1 can handle any arbitrary number of points. In fact, one of the most common uses of cubic or spline interpolation is to smooth data. That is, given a set of data, use interpolation to evaluate the data at a finer interval:

```
>> Hzi = linspace(2e3,5e3); % look closely near minimum
>> spli = interp1(Hz,spl,Hzi,'spline'); % interpolate near minimum
>> i = find(Hz>=2e3 & Hz<=5e3); % find original data indices near minimum

>> semilogx(Hz(i),spl(i),'--o',Hzi,spli) % plot old and new data
>> xlabel('Frequency, Hz')
>> ylabel('Relative Sound Pressure Level, dB')
>> title('Figure 18.3: Threshold of Human Hearing')
>> grid on
```

In the plot, the dashed line is the linear interpolation, the solid line is the cubic interpolation, and the original data are marked with 'o'. By asking for a finer resolution on the frequency axis, and by using spline interpolation, we have a smoother estimate of the sound pressure level. In particular, note how the slope of the spline solution does not change abruptly at the data points.

With these data, we can make a better estimate of the frequency of greatest sensitivity, using, for example, the following code:

```
>> [spl_min,i] = min(spli) % minimum and index of minimum
spl_min =
      -8.4245
i =
   45
>> Hz_min = Hzi(i) % frequency at minimum
```

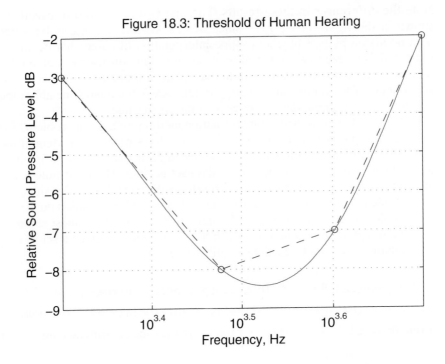

Figure 18.3: Threshold of Human Hearing

```
Hz_min =

        3333.3
```

According to this analysis, the human ear is most sensitive to tones near 3.33 kHz.

It is important to recognize the major restriction enforced by `interp1`, namely, that the independent variable must be monotonic. That is, the first variable must always increase or decrease. In our example, `Hz` is monotonic.

Finally, it is possible to interpolate more than one data set at a time, because the function `interp1` supports multidimensional input (e.g., if `y` is a column-oriented data array). That is, if `x` is a vector, either `y` can be a vector, as shown previously, or can be an array having `length(x)` rows and any number of columns. For example, in the code

```
>> x = linspace(0,2*pi,11)'; % example data
>> y = [sin(x) cos(x) tan(x)];
>> size(y) % three columns
ans =
     11    3
>> xi = linspace(0,2*pi); % interpolate on a finer scale
>> yi = interp1(x,y,xi,'cubic');
```

```
>> size(yi) % result is all three columns interpolated
ans =
   100    3
```

$\sin(x)$, $\cos(x)$, and $\tan(x)$ are all interpolated at the points in xi.

18.2 TWO-DIMENSIONAL INTERPOLATION

Two-dimensional interpolation is based on the same underlying ideas as 1-D interpolation. However, as the name implies, 2-D interpolation interpolates functions of two variables: $z = f(x,y)$. To illustrate this added dimension, consider the following example. An exploration company is using sonar to map the ocean floor. At points every 0.5 km on a rectangular grid, the ocean depth in meters is recorded for later analysis. A portion of the data collected is entered into MATLAB in the script M-file ocean.m, as shown in the following:

```
% ocean.m, example test data

% ocean depth data

x = 0:.5:4; % x-axis (varies across the rows of z)
y = 0:.5:6; % y-axis (varies down the columns of z)
z = [100    99   100    99   100    99    99    99   100
     100    99    99    99   100    99   100    99    99
      99    99    98    98   100    99   100   100   100
     100    98    97    97    99   100   100   100    99
     101   100    98    98   100   102   103   100   100
     102   103   101   100   102   106   104   101   100
      99   102   100   100   103   108   106   101    99
      97    99   100   100   102   105   103   101   100
     100   102   103   101   102   103   102   100    99
     100   102   103   102   101   101   100    99    99
     100   100   101   101   100   100   100    99    99
     100   100   100   100   100    99    99    99    99
     100   100   100    99    99   100    99   100    99];
```

A plot of this data can be displayed by entering

```
>> mesh(x,y,z)
>> xlabel('X-axis, km')
>> ylabel('Y-axis, km')
>> zlabel('Ocean Depth, m')
>> title('Figure 18.4: Ocean Depth Measurements')
```

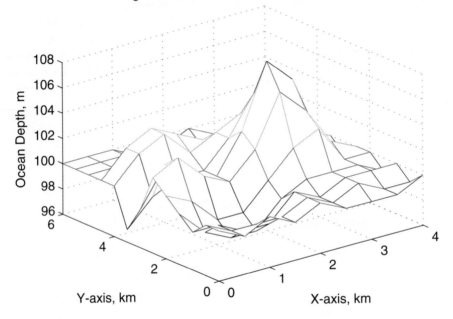

Figure 18.4: Ocean Depth Measurements

With these data, the depth at arbitrary points within the rectangle can be found by using the function interp2, as in the following code:

```
>> zi = interp2(x,y,z,2.2,3.3)
zi =
       103.92
>> zi = interp2(x,y,z,2.2,3.3,'linear')
zi =
       103.92
>> zi = interp2(x,y,z,2.2,3.3,'cubic')
zi =
       104.19
```

```
>> zi = interp2(x,y,z,2.2,3.3,'spline')
zi =
        104.3
>> zi = interp2(x,y,z,2.2,3.3,'nearest')
zi =
    102
```

As was the case with 1-D interpolation, several interpolation methods are available, with the default method being linear.

Once again, we can interpolate on a finer scale, or mesh, to smooth the plot:

```
>> xi = linspace(0,4,30);    % finer x-axis
>> yi = linspace(0,6,40);    % finer y-axis
```

For each value in xi, we wish to interpolate at all values in yi. That is, we wish to create a grid of all combinations of the values of xi and yi, and then interpolate at all of these points. The function meshgrid accepts two vectors and produces two arrays, each containing duplicates of its inputs, so that all combinations of the inputs are considered:

```
>> xtest = 1:5
xtest =
    1   2   3   4   5
>> ytest = 6:9
ytest =
    6   7   8   9
>> [xx,yy] = meshgrid(xtest,ytest)
xx =
    1   2   3   4   5
    1   2   3   4   5
    1   2   3   4   5
    1   2   3   4   5
yy =
    6   6   6   6   6
    7   7   7   7   7
    8   8   8   8   8
    9   9   9   9   9
```

As shown in this code, xx has length(ytest) rows, each containing xtest, and yy has length(xtest) columns, each containing ytest. With this structure, xx(i,j) and yy(i,j) for all i and j cover all combinations of the original vectors, xtest and ytest.

Applying meshgrid to our ocean depth example produces

```
>> [xxi,yyi] = meshgrid(xi,yi); % grid of all combinations of xi and yi
>> size(xxi) % xxi has 40 rows each containing xi
ans =
   40    30
>> size(yyi) % yyi has 30 columns each containing yi
ans =
   40    30
```

Given xxi and yyi, the ocean depth can now be interpolated on the finer scale by entering

```
>> zzi = interp2(x,y,z,xxi,yyi,'cubic');  % interpolate
>> size(zzi) % zzi is the same size as xxi and yyi
ans =
   40    30
>> mesh(xxi,yyi,zzi)  % plot smoothed data
>> hold on
>> [xx,yy] = meshgrid(x,y); % grid original data
>> plot3(xx,yy,z+0.1,'ok') % plot original data up a bit to show nodes
>> hold off

>> xlabel('X-axis, km')
>> ylabel('Y-axis, km')
>> zlabel('Ocean Depth, m')
>> title('Figure 18.5: 2-D Smoothing')
```

Using these data, we can now estimate the peak and its location:

```
>> zmax = max(max(zzi))
zmax =
      108.05
>> [i,j] = find(zmax==zzi);
```

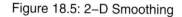

Figure 18.5: 2–D Smoothing

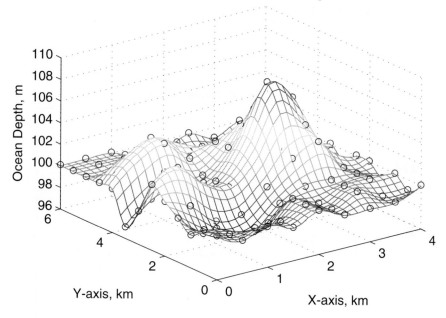

```
>> xmax = xi(i)
xmax =
      2.6207
>> ymax = yi(j)
ymax =
      2.9231
```

The concepts discussed in the first two sections extend naturally to higher dimensions, where `ndgrid`, `interp3`, and `interpn` apply. The function `ndgrid` is the multidimensional equivalent of `meshgrid`. Multidimensional interpolation uses multidimensional arrays in a straightforward way to organize the data and perform the interpolation. The function `interp3` performs interpolation in 3-D space, and `interpn` performs interpolation in higher-order dimensions. Both `interp3` and `interpn` offer method choices of `'linear'`, `'cubic'`, `'spline'`, and `'nearest'`. (For more information regarding these functions, see the MATLAB documentation and the online help.)

18.3 TRIANGULATION AND SCATTERED DATA

In a number of applications, such as those involving geometric analysis, data points are often scattered, rather than appearing on a rectangular grid, like the ocean data in the example discussed in the last section. For example, consider the following 2-D random data:

```
>> x = randn(12,1); % use column vector format
>> y = randn(12,1);
>> plot(x,y,'o')
>> title('Figure 18.6: Random Data')
```

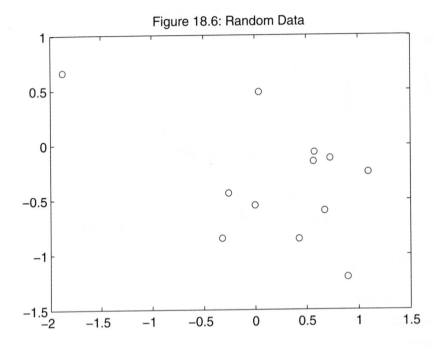

Figure 18.6: Random Data

Given these scattered data, it is common to apply ***Delaunay triangulation***, which returns a set of triangles connecting the data points such that no data points are contained within any triangle. The MATLAB class constructor `DelaunayTri` accepts data points and returns a structure which includes a list of indices that identify the triangle vertices. For the random data, `DelaunayTri` returns

```
>> tri = DelaunayTri(x,y)
tri =
  DelaunayTri

  Properties:
      Constraints: []
                X: [12x2 double]
    Triangulation: [17x3 double]
  Methods, Superclasses
```

where `tri.X` contains the data points [x y]

```
>> tri.X
ans =
        0          -0.5465
   -0.3179         -0.8468
    1.0950         -0.2463
   -1.8740          0.6630
    0.4282         -0.8542
    0.8956         -1.2013
    0.7310         -0.1199
    0.5779         -0.0653
    0.0403          0.4853
    0.6771         -0.5955
    0.5689         -0.1497
   -0.2556         -0.4348
```

and `tri.Triangulation` contains the array of indices identifying the triangles:

```
>> tri.Triangulation
ans =
    1    8   12
    1    2    5
    8    1   11
   10    1    5
    2   12    4
    2    1   12
    4   12    9
   10   11    1
    8    9   12
    7    8   11
    7    9    8
   10    7   11
    3    9    7
    6   10    5
```

$$
\begin{array}{ccc}
6 & 3 & 10 \\
2 & 6 & 5 \\
10 & 3 & 7
\end{array}
$$

Each row contains indices into `tri.X` that identify triangle vertices. For example, the first triangle is described by the data points in `tri.X([1 8 12],:)`. The triangles can be plotted by using the function `triplot`:

```
>> hold on, triplot(tri), hold off
>> title('Figure 18.7: Delaunay Triangulation')
```

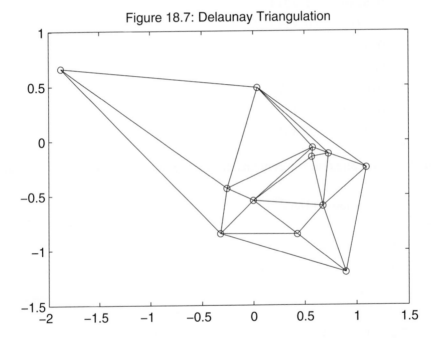

Figure 18.7: Delaunay Triangulation

Once the Delaunay triangulation is known, the methods `pointLocation` and `nearestNeighbor` can be used to interpolate it. For example, the triangle enclosing the origin is

```
>> tri.pointLocation(0,0) % find row of tri.Triangulation closest to (0,0)
ans =
         9
>> tri.Triangulation(ans,:) % vertices of triangle closest to (0,0)
ans =
      8    9    12
```

Naturally, the pointLocation method accepts multiple values. For example,

```
>> testpoints = [0,0; -0.5,0.1; 1,0.5];
>> tri.pointLocation(testpoints)
ans =
     9
     7
     NaN
```

Here, triangle 9 encloses the point $(0, 0)$, triangle 7 encloses the point $(-0.5, 0.1)$, and no triangle encloses the point $(1, 0.5)$.

Rather than returning the triangle enclosing one or more data points, the method nearestNeighbor returns the indices into tri.X that are closest to the desired points:

```
>> tri.nearestNeighbor(testpoints)
ans =
     9
     12
     7
```

Here, the point (tri.X(9,:)) is closest to the point $(0, 0)$, the point (tri.X(12,:)) is closest to the point $(-0.5, 0.1)$, and the point (tri.X(7,:)) is closest to $(1, 0.5)$.

In addition to interpolating the data, it is often useful to know which points form the outer boundary or **convex hull** for the set. The function convhull returns indices into x and y that describe the convex hull, as in the following example:

```
>> [k,a] = convhull(x,y); % use the original data points
>> k'
k =
     2   6   3   9   4   2
>> a % optional 2nd argument gives area enclosed
a =
        2.7551
>> [k,a] = convexHull(tri); % use the DelaunayTri/convexHull method
>> k'
```

```
k =

    2    6    3    9    4    2
>> a

a =

      2.7551
```

Note that convhull and convexHull return the indices of a closed curve, since the first and last index values are the same. In addition, the optional second output argument returns the area enclosed by the convex hull. Based on the data returned by convhull, the boundary can be drawn with the following code:

```
>> plot(x,y,'o',x(k),y(k))
>> title('Figure 18.8: Convex Hull')
```

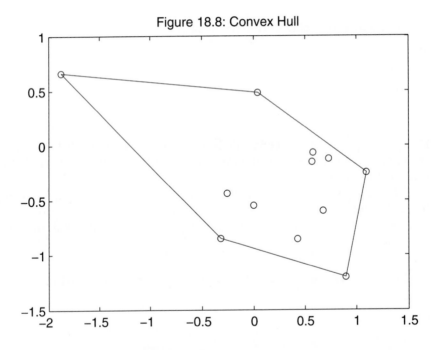

Figure 18.8: Convex Hull

It is also possible to define and draw the lines that separate regions in the plane that are closest to a particular data point. These lines form what is called a ***Voronoi polygon***. In MATLAB, these lines are drawn by the function voronoi:

```
>> voronoi(tri)
>> title('Figure 18.9: Voronoi Diagram')
```

Figure 18.9: Voronoi Diagram

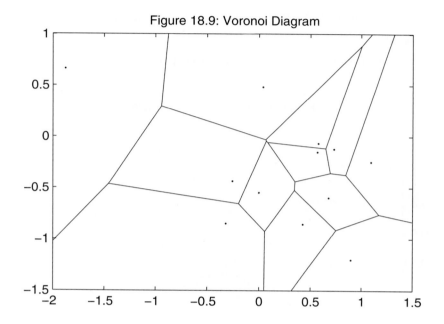

Finally, it is possible to interpolate a Delaunay triangulation to produce interpolated points on a rectangular grid by using the function griddata. In particular, this step is required to use functions such as surf and other standard plotting routines. These plotting routines require data that contain information at a complete sequence of points along two coordinate axes, rather than the scattered data that triplot uses. Think about a map analogy. Delaunay triangulation allows you to identify specific scattered points on a map. The function griddata uses this information to construct an approximation to the rest of the map, filling in data in a user-specified rectangular region in two coordinate directions:

```
>> z = rand(12,1); % now use some random z axis data
>> xi = linspace(min(x),max(x),30); % x interpolation points
>> yi = linspace(min(y),max(y),30); % y interpolation points

>> [Xi,Yi] = meshgrid(xi,yi);        % create grid of x and y

>> Zi = griddata(x,y,z,Xi,Yi);       % grid the data at Xi,Yi points

>> mesh(Xi,Yi,Zi)
>> hold on
```

```
>> plot3(x,y,z,'ko') % show original data as well
>> hold off
>> title('Figure 18.10: Griddata Example')
```

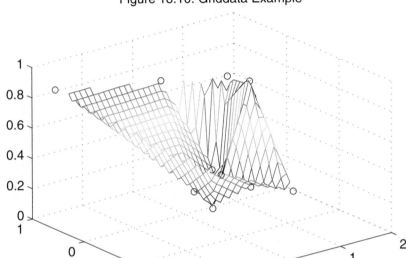

Figure 18.10: Griddata Example

In this example, the information at the 12 scattered data points was interpolated to produce an array of data on a 30-by-30 grid in the x–y plane. The variable Zi contains a 30-by-30 array of points linearly interpolated from the triangulation of the data in x, y, and z. Just as interp1 and interp2 support other interpolations, griddata also supports others:

```
>> Zi = griddata(x,y,z,Xi,Yi,'linear') % same as above (default)
>> Zi = griddata(x,y,z,Xi,Yi,'cubic')   % triangle based cubic interpolation
>> Zi = griddata(x,y,z,Xi,Yi,'nearest') % triangle based nearest neighbor
```

The functions discussed in this section also support higher dimensions. The DelaunayTri(X) form of the constructor supports a single multidimensional array as input. The functions trimesh and trisurf are the multidimensional analogs of triplot, and voronoin is the multidimensional version of voronoi. (For more information regarding these functions and the TriRep class for triangular representation in 2-D and 3-D space, see the MATLAB documentation and the online help.)

18.4 SUMMARY

In MATLAB, the triangulation functions discussed in this chapter have n-dimensional forms as well. Many of the data interpolation functions have other features and options that cannot be covered here. The data interpolation functions and classes in MATLAB are summarized in the following table:

Function	Description
convhull	Convex hull in 2-D or 3-D space
convhulln	n-D convex hull
delaunay	2-D and 3-D Delaunay triangulation
delaunay3	3-D Delaunay tetrahedra creation (depreciated in favor of DelaunayTri)
delaunayn	n-D Delaunay tesselation
dsearch	Nearest point search in Delaunay triangulation (depreciated in favor of DelaunayTri/nearestNeighbor)
griddata	2-D rectangular gridding
griddata3	3-D rectangular gridding (depreciated in favor of TriScatteredInterp)
griddatan	n-D rectangular gridding
interp1	1-D interpolation
interp1q	1-D quick interpolation (no error checking)
interp2	2-D interpolation
interp3	3-D interpolation
interpft	1-D interpolation using FFT method
interpn	n-D interpolation
meshgrid	Generates X and Y matrices for 3-D functions
ndgrid	Generates arrays for multidimensional functions
tetramesh	Tetrahedron mesh plot
trimesh	Triangular mesh plot
triplot	2-D triangular plot
trisurf	Triangular surface plot
tsearch	Closest triangle search (depreciated in favor of DelaunayTri/pointLocation)

Function	Description
tsearchn	n-D closest simplex search
voronoi	Voronoi diagram
voronoin	n-D Voronoi diagram
DelaunayTri	Delaunay triangulation representation class constructor (and methods)
TriRep	Triangulation representation class constructor (and methods)
TriScatteredInterp	Scattered data interpolant

Polynomials

MATLAB provides a number of functions for manipulating polynomials. Polynomials are easily differentiated and integrated, and it is straightforward to find polynomial roots. However, higher-order polynomials pose numerical difficulties in a number of situations and therefore should be used with caution.

19.1 ROOTS

Finding the roots of a polynomial—that is, the values for which the polynomial is zero—is a problem common to many disciplines. MATLAB solves this problem and provides other polynomial manipulation tools as well. In MATLAB, a polynomial is represented by a row vector of its coefficients in descending order. For example, the polynomial $2x^4 - 6x^3 + 2x^2 + 0x + 86$ is entered as

```
>> p = [2 -6 2 0 86]
p =
     2    -6    2    0    86
```

Note that terms with zero coefficients must be included. MATLAB has no way of knowing which terms are zero unless you specifically identify them. Given this form, the roots of a polynomial are found by using the function roots:

```
>> r = roots(p)
r =
    2.7147 + 1.5601i
    2.7147 - 1.5601i
```

```
-1.2147 + 1.7062i
-1.2147 - 1.7062i
```

Since in MATLAB both a polynomial and its roots are vectors, MATLAB adopts the convention that *polynomials are row vectors and roots are column vectors*.

Given the roots of a polynomial, it is also possible to construct the associated polynomial. In MATLAB, the command poly performs this task:

```
>> pp = poly(r)
pp =
        1.0000    -3.0000     1.0000    -0.0000    43.0000
>> pp(abs(pp)<.1) = 0 % change small element appearing to be -0 to zero!
pp =
        1.0000    -3.0000     1.0000         0    43.0000
```

Because of truncation errors, it is not uncommon for the results of poly to have near-zero components or to have components with small imaginary parts. As already shown, near-zero components can be corrected by array manipulation. Similarly, eliminating spurious imaginary parts is simply a matter of using the function real to extract the real part of the result.

19.2 MULTIPLICATION

Polynomial multiplication is supported by the function conv (which performs the convolution of two arrays). Consider the product of the two polynomials $a(x) = 2x^3 + x^2 - 3x + 1$ and $b(x) = x^3 - 4x^2 + 3x + 23$:

```
>> a = [2 1 -3 1]; b = [1 -4 3 23];
>> c = conv(a,b)
c =
      1      -7      -1      62      10      -66      23
```

This result is $c(x) = 2x^6 - 7x^5 - x^4 + 62x^3 + 10x^2 - 66x + 23$. Multiplication of more than two polynomials requires repeated use of conv.

19.3 ADDITION

MATLAB does not provide a direct function for adding polynomials. Standard array addition works if both polynomial vectors are of the same order, as in the code

```
>> d = a + b
```

d =

    3   -3   0   24

which represents $d(x) = 3x^3 - 3x^2 + 0x + 24$. When two polynomials are of different orders, the polynomial of lower order must be padded with leading zeros so that it has the same effective order as the higher-order polynomial. Consider the addition of the preceding polynomials c and d:

```
>> e = c + [0 0 0 d]
```

e =

    2   -7   -1   65   7   -66   47

The resulting polynomial is $e(x) = 2x^6 - 7x^5 - 1x^4 + 65x^3 + 7x^2 + 66x + 47$. Leading zeros, rather than trailing zeros, are required, because coefficients associated with like powers of x must line up. The following M-file functions automate polynomial simplification:

```
function y = mmpsim(x,tol)
%MMPSIM Polynomial Simplification
% Strip leading zero terms and small coefficients.

if nargin<1 | ~isnumeric(x)
error('First Input Must be Numeric')
end
x = x(:).';                     % make sure input is a row
if nargin<2
   tol = max(abs(x))*100*eps;   % default tolerance
else
   tol = max(tol(1),eps);       % check user tolerance
end
i = find(abs(x)<.99 & abs(x)<tol); % find insignificant indices
x(i) = 0;                       % set them to zero
i = find(x~=0);                 % find significant indices
if isempty(i)
   y = 0;                       % the extreme case: nothing left!
else
y = x(i(1):end);                % start with first significant term
end
```

and addition:

```
function p = mmpadd(a,b)
%MMPADD Polynomial Addition.
% MMPADD(A,B) adds the polynomials A and B.

if nargin<2
   error('Not Enough Input Arguments.')
end

a = reshape(a,1,[]); % make sure inputs are polynomial row vectors
b = b(:).';          % this makes a row as well

na = length(a); % find lengths of a and b
nb = length(b);

p = [zeros(1,nb-na) a]+[zeros(1,na-nb) b]; % pad with zeros as necessary
```

To illustrate the use of mmpadd, consider again the preceding example

```
>> f = mmpadd(c,d)
f =
      2    -7    -1    65     7   -66    47
```

which is the same as our earlier e. Of course, mmpadd can also be used for subtraction, as in

```
>> g = mmpadd(c,-d)
g =
      2    -7    -1    59    13   -66    -1
```

The resulting polynomial is $g(x) = 2x^6 - 7x^5 - x^4 + 59x^3 + 13x^2 + 66x - 1$.

19.4 DIVISION

In some special cases, it is necessary to divide one polynomial into another. In MATLAB, this is accomplished with the function deconv:

```
>> [q,r] = deconv(c,b)
q =
    2    1   -3    1
r =
    0    0    0    0    0    0    0
```

This result says that b divided into c gives the quotient polynomial q and the remainder r, which is zero in this case, since the product of b and q is exactly c. Another example gives a remainder:

```
>> [q,r] = deconv(e,b)
q =
    2    1   -3    4
r =
    0    0    0    0    9   -9   -45
```

Here, b divided into e gives the quotient polynomial q and the remainder r. The leading zeros in r simply make r the same length as f. In this case, the quotient is $q(x) = 2x^3 + 1x^2 - 3x + 4$ and the remainder term is $r(x) = 9x^2 - 9x - 45$.

19.5 DERIVATIVES AND INTEGRALS

Because differentiation of a polynomial is simple to express, MATLAB offers the function polyder for polynomial differentiation:

```
>> a % recall polynomial
a =
    2    1   -3    1

>> h = polyder(a)
h =
    6    2   -3
```

Similarly, the integral of a polynomial is easy to express. Given an integration constant, the function polyint returns the integral:

```
>> polyint(h,1) % get g back from h = polyder(a)
ans =
    2    1   -3    1
```

19.6 EVALUATION

Given that you can add, subtract, multiply, divide, and differentiate polynomials on the basis of row vectors of their coefficients, you should be able to evaluate them also. In MATLAB, this is accomplished with the function polyval, as in the following example:

```
>> p = [1 4 -7 -10];     % the polynomial
>> x = linspace(-1,3);   % evaluation points
>> v = polyval(p,x);     % evaluate p at points in x
>> plot(x,v)             % plot results
>> title('Figure 19.1:   x{^3} + 4x{^2} - 7x -10')
>> xlabel('x')
```

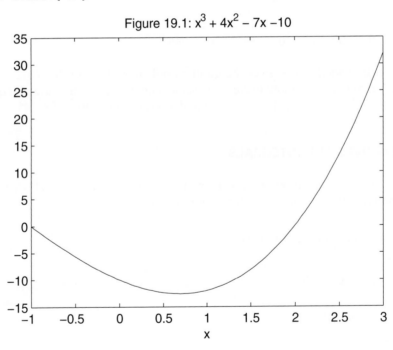

Figure 19.1: $x^3 + 4x^2 - 7x - 10$

19.7 RATIONAL POLYNOMIALS

Sometimes one encounters ratios of polynomials—for example, transfer functions and Pade approximations to functions. In MATLAB, these are manipulated by considering the numerator and denominator polynomials separately, as in the following example:

```
>> n = [30 2 110] % a numerator
n =
```

```
      30    2    110
>> d = [2 1 20 3] % a denominator
d =
    2    1    20    3

>> z = roots(n) % the zeros of n(x)/d(x)
z =
           -0.0333 + 1.9146i
           -0.0333 - 1.9146i

>> p = roots(d) % the poles of n(x)/d(x)
p =
          -0.1746 + 3.1491i
          -0.1746 - 3.1491i
          -0.1508
```

The derivative of this rational polynomial with respect to x is found by using polyder:

```
>> [nd,dd] = polyder(n,d)
nd =
     -60      -8      -62      -40      -2194
dd =
       4      4      81      52      406    120    9
```

Here, nd and dd are the respective numerator and denominator polynomials of the derivative.

Another common operation is to find the partial-fraction expansion of a rational polynomial. Consider the following data:

```
>> [r,p,k] = residue(n,d)
r =
   4.7175 +     0.6372i
   4.7175 -     0.6372i
   5.5650
```

p =

$$-0.1746 +   3.1491i$$
$$-0.1746 -   3.1491i$$
$$0.1508$$

k =

[]

In this case, the `residue` function returns the residues or partial-fraction expansion coefficients r, their associated poles p, and the direct term polynomial k. Since the order of the numerator is less than that of the denominator, there are no direct terms. For this example, the partial-fraction expansion of the rational polynomial is

$$\frac{n(x)}{d(x)} = \frac{4.7175 + 0.6372i}{x + 0.1746 - 3.1491i} + \frac{4.7175 - 0.6372i}{x + 0.1746 + 3.1491i} + \frac{5.5650}{x + 0.1508}$$

Given this information, the original rational polynomial is found by using `residue` yet again:

```
>> [nn,dd] = residue(r,p,k) % Both polynomials are divided by 2
nn =
```
             15.000   1.0000   55.0000
```
dd =
```
          1.0000   0.5000   10.0000   1.5000

So, in this case, the function `residue` performs two operations that are inverses of one another depending on how many input and output arguments are used.

19.8 CURVE FITTING

In numerous application areas, you are faced with the task of fitting a curve to measured data. Sometimes the chosen curve passes through the data points, but at other times the curve comes close to, but does not necessarily pass through, the data points. In the most common situation, the curve is chosen so that the sum of the squared errors at the data points is minimized. This choice results in a *least-squares* curve fit. While least-squares curve fitting can be done by using any set of basis functions, it is straightforward and common to use a truncated power series—that is, a polynomial.

In MATLAB, the function `polyfit` solves the least-squares polynomial curve-fitting problem. To illustrate the use of this function, let's start with the following data:

```
>> x = [0 .1 .2 .3 .4 .5 .6 .7 .8 .9 1];
>> y = [-.447 1.978 3.28 6.16 7.08 7.34 7.66 9.56 9.48 9.30 11.2];
```

To use `polyfit`, we must supply the data and the order, or degree, of the polynomial we wish to best fit to the data. If we choose $n = 1$ as the order, the best straight-line approximation will be found. This is often called ***linear regression***. On the other hand, if we choose $n = 2$ as the order, a quadratic polynomial will be found. For now, let's choose a quadratic polynomial:

```
>> n = 2;
>> p = polyfit(x,y,n)
p =
     -9.8108     20.129     -0.031671
```

The output of `polyfit` is a row vector of the polynomial coefficients. Here the solution is $y(x) = -9.8108x^2 + 20.129x - 0.031671$. To compare the curve-fit solution to the data points, let's plot both:

```
>> xi = linspace(0,1,100);
>> yi = polyval(p,xi);

>> plot(x,y,'-o',xi,yi,'--')
>> xlabel('x'), ylabel('y = f(x)')
>> title('Figure 19.2: Second Order Curve Fitting')
```

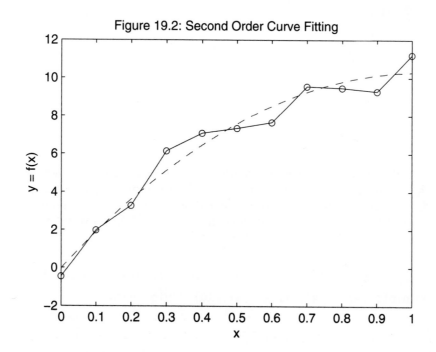

This plot contains the original data x and y, marking the data points with 'o' and connecting them with straight lines. In addition, the evaluated polynomial data xi and yi are plotted with a dashed line ('--').

The choice of polynomial order is somewhat arbitrary. It takes two points to define a straight line or first-order polynomial. (If this isn't clear to you, mark two points and draw a straight line between them.) It takes three points to define a quadratic, or second-order, polynomial. Following this progression, it takes $n + 1$ data points to uniquely specify an nth-order polynomial. Thus, in the preceding case, where there are 11 data points, we could choose up to a 10th-order polynomial. However, given the poor numerical properties of higher-order polynomials, you should not choose a polynomial order that is any higher than necessary. In addition, as the polynomial order increases, the approximation becomes less smooth, since higher-order polynomials can be differentiated more times before they become zero. For example, consider choosing a 10th-order polynomial:

```
>> pp = polyfit(x,y,10);
>> pp.' % display polynomial coefficients as a column
ans =
  -4.6436e+05
   2.2965e+06
  -4.8773e+06
   5.8233e+06
  -4.2948e+06
   2.0211e+06
  -6.0322e+05
   1.0896e+05
       -10626
       435.99
       -0.447
```

Note the size of the polynomial coefficients in this case, compared with those of the earlier quadratic fit. Note also the seven orders of magnitude difference between the smallest (-0.447) and largest (-4.6436e+005) coefficients and the alternating signs on the coefficients. To see how this polynomial differs from the quadratic fit shown earlier, consider a plot of both:

```
>> y10 = polyval(pp,xi);  % evaluate 10th order polynomial
>> plot(x,y,'o',xi,yi,'--',xi,y10) % plot data
```

```
>> xlabel('x'), ylabel('y = f(x)')
>> title('Figure 19.3: 2nd and 10th Order Curve Fitting')
```

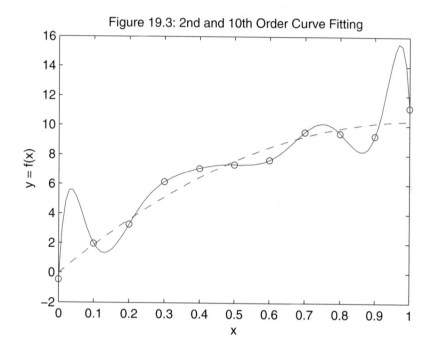

Figure 19.3: 2nd and 10th Order Curve Fitting

In this plot, the original data are marked with 'o', the quadratic curve fit with a dashed line, and the 10th-order fit with a solid line. Note the wavelike ripples that appear between the data points at the left and right extremes in the 10th-order fit. This example clearly demonstrates the difficulties with higher-order polynomials.

19.9 POLYNOMIAL FUNCTIONS

The polynomial functions in MATLAB are summarized in the following table

Polynomial Function	Description
conv	Convolution and polynomial multiplication
dconv	Deconvolution and polynomial division
poly	Polynomial with specified roots
polyder	Polynomial derivative

Polynomial Function	Description
polyeig	Polynomial eigenvalue problem solver
polyfit	Polynomial curve fitting
polyint	Polynomial integration
polyval	Polynomial evaluation
polyvalm	Polynomial matrix evaluation
residue	Partial-fraction expansion
roots	Finds polynomial roots

20

Cubic Splines

It is well known that interpolation using high-order polynomials often produces ill-behaved results. There are numerous approaches to eliminating this poor behavior. Of these approaches, cubic splines are very popular. In MATLAB, basic cubic splines interpolation is accomplished by the functions `spline`, `ppval`, `mkpp`, and `unmkpp`. Of these, only `spline` appears in earlier MATLAB documentation. However, help text is available for all of these functions. In the following sections, the basic features of cubic splines, as implemented in these M-file functions, are demonstrated. Also considered is an alternative to cubic splines called a piecewise cubic Hermite interpolating polynomial. This piecewise polynomial is computed by the function `pchip` and returns a piecewise polynomial, just as `spline` does.

20.1 BASIC FEATURES

In cubic splines, cubic polynomials are found to approximate the curve between each pair of data points. In the language of splines, these data points are called breakpoints. Since a straight line is uniquely defined by two points, an infinite number of cubic polynomials can be used to approximate a curve between two points. Therefore, in cubic splines, additional constraints are placed on the cubic polynomials to make the result unique. By constraining the first and second derivatives of each cubic polynomial to match at the breakpoints, all internal cubic polynomials are well defined. Moreover, both the slope and curvature of the approximating polynomials are continuous across the breakpoints. However, the first and last cubic polynomials do not have adjoining cubic polynomials beyond the first and last breakpoints. As a result, the remaining constraints must be determined by some other means. The most common approach, which is used by the function `spline`, is to adopt a *not-a-knot* condition if x and y are the same length. This condition forces the third derivative of the first and second cubic polynomials to be identical, and likewise for the last and second-to-last cubic polynomials. However, if y contains two more values than x, the first and last values in y are used as the slopes of the ends of the cubic spline.

Based on this description, you could guess that finding cubic spline polynomials requires solving a large set of linear equations. In fact, given n breakpoints, there are $n - 1$ cubic polynomials to be found, each having 4 unknown coefficients. Thus, the set of equations to be solved involves $4(n - 1)$ unknowns. By writing each cubic polynomial in a special form and by applying the constraints, the cubic polynomials can be found by solving a reduced set of n equations in n unknowns. Thus, if there are 50 breakpoints, there are 50 equations in 50 unknowns. Luckily, these equations can be concisely written and solved using sparse matrices, which describe what the function spline uses to compute the unknown coefficients.

20.2 PIECEWISE POLYNOMIALS

In its most simple use, spline takes data x and y and desired values xi, finds the cubic spline interpolation polynomials that fit x and y, and then evaluates the polynomials to find the corresponding yi values for each xi value. This approach matches the use of yi = interp1(x,y,xi,'spline'):

```
>> x = 0:12;
>> y = tan(pi*x/25);
>> xi = linspace(0,12);

>> yi = spline(x,y,xi);

>> plot(x,y,'o',xi,yi)
>> title('Figure 20.1: Spline Fit')
```

This approach is appropriate if only one set of interpolated values is required. However, if more values are needed from the same set of data, it doesn't make sense to recompute the same set of cubic spline coefficients a second time. In this situation, one can call spline with only the first two arguments:

```
>> pp = spline(x,y)
pp =
      form: 'pp'
    breaks: [0 1 2 3 4 5 6 7 8 9 10 11 12]
     coefs: [12x4 double]
    pieces: 12
     order: 4
       dim: 1
```

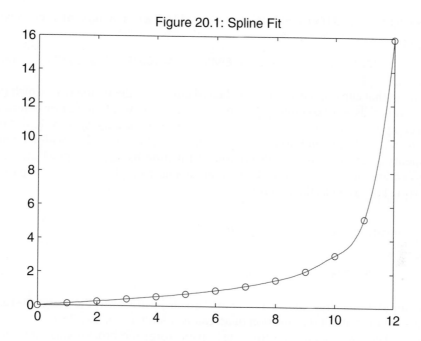

Figure 20.1: Spline Fit

When called in this way, `spline` returns a structure containing the ***pp-form***, or piecewise polynomial form, of the cubic splines. This structure contains all of the information necessary to evaluate the cubic splines for any set of desired interpolation values. The pp-form structure is also compatible with the optional *Spline Toolbox* available with MATLAB. Given the pp-form, the function `ppval` evaluates the cubic splines. For example,

```
>> yi = ppval(pp,xi);
```

computes the same `yi` values computed earlier. Similarly,

```
>> xi2 = linspace(10,12);
>> yi2 = ppval(pp,xi2);
```

uses the pp-form again to evaluate the cubic splines over a finer spacing, restricted to the region between 10 and 12. The code

```
>> xi3 = 10:15;

>> yi3 = ppval(pp,xi3)
yi3 =
      3.0777     5.2422     15.8945     44.0038     98.5389     188.4689
```

```
>> yi4 = ppval(xi3,pp) % can be called with arguments reversed
yi4 =
    3.0777    5.2422    15.8945    44.0038    98.5389    188.4689
```

shows that cubic splines can be evaluated outside of the region over which the cubic polynomials were computed. When data appears beyond the last or before the first breakpoint, the last and first cubic polynomials are used, respectively, to find interpolated values. In addition, this shows that ppval can be called with its input arguments reversed. This permits creating a function handle for ppval and passing it as an argument to functions that evaluate a user-supplied function as part of their work. For example, the function

```
>> quad(@ppval,0,10,[],[],pp)

ans =
        9.3775
```

computes the area under the spline given by pp over the range 0–10. (More information about the function quad can be found in Chapter 23.)

The cubic splines pp-form just given stores the breakpoints and polynomial coefficients, as well as other information regarding the cubic splines representation. This form is a convenient data structure in MATLAB, since all information is stored in a single structure. When a cubic spline representation is evaluated, the various fields in the pp-form must be extracted. In MATLAB, this process is conveniently performed by the function unmkpp. Using this function on the preceding pp-form gives

```
>> [breaks,coefs,npolys,ncoefs,dim] = unmkpp(pp)
breaks =
0   1   2   3   4   5   6   7   8   9   10   11   12
coefs =
        0.0007   -0.0001    0.1257         0
        0.0007    0.0020    0.1276    0.1263
        0.0010    0.0042    0.1339    0.2568
        0.0012    0.0072    0.1454    0.3959
        0.0024    0.0109    0.1635    0.5498
        0.0019    0.0181    0.1925    0.7265
        0.0116    0.0237    0.2344    0.9391
       -0.0083    0.0586    0.3167    1.2088
        0.1068    0.0336    0.4089    1.5757
       -0.1982    0.3542    0.7967    2.1251
```

```
       1.4948    -0.2406     0.9102     3.0777
       1.4948     4.2439     4.9136     5.2422
npolys =
    12
ncoefs =
    4
dim =
    1
```

Here, breaks contains the breakpoints, coefs is a matrix whose *i*th row is the *i*th cubic polynomial, npolys is the number of polynomials, ncoefs is the number of coefficients per polynomial, and dim is the spline dimension. Note that this pp-form is sufficiently general that the spline polynomials need not be cubic. This fact is useful when the spline is integrated or differentiated.

In MATLAB prior to version 6, the pp-form was stored in a single numerical array rather than in a structure. As a result, unmkpp was valuable in separating the parts of the pp-form from the numerical array. Given the simplicity of the structure form, you can easily address the fields directly and avoid using unmkpp entirely. However, unmkpp continues to support the prior numerical array pp-form, thereby making the process of extracting the parts of a pp-form transparent to the user.

Given the broken-apart form, the function mkpp restores the pp-form:

```
>> pp = mkpp(breaks,coefs)
pp =

      form: 'pp'
    breaks: [0 1 2 3 4 5 6 7 8 9 10 11 12]
     coefs: [12x4 double]
    pieces: 12
     order: 4
       dim: 1
```

Since the size of the matrix coefs determines npolys and ncoefs, they are not needed by mkpp to reconstruct the pp-form.

20.3 CUBIC HERMITE POLYNOMIALS

When the underlying data to be interpolated represents a smooth function, cubic splines return appropriate values. However, when the underlying data are not so smooth, cubic splines can predict minima and maxima that do not exist and can destroy monotonicity. Therefore, for nonsmooth data, a different piecewise polynomial interpolation is called for. In MATLAB, the function pchip returns a piecewise cubic polynomial that has the properties described in the following help text:

```
>> help pchip

  PCHIP Piecewise Cubic Hermite Interpolating Polynomial.
  PP = PCHIP(X,Y) provides the piecewise polynomial form of a certain
  shape-preserving piecewise cubic Hermite interpolant, to the values
  Y at the sites X, for later use with PPVAL and the spline utility UNMKPP.
  X must be a vector.
  If Y is a vector, then Y(j) is taken as the value to be matched at X(j),
  hence Y must be of the same length as X.
  If Y is a matrix or ND array, then Y(:,...,:,j) is taken as the value to
  be matched at X(j), hence the last dimension of Y must equal length(X).

  YY = PCHIP(X,Y,XX) is the same as YY = PPVAL(PCHIP(X,Y),XX), thus
  providing, in YY, the values of the interpolant at XX.

  The PCHIP interpolating function, p(x), satisfies:
  On each subinterval, X(k) <= x <= X(k+1), p(x) is the cubic Hermite
     interpolant to the given values and certain slopes at the two endpoints.
  Therefore, p(x) interpolates Y, i.e., p(X(j)) = Y(:,j), and
     the first derivative, Dp(x), is continuous, but
     D^2p(x) is probably not continuous; there may be jumps at the X(j).
  The slopes at the X(j) are chosen in such a way that
     p(x) is "shape preserving" and "respects monotonicity". This means that,
  on intervals where the data is monotonic, so is p(x);
  at points where the data have a local extremum, so does p(x).
  Comparing PCHIP with SPLINE:
  The function s(x) supplied by SPLINE is constructed in exactly the same way,
  except that the slopes at the X(j) are chosen differently, namely to make
  even D^2s(x) continuous. This has the following effects.
  SPLINE is smoother, i.e., D^2s(x) is continuous.
  SPLINE is more accurate if the data are values of a smooth function.
  PCHIP has no overshoots and less oscillation if the data are not smooth.
  PCHIP is less expensive to set up.
```

The two are equally expensive to evaluate.

Example:

```
x = -3:3;
y = [-1 -1 -1 0 1 1 1];
t = -3:.01:3;
plot(x,y,'o',t,[pchip(x,y,t); spline(x,y,t)])
legend('data','pchip','spline',4)
```

Class support for inputs x, y, xx:
    float: double, single

See also interp1, spline, ppval, unmkpp.

Reference page in Help browser
    doc pchip

The following example demonstrates the similarities and differences between spline and pchip:

```
>> x = [0 2 4 5 7.5 10];% sample data
>> y = exp(-x/6).*cos(x);

>> cs = spline(x,y);     % cubic spline
>> ch = pchip(x,y);      % cubic Hermite

>> xi = linspace(0,10);
>> ysi = ppval(cs,xi);   % interpolate spline
>> yhi = ppval(ch,xi);   % interpolate Hermite

>> plot(x,y,'o',xi,ysi,':',xi,yhi)
>> legend('data','spline','hermite')
>> title('Figure 20.2: Spline and Hermite Interpolation')
```

20.4 INTEGRATION

In many situations, it is desirable to know the area under a function described by piecewise polynomials as a function of the independent variable x. That is, if the piecewise polynomials are denoted $y = s(x)$, we are interested in computing

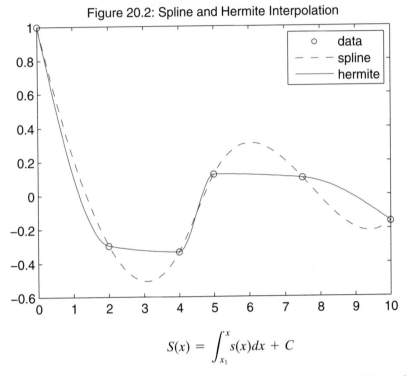

Figure 20.2: Spline and Hermite Interpolation

$$S(x) = \int_{x_1}^{x} s(x)dx + C$$

where x_1 is the first breakpoint and C is the integration constant. Since $s(x)$ is composed of connected cubic polynomials, with the kth cubic polynomial being

$$s_k(x) = a_k(x - x_k)^3 + b_k(x - x_k)^2 + c_k(x - x_k) + d_k, \quad x_k \leq x \leq x_{k+1}$$

and whose integral or area over the range $[x_k, x]$, where $x_k \leq x \leq x_{k+1}$, is

$$S_k(x) = \int_{x_k}^{x} s_k(x)dx = \frac{a_k}{4}(x - x_k)^4 + \frac{b_k}{3}(x - x_k)^3 + \frac{c_k}{2}(x - x_k)^2 + d_k(x - x_k)$$

the area under a piecewise polynomial is easily computed as

$$S(x) = \sum_{i=1}^{k-1} S_i(x_{i+1}) + S_k(x)$$

where $x_k \leq x \leq x_{k+1}$. The summation term is the cumulative sum of the areas under all preceding cubic polynomials. As such, it is readily computed and forms the constant term in the polynomial describing $S(x)$, since $S_k(x)$ is a polynomial. With this understanding, the integral itself can be written as a piecewise polynomial. In this case, it is a quartic piecewise polynomial, because the individual polynomials are of order four.

Because the pp-form used in MATLAB can support piecewise polynomials of any order, the preceding piecewise polynomial integration is embodied in the function mmppint. The body of this function is as follows:

```
function ppi = mmppint(pp,c)
%MMPPINT Cubic Spline Integral Interpolation.
% PPI = MMPPINT(PP,C) returns the piecewise polynomial vector PPI
% describing the integral of the cubic spline described by
% the piecewise polynomial in PP and having integration constant C.

if prod(size(c))~=1
    error('C Must be a Scalar.')
end
[br,co,npy,nco] = unmkpp(pp);      % take apart pp
sf = nco:-1:1;                     % scale factors for integration
ico = [co./sf(ones(npy,1),:) zeros(npy,1)]; % integral coefficients
nco = nco+1;                       % integral spline has higher order
ico(1,nco) = c;                    % integration constant
for k = 2:npy                      % find constant terms in polynomials
     ico(k,nco) = polyval(ico(k-1,:),br(k)-br(k-1));
end
ppi = mkpp(br,ico);                % build pp form for integral
```

Consider the following example using mmppint:

```
>> x = (0:.1:1)*2*pi;
>> y = sin(x); % create rough data
>> pp = spline(x,y);         % pp-form fitting rough data
>> ppi = mmppint(pp,0);      % pp-form of integral
>> xi = linspace(0,2*pi);    % finer points for interpolation
>> yi = ppval(pp,xi);        % evaluate curve
>> yyi = ppval(ppi,xi);      % evaluate integral
>> plot(x,y,'o',xi,yi,xi,yyi,'--') % plot results
>> title('Figure 20.3: Spline Integration')
```

Figure 20.3: Spline Integration

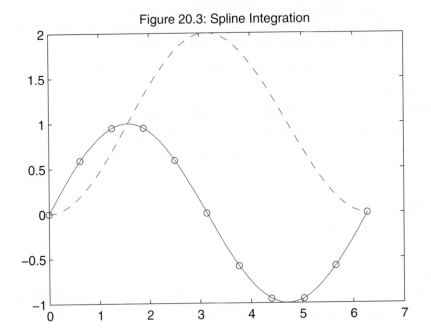

Note that this plot qualitatively shows the identity

$$\int_0^x \sin(x)dx = 1 - \cos(x)$$

20.5 DIFFERENTIATION

Just as you may be interested in piecewise polynomial integration, you may find that the derivative or slope of a function described by piecewise polynomials is also useful. Given that the kth cubic polynomial is

$$s_k(x) = a_k(x - x_k)^3 + b_k(x - x_k)^2 + c_k(x - x_k) + d_k, \quad x_k \le x \le x_{k+1}$$

the derivative of $s_k(x)$ is easily written as

$$\frac{ds_k(x)}{dx} = 3a_k(x - x_k)^2 + 2b_k(x - x)c_k$$

where $x_k \le x \le x_{k+1}$. As with integration, the derivative is also a piecewise polynomial. However, in this case, it is a quadratic piecewise polynomial, since the order of the polynomial is two.

Based on this expression, the function mmppder performs piecewise polynomial differentiation. The body of this function is as follows:

```
function ppd = mmppder(pp)
%MMPPDER Cubic Spline Derivative Interpolation.
% PPD = MMPPDER(PP) returns the piecewise polynomial vector PPD
% describing the cubic spline derivative of the curve described
% by the piecewise polynomial in PP.
[br,co,npy,nco] = unmkpp(pp);        % take apart pp
sf = nco-1:-1:1;                     % scale factors for differentiation
dco = sf(ones(npy,1),:).*co(:,1:nco-1); % derivative coefficients
ppd = mkpp(br,dco);                  % build pp form for derivative
```

To demonstrate the use of mmppder, consider the following example:

```
>> x = (0:.1:1)*2*pi;      % same data as earlier
>> y = sin(x);

>> pp = spline(x,y);       % pp-form fitting rough data
>> ppd = mmppder(pp);      % pp-form of derivative

>> xi = linspace(0,2*pi);  % finer points for interpolation
>> yi = ppval(pp,xi);      % evaluate curve
>> yyd = ppval(ppd,xi);    % evaluate derivative

>> plot(x,y,'o',xi,yi,xi,yyd,'--') % plot results
>> title('Figure 20.4: Spline Differentiation')
```

Note that this plot qualitatively shows the identity

$$\frac{d}{dx}\sin(x) = \cos(x)$$

20.6 SPLINE INTERPOLATION ON A PLANE

Spline interpolation, as implemented by the function spline, assumes that the independent variable is monotonic. That is, the spline $y = s(x)$ describes a continuous function. When it is not monotonic, there is no one-to-one relationship between x and y, and the function ppval has no way of knowing what y value to return for a given x. A common situation where this occurs is a curve defined on a plane. For example,

```
>> t = linspace(0,3*pi,15);
>> x = sqrt(t).*cos(t);
```

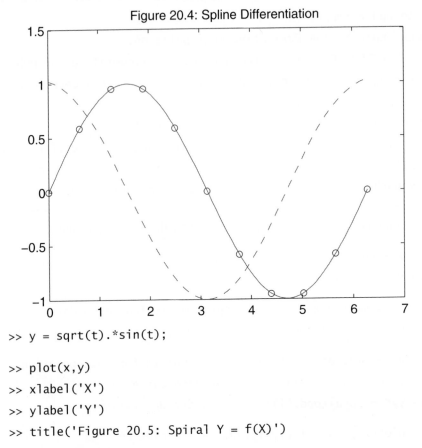

Figure 20.4: Spline Differentiation

```
>> y = sqrt(t).*sin(t);

>> plot(x,y)

>> xlabel('X')

>> ylabel('Y')

>> title('Figure 20.5: Spiral Y = f(X)')
```

It is not possible to compute a cubic spline for the spiral as a function of x, since there are multiple y values for each x near the origin. However, it is possible to compute a spline for each axis with respect to the variable or parameter t. This can be accomplished in two ways in MATLAB. First, you could make two calls to spline to fit a spline to $x(t)$ and then make another call to fit a spline to $y(t)$. Alternatively, the spline function can fit both splines simultaneously and return a single pp-form structure containing both fits. The following example demonstrates the latter approach:

```
>> ppxy = spline(t,[x;y])

ppxy =

      form: 'pp'

    breaks: [1x15 double]

     coefs: [28x4 double]

    pieces: 14

     order: 4

       dim: 2
```

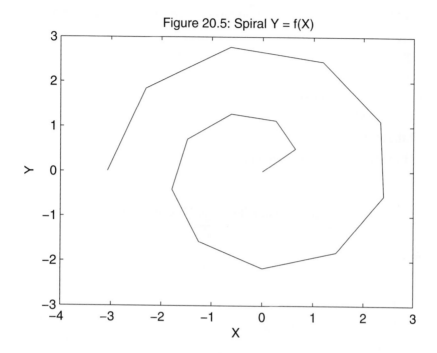

Figure 20.5: Spiral Y = f(X)

Here, the second argument to spline is an array containing two **rows**, each of which is fit with a spline by using the independent variable t, which is monotonic. Elsewhere in MATLAB, data arrays are column-oriented, with different columns representing different variables. However, the function spline adopts a row-oriented approach in which different rows represent different variables. Not recognizing this subtle fact can lead to errors. For example,

```
>> ppz = spline(t,[x;y]') % try "normal" column oriented data
??? Error using ==> chckxy at 89
The number of sites, 15, is incompatible with the number of values, 2.

Error in ==> spline at 55
[x,y,sizey,endslopes] = chckxy(x,y);
```

In addition, the pp-form structure returned now identifies ppxy.dim = 2, meaning that ppxy describes a 2-D spline.

Given this spline fit, the original data can be interpolated as desired, as in the following example.

```
>> ti = linspace(0,3*pi); % total range, 100 points
>> xy = ppval(ppxy,ti);   % evaluate both splines
```

```
>> size(xy)                    % results are row-oriented too!
ans =
   2 100

>> plot(x,y,'d',xy(1,:),xy(2,:))
>> xlabel('X')
>> ylabel('Y')
>> title('Figure 20.6: Interpolated Spiral Y = f(X)')
```

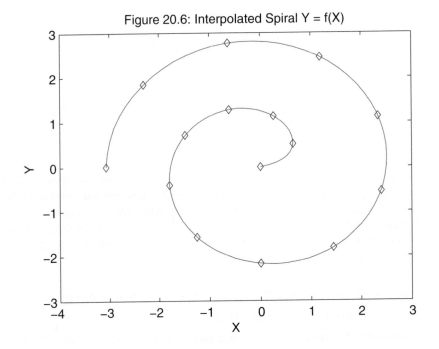

Figure 20.6: Interpolated Spiral Y = f(X)

In the example, the results of ppval are also row-oriented, the first row being associated with the first spline, and so on. Therefore, to plot *y* versus *x*, the second row xy(2,:) is plotted versus the first row xy(1,:).

Finally, the preceding approach is not limited to two dimensions. The pp-form structure and MATLAB piecewise polynomial functions all handle *n*-dimensional splines.

21

Fourier Analysis

Frequency-domain tools, such as Fourier series, Fourier transforms, and their discrete-time counterparts, form a cornerstone in signal processing. These transforms decompose a signal into a sequence or continuum of sinusoidal components that identify the frequency-domain content of the signal. MATLAB provides the functions fft, ifft, fft2, ifft2, fftn, ifftn, fftshift, and ifftshift for Fourier analysis. This collection of functions performs the discrete Fourier transform and its inverse in one or more dimensions. (More extensive signal-processing tools are available in the optional *Signal Processing Toolbox.*)

Because signal processing encompasses such a diverse area, it is beyond the scope of this text to illustrate even a small sample of the type of problems that can be solved using the discrete Fourier transform functions in MATLAB. Therefore, only one example of using the function fft to approximate the Fourier transform of a continuous time signal is illustrated. In addition, the use of the function fft to approximate the Fourier series of a periodic continuous time signal is demonstrated.

21.1 DISCRETE FOURIER TRANSFORM

In MATLAB, the function fft computes the discrete Fourier transform of a signal. In cases where the length of the data is a power of 2, or a product of prime factors, fast Fourier transform (FFT) algorithms are employed to compute the discrete Fourier transform.

> Because of the substantial increase in computational speed that occurs when data length is a power of 2, whenever possible it is important to choose data lengths equal to a power of 2, or to pad data with zeros to give it a length equal to a power of 2.

The fast Fourier transform implemented in MATLAB follows that commonly used in engineering texts:

$$F(k) = FFT\{f(n)\} = \sum_{n=0}^{N-1} f(n)e^{-j2\pi nk/N} \quad k = 0,1,\ldots,N-1$$

Since MATLAB does not support zero indices, the values are shifted by one index value to

$$F(k) = FFT\{f(n)\} = \sum_{n=1}^{N} f(n)e^{-j2\pi(n-1)(k-1)/N} \quad k = 1,2,\ldots,N$$

The inverse transform follows accordingly as

$$f(n) = FFT^{-1}\{F(k)\} = \frac{1}{N}\sum_{k=1}^{N} F(k)e^{j2\pi(n-1)(k-1)/N} \quad n = 1,2,\ldots,N$$

Specific details on the use of the `fft` function are described in its help text:

```
>> help fft
 FFT Discrete Fourier transform.
    FFT(X) is the discrete Fourier transform (DFT) of vector X. For
    matrices, the FFT operation is applied to each column. For N-D
    arrays, the FFT operation operates on the first non-singleton
    dimension.

    FFT(X,N) is the N-point FFT, padded with zeros if X has less
    than N points and truncated if it has more.

    FFT(X,[],DIM) or FFT(X,N,DIM) applies the FFT operation across the
    dimension DIM.

    For length N input vector x, the DFT is a length N vector X,
    with elements
                        N
        X(k) =         sum  x(n)*exp(-j*2*pi*(k-1)*(n-1)/N), 1 <= k <= N.
                       n = 1

    The inverse DFT (computed by IFFT) is given by
                        N
        x(n) = (1/N) sum X(k)*exp( j*2*pi*(k-1)*(n-1)/N), 1 <= n <= N.
                       k = 1

    See also fft2, fftn, fftshift, fftw, ifft, ifft2, ifftn.
```

Overloaded methods:
    uint8/fft
    uint16/fft

Reference page in Help browser
    doc fft

To illustrate the use of the FFT, consider the problem of estimating the continuous Fourier transform of the signal

$$f(t) = 2e^{-3t} \quad t \geq 0$$

Analytically, the Fourier transform of $f(t)$ is given by

$$F(\omega) = \frac{2}{3 + j\omega}$$

Although using the FFT has little real value in this case (since the analytical solution is known), this example illustrates an approach to estimating the Fourier transform of less common signals, especially those whose Fourier transform is not readily found analytically. The following MATLAB statements estimate $|F(\omega)|$ by using the FFT and graphically compare it to the preceding analytical expression:

```
>> N = 128;              % choose a power of 2 for speed
>> t = linspace(0,3,N);  % time points for function evaluation
>> f = 2*exp(-3*t);      % evaluate function, minimize aliasing: f(3) ~ 0

>> Ts = t(2) - t(1);     % the sampling period
>> Ws = 2*pi/Ts;         % the sampling frequency in rad/sec

>> F = fft(f);           % compute the fft
>> Fc = fftshift(F)*Ts;  % shift and scale

>> W = Ws*(-N/2:(N/2)-1)/N; % frequency axis
>> Fa = 2./(3+j*W);      % analytical Fourier transform

>> plot(W,abs(Fa),W,abs(Fc),'.') % generate plot, 'o' marks fft
>> xlabel('Frequency, Rad/s')
>> ylabel('|F(\omega)|')
>> title('Figure 21.1: Fourier Transform Approximation')
```

The function fftshift flips the halves of F so that the $(N/2) + 1$ element of Fc is the DC component of the result. Elements less than this are negative frequency

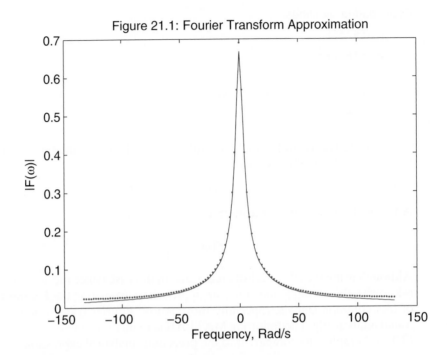

Figure 21.1: Fourier Transform Approximation

components, while those greater are positive frequency components. Using this fact, W creates the appropriate analog frequency axis with W(N/2 + 1) = 0. Graphically, the FFT approximation is good at low frequencies, but demonstrates some aliasing at higher frequencies near the Nyquist frequency.

The FFT-related functions in MATLAB include those listed in the following table:

Function	Description
conv	Convolution
conv2	2-D convolution
convn	n-D convolution
deconv	Deconvolution
filter	Discrete-time filter
filter2	2-D discrete-time filter
fft	Discrete Fourier transform
fft2	2-D discrete Fourier transform
fftn	n-D discrete Fourier transform
fftw	Tunes FFTW library planner method used by subsequent FFT functions
ifft	Inverse discrete Fourier transform

`ifft2`	2-D inverse discrete Fourier transform
`ifftn`	n-D inverse discrete Fourier transform
`fftshift`	Shifts FFT results so that negative frequencies appear first
`ifftshift`	Undo actions performed by `fftshift`
`abs`	Magnitude of complex array
`angle`	Radian angle of complex array
`unwrap`	Remove phase angle jumps
`cplxpair`	Sorts vector into complex conjugate pairs
`nextpow2`	Next higher power of 2

21.2 FOURIER SERIES

MATLAB itself offers no functions specifically tailored to Fourier series analysis and manipulation. However, these functions can be easily added when you understand the relationship between the discrete Fourier transform of samples of a periodic signal and its Fourier series.

The Fourier series representation of a real-valued periodic signal $f(t)$ can be written in complex exponential form as

$$f(t) = \sum_{n=-\infty}^{\infty} F_n e^{jn\omega_o t}$$

where the Fourier series coefficients are given by

$$F_n = \frac{1}{T_o} \int_t^{t+T_o} f(t)e^{-jn\omega_o t}dt$$

and the fundamental frequency is $\omega_o = 2\pi/T_o$, where T_o is the period. The complex exponential form of the Fourier series can be rewritten in trigonometric form as

$$f(t) = A_o + \sum_{n=1}^{\infty}\{A_n\cos(n\omega_o t) + B_n \sin(n\omega_o t)\}$$

where the coefficients are given by

$$A_o = \frac{1}{T_o} \int_t^{t+T_o} f(t)dt$$

$$A_n = \frac{2}{T_o} \int_t^{t+T_o} f(t)\cos(n\omega_o t)dt$$

$$B_n = \frac{2}{T_o} \int_t^{t+T_o} f(t)\sin(n\omega_o t)dt$$

Of these two forms, the complex exponential Fourier series is generally easier to manipulate analytically, whereas the trigonometric form provides a more intuitive understanding because it makes it easier to visualize sine and cosine waveforms. The relationships between the coefficients of the two forms are

$$A_o = F_o$$
$$A_n = 2\,\mathrm{Re}\{F_n\}$$
$$B_n = -2\,\mathrm{Im}\{F_n\}$$
$$F_n = F_{-n}^* = (A_n - jB_n)/2$$

Using these relationships, you can use the complex exponential form analytically and then convert results to the trigonometric form for display.

The discrete Fourier transform can be used to compute the Fourier series coefficients, provided that the time samples are appropriately chosen and the transform output is scaled. For example, consider computing the Fourier series coefficients of the sawtooth waveform shown next.

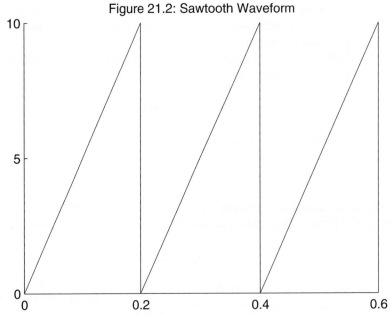

Figure 21.2: Sawtooth Waveform

First, you must create a function to evaluate the sawtooth at arbitrary points:

```
function f = sawtooth(t,To)
%SAWTOOTH Sawtooth Waveform Generation.
% SAWTOOTH(t,To) computes values of a sawtooth having
```

```
% a period To at the points defined in the vector t.
f = 10*rem(t,To)/To;
f(f==0 | f==10) = 5; % must average value at discontinuity!
```

To minimize aliasing, it is necessary to compute enough harmonics so that the highest harmonic amplitude is negligible. In this case, choose

```
>> N = 25;    % number of harmonics
>> To = 0.2; % choose period
```

The number of terms to consider in the discrete Fourier transform is twice the number of harmonics, since the discrete Fourier transform computes both positive and negative harmonics:

```
>> n = 2*N;
```

The function must be evaluated at n points over one period in such a manner that the $(n + 1)$th point is one period away from the first point:

```
>> t = linspace(0,To,n+1);  % (n+1)th point is one period away
>> t(end) = [];             % throw away undesired last point
>> f = sawtooth(t,To);      % compute sawtooth
```

We are now ready to compute the transform, rearrange the components, and scale the results:

```
>> Fn = fft(f);                                % compute FFT
>> Fn = [conj(Fn(N+1)) Fn(N+2:end) Fn(1:N+1)]; % rearrange values
>> Fn = Fn/n;                                  % scale results
```

The vector Fn now contains the complex exponential Fourier series coefficients in ascending order. That is, Fn(1) is F_{-25}; Fn(N+1) is F_0, the DC component; and Fn(2*N+1) is F_{25}, the 25th harmonic component.

From these data, the trigonometric Fourier series coefficients are as follows:

```
>> A0 = Fn(N+1) % DC component
A0 =
      5
```

```
>> An = 2*real(Fn(N+2:end)) % Cosine terms
An =

   1.0e-015 *
   Columns 1 through 7
    -0.1176   -0.0439   -0.2555    0.3814    0.0507   -0.2006    0.1592
   Columns 8 through 14
    -0.1817    0.0034         0    0.0034   -0.1141   -0.1430   -0.0894
   Columns 15 through 21
    -0.0685   -0.0216    0.0537   -0.0496   -0.0165         0   -0.0165
   Columns 22 through 25
    -0.0079    0.2405    0.3274    0.2132

>> Bn = -2*imag(Fn(N+2:end)) % Sine terms
Bn =

   Columns 1 through 7
    -3.1789   -1.5832   -1.0484   -0.7789   -0.6155   -0.5051   -0.4250
   Columns 8 through 14
    -0.3638   -0.3151   -0.2753   -0.2418   -0.2130   -0.1878   -0.1655
   Columns 15 through 21
    -0.1453   -0.1269   -0.1100   -0.0941   -0.0792   -0.0650   -0.0514
   Columns 22 through 25
    -0.0382   -0.0253   -0.0126         0
```

Since the sawtooth waveform has odd symmetry (except for its DC component), it makes sense that the cosine coefficients An are negligible. (Note that they are scaled by 10^{-15}.) Comparing the actual Fourier series coefficients for this sawtooth waveform with the preceding Bn terms gives a relative error of

```
>> idx = -N:N;           % harmonic indices
>> Fna = 5j./(idx*pi); % complex exponential terms
>> Fna(N+1) = 5;
>> Bna = -2*imag(Fna(N+2:end)); % sine terms

>> Bn_error = (Bn-Bna)./Bna     % relative error
Bn_error =
```

```
Columns 1 through 7
 -0.0013   -0.0053   -0.0119  -0.0211  -0.0331   -0.0478  -0.0653
Columns 8 through 14
 -0.0857   -0.1089   -0.1352  -0.1645  -0.1971   -0.2330  -0.2723
Columns 15 through 21
 -0.3152   -0.3620   -0.4128  -0.4678  -0.5273   -0.5917  -0.6612
Columns 22 through 25
 -0.7363   -0.8174   -0.9051  -1.0000
```

As with the earlier Fourier transform example, aliasing causes errors that increase with increasing frequency. Since all practical signals are not band-limited, aliasing is inevitable, and a decision must be made about the degree of aliasing that can be tolerated in a given application. As the number of requested harmonics increases, the degree of aliasing decreases. Therefore, to minimize aliasing, you can request a larger number of harmonics and then choose a subset of them to view and further manipulate.

Finally, the line spectra of the complex exponential Fourier series can be plotted using the stem function:

```
>> stem(idx,abs(Fn))
>> xlabel('Harmonic Index')
>> title('Figure 21.3: Sawtooth Harmonic Content')
>> axis tight
```

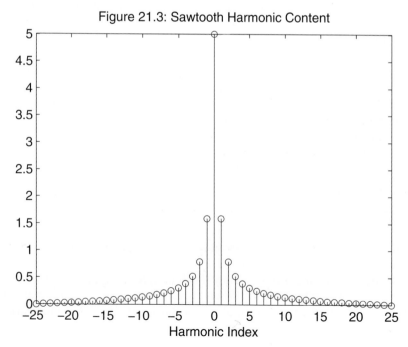

Figure 21.3: Sawtooth Harmonic Content

22

Optimization

Optimization in the context of this chapter refers to the process of determining where a function $y = g(x)$ takes on either specific or extreme values. When a function is defined simply, the corresponding inverse function $x = g^{-1}(y)$ can often be found, in which case you can determine what x values produce a given y by evaluating the inverse function. On the other hand, many functions, including simple ones, have no inverse. When this is the case, you must estimate the x that produces a known y by some iterative procedure. In practice, this iterative procedure is called **zero finding**, because finding x such that $y = g(x)$ for some y is equivalent to finding x such that $f(x) = 0$, where $f(x) = y - g(x)$.

In addition to knowing where a function takes on specific values, it is also common to know its extreme values—that is, where it achieves maximum or minimum values. As before, there are numerous times when these extreme values must be estimated by some iterative procedure. Since a function maximum is the minimum of its negative (i.e., $\max f(x) = \min \{ - f(x) \}$), iterative procedures for finding extreme values typically find only minimum values, and the procedures are called **minimization** algorithms.

In this chapter, the optimization functions available in basic MATLAB are covered. (Many more functions are available in the optional *Optimization Toolbox*.)

22.1 ZERO FINDING

Finding a zero of a function can be interpreted in a number of ways, depending on the function. When the function is 1-D, the MATLAB function `fzero` can be used to find a zero. The algorithm used by this function is a combination of bisection and inverse quadratic interpolation. When the function is multidimensional—that is, the function definition consists of multiple scalar functions of a vector variable—you

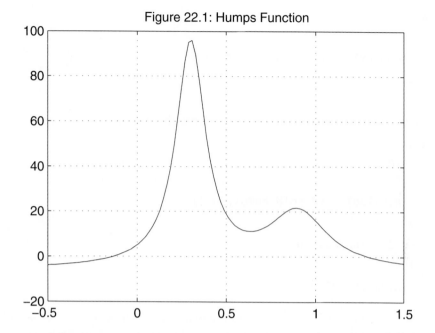

must look beyond basic MATLAB for a solution. The *Optimization Toolbox* or other third-party toolbox is required to solve the multidimensional case.

To illustrate the use of the function fzero, consider the function humps:

```
>> x = linspace(-.5,1.5);
>> y = humps(x);
>> plot(x,y)
>> grid on
>> title('Figure 22.1: Humps Function')
```

The humps M-file evaluates the function

$$\text{humps}(x) = \frac{1}{(x - 0.3)^2 + 0.01} + \frac{1}{(x - 0.9)^2 + 0.04} - 6$$

which crosses zero near $x = -0.2$ and $x = 1.3$. The function fzero provides a way to find a better approximation to these zero crossings:

```
>> format long % display more precision

>> H_humps = @humps; % create function handle to humps.m function.

>> x = fzero(H_humps,1.3)
```

```
x =
    1.299549682584822
>> humps(x) % how close is it to 0?
ans =
    0
>> H_humps(x) % evaluate humps through its handle as well
ans =
    0

>> [x,value] = fzero(H_humps,-0.2)
x =
   -0.131618018099606
value =
    8.881784197001252e-16
```

Here, the two zeros of the function were found. The first zero is very close to 1.3, and evaluation of function at the zero produced zero. The second zero was found to be close to –0.13. In this call to fzero, a second output argument was supplied, which returned the function evaluated at the zero. Therefore, it wasn't necessary to call humps(x) to check the accuracy of the solution found by fzero. It is important to note that fzero returns just one zero—the zero found closest to the initial guess. So, if a function has more than one zero, it is up to the user to call fzero multiple times with different initial guesses.

When initially called, the function fzero searches on either side of the initial guess for a sign change in the function. When a sign change is found, the two endpoints that produced the sign change form a **bracket** on the number line. If a function is continuous, it must cross through zero somewhere in the bracket. Knowing this, the function fzero then searches for the zero crossing and returns the value of *x* that comes closest to making this happen.

In many cases, supplying an initial guess or estimate of the zero is unnecessary, because a bracket is already known from the properties of the problem to be solved. When this occurs, you can simply supply fzero with the bracket rather than with an initial guess of the zero location:

```
>> [x,value] = fzero(H_humps,[-2 0])
x =
   -0.131618018099606
value =
    0

>> [x,value] = fzero(H_humps,[0 1.2])
```

```
??? Error using ==> fzero at 283
The function values at the interval endpoints must differ in sign.
```

In the first example, [-2 0] is a bracket around the zero near −0.13. As a result, fzero finds the zero. In the second example, [0 1.2] is not a bracket around a zero, forcing fzero to report an error and abort its search. So, if a two-element array is supplied to fzero, it must bracket a zero, or else the function terminates without doing a zero search.

In the preceding examples, the function to be searched was provided to fzero as a function handle, which uniquely identifies the function M-file. As discussed in Chapter 12, the function to be searched can also be supplied as an *anonymous function*, an *in-line function* object, or a *string expression*. While all three of these work, the use of function handles is encouraged. In-line function objects and string expression definitions are obsolete, but remain operable in MATLAB for the time being. (To learn more about these alternatives, see the MATLAB documentation.)

All functions in this chapter have various settable parameters. These functions, as well as those in the *Optimization Toolbox*, share the same format for managing parameters. The functions optimset and optimget are used, respectively, to set and get parameters for all functions. For fzero, there are two settable parameters, 'Display' and 'TolX'. The first parameter controls the amount of detail returned while the function is working; the second sets a tolerance range for accepting the final answer. The following code is illustrative:

```
>> options = optimset('Display','iter'); % show iteration history
>> [x,value] = fzero(H_humps,[-2 0],options)
```

Func-count	x	f(x)	Procedure
2	0	5.17647	initial
3	-0.952481	-5.07853	interpolation
4	-0.480789	-3.87242	interpolation
5	-0.240394	-1.94304	bisection
6	-0.120197	0.28528	bisection
7	-0.135585	-0.0944316	interpolation
8	-0.131759	-0.00338409	interpolation
9	-0.131618	1.63632e-06	interpolation
10	-0.131618	-7.14819e-10	interpolation
11	-0.131618	0	interpolation

```
Zero found in the interval [-2, 0]
x =
  -0.131618018099606
```

```
value =
     0

>> options = optimset('Display','final'); % display successful interval
>> [x,value] = fzero(H_humps,[-2 0],options)

Zero found in the interval: [-2, 0].
x =
  -0.131618018099606
value =
     0

>> options = optimset('TolX',0.1);

>> [x,value] = fzero(H_humps,[-2 0],options)
x =
  -0.240394472507622
value =
  -1.943038259725652

>> options = optimset('Display','iter','TolX',0.1); % set both
>> [x,value] = fzero(H_humps,[-2 0],options)
```

Func-count	x	f(x)	Procedure
2	0	5.17647	initial
3	-0.952481	-5.07853	interpolation
4	-0.480789	-3.87242	interpolation
5	-0.240394	-1.94304	bisection

```
Zero found in the interval [-2, 0]
x =
  -0.240394472507622
value =
  -1.943038259725652
```

In the preceding code, an options structure was created with the desired parameters and then passed as a third argument to fzero. The 'Display' option has four settings: 'final', 'iter', 'notify', and 'off'. The setting 'notify' is the default, and it means "Display information only if no solution is found." The 'TolX' option sets the tolerance for the final answer, which is equal to eps by default. (See the online help for optimset and optimget for more information regarding parameters for MATLAB optimization functions.)

22.2 MINIMIZATION IN ONE DIMENSION

In addition to the visual information provided by plotting, it is often necessary to determine other, more specific attributes of a function. Of particular interest in many applications are function extremes, that is, the function's maxima (peaks) and minima (valleys). Mathematically, these extremes are found analytically by determining where the derivative (slope) of a function is zero. This idea can be readily understood by inspecting the slope of the humps plot at its peaks and valleys. Clearly, when a function is simply defined, this process often works. However, even for many simple functions that can be differentiated readily, it is often impossible to find where the derivative is zero. In these cases and in cases where it is difficult or impossible to find the derivative analytically, it is necessary to search for function extremes numerically. MATLAB provides two functions that perform this task, fminbnd and fminsearch. These two functions find minima of 1-D and n-D functions, respectively. The function fminbnd employs a combination of golden-section search and parabolic interpolation. Since a maximum of $f(x)$ is equal to a minimum of $-f(x)$, fminbnd and fminsearch can be used to find both minima and maxima. If this notion is not clear, visualize the preceding humps(x) plot flipped upside down. In the upside-down state, peaks become valleys and valleys become peaks.

To illustrate 1-D minimization and maximization, consider the preceding humps(x) example once again. From the figure, there is a maximum near $x = 0.3$ and a minimum near $x = 0.6$. With fminbnd, these extremes can be found with more accuracy:

```
>> H_humps = @humps; % create handle to humps.m function.

>> [xmin,value] = fminbnd(H_humps,0.5,0.8)
xmin =
   0.637008211963619
value =
  11.252754125877694
>> options = optimset('Display','iter');
>> [xmin,value] = fminbnd(H_humps,0.5,0.8,options)
```

Func-count	x	f(x)	Procedure
1	0.61459	11.4103	initial
2	0.68541	11.9288	golden

3	0.57082	12.7389	golden
4	0.638866	11.2538	parabolic
5	0.637626	11.2529	parabolic
6	0.637046	11.2528	parabolic
7	0.637008	11.2528	parabolic
8	0.636975	11.2528	parabolic

```
Optimization terminated:
the current x satisfies the termination criteria using OPTIONS.TolX
of 1.000000e-04

xmin =
    0.637008211963619
value =
    11.252754125877694
```

In this example, in the two calls to fminbnd, 0.5 and 0.8 denote the range over which to search for minimum. In the second case, options were set to display the iterations performed by fminbnd.

To find the maximum near $x = 0.3$, we can either modify the humps.m file to negate the expression, or we can create an anonymous function. The following example demonstrates the latter approach:

```
>> AH_humps = @(x) -1./((x-.3).^2 +.01) - 1./((x-.9).^2 +.04) + 6;
>> [xmax,value] = fminbnd(AH_humps,0.2,0.4,options)
```

Func-count	x	f(x)	Procedure
1	0.276393	-91.053	initial
2	0.323607	-91.4079	golden
3	0.352786	-75.1541	golden
4	0.300509	-96.5012	parabolic
5	0.300397	-96.5014	parabolic
6	0.300364	-96.5014	parabolic
7	0.300331	-96.5014	parabolic

```
Optimization terminated:
the current x satisfies the termination criteria using OPTIONS.TolX
of 1.000000e-04
```

```
xmax =
    0.300364137900245
value =
    -96.501407243870503
```

On termination, the maximum is found to be very close to 0.3, and the peak has an amplitude of $+96.5$. The value returned by fminbnd is the negative of the actual value, because fminbnd computes the minimum of $-\text{humps}(x)$.

22.3 MINIMIZATION IN HIGHER DIMENSIONS

As described previously, the function fminsearch provides a simple algorithm for minimizing a function of several variables. That is, fminsearch attempts to find the minimum of $f(x)$, where $f(x)$ is a scalar function of a vector argument x. The function fminsearch implements the Nelder–Mead simplex search algorithm, which modifies the components of x to find the minimum of $f(x)$. This algorithm is not as efficient on smooth functions as some other algorithms are, but, on the other hand, it does not require gradient information that is often expensive to compute. It also tends to be more robust on functions that are not smooth, where gradient information is less valuable. If the function to be minimized is inexpensive to compute, the Nelder–Mead algorithm usually works very well.

To illustrate usage of fminsearch, consider the banana function, also called Rosenbrock's function:

$$f(x) = 100\,(x_2 - x_1^2)^2 + (1 - x_1)^2$$

This function can be visualized by creating a 3-D mesh plot with x_1 as the x-dimension and x_2 as the y-dimension:

```
x = [-1.5:0.125:1.5];   % range for x1 variable
y = [-.6:0.125:2.8];    % range for x2 variable

[X,Y] = meshgrid(x,y); % grid of all x and y
Z = 100.*(Y-X.*X).^2 + (1-X).^2; % evaluate banana

mesh(X,Y,Z)
hidden off
xlabel('x(1)')
ylabel('x(2)')
title('Figure 22.2: Banana Function')
```

```
hold on
plot3(1,1,1,'k.','markersize',30)
hold off
```

Figure 22.2: Banana Function

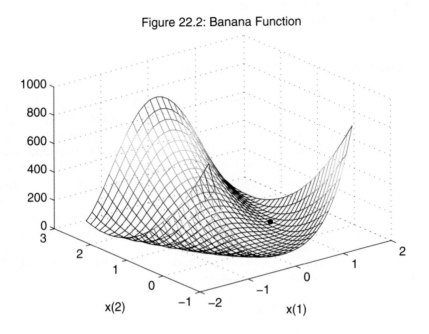

As shown in the plot, the banana function has a unique minimum of zero at $x = [1; 1]$.
To find the minimum of this function, it must be rewritten in terms of $x_1 = x(1)$ and
$x_2 = x(2)$, as shown before mathematically. It can be entered as the M-file

```
function f = banana(x)
% Rosenbrock's banana function
    f = 100*(x(2)-x(1)^2)^2 + (1-x(1))^2;
```

or as the anonymous function

```
>> AH_banana = @(x) 100*(x(2)-x(1)^2)^2 + (1-x(1))^2;
```

Using either of these representations, `fminsearch` produces

```
>> [xmin,value,flag,output] = fminsearch(AH_banana,[-1.9,2])
xmin =
    1.000016668894802   1.000034473862770
value =
    4.068551535063419e-10
flag =
     1
output =
    iterations: 114
    funcCount: 210
    algorithm: 'Nelder-Mead simplex direct search'
      message: [1x194 char]
```

Here, four output parameters are shown: the minimum found, the function evaluated at the minimum, a flag signifying success, and finally an algorithm statistics structure. Finding the minimum with a tolerance of 1e–4 required 114 iterations and 210 banana function evaluations. If less output is desired, it is simply a matter of providing fewer output variables.

As with fminbnd, fminsearch accepts an options structure. Although it is possible to set over 65 options in an options structure, those used by fminsearch are listed in the next table. As shown earlier, preferences are set by calling optimset as options = optimset('Name',value,'Name',value,...). Setting 'Display' to 'iter' in fminsearch can lead to a tremendous amount of output to the *Command* window.

Option Name	Description	Default Value
'Display'	Displays frequency, 'iter','final','notify', or 'off'	'notify' (displays information only if no solution is found)
'FunValCheck'	Checks validity of objective function values 'on' or 'off'	'off' (do not display an error if the objective function returns a complex value or a NaN)
'MaxFunEvals'	Maximum function evaluations	200*length(x)
'MaxIter'	Maximum algorithm iterations	200*length(x)
'PlotFcns'	Plot function called by an optimization function at each iteration	[] (empty array)
'OutputFcn'	User-defined function called by an optimization function at each iteration	[] (empty array)
'TolFun'	Function solution tolerance	1.00E-04
'TolX'	Variable solution tolerance	1.00E-04

To demonstrate how to use the options shown in the table, consider finding the solution to the previous problem with tighter function and variable tolerances:

```
>> options = optimset('TolFun',1e-8,'TolX',1e-8);
>> [xmin,value,flag,output] = fminsearch(AH_banana,[-1.9,2],options)
xmin =
    1.000000001260770  1.000000002307897
value =
    6.153858843361103e-18
flag =
    1
output =
    iterations: 144
     funcCount: 266
     algorithm: 'Nelder-Mead simplex direct search'
       message: [1×194 char]
```

With tighter tolerances, xmin is now within 1e–8 of the actual minimum, the function evaluates to well within eps of the minimum of zero, and the number of algorithm iterations and function evaluations increases by approximately 26 percent. From this information, it is clear that fminsearch requires a lot of function evaluations and therefore can be slow for functions that are computationally expensive.

22.4 PRACTICAL ISSUES

Iterative solutions, such as those found by fzero, fminbnd, and fminsearch, all make some assumptions about the function to be iterated. Since there are essentially no limits to the function provided, it makes sense that these *function functions* may not converge or may take many iterations to converge. At worst, these functions can produce a MATLAB error that terminates the iteration without producing a result. And even if they do terminate promptly, there is no guarantee that they have stopped at the desired result. To make the most efficient use of these function functions, consider applying the following points:

1. Start with a good initial guess. This is the most important consideration. A good guess keeps the problem in the neighborhood of the solution, where its numerical properties are hopefully stable.
2. If components of the solution (e.g., in fminsearch) are separated by several orders of magnitude or more, consider scaling them to improve iteration efficiency and accuracy. For example, if x(1) is known to be near 1, and x(2) is known to be near 1e6, scale x(2) by 1e-6 in the function definition and then

scale the returned result by 1e6. When you do so, the search algorithm uses numbers that are all around the same order of magnitude.

3. If the problem is complicated, look for ways to simplify it into a sequence of simpler problems that have fewer variables.

4. Make sure your function cannot return complex numbers, Inf, or NaN, which usually results in convergence failure. The functions isreal, isfinite, and isnan can be used to test results before returning them.

5. Avoid functions that are discontinuous. Functions such as abs, min, and max all produce discontinuities that can lead to divergence.

6. Constraints on the allowable range of x can be included by adding a penalty term to the function to be iterated, such that the algorithm is persuaded to avoid out-of-range values.

23

Integration and Differentiation

Integration and differentiation are fundamental tools in calculus. Integration computes the area under a function, and differentiation describes the slope or gradient of a function. MATLAB provides functions for numerically approximating the integral and slope of a function. Functions are provided for making approximations when functions exist as M-files and as anonymous functions, as well as when functions are tabulated at uniformly spaced points over the region of interest.

23.1 INTEGRATION

MATLAB provides seven functions for computing integrals of functions: quad, quadl, quadgk, quadv, quad2d, dblquad, and triplequad.

To illustrate integration, consider the function humps(x), as shown in Figure 23.1. As is apparent in the figure, the sum of the trapezoidal areas approximates the integral of the function. Clearly, as the number of trapezoids increases, the fit between the function and the trapezoids gets better, leading to a better integral or area approximation.

Using tabulated values from the humps function, the MATLAB function trapz approximates the area using the trapezoidal approximation. Duplicating the trapezoids shown in the figure produces the following result:

```
>> x = -1:.17:2;
>> y = humps(x);
>> area = trapz(x,y)
```

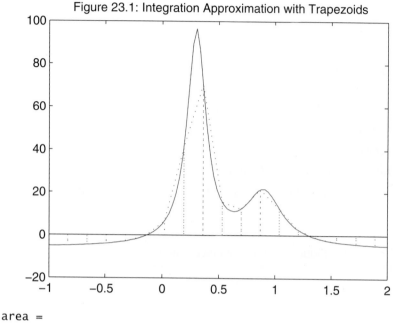

Figure 23.1: Integration Approximation with Trapezoids

```
area =

        25.917
```

In comparison to the figure, this is probably not a very accurate estimate of the area. However, when a finer discretization is used, more accuracy is achieved:

```
>> x = linspace(-1,2,100);
>> y = humps(x);
>> format long
>> area = trapz(x,y)
area =

        26.344731195245956
```

This area agrees with the analytical integral through five significant digits.

Sometimes, we are interested in the integral as a function of x—that is,

$$\int_{x_1}^{x} f(x)\,dx$$

where x_1 is a known lower limit of integration. The definite integral from x_1 to any point x is then found by evaluating the function at x. Using the trapezoidal rule, tabulated values of the cumulative integral are computed using the function cumtrapz, as in the following example:

```
>> x = linspace(-1,2,100);
>> y = humps(x);
```

```
>> z = cumtrapz(x,y);
>> size(z)
ans =
     1    100

>> plotyy(x,y,x,z)
>> grid on
>> xlabel('x')
>> ylabel('humps(x) and integral of humps(x)')
>> title('Figure 23.2: Cumulative Integral of humps(x)')
```

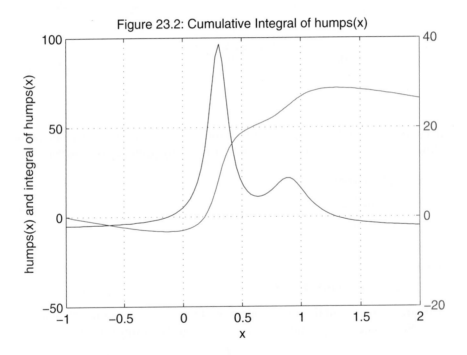

Depending on the properties of the function at hand, it may be difficult to determine an optimum trapezoidal width. Clearly, if you could somehow vary the individual trapezoid widths to match the characteristics of the function, much greater accuracy could be achieved.

The MATLAB functions quad, quadl and quadgk, which are based on a mathematical concept called quadrature, take this approach. These integration functions operate in the same way. They all evaluate the function to be integrated at whatever intervals are necessary to achieve accurate results. Moreover, all three functions make higher-order approximations than a simple trapezoid, with quadl being more rigorous than quad and quadgk being the most rigorous. As an example, consider computing the integral of the humps function again:

```
>> z(end)              % cumtrapz result
ans =
   26.344731195245959

>> H_humps = @humps;   % create function handle

>> quad(H_humps,-1,2)  % use adaptive Simpson quadrature
ans =
   26.344960501201232

>> quadl(H_humps,-1,2) % use adaptive Lobatto quadrature
ans =
   26.344960471378968

>> quadgk(H_humps,-1,2)% use adaptive Gauss-Kronrod quadrature
ans =
   26.344960471378776
```

For this example, quad, quadl, and quadgk all return the same result to eight significant digits, exhibiting eight-digit accuracy with respect to the true solution. On the other hand, cumtrapz achieves only five significant digit accuracy.

The function to be integrated (i.e., the integrand) must support a vector input argument. That is, the integrand must return a vector of outputs for a vector of inputs. Doing so means using dot-arithmetic operators, .*, ./, .\, and .^. For example, the function in humps.m is given by the statement

```
y = 1 ./ ((x-.3).^2 + .01) + 1 ./ ((x-.9).^2 + .04) - 6;
```

The functions quad and quadl also allow you to specify an absolute error tolerance as a fourth input argument, with the default absolute tolerance being 10^{-6}. The quadgk function allows you to specify four parameters, as shown in the following table.

Parameter	Description	Default Value
'AbsTol'	Absolute error tolerance	1.0e–10
'RelTol'	Relative error tolerance	1.0e–6
'Waypoints'	Vector of integration waypoints	[] (empty array)
'MaxIntervalCount'	Maximum number of intervals	650

In addition to 1-D integration, MATLAB supports 2-D integration with the function dblquad. That is, dblquad approximates the integral

$$\int_{y\,min}^{y\,max} \int_{x\,min}^{x\,max} f(x,y)\,dx\,dy$$

To use dblquad, you must first create a function that evaluates $f(x,y)$. For example, consider the function myfun.m:

```
function z = myfun(x,y)
%MYFUN(X,Y) an example function of two variables
z = sin(x).*cos(y) + 1; % must handle vector x input
```

This function can be plotted by issuing the commands

```
>> x = linspace(0,pi,20);    % xmin to xmax
>> y = linspace(-pi,pi,20); % ymin to ymax

>> [xx,yy] = meshgrid(x,y); % create grid of point to evaluate at

>> zz = myfun(xx,yy);         % evaluate at all points

>> mesh(xx,yy,zz)
>> xlabel('x'), ylabel('y')
>> title('Figure 23.3: myfun.m plot')
```

The volume under this function is computed by calling dblquad as

```
>> area = dblquad(@myfun,0,pi,-pi,pi)
area =
   19.739208806091021
>> relerr = (area-2*pi^2)/(2*pi^2)
relerr =
   1.981996941074027e-10
```

Here, dblquad is called as dblquad(Fname,xmin,xmax,ymin,ymax). Based on the relative error computed above, we can say that the results produced by dblquad are highly accurate, even though the function quad is called by dblquad to do the actual integration.

The function triplequad extends the above quadrature integration schemes to triple integration, or volume integration. In this case, triplequad(Fname,xmin,

Figure 23.3: myfun.m plot

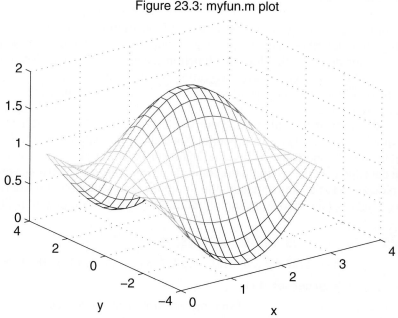

xmax,ymin,ymax,zmin,zmax) integrates the function *Fname*(*x*,*y*,z) over the limits specified by the remainder of the triplequad arguments.

The integration functions in MATLAB are summarized in the following table.

Function	Description
trapz	Trapezoidal numerical integration
cumtrapz	Cumulative trapezoidal numerical integration
quad	Numerical integration using adaptive Simpson quadrature
quadl	Numerical integration using adaptive Lobatto quadrature
quadgk	Numerical integration using adaptive Gauss–Kronrod quadrature
quadv	Vectorized quad
dblquad	Numerical double integration
quad2d	Numerical double integration over a planar region
triplequad	Numerical triple integration

For further information regarding these integration functions, see the MATLAB documentation.

23.2 DIFFERENTIATION

As opposed to integration, numerical differentiation is much more difficult. Integration describes an overall or macroscopic property of a function, whereas differentiation describes the slope of a function at a point, which is a microscopic property of a function. As a result, integration is not sensitive to minor changes in the shape of a function, whereas differentiation is. Any small change in a function can easily create large changes in its slope in the neighborhood of the change.

Because of this inherent sensitivity in differentiation, numerical differentiation is avoided whenever possible, especially if the data to be differentiated is obtained experimentally. In this case, it is best to perform a least-squares curve fit to the data and then differentiate the resulting polynomial. Alternatively, you could fit cubic splines to the data and then find the spline representation of the derivative, as discussed in Chapter 20. For example, consider again the example from Chapter 19:

```
>> x = [0 .1 .2 .3 .4 .5 .6 .7 .8 .9 1];
>> y = [-.447 1.978 3.28 6.16 7.08 7.34 7.66 9.56 9.48 9.30 11.2];
>> n = 2; % order of fit
>> p = polyfit(x,y,n) % find polynomial coefficients
p =
    -9.8108    20.1293    -0.0317

>> xi = linspace(0,1,100);
>> yi = polyval(p,xi); % evaluate polynomial

>> plot(x,y,'-o',xi,yi,'--')
>> xlabel('x'), ylabel('y = f(x)')
>> title('Figure 23.4: Second Order Curve Fitting')
```

The derivative in this case is found by using the polynomial derivative function `polyder`:

```
>> dp = polyder(p)
dp =
    -19.6217    20.1293
```

The derivative of $y(x) = -9.8108x^2 + 20.1293x - 0.0317$ is $dy/dx = -19.6217x + 20.1293$. Since the derivative of a polynomial is yet another polynomial of the next-lowest order, the derivative can also be evaluated at any point. In this case, the polynomial fit is second order, making the resulting derivative first order. As a result, the derivative is a straight line, meaning that it changes linearly with x.

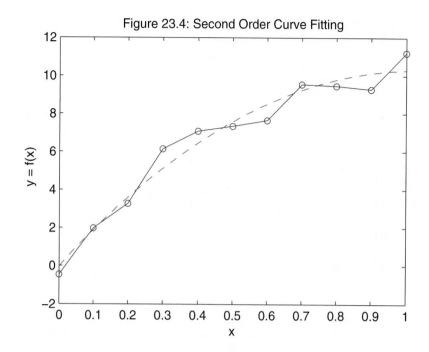

Figure 23.4: Second Order Curve Fitting

MATLAB provides a function for computing an approximate derivative, given tabulated data describing some function. This function, named `diff`, computes the difference between elements in an array. Since differentiation is defined as

$$\frac{dy}{dx} = \lim_{\Delta x \to 0} \frac{f(x + \Delta x) - f(x)}{\Delta x}$$

the derivative of $y = f(x)$ can be approximated by

$$\frac{dy}{dx} \approx \frac{\Delta y}{\Delta x} = \frac{f(x + \Delta x) - f(x)}{\Delta x}$$

which is the *forward* finite difference in y divided by the finite difference in x. Since `diff` computes differences between array elements, differentiation can be approximated in MATLAB. Continuing with the previous example, we have

```
>> dyp = polyval(dp,x);      % poly derivative for comparison

>> dy = diff(y)./diff(x);    % compute differences and use array division
>> xd = x(1:end-1);          % new x axis array since dy is shorter than y

>> plot(xd,dy,x,dyp,':')
>> ylabel('dy/dx'), xlabel('x')
>> title('Figure 23.5: Forward Difference Derivative Approximation')
```

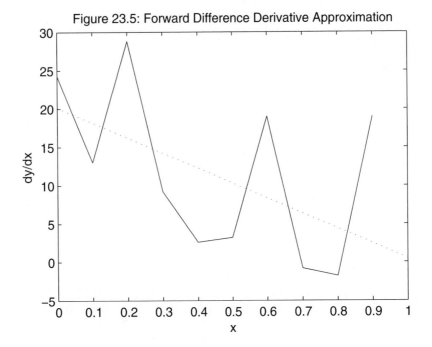

Figure 23.5: Forward Difference Derivative Approximation

     The resulting output produced by diff contains one less element than the original array. Thus, to plot the derivative, one element of the x array must be thrown out. When the first element of x is thrown out, the procedure gives a backward difference approximation, which uses information at x(n-1) and x(n) to approximate the derivative at x(n). On the other hand, throwing out the last element gives a forward difference approximation, which uses x(n+1) and x(n) to compute results at x(n). Comparing the derivative found using diff with that found from polynomial approximation, it is overwhelmingly apparent that approximating the derivative by finite differences can lead to poor results, especially if the data originates from experimental or noisy measurements. When the data used do not have uncertainty, the results of using diff can be acceptable, especially for visualization purposes, as in this example:

```
>> x = linspace(0,2*pi);
>> y = sin(x);

>> dy = diff(y)/(x(2)-x(1));
>> xd = x(2:end);

>> plot(x,y,xd,dy)
>> axis tight
>> xlabel('x'), ylabel('sin(x) and cos(x)')
>> title('Figure 23.6: Backward Difference Derivative Approximation')
```

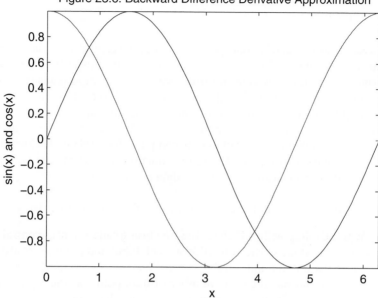

Figure 23.6: Backward Difference Derivative Approximation

Here, x was linearly spaced, so dividing by `x(2)-x(1)` gives the same answer as `diff(x)`, which is required if x is not linearly spaced. In addition, the first element in x is thrown out, making the result a backward difference derivative approximation. Visually, the derivative in this example is quite accurate. In fact, the maximum error is

```
>> max(abs(cos(xd)-dy))
ans =
    0.0317
```

When forward or backward difference approximations are not sufficient, *central* differences can be computed by performing the required array operations directly. The first central difference for equally spaced data is given by

$$\frac{dy(x_n)}{dx} \approx \frac{f(x_{n+1}) - f(x_{n-1})}{x_{n+1} - x_{n-1}}$$

Therefore, the slope at x_n is a function of its neighboring data points. Repeating the previous example gives

```
>> dy = (y(3:end)-y(1:end-2)) / (x(3)-x(1));
>> xd = x(2:end-1);
>> max(abs(cos(xd)-dy))
```

ans =

    0.00067086

In this case, the first and last data points do not have a central difference approximation, because there are no data at n = 0 and n = 101, respectively. However, at all intermediate points, the central difference approximation is nearly two orders of magnitude more accurate than the forward or backward difference approximation. The MATLAB function `gradient` implements this approach for estimating the derivative of tabulated 1-D data. Since central differences cannot be computed for the first or last data point, `gradient` simply computes forward and backward differences for these points respectively. Using the function `gradient`, the derivative of the preceding data can be estimated at each data point in x as

```
>> dy = gradient(y,x); % derivative of y at points in x.
```

When dealing with 2-D data, the function `gradient` uses central differences to compute the slope in each direction at each tabulated point. Forward differences are used at the initial points, and backward differences are used at the final points, so that the output has the same number of data points as the input. The function `gradient` is used primarily for graphical data visualization:

```
>> [x,y,z] = peaks(20);  % simple 2-D function
>> dx = x(1,2) - x(1,1); % spacing in x direction
>> dy = y(2,1) - y(1,1); % spacing in y direction
>> [dzdx,dzdy] = gradient(z,dx,dy);
```

Figure 23.7: Gradient Arrow Plot

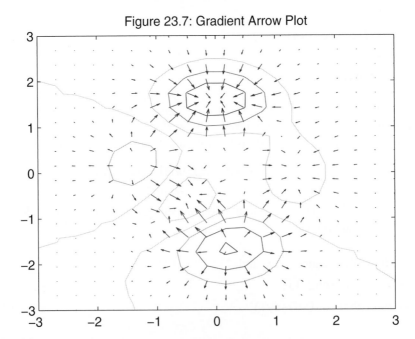

```
>> contour(x,y,z)
>> hold on
>> quiver(x,y,dzdx,dzdy)
>> hold off
>> title('Figure 23.7: Gradient Arrow Plot')
```

In this example, gradient computes dz/dx and dz/dy from the tabulated data output of peaks. These data are supplied to the function quiver, which draws arrows normal to the underlying surface with lengths scaled by the slope at each point.

   In addition to the gradient, it is sometimes useful to know the curvature of a surface. The curvature, or change in slope, at each point is calculated by the function del2, which computes the discrete approximation to the Laplacian:

$$\nabla^2 z(x,y) = \frac{d^2 z}{dx^2} + \frac{d^2 z}{dy^2}$$

In its simplest form, this value is computed by taking each surface element and subtracting from it the average of its four neighbors. If the surface is flat in each direction at a given point, the element does not change. Since MATLAB version 5, central second differences have been used at interior points to produce more accurate results. In the following example, the absolute surface curvature influences the color of the surface:

```
>> [x,y,z] = peaks;       % default output of peaks
>> dx = x(1,2) - x(1,1); % spacing in x direction
>> dy = y(2,1) - y(1,1); % spacing in y direction

>> L = del2(z,dx,dy);
>> surf(x,y,z,abs(L))
>> shading interp
>> title('Figure 23.8: Discrete Laplacian Color')
```

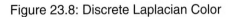

Figure 23.8: Discrete Laplacian Color

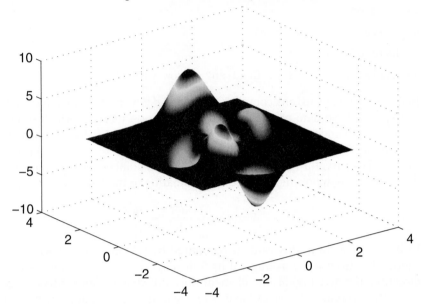

For further information regarding diff, gradient, and del2, see the MATLAB documentation.

24

Differential Equations

In 1995, MATLAB introduced a collection of M-files called the MATLAB ODE suite for solving ordinary differential equations (ODEs). With the introduction of MATLAB 5, the MATLAB ODE suite became a standard part of MATLAB. In MATLAB 6, the ODE suite added two initial value problem (IVP) solvers. In addition, functions have been added in MATLAB 6 to solve boundary value problems (BVPs) and partial differential equations (PDEs). Since MATLAB 6, the ODE suite has added the ability to solve delay differential equations (DDEs) and to solve implicit differential equations.

Taken as a whole, MATLAB now has the capability to solve a wide variety of problems involving differential equations. However, discussing each of these is beyond the scope of this text. Because initial value problems appear most often in applications, they are discussed here.

24.1 IVP FORMAT

The initial value problem solvers in MATLAB compute the time history of a set of coupled first-order differential equations with known initial conditions. In mathematical terms, these problems have the form

$$\dot{y} = f(t, y) \quad y(t_o) = y_o$$

which is vector notation for the set of differential equations

$$\dot{y}_1 = f_1(t, y_1, y_2, \ldots, y_n) \quad y_1(t_o) = y_{1o}$$
$$\dot{y}_2 = f_2(t, y_1, y_2, \ldots, y_n) \quad y_2(t_o) = y_{2o}$$
$$\vdots \qquad\qquad\qquad \vdots$$
$$\dot{y}_n = f_n(t, y_1, y_2, \ldots, y_n) \quad y_n(t_o) = y_{no}$$

where $\dot{y}_i = dy_i/dt$, n is the number of first-order differential equations, and y_{io} is the initial condition associated with the ith equation. When an initial value problem is not specified as a set of first-order differential equations, it must be rewritten as one. For example, consider the classic van der Pol equation

$$\ddot{x} - \mu(1 - x^2)\dot{x} + x = 0$$

where μ is a parameter greater than zero. If we choose $y_1 = x$ and $y_2 = dx/dt$, the van der Pol equation becomes

$$\dot{y}_1 = y_2$$
$$\dot{y}_2 = \mu(1 - y_1^2)y_2 - y_1$$

This initial value problem is used throughout this chapter to demonstrate aspects of the IVP solvers in MATLAB.

24.2 ODE SUITE SOLVERS

The MATLAB ODE suite offers eight initial value problem solvers. Each has characteristics appropriate for different initial value problems. The calling syntax for each solver is identical, making it relatively easy to change solvers for a given problem. A description of each solver is given in the following table:

Solver	Description
ode23	An explicit one-step Runge–Kutta low-order (2nd- to 3rd order) solver. Suitable for problems that exhibit mild stiffness, problems where lower accuracy is acceptable, or problems where $f(t,y)$ is not smooth (e.g., discontinuous).
ode23s	An implicit one-step modified Rosenbrock solver of order two. Suitable for stiff problems where lower accuracy is acceptable or where $f(t,y)$ is discontinuous. ***Stiff problems are generally described as problems in which the underlying time constants vary by several orders of magnitude or more.***
ode23t	An implicit, one-step trapezoidal rule using a *free* interpolant. Suitable for moderately stiff problems. Can be used to solve differential-algebraic equations (DAEs).
ode23tb	An implicit trapezoidal rule followed by a backward differentiation of order two. Similar to ode23s. Can be more efficient than ode15s for crude tolerances.
ode45	An explicit one-step Runge–Kutta medium-order (4th- to 5th order) solver. Suitable for nonstiff problems that require moderate accuracy. ***This is typically the first solver to try on a new problem.***
ode113	A multistep Adams–Bashforth–Moulton PECE solver of varying order (1st- to 13th order). Suitable for nonstiff problems that require moderate to high accuracy or where $f(t,y)$ is expensive to compute. Not suitable for problems where $f(t,y)$ is not smooth.

Solver	Description
ode15s	An implicit, multistep numerical differentiation solver of varying order (1st- to 5th order). Suitable for stiff problems that require moderate accuracy. ***This is typically the solver to try if ode45 fails or is too inefficient.***
ode15i	A solver of varying order (1st- to 5th order) for solving fully implicit differential equations.

This table uses terminology—for example, *explicit*, *implicit*, and *stiff*—that requires a substantial theoretical background to understand. If you understand the terminology, the table describes the basic properties of each solver. If you don't understand the terminology, just follow the guidelines presented in the table, and apply ode45 and ode15s, respectively, as the first and second solvers to be tried on a given problem.

It is important to note that the MATLAB ODE suite is provided as a set of M-files that can be viewed. In addition, these same solvers are included internally in SIMULINK for the simulation of dynamic systems.

24.3 BASIC USE

Before a set of differential equations can be solved, they must be coded in a function M-file as ydot = odefile(t,y). That is, the file must accept a time t and a solution y and return values for the derivatives. For the van der Pol equation, this ODE file can be written as follows:

```
function ydot = vdpol(t,y)
%VDPOL van der Pol equation.
% Ydot = VDPOL(t,Y)
% Ydot(1) = Y(2)
% Ydot(2) = mu*(1-Y(1)^2)*Y(2)-Y(1)
% mu = 2

mu = 2;
ydot = [y(2); mu*(1-y(1)^2)*y(2)-y(1)];
```

Note that the input arguments are t and y, but that this particular function does not use t. Note also that the output ydot must be a column vector.

Given the preceding ODE file, this set of ODEs is solved by using the following commands:

```
>> tspan = [0 20];    % time span to integrate over

>> yo = [2; 0];       % initial conditions (must be a column)
>> [t,y] = ode45(@vdpol,tspan,yo);
>> size(t)            % number of time points
ans =

   333    1
>> size(y)            % (i)th column is y(i) at t(i)
ans =

   333    2
>> plot(t,y(:,1),t,y(:,2),'--')
>> xlabel('time')
>> title('Figure 24.1: van der Pol Solution')
```

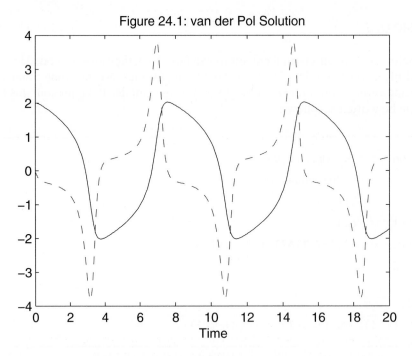

Figure 24.1: van der Pol Solution

Note that the use of function handles is appropriate in this case, since differential equations are written as M-files, and they are evaluated many times in the process of generating a solution over a reasonable time span.

By default, if a solver is called with no output arguments (e.g., ode45 (@vdpol,tspan,yo)) the solver generates no output variables, but generates a time plot similar to that in Figure 24.1. Alternatively, if a solver is called with one output argument, that argument is a structure containing all of the information needed to evaluate the solution at arbitrary time points by using the function deval:

```
>> sol = ode45(@vdpol,tspan,yo)
sol =
    solver: 'ode45'
   extdata: [1×1 struct]
         x: [1×84 double]
         y: [2×84 double]
     stats: [1×1 struct]
     idata: [1×1 struct]
```

The help text for deval provides information about the contents of the output variable sol. It is also possible to use this solution structure to extend or extrapolate the solution by using the function odextend.

In addition to specifying the initial and final time points in tspan, you can identify the desired specific solution time points by simply adding them to tspan, as in the following example:

```
>> yo = [2; 0];
>> tspan = linspace(0,20,100);

>> [t,y] = ode45(@vdpol,tspan,yo);

>> size(t)
ans =
   100      1
>> size(y)
ans =
   100      2
```

Here, 100 points are gathered over the same time interval as in the earlier example. When called in this way, the solver still uses automatic step-size control to maintain accuracy. ***The solver does not use fixed step integration.*** To gather the solution at the desired points, the solver interpolates its own solution in an efficient way that does not deteriorate the solution's accuracy.

Sometimes the set of differential equations to be solved contains a set of parameters that the user wishes to vary. Rather than open the ODE file and change the parameters before each call to a solver, the parameters can be added to the solver and ODE file input arguments:

```
function ydot = vdpol(t,y,mu)
%VDPOL van der Pol equation.
% Ydot = VDPOL(t,Y,mu)
% Ydot(1) = Y(2)
% Ydot(2) = mu*(1-Y(1)^2)*Y(2)-Y(1)

% mu = ?; now passed as an input argument

if nargin<3 % supply default if not given
   mu = 2;
end
ydot = [y(2); mu*(1-y(1)^2)*y(2)-y(1)];
```

```
>> mu = 10 % set mu in Command window
mu =
     10
>> ode45(@vdpol,tspan,yo,[],mu)
>> title('Figure 24.2: van der Pol Solution, \mu=10')
```

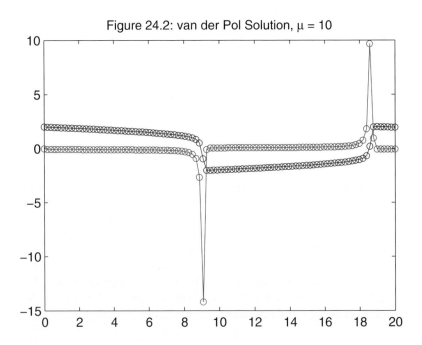

Figure 24.2: van der Pol Solution, $\mu = 10$

The preceding code solves the van der Pol equation with $\mu = 10$. Here, mu is added as a third input argument to vdpol.m. Then the differential equations are solved by adding two input arguments to the ode45 function call. The first added argument is an empty array; this tells the function to use default solver options. The second added argument contains the value of the parameter. Since no output arguments were specified, the plotted solution is automatically created. While the above example describes the use of one parameter, any number of parameters can be added after the first one shown. In addition, each parameter can be an array of any data type.

24.4 SETTING OPTIONS

Up to this point, we have just accepted all default tolerances and options. When these are not sufficient, an options structure can be passed to a solver as a fourth input argument. The MATLAB ODE suite contains the functions odeset and odeget to manage this options structure. The function odeset works similarly to the Handle Graphics set function, in that parameters are specified in name/value pairs—for example, options = odeset('Name1',Value1,'Name2',Value2,...);. The available parameter names and values are described in the online help for odeset:

```
>> help odeset
 ODESET Create/alter ODE OPTIONS structure.
    OPTIONS = ODESET('NAME1',VALUE1,'NAME2',VALUE2,...) creates an integrator
    options structure OPTIONS in which the named properties have the
    specified values. Any unspecified properties have default values. It is
    sufficient to type only the leading characters that uniquely identify the
    property. Case is ignored for property names.
    OPTIONS = ODESET(OLDOPTS,'NAME1',VALUE1,...) alters an existing options
    structure OLDOPTS.
    OPTIONS = ODESET(OLDOPTS,NEWOPTS) combines an existing options structure
    OLDOPTS with a new options structure NEWOPTS. Any new properties
    overwrite corresponding old properties.
    ODESET with no input arguments displays all property names and their
    possible values.
 ODESET PROPERTIES
 RelTol - Relative error tolerance [ positive scalar {1e-3} ]
    This scalar applies to all components of the solution vector, and
    defaults to 1e-3 (0.1% accuracy) in all solvers. The estimated error in
    each integration step satisfies e(i) <= max(RelTol*abs(y(i)),AbsTol(i)).
```

AbsTol - Absolute error tolerance [positive scalar or vector {1e-6}]

    A scalar tolerance applies to all components of the solution vector.
Elements of a vector of tolerances apply to corresponding components of
the solution vector. AbsTol defaults to 1e-6 in all solvers. See RelTol.

NormControl - Control error relative to norm of solution [on | {off}]

    Set this property 'on' to request that the solvers control the error in
each integration step with norm(e) <= max(RelTol*norm(y),AbsTol). By
default the solvers use a more stringent component-wise error control.

Refine - Output refinement factor [positive integer]

    This property increases the number of output points by the specified
factor producing smoother output. Refine defaults to 1 in all solvers
except ODE45, where it is 4. Refine does not apply if length(TSPAN) > 2
or the ODE solver returns the solution as a structure.

OutputFcn - Installable output function [function_handle]

    This output function is called by the solver after each time step. When
a solver is called with no output arguments, OutputFcn defaults to
@odeplot. Otherwise, OutputFcn defaults to [].

OutputSel - Output selection indices [vector of integers]

    This vector of indices specifies which components of the solution vector
are passed to the OutputFcn. OutputSel defaults to all components.

Stats - Display computational cost statistics [on | {off}]

Jacobian - Jacobian function [function_handle | constant matrix]

    Set this property to @FJac if FJac(t,y) returns dF/dy, or to
the constant value of dF/dy.

    For ODE15I solving F(t,y,y') = 0, set this property to @FJac if
[dFdy, dFdyp] = FJac(t,y,yp), or to a cell array of constant
values {dF/dy,dF/dyp}.

JPattern - Jacobian sparsity pattern [sparse matrix]

    Set this property to a sparse matrix S with S(i,j) = 1 if component i of
F(t,y) depends on component j of y, and 0 otherwise.

    For ODE15I solving F(t,y,y') = 0, set this property to
{dFdyPattern,dFdypPattern}, the sparsity patterns of dF/dy and
dF/dy', respectively.

Vectorized - Vectorized ODE function [on | {off}]

    Set this property 'on' if the ODE function F is coded so that

```
F(t,[y1 y2 ...]) returns [F(t,y1) F(t,y2) ...].
For ODE15I solving F(t,y,y') = 0, set this property to
{yVect,ypVect}. Setting yVect 'on' indicates that
F(t,[y1 y2 ...],yp) returns [F(t,y1,yp) F(t,y2,yp) ...].
Setting ypVect 'on' indicates that F(t,y,[yp1 yp2 ...])
returns [F(t,y,yp1) F(t,y,yp2) ...].
```

Events - Locate events [function_handle]

To detect events, set this property to the event function.

Mass - Mass matrix [constant matrix | function_handle]

For problems M*y' = f(t,y) set this property to the value of the constant mass matrix. For problems with time- or state-dependent mass matrices, set this property to a function that evaluates the mass matrix.

MStateDependence - Dependence of the mass matrix on y [none | {weak} | strong]

Set this property to 'none' for problems M(t)*y' = F(t,y). Both 'weak' and 'strong' indicate M(t,y), but 'weak' will result in implicit solvers using approximations when solving algebraic equations.

MassSingular - Mass matrix is singular [yes | no | {maybe}]

Set this property to 'no' if the mass matrix is not singular.

MvPattern - dMv/dy sparsity pattern [sparse matrix]

Set this property to a sparse matrix S with S(i,j) = 1 if for any k, the (i,k) component of M(t,y) depends on component j of y, and 0 otherwise.

InitialSlope - Consistent initial slope yp0 [vector]

yp0 satisfies M(t0,y0)*yp0 = F(t0,y0).

InitialStep - Suggested initial step size [positive scalar]

The solver will try this first. By default the solvers determine an initial step size automatically.

MaxStep - Upper bound on step size [positive scalar]

MaxStep defaults to one-tenth of the tspan interval in all solvers.

NonNegative - Non-negative solution components [vector of integers]

This vector of indices specifies which components of the solution vector must be non-negative. NonNegative defaults to []. This property is not available in ODE23S, ODE15I. In ODE15S, ODE23T, and ODE23TB, the property is not available for problems where there is a mass matrix.

BDF - Use Backward Differentiation Formulas in ODE15S [on | {off}]

This property specifies whether the Backward Differentiation Formulas (Gear's methods) are to be used in ODE15S instead of the default Numerical Differentiation Formulas.

MaxOrder - Maximum order of ODE15S and ODE15I [1 | 2 | 3 | 4 | {5}]

See also odeget, ode45, ode23, ode113, ode15i, ode15s, ode23s, ode23t, ode23tb, function_handle.

NOTE:

Some of the properties available through ODESET have changed in MATLAB 6.0. Although we still support the v5 properties when used with the v5 syntax of the ODE solvers, any new functionality will be available only with the new syntax. To see the properties available in v5, type in the command line more on, type odeset, more off

Reference page in Help browser

doc odeset

The following examples use the preceding options:

```
>> odeset
            AbsTol: [ positive scalar or vector {1e-6} ]
            RelTol: [ positive scalar {1e-3} ]
       NormControl: [ on | {off} ]
       NonNegative: [ vector of integers ]
         OutputFcn: [ function_handle ]
         OutputSel: [ vector of integers ]
            Refine: [ positive integer ]
             Stats: [ on | {off} ]
       InitialStep: [ positive scalar ]
           MaxStep: [ positive scalar ]
               BDF: [ on | {off} ]
          MaxOrder: [ 1 | 2 | 3 | 4 | {5} ]
          Jacobian: [ matrix | function_handle ]
          JPattern: [ sparse matrix ]
        Vectorized: [ on | {off} ]
              Mass: [ matrix | function_handle ]
   MStateDependence: [ none | {weak} | strong ]
         MvPattern: [ sparse matrix ]
```

```
            MassSingular:  [ yes | no | {maybe} ]
            InitialSlope:  [ vector ]
                  Events:  [ function_handle ]
```

Invoking odeset without input or output arguments returns an option listing, their possible values, and default values in braces:

```
>> tspan = [0 20]; % set time span to solve
>> yo = [2; 0];    % intial conditions
>> mu = 10;        % parameter mu
>> options = odeset('AbsTol',1e-12,'RelTol',1e-6);

>> [t,y] = ode45(@vdpol,tspan,yo,[],mu); % default tolerances
>> length(t)
ans =
   593
>> [t,y] = ode45(@vdpol,tspan,yo,options,mu); % tight tolerances
>> length(t)
ans =
            1689

>> [t,y] = ode15s(@vdpol,tspan,yo,[],mu); % new solver, default tols
>> length(t)
ans =
   232
>> [t,y] = ode15s(@vdpol,tspan,yo,options,mu); % new solver, tight tols
>> length(t)
ans =
   651
```

Here, the steps needed to integrate the first 20 seconds are shown. Default tolerances (i.e., AbsTol = 1e-6 and RelTol = 1e-3) force ode45 to take 593 time steps. Decreasing the tolerances to AbsTol = 1e-12 and RelTol = 1e-6 takes 1689 time steps. However, changing to the stiff solver ode15s requires only 651 steps at the tighter tolerances.

The stiff solvers ode15s, ode23s, ode23t, and ode23tb allow you to specify an analytical Jacobian, rather than a numerically computed approximation, which is the default. The Jacobian is a matrix of partial derivatives that has the form

$$\begin{bmatrix} \dfrac{\partial f_1}{\partial y_1} & \dfrac{\partial f_1}{\partial y_2} & \cdots & \dfrac{\partial f_1}{\partial y_n} \\ \dfrac{\partial f_2}{\partial y_1} & \dfrac{\partial f_2}{\partial y_2} & \cdots & \dfrac{\partial f_2}{\partial y_n} \\ \vdots & \vdots & \ddots & \vdots \\ \dfrac{\partial f_n}{\partial y_1} & \dfrac{\partial f_n}{\partial y_2} & \cdots & \dfrac{\partial f_n}{\partial y_n} \end{bmatrix}$$

This matrix or a numerical approximation of it is used by the stiff solvers to compute the solution of a set of nonlinear equations at each time step. If at all possible, an analytical Jacobian should be supplied, as shown in the following code:

```
function jac = vdpoljac(t,y,mu)
%VDPOLJAC van der Pol equation Jacobian.

% mu = ?; passed as an input argument
if nargin<3 % supply default if not given
   mu = 2;
end
jac = [        0                      1
        (-2*mu*y(1)*y(2)-1) (mu*(1-y(1)^2))];
```

```
>> options = odeset(options,'Jacobian',@vdpoljac);
>> [t,y] = ode15s(@vdpol,tspan,yo,options,mu);
>> length(t)
ans =
    670
```

In this example, the function handle of the M-file that computes the Jacobian is added to the options structure set discussed earlier. Running the solver now shows that it takes 670 steps. While this is more than the 651 steps shown earlier, providing the analytical Jacobian still significantly increases execution speed.

When it is not possible to supply an analytical Jacobian, it is beneficial to supply a *vectorized* ODE file. Vectorizing the ODE file usually means replacing y(i) with y(i,:) and using array operators. Doing so allows computation of the numerical Jacobian to proceed as fast as possible, as in this example:

```
function ydot = vdpol(t,y,mu)
%VDPOL van der Pol equation.
% Ydot = VDPOL(t,Y)
% Ydot(1) = Y(2)
% Ydot(2) = mu*(1-Y(1)^2)*Y(2)-Y(1)

% mu = ?; now passed as an input argument
if nargin<3 % supply default if not given
   mu = 2;
end
ydot = [y(2,:); mu*(1-y(1,:)^2)*y(2,:)-y(1,:)];
```

The 'Refine' property determines how much output data to generate. It does not affect the step sizes chosen by the solvers, or the solution accuracy. It merely dictates how many intermediate points to interpolate the solution at within each integration step, as in the following example:

```
>> options = odeset('Refine',1);
>> [t,y] = ode45(@vdpol,tspan,yo,options,mu);
>> length(t) % # of time points
ans =
    149
>> options = odeset('Refine',4);
>> [t,y] = ode45(@vdpol,tspan,yo,options,mu);
>> length(t) % # of time points
ans =
    593
```

With 'Refine' set to 1, 149 time points are returned over the 20-second time span. Setting 'Refine' to 4 increases the number of time points returned to 593.

The 'Events' property allows you to flag one or more events that occur as an ODE solution evolves in time. For example, a simple event could be when some solution component reaches a maximum, a minimum, or crosses through zero. Optionally, the occurrence of an event can force a solver to stop integrating. To make use of this feature, you must simply supply a function handle that computes the values of the events to be tracked. At each time step, the solver computes the events and marks in time those that cross through zero. For example,

```
function [value,isterminal,direction] = vdpolevents(t,y,mu)
%VDPOLEVENTS van der Pol equation events.
value(1) = abs(y(2))-2; % find where |y(2)|=2
isterminal(1) = 0;        % don't stop integration
direction(1) = 0;         % don't care about crossing direction
```

```
>> mu = 2;
>> options = odeset('Events',@vdpolevents);
>> [t,y,te,ye] = ode45(@vdpol,tspan,yo,options,mu);
>> plot(t,y,te,ye(:,2),'o')
>> title('Figure 24.3: van der Pol Solution, |y(2)|=2')
```

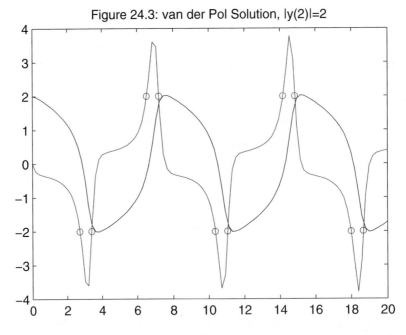

The function vdpolevents receives the same three arguments as the other functions and returns three numerical vectors. The first is the value of the events, the second is a logical array dictating whether the solver should terminate execution on a zero crossing in one or more of the computed events, and the third gives the user the ability to specify whether the direction of event crossing should be considered. In the preceding example, the points where $abs(y_2(t)) = 2$ are selected as the only event. The solver is told to not halt on sensing events and to not care about the direction of event crossing. The generated plot shows the system solution and the points where the chosen events occur.

24.5 BVPs, PDEs, AND DDEs

In addition to the eight MATLAB solvers for solving initial value ordinary differential equations, MATLAB includes functions for the solution of boundary value problems, partial differential equations, and delay differential equations. The functions used to solve BVPs are shown in the following table:

Function	Description
bvp4c	BVP solver
bvp5c	BVP solver (more efficient for small error tolerances)
bvpextend	Forms a guess structure for extending a BVP solution
bvpget	Gets BVP options structure
bvpinit	Forms the initial solution guess, which is refined by bvp4c or bvp5c
bvpset	Sets the BVP options structure
deval	Evaluates/interpolates the solution found using bvp4c or bvp5c

(Further information regarding the solution of BVPs can be found in the MATLAB documentation.)

The functions used to solve PDEs are shown in this table:

Function	Description
pdepe	Solves IVPs for parabolic-elliptic PDEs in one dimension
pdeval	Evaluates/interpolates the solution found using pdepe

(Further information regarding the solution of PDEs can be found in the online documentation.)

The functions used to solve DDEs are shown in this table:

Function	Description
dde23	Solves DDE initial value problems with constant delays
ddesd	Solves DDE initial value problems with general delays
deval	Evaluates/interpolates the solution found using dde23 or ddesd
ddeget	Gets DDE options from options structure
ddeset	Creates or alters options structure for dde23 and ddesd

(Further information regarding the solution of DDEs can be found in the MATLAB documentation.)

25

Two-Dimensional Graphics

Throughout this text, several of MATLAB's graphics features have been introduced. In this and the next several chapters, the graphics features in MATLAB are more rigorously illustrated. Many of the features and capabilities illustrated here are available as menu items from the top of a *Figure* window. They are also available as buttons on the *Figure* or *Camera* toolbar, which appears by default when plots are generated or can be chosen via the **View** menu in a *Figure* window.

The general rule is to use the toolbar and menu features of *Figure* windows if you want to customize a single *Figure*. Otherwise, use *Command* window functions to automate the process of customizing plots. This text concentrates on the *Command* window functions, since they perform the actions taken when menu and toolbar items are used.

25.1 THE plot FUNCTION

As you have seen in earlier examples, the most common function for plotting 2-D data is the plot function. This versatile function plots sets of data arrays on appropriate axes and connects the points with straight lines:

```
>> x = linspace(0,2*pi,30);
>> y = sin(x);
>> plot(x,y), title('Figure 25.1: Sine Wave')
```

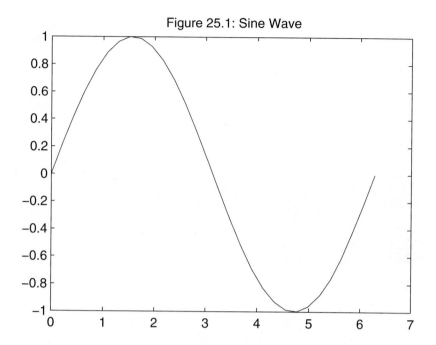

This example creates 30 data points over $0 \leq x \leq 2\pi$ to form the horizontal axis of the plot and creates another vector y containing the sine of the data points in x. The plot function opens a graphics window, called a *Figure* window, scales the axes to fit the data, plots the points, and then connects the points with straight lines. It also adds numerical scales and tick marks to the axes automatically. If a *Figure* window already exists, plot generally clears the current *Figure* window and draws a new plot.

The following code plots more than one curve or line:

```
>> z = cos(x);
>> plot(x,y,x,z)
>> title('Figure 25.2: Sine and Cosine')
```

Just giving plot another pair of arguments instructs it to generate a second line. This time, $\sin(x)$ versus x, and $\cos(x)$ versus x were plotted on the same plot. Although the figure doesn't show color, plot automatically draws the second curve in a different color. The function plot generates as many curves as it receives pairs of input arguments.

If one of the arguments is a matrix and the other a vector, the plot function plots each column of the matrix versus the vector. For example, the code

```
>> W = [y;z]; % create a matrix of the sine and cosine
>> plot(x,W) % plot the columns of W vs. x
```

reproduces the preceding plot.

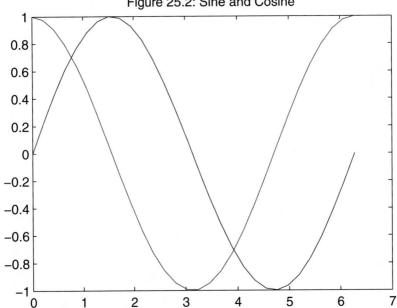

Figure 25.2: Sine and Cosine

If you change the order of the arguments, the orientation of the plot changes accordingly:

```
>> plot(W,x) % plot x vs. the columns of W
title('Figure 25.3: Change Argument Order')
```

When the plot function is called with only one argument (e.g., plot(Y)) the function acts differently, depending on the data contained in Y. If Y is a ***complex-valued*** vector, plot(Y) is interpreted as plot(real(Y),imag(Y)). In all other cases, the imaginary components of the input vectors are *ignored*. On the other hand, if Y is ***real-valued***, then plot(Y) is interpreted as plot(1:length(Y),Y); that is, Y is plotted versus an index of its values. When Y is a matrix, this interpretation is applied to each column of Y.

25.2 LINESTYLES, MARKERS, AND COLORS

In the previous examples, MATLAB chose the solid linestyle, and the colors blue and green for the plots. You can specify your own colors, markers, and linestyles by giving plot a third argument after each pair of data arrays. This optional argument is a character string consisting of one or more characters from the following table:

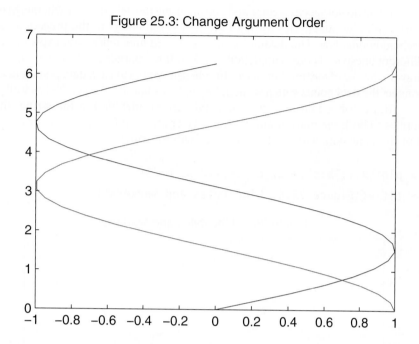

Figure 25.3: Change Argument Order

Symbol	Color	Symbol	Marker	Symbol	Linestyle
b	Blue	.	Point	–	Solid line
g	Green	o	Circle	:	Dotted line
r	Red	x	Cross	-.	Dash-dot line
c	Cyan	+	Plus sign	––	Dashed line
m	Magenta	*	Asterisk	none	No line
y	Yellow	s	Square		
k	Black	d	Diamond		
w	White	v	Triangle (down)		
		^	Triangle (up)		
		<	Triangle (left)		
		>	Triangle (right)		
		p	Pentagram		
		h	Hexagram		
		none	No marker		

If you do not specify a color and you are using the default color scheme, MATLAB starts with blue and cycles through the first seven colors in the preceding table for each additional line. The default linestyle is a solid line, unless you explicitly specify a different linestyle. There is no default marker. If no marker is selected, no markers are drawn. The use of any marker places the chosen symbol at each data point, but does not connect the data points with a straight line unless a linestyle is specified as well.

If a color, a marker, and a linestyle are all included in the string, the color applies to both the marker and the line. To specify a different color for the marker, plot the same data with a different specification string:

```
>> plot(x,y,'b:p',x,z,'c-',x,1.2*z,'m+')
>> title('Figure 25.4: Linestyles and Markers')
```

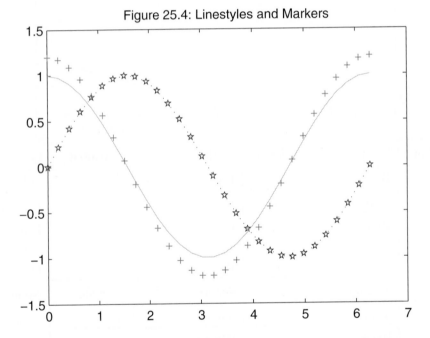

Figure 25.4: Linestyles and Markers

As with many of the plots in this section, your computer displays color, but the figures shown here do not. If you are following along in MATLAB, just enter the commands listed in the examples to see the effects of color.

25.3 PLOT GRIDS, AXES BOX, AND LABELS

The grid on command adds grid lines to the current plot at the tick marks. The grid off command removes the grid. The command grid with no arguments alternately turns the grid lines on and off—that is, it *toggles* them. By default, MATLAB starts up with grid off for most plots. If you like to have grid lines on all of your plots by default, add the following lines to your startup.m file:

```
set(0,'DefaultAxesXgrid','on')
set(0,'DefaultAxesYgrid','on')
set(0,'DefaultAxesZgrid','on')
```

These lines illustrate the use of Handle Graphics features in MATLAB and the setting of default behavior. (More information on these topics can be found in Chapter 30.)

Normally, 2-D axes are fully enclosed by solid lines called an ***axes box***. This box can be turned off with box off. The command box on restores the axes box. The box command toggles the state of the axes box. Horizontal and vertical axes can be labeled with the xlabel and ylabel functions, respectively. The title function adds a line of text at the top of the plot. The following example is illustrative:

```
>> x = linspace(0,2*pi,30);
>> y = sin(x);
>> z = cos(x);
>> plot(x,y,x,z)
>> box off                              % turn off the axes box
>> xlabel('Independent Variable X')      % label horizontal axis
>> ylabel('Dependent Variables Y and Z')  % label vertical axis
>> title('Figure 25.5: Sine and Cosine Curves, No Box') % title
```

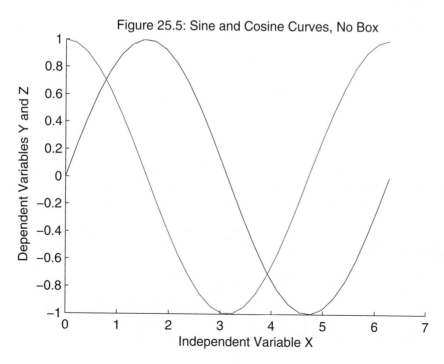

Figure 25.5: Sine and Cosine Curves, No Box

You can add a label or any other text string to any specific location on your plot with the text function. The syntax for text is text(x,y,'string'), where (x,y) represents the coordinates of the center left edge of the text string in units taken from the plot axes. For example, the following code segment places the text 'sin(x)' at the location $x = 2.5$, $y = 0.7$:

```
>> grid on, box on % turn axes box and grid lines on
>> text(2.5,0.7,'sin(x)')
>> title('Figure 25.6: Sine and Cosine Curves, Added Label')
```

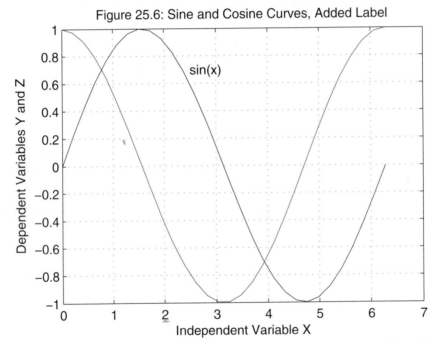

Figure 25.6: Sine and Cosine Curves, Added Label

If you want to add a label, but don't want to stop to figure out the coordinates to use, you can place a text string with the mouse. The gtext('text') function switches to the current *Figure* window, puts up a crosshair that follows the mouse, and waits for a mouse click or key press. When either one occurs, the string argument to gtext is placed with the lower left corner of the first character at that location.

25.4 CUSTOMIZING PLOT AXES

MATLAB gives you complete control over the scaling and appearance of both the horizontal and vertical axes of your plot with the axis command. Because this command has so many features, only the most useful are described here. The primary features of the axis command are given in the following table:

Command	Description
`axis([xmin xmax ymin ymax])`	Sets axis limits on the current plot
`V = axis`	Returns a row vector containing the current axis limits
`axis auto`	Returns axis scaling to automatic defaults
`axis manual`	Freezes axis scaling so that if `hold` is on, subsequent plots use the same axis limits
`axis tight`	Sets axis limits to the range of the plotted data
`axis fill`	Sets the axis limits and aspect ratio so that the axis fills the allotted space. This option has an effect only if `PlotBoxAspectRatio` or `DataAspectRatioMode` is `'manual'`
`axis ij`	Puts axis in *matrix* mode. The horizontal axis increases from left to right. The vertical axis increases from top to bottom
`axis xy`	Puts axis in Cartesian mode. The horizontal axis increases from left to right. The vertical axis increases from bottom to top
`axis equal`	Sets the aspect ratio so that equal tick mark increments on each axis are equal in size
`axis image`	Sets axis limits appropriate for displaying an image
`axis square`	Makes the axis box square
`axis normal`	Restores the current axis box to full size and removes any restrictions on unit scaling
`axis vis3d`	Freezes the aspect ratio to enable rotation of 3-D objects without axis size changes
`axis off`	Turns off all axis labeling, tick marks, and background
`axis on`	Turns on all axis labeling, tick marks, and background

Multiple commands to `axis` can be given at once. For example, `axis auto on xy` is the default axis scaling. The `axis` command affects only the current plot. Therefore, it is issued after the `plot` command, just as `grid`, `xlabel`, `ylabel`, `title`, `text`, and so on, are issued after the plot is on the screen:

```
>> x = linspace(0,2*pi,30);
>> y = sin(x);
>> plot(x,y)
>> title('Figure 25.7: Fixed Axis Scaling')
>> axis([0 2*pi -1.5 2]) % change axis limits
```

Figure 25.7: Fixed Axis Scaling

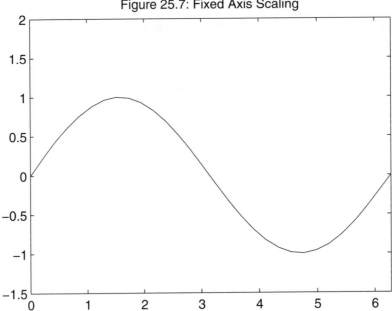

Note that by specifying the maximum x-axis value to be 2*pi, the plot axis ends at exactly 2*pi rather than rounding the axis limit up to 7. The simplest way to see what the various axis command arguments do is to generate a simple plot, then issue multiple axis commands, and, finally, view the resulting changes.

When you simply want to change the axis limits on a single axis, the axis command is cumbersome because it requires you to enter limits for all axes. To solve this problem, MATLAB provides the functions xlim, ylim, and zlim, which are described by the help text for xlim, with the obvious change for ylim and zlim:

```
>> help xlim
XLIM X limits.
    XL = XLIM                gets the x limits of the current axes.
    XLIM([XMIN XMAX])        sets the x limits.
    XLMODE = XLIM('mode')    gets the x limits mode.
    XLIM(mode)               sets the x limits mode.
                             (mode can be 'auto' or 'manual')
    XLIM(AX,...)             uses axes AX instead of current axes.
    XLIM sets or gets the XLim or XLimMode property of an axes.
    See also pbaspect, daspect, ylim, zlim.
    Reference page in Help browser
    doc xlim
```

25.5 MULTIPLE PLOTS

You can add new plots to an existing plot by using the `hold` command. When you enter `hold on`, MATLAB does not remove the existing axes when new `plot` functions are issued. Instead, it adds new curves to the current axes. However, if the new data does not fit within the current axes limits, the axes are rescaled. Entering `hold off` releases the current *Figure* window for new plots. The `hold` command, without arguments, toggles the hold setting:

```
>> x = linspace(0,2*pi,30);
>> y = sin(x);
>> z = cos(x);
>> plot(x,y)
>> hold on
>> ishold % return 1 (True) if hold is ON
ans =
     1
>> plot(x,z,'m')
>> hold off
>> ishold % hold is no longer ON
ans =
     0
>> title 'Figure 25.8: Use of hold command'
```

Notice that this example specifies the color of the second curve. Since there is only one set of data arrays in each `plot` function, the line color for each `plot` function would otherwise default to the first color in the color order list, resulting in two lines plotted in the same color. Note also that the title text is not enclosed in parentheses, but the effect remains unchanged. In the alternative form shown, `title` is interpreted as a command rather than a function. In addition, the command `hold all` holds both the current plot and the next line color to be used. Therefore, if `hold on` in the previous example is replaced by `hold all`, and no explicit color is specified for the second plot, then, by default, `plot(x,z)` would have produced a green line.

25.6 MULTIPLE FIGURES

It is possible to create multiple *Figure* windows and plot different data sets in different ways in each one. To create new *Figure* windows, use the `figure` command in the *Command* window or the **New Figure** selection from the **File** menu in the *Command* or *Figure* window. You can choose a specific *Figure* window to be the active, or current, figure by clicking on it with the mouse or by using `figure(n)`,

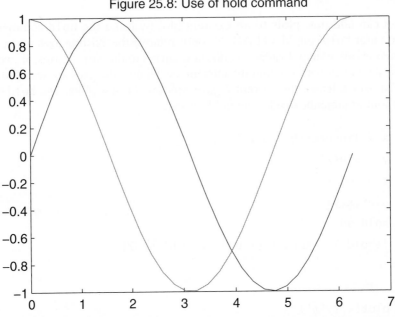

Figure 25.8: Use of hold command

where n is the number of the window. The current *Figure* window is the window that is active for subsequent plotting functions.

Every time a new *Figure* window is created, a number identifying it—that is, its **handle**—is returned and stored for future use. The figure handle is also displayed in the *Figure* window title bar. When a new *Figure* window is created, it is placed in the default figure position on the screen. As a result, when more than one *Figure* window is created, each new window covers all preceding *Figure* windows. To see the windows simultaneously, simply drag them around by using the mouse on the *Figure* window title bar.

To reuse a *Figure* window for a new plot, it must be made the active, or current, figure. Clicking on the figure of choice with the mouse makes it the current figure. From within MATLAB, figure(h), where h is the figure handle, makes the corresponding figure active or current. Only the current figure is responsive to the axis, hold, xlabel, ylabel, title, and grid commands.

Figure windows can be deleted by closing them with the mouse, similar to the way you may close windows on your computer. Alternatively, the command close can be issued. For example,

```
>> close
```

closes the current *Figure* window,

```
>> close(h)
```

closes the *Figure* window having handle h, and

```
>> close all
```

closes all *Figure* windows.

If you simply want to erase the contents of a *Figure* window without closing it, use the command clf. For example,

```
>> clf
```

clears the current *Figure* window, and

```
>> clf reset
```

clears the current *Figure* window and resets all properties, such as hold, to their default states.

25.7 SUBPLOTS

One *Figure* window can hold more than one set of axes. The subplot(m,n,p) command subdivides the current *Figure* window into an m-by-n matrix of plotting areas and chooses the pth area to be active. The subplots are numbered left to right along the top row, then along the second row, and so on. The following code is illustrative:

```
>> x = linspace(0,2*pi,30);
>> y = sin(x);
>> z = cos(x);
>> a = 2*sin(x).*cos(x);
>> b = sin(x)./(cos(x)+eps);

>> subplot(2,2,1) % pick the upper left of a 2-by-2 grid of subplots
>> plot(x,y), axis([0 2*pi -1 1]), title('Figure 25.9a: sin(x)')

>> subplot(2,2,2) % pick the upper right of the 4 subplots
>> plot(x,z), axis([0 2*pi -1 1]), title('Figure 25.9b: cos(x)')

>> subplot(2,2,3) % pick the lower left of the 4 subplots
>> plot(x,a), axis([0 2*pi -1 1]), title('Figure 25.9c: 2sin(x)cos(x)')

>> subplot(2,2,4) % pick the lower right of the 4 subplots
>> plot(x,b), axis([0 2*pi -20 20]), title('Figure 25.9d: sin(x)/cos(x)')
```

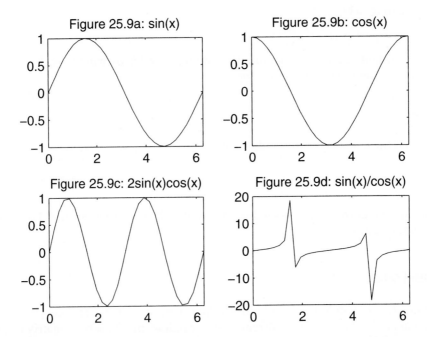

Note that when a particular subplot is active, it is the only subplot or axis that is responsive to the axis, hold, xlabel, ylabel, title, grid, and box commands. The other subplots are not affected. In addition, the active subplot remains active until another subplot or figure command is issued. When a new subplot command changes the number of subplots in the *Figure* window, previous subplots are erased to make room for the new orientation. To return to the default mode and use the entire *Figure* window for a single set of axes, use the command subplot(1,1,1). When you print a *Figure* window containing multiple plots, all of them are printed on the same page. For example, when the current *Figure* window contains four subplots and the orientation is landscape mode, each of the plots uses one-quarter of the printed page.

25.8 INTERACTIVE PLOTTING TOOLS

Before the *Figure* window menu bar and toolbars existed, MATLAB offered several functions for annotating plots interactively. These functions are described in this section, and most are available from the *Figure* menu bar and toolbars.

Rather than using individual text strings, a legend can be used to identify the data sets on your plot. The legend command creates a legend box on the plot, keying any text you supply to each line in the plot. If you wish to move the legend, simply click and hold down the mouse button on the legend and drag the legend to the desired location. The command legend off deletes the legend. The following example is illustrative:

```
>> close % close figure containing subplots
>> x = linspace(0,2*pi,30);
```

```
>> y = sin(x);
>> z = cos(x);
>> plot(x,y,x,z)
>> legend('sin(x)','cos(x)')
>> title('Figure 25.10: Legend Example')
```

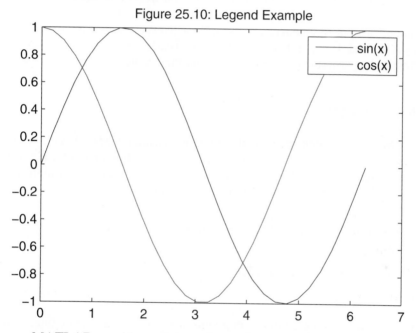

Figure 25.10: Legend Example

MATLAB provides an interactive tool to expand sections of a 2-D plot to see more detail or to ***zoom in*** on a region of interest. The command zoom on turns on the zoom mode. Clicking the left mouse button within the *Figure* window expands the plot by a factor of 2 centered around the point under the mouse pointer. Each time you click, the plot expands. Double-clicking zooms back out. Click the right mouse button to open a contextual menu to zoom out or select other properties. You can also click and drag a rectangular area to zoom in on a specific area. The zoom out command returns the plot to its initial state. The command zoom off turns off the zoom mode. Just zoom with no arguments toggles the zoom state of the active *Figure* window. The *Figure* toolbar and *Figure* window menus offer a GUI approach to implementing this feature as well.

In some situations, it is convenient to select coordinate points from a plot in a *Figure* window. In MATLAB, this feature is embodied in the ginput function. The form [x,y] = ginput(n) gets n points from the current plot or subplot based on mouse click positions within the plot or subplot. If you press the **Return** or **Enter** key before all n points are selected, ginput terminates with fewer points. The points returned in the vectors x and y are the respective *x* and *y* data coordinate points selected. The returned data are not necessarily points from the plotted data, but rather the explicit *x*- and *y*-coordinate values where the mouse was clicked. If points are selected outside of the plot or subplot axes limits—for example, outside the plot box—the points returned are extrapolated values.

This function can be somewhat confusing when used in a *Figure* window containing subplots. The data returned is with respect to the current or active subplot. Thus, if ginput is issued after a subplot(2,2,3) command, the data returned is with respect to the axes of the data plotted in subplot(2,2,3). If points are selected from other subplots, the data is still with respect to the axes of the data in subplot(2,2,3). When an unspecified number of data points are desired, the form [x,y] = ginput without an input argument can be used. Here, data points are gathered until the **Return** key is pressed. The gtext function (described earlier in this chapter) uses the function ginput along with the function text for placing text with the mouse.

25.9 SCREEN UPDATES

Because screen rendering is relatively time-consuming, MATLAB does not always update the screen after each graphics command. For example, if the following commands are entered at the MATLAB prompt, MATLAB updates the screen after each graphics command (plot, axis, and grid):

```
>> x = linspace(0,2*pi); y = sin(x);
>> plot(x,y)
>> axis([0 2*pi -1.2 1.2])
>> grid
```

However, if the same graphics commands are entered on a single line, such as

```
>> plot(x,y), axis([0 2*pi -1.2 1.2]), grid
```

MATLAB renders the figure only once—when the MATLAB prompt reappears. A similar procedure occurs when graphics commands appear as part of a script or function M-file. In this case, even if the commands appear on separate lines in the file, the screen is rendered only once—when all commands are completed and the MATLAB prompt reappears.

In general, six events cause MATLAB to render the screen:

1. A return to the MATLAB prompt
2. Encountering a function that temporarily stops execution, such as pause, keyboard, input, ginput, waitfor, and waitforbuttonpress
3. Execution of a getframe command
4. Execution of a drawnow command
5. Execution of a figure command
6. Resizing of a *Figure* window

Of these, only the `drawnow` command specifically allows one to force MATLAB to update the screen at arbitrary times. The `refresh` command completely erases and redraws the current figure.

25.10 SPECIALIZED 2-D PLOTS

Up to this point, the basic plotting function `plot` has been illustrated. In many situations, plotting lines or points on linearly scaled axes does not convey the desired information. As a result, MATLAB offers other basic 2-D plotting functions, as well as specialized plotting functions that are embodied in function M-files.

In addition to `plot`, MATLAB provides the functions `semilogx` for plotting with a logarithmically scaled x-axis, `semilogy` for plotting with a logarithmically scaled y-axis, and `loglog` for plotting with both axes logarithmically spaced. All of the features discussed previously with respect to the function `plot` apply to these functions as well.

The `area` function is useful for building a stacked area plot. The function `area(x,y)` is the same as `plot(x,y)` for vectors x and y, except that the area under the plot is filled in with color. The lower limit for the filled area may be specified, but defaults to zero. To stack areas, use the form `area(X,Y)`, where Y is a matrix and X is a matrix or vector whose length equals the number of rows in Y. If X is omitted, as in the following example, `area` uses the default value X = `1:size(Y,1)`:

```
>> z = -pi:pi/5:pi;
>> area([sin(z);cos(z)])
>> title('Figure 25.11: Stacked Area Plot')
```

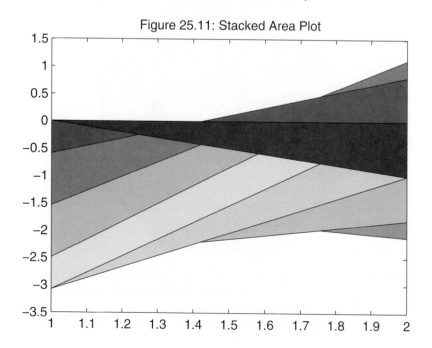

Figure 25.11: Stacked Area Plot

Filled polygons can be drawn by using the `fill` function. The function `fill(x,y,'c')` fills the 2-D polygon defined by the column vectors x and y with the color specified by c. The vertices of the polygon are specified by the pairs (xi,yi). If necessary, the polygon is closed by connecting the last vertex to the first. Like the `plot` function, `fill` can have any number of pairs of vertices and associated colors. Moreover, when x and y are matrices of the same dimension, the columns of x and y are assumed to describe separate polygons. In the code

```
>> t = (1:2:15)'*pi/8;
>> x = sin(t);
>> y = cos(t);
>> fill(x,y,'r') % a filled red circle using only 8 data points
>> axis square off
>> text(0,0,'STOP', ...
   'Color',[1 1 1], ...
   'FontSize',80, ...
   'FontWeight','bold', ...
   'HorizontalAlignment','center')
>> title('Figure 25.12: Stop Sign')
```

Figure 25.12: Stop Sign

the `text(x,y,'string')` function is used with extra arguments. The `Color`, `FontSize`, `FontWeight`, and `HorizontalAlignment` arguments tell MATLAB to use Handle Graphics to modify the text. Handle Graphics is the name of MATLAB's underlying graphics functions. You can access this rich set of powerful, versatile graphics functions yourself. (See Chapter 30 for more information on these features.)

Standard pie charts can be created by using the `pie(a,b)` function, where a is a vector of values and b is an optional logical vector describing a slice or slices to be pulled out of the pie chart. The `pie3` function renders the pie chart with a 3-D appearance:

```
>> a = [.5 1 1.6 1.2 .8 2.1];
>> pie(a,a==max(a));    % chart a and pull out the biggest slice
>> title('Figure 25.13: Example Pie Chart')
```

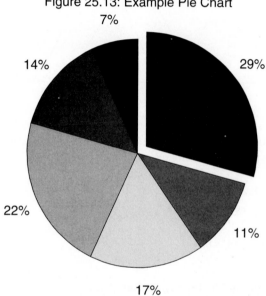

Figure 25.13: Example Pie Chart

Sometimes it is desirable to plot two different functions on the same axes by using different y-axis scales. The function `plotyy` does just that:

```
>> x = -2*pi:pi/10:2*pi;
>> y = sin(x);
>> z = 3*cos(x);
>> subplot(2,1,1), plot(x,y,x,z)
>> title('Figure 25.14a: Two plots on the same scale.');
>> subplot(2,1,2), plotyy(x,y,x,z)
>> title('Figure 25.14b: Two plots on different scales.');
```

Figure 25.14a: Two plots on the same scale.

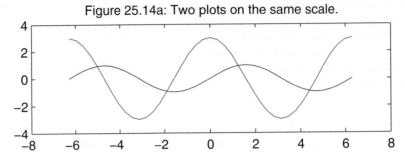

Figure 25.14b: Two plots on different scales.

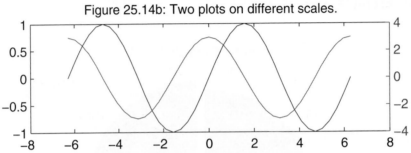

Bar and stair plots can be generated by using the bar, barh, and stairs plotting functions. The bar3 and bar3h functions render the bar charts with a 3-D appearance, as in the following example:

```
>> x = -2.9:0.2:2.9;
>> y = exp(-x.*x);

>> subplot(2,2,1)
>> bar(x,y)
>> title('Figure 25.15a: 2-D Bar Chart')

>> subplot(2,2,2)
>> bar3(x,y,'r')
>> title('Figure 25.15b: 3-D Bar Chart<')

>> subplot(2,2,3)
>> stairs(x,y)
>> title('Figure 25.15c: Stair Chart')

>> subplot(2,2,4)
>> barh(x,y)
>> title('Figure 25.15d: Horizontal Bar Chart')
```

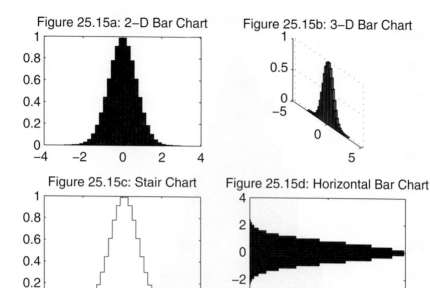

The various bar functions accept a single-color argument for all bars. Bars can be grouped or stacked as well. The form bar(x,Y) for vector x and matrix Y draws groups of bars corresponding to the columns of Y. The form bar(x,Y,'stacked') draws the bars stacked vertically. The functions barh, bar3, and bar3h have similar options.

Histograms illustrate the distribution of values in a vector. The function hist(y) draws a 10-bin histogram for the data in vector y. The function hist(y,n), where n is a scalar, draws a histogram with n bins. The function hist(y,x), where x is a vector, draws a histogram using the bins specified in x:

```
>> x = -2.9:0.2:2.9;    % specify the bins to use
>> y = randn(5000,1);   % generate 5000 random data points

>> hist(y,x)            % draw the histogram
>> title('Figure 25.16: Histogram of Gaussian Data')
```

Discrete sequence data can be plotted by using the stem function. The function stem(z) creates a plot of the data points in vector z connected to the horizontal axis by a line. An optional character-string argument can be used to specify linestyle, as in the following code:

```
>> z = randn(30,1);     % create some random data
>> stem(z,'--')         % draw a stem plot using dashed linestyle
>> set(gca,'YGrid','on') % turn grid on Y-axis only
>> title('Figure 25.17: Stem Plot of Random Data')
```

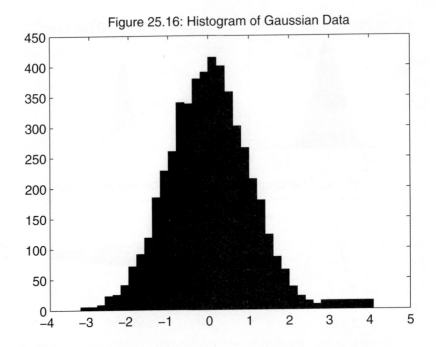

Figure 25.16: Histogram of Gaussian Data

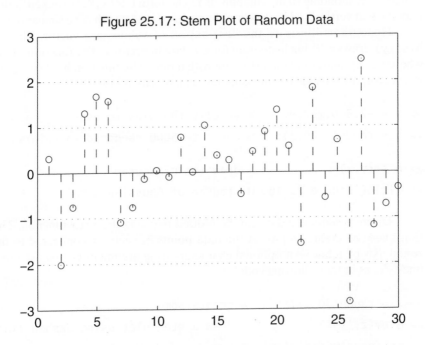

Figure 25.17: Stem Plot of Random Data

The function stem(x,z) plots the data points in z at the values specified in x.

A plot can include error bars at the data points. The function errorbar(x,y,e) plots the graph of vector x versus vector y with error bars specified by vector e:

```
>> x = linspace(0,2,21);   % create a vector
>> y = erf(x);             % y is the error function of x
>> e = rand(size(x))/10;   % e contains random error values

>> errorbar(x,y,e)         % create the plot
>> title('Figure 25.18: Errorbar Plot')
```

Note that all vectors must be the same length. For each data point (xi,yi), an error bar is drawn a distance ei above and ei below the point:

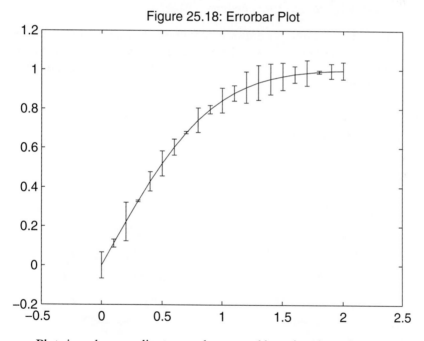

Figure 25.18: Errorbar Plot

Plots in polar coordinates can be created by using the polar(t,r,S) function, where t is the angle vector in radians, r is the radius vector, and S is an optional character string describing color, marker symbol, and/or linestyle:

```
>> t = linspace(0,2*pi);
>> r = sin(2*t).*cos(2*t);
>> subplot(2,2,1)
>> polar(t,r), title('Figure 25.19a: Polar Plot')
```

Complex data can be plotted by using compass and feather. The function compass(z) draws a plot that displays the angle and magnitude of the complex elements of z as arrows emanating from the origin. The function feather(z) plots the same data by using arrows emanating from equally spaced points on a horizontal line. The functions compass(x,y) and feather(x,y) are equivalent to compass(x+i*y) and feather(x+i*y), respectively. The following example is illustrative:

```
>> z = eig(randn(20));
>> subplot(2,2,2)
>> compass(z)
>> title('Figure 25.19b: Compass Plot')
>> subplot(2,2,3)
>> feather(z)
>> title('Figure 25.19c: Feather Plot')
```

The function rose(v) draws a 20-bin polar histogram for the angles in vector v. The function rose(v,n), where n is a scalar, draws a histogram with n bins. The function rose(v,x), where x is a vector, draws a histogram using the bins specified in x. The following code illustrates the first of these options:

```
>> subplot(2,2,4)
>> v = randn(1000,1)*pi;
>> rose(v)
>> title('Figure 25.19d: Angle Histogram')
```

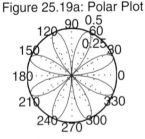

Figure 25.19a: Polar Plot

Figure 25.19b: Compass Plot

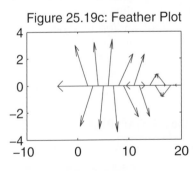

Figure 25.19c: Feather Plot

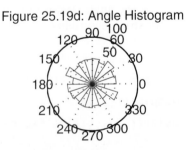

Figure 25.19d: Angle Histogram

The function scatter generates a scatter plot—that is, a plot of circles at data points, where the circle size or color can vary, point by point:

```
>> x = rand(40,1);
>> y = randn(40,1);
>> area = 20+(1:40);
>> scatter(x,y,area)
>> box on
>> title('Figure 25.20: A scatter plot')
```

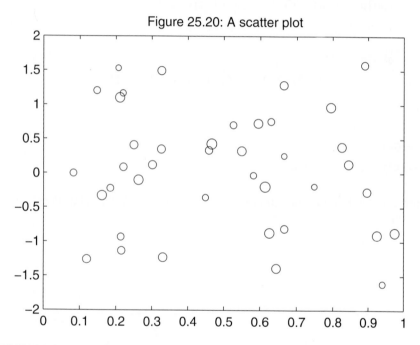

Figure 25.20: A scatter plot

25.11 EASY PLOTTING

When you don't want to take the time to specify the data points explicitly for a plot, MATLAB provides the functions fplot, ezplot, and ezpolar. The function fplot plots functions defined by M-file names or function handles. The functions ezplot and ezpolar plot functions defined by function handles, string expressions, or symbolic math objects, with the obvious difference in plot type. These functions simply spare the user from having to define the data for the independent variable, as the following code indicates:

```
>> subplot(2,2,1)
>> fplot(@humps,[-.5 3])
>> title('Figure 25.21a: Fplot of the Humps Function')
```

```
>> xlabel('x')
>> ylabel('humps(x)')

>> subplot(2,2,2)
>> f_hdl = @(x) sin(x)/x;
>> ezplot(f_hdl,[-15,15])
>> title('Figure 25.21b: sin(x)/x')

>> subplot(2,2,3)
>> ezpolar('sin(3*t).*cos(3*t)',[0 pi])
>> title('Figure 25.21c: ezpolar plot')

>> subplot(2,2,4)
>> istr = '(x-2)^2/(2^2) + (y+1)^2/(3^2) - 1';
>> ezplot(istr,[-2 6 -5 3])
>> axis square
>> grid
>> title(['Figure 25.21d: ' istr])
```

Figure 25.21a: Fplot of the Humps Function   Figure 25.21b: sin(x)/(x)

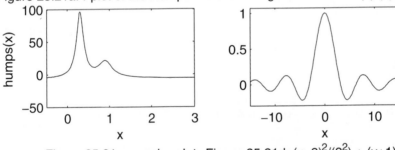

Figure 25.21c: ezpolar plot Figure 25.21d: $(x-2)^2/(2^2) + (y+1)^2/(3^2) - 1$

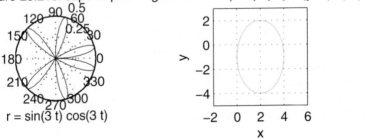

The last example shows that `ezplot` can be used to plot implicit functions. In this case, the string expression is that of an ellipse centered at $(2, -1)$.

25.12 TEXT FORMATTING

Multiline text can be used in any text string, including titles and axis labels, as well as the `text` and `gtext` functions. Simply use string arrays or cell arrays for multiline text. For example,

```
>> xlabel({'This is the first line','and this is the second.'});
```

labels the *x*-axis with two lines of text. Note that the string separator can be a space, a comma, or a semicolon; each style produces the same result. (See Chapter 9 for more details on cell arrays of strings.)

A selection of more than 100 symbols, including Greek letters and other special characters, can be included in MATLAB text strings by embedding commands within the string. The available symbols and the character strings used to define them are listed in the following table (this information can also be found by viewing the `string` property of the *text* Handle Graphics object in the online documentation):

Character Sequence	Symbol	Character Sequence	Symbol	Character Sequence	Symbol
\alpha	α	\pi	π	\upsilon	υ
\angle	∠	\rho	ρ	\phi	Φ
\ast	*	\sigma	σ	\chi	χ
\beta	β	\varsigma	ς	\psi	ψ
\gamma	γ	\tau	τ	\omega	ω
\delta	δ	\equiv	≡	\Gamma	Γ
\epsilon	ε	\Im	ℑ	\Delta	Δ
\zeta	ζ	\otimes	⊗	\Theta	Θ
\eta	η	\cap	∩	\Lambda	Λ
\theta	Θ	\supset	⊃	\Xi	Ξ
\vartheta	ϑ	\int	∫	\Pi	Π
\iota	ι	\rfloor	⌋	\Sigma	Σ
\kappa	κ	\lfloor	⌊	\Upsilon	ϒ
\lambda	λ	\perp	⊥	\Phi	Φ
\mu	μ	\wedge	∧	\Psi	Ψ
\nu	ν	\rceil	⌉	\Omega	Ω
\xi	ξ	\vee	∨	\forall	∀

Character Sequence	Symbol	Character Sequence	Symbol	Character Sequence	Symbol
\exists	∃	\sim	~	\propto	∝
\ni	∋	\leq	≤	\partial	∂
\cong	≅	\infty	∞	\bullet	•
\approx	≈	\clubsuit	♣	\div	÷
\Re	ℜ	\diamondsuit	♦	\neq	≠
\oplus	⊕	\heartsuit	♥	\aleph	ℵ
\cup	∪	\spadesuit	♠	\wp	℘
\subseteq	⊆	\leftrightarrow	↔	\oslash	∅
\in	∈	\leftarrow	←	\supseteq	⊇
\lceil	⌈	\Leftarrow	⇐	\subset	⊂
\cdot	•	\uparrow	↑	\o	o
\neg	¬	\rightarrow	→	\nabla	∇
\times	ξ	\Rightarrow	⇒	\ldots	…
\surd	√	\downarrow	↓	\prime	′
\varpi	ϖ	\circ	°	\0	∅
\rangle	⟩	\pm	±	\mid		
\langle	⟨	\geq	≥	\copyright	©

In several previous versions of MATLAB, only a limited subset of TEX formatting commands was available. In MATLAB 7 and above, all TEX formatting commands are available. Complete information regarding these commands can be found online. Some of the common and important commands include the following: Superscripts and subscripts are specified by ^ and _ respectively; text font and size are chosen by using the \fontname and \fontsize commands; and a font style is specified by using the \bf, \it, \sl, or \rm command to select a boldface, italic, oblique (or slant), or normal Roman font, respectively. To change the color of the text, use \color{colorname} or \color[rgb]{r g b} to specify the color. To print the special characters used to define TEX strings, prefix them with the backslash (\) character. The characters affected are the backslash (\), left and right curly braces { }, underscore (_), and carat (^). The following example illustrates the use of TEX formatting commands:

```
>> close % close last Figure window and start over
>> axis([0 1 0 0.5])
```

```
>> text(0.2,0.1,'\color{blue}\itE = M\cdotC^{\rm2}')
>> text(0.2,0.2,'\fontsize{16} \nabla \times H = J + \partialD/\partialt')
>> text(0.2,0.3,'\fontname{courier}\fontsize{16}\bf x_{\alpha}+y^{2\pi}')
>> fsstr = 'f(t) = A_o + \fontsize{30}_\Sigma\fontsize{10}';
>> text(0.2,0.4,[fsstr '[A_ncos(n\omega_ot) + B_nsin(n\omega_ot)]'])
>> title('Figure 25.22: TeX Formatting Examples')
```

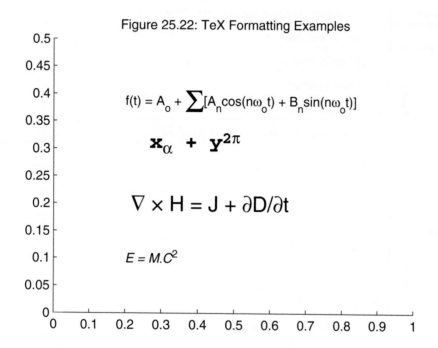

Figure 25.22: TeX Formatting Examples

25.13 SUMMARY

The following table lists MATLAB functions for 2-D plotting:

Function	Description
plot	Linear plot
loglog	Log–log plot
semilogx	Semilog x-axis plot
semilogy	Semilog y-axis plot
polar	Polar coordinate plot
plotyy	Linear plot with two y-axes
axis	Control axis scaling and appearance

Function	Description
xlim	*x*-axis limits
ylim	*y*-axis limits
zlim	*z*-axis limits
daspect	Sets and gets data aspect ratio, that is, `axis equal`
pbaspect	Sets and gets plot box aspect ratio, that is, `axis square`
zoom	Zooms in and out
grid	Grid line visibility
box	Axis box visibility
hold	Holds current plot
subplot	Creates multiple axes in *Figure* window
figure	Creates *Figure* windows
legend	Adds legend
title	Title at top of plot
xlabel	*x*-axis label
ylabel	*y*-axis label
text	Places text on plot
gtext	Places text with mouse
ginput	Gets coordinates at cursor
area	Filled area plot
bar	Bar graph
barh	Horizontal bar graph
bar3	3-D bar graph
bar3h	3-D horizontal bar graph
comet	2-D animated comet plot
compass	Compass graph
errorbar	Linear plot with error bars
ezplot	Easy line plot of string expression
ezpolar	Easy polar plot of string expression
feather	Feather plot

`fill`	Filled 2-D polygons
`fplot`	Plot function
`hist`	Histogram
`pareto`	Pareto chart
`pie`	Pie chart
`pie3`	3-D pie chart
`plotmatrix`	Scatter plot matrix
`rectangle`	Creates rectangle, rounded-rectangle, or ellipse
`ribbon`	Linear plot with 2-D lines as ribbons
`scatter`	Scatter plot
`stem`	Discrete sequence or stem plot
`stairs`	Stairstep plot

26

Three-Dimensional Graphics

MATLAB provides a variety of functions to display 3-D data. Some functions plot lines in three dimensions, while others draw surfaces and wire frames. In addition, color can be used to represent a fourth dimension. When color is used in this manner, it is called *pseudocolor*, since color is not an inherent or natural property of the underlying data in the way that color in a photograph is a natural characteristic of the image. To simplify the discussion of 3-D graphics, the use of color is postponed until the next chapter. In this chapter, the fundamental concepts of producing useful 3-D plots are discussed.

26.1 LINE PLOTS

The plot function from the 2-D world is extended into three dimensions with plot3. The format is the same as the 2-D plot, except that the data is supplied in triplets rather than in pairs. The general format of the function call is plot3(x1,y1,z1,S1, x2,y2,z2,S2,...), where xn, yn, and zn are vectors or matrices and Sn are optional character strings specifying color, marker symbol, and/or linestyle. The function plot3 is commonly used to plot a 3-D function of a single variable, as in the following example:

```
>> t = linspace(0,10*pi);
>> plot3(sin(t),cos(t),t)
>> xlabel('sin(t)'), ylabel('cos(t)'), zlabel('t')
>> text(0,0,0,'Origin')
>> grid on
```

```
>> title('Figure 26.1: Helix')
>> v = axis
v =
      -1    1    -1    1    0    35
```

Figure 26.1: Helix

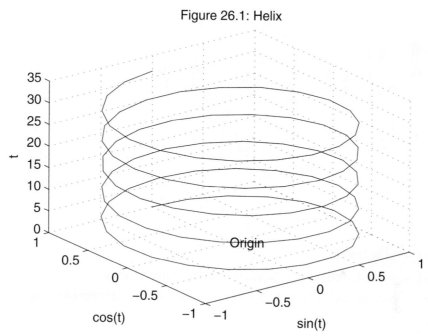

From this simple example, it is apparent that all of the basic features of 2-D graphics exist in 3-D graphics also. The axis command extends to 3-D by returning the *z*-axis limits (0 and 35) as two additional elements in the axis vector. There is a zlabel function for labeling the *z*-axis. The grid command toggles a 3-D grid underneath the plot, and the box command creates a 3-D box around the plot. The defaults for plot3 are grid off and box off. The function text(x,y,z,'string') places a character string at the position identified by the triplet x,y,z. In addition, subplots and multiple *Figure* windows apply directly to 3-D graphics functions.

In the last chapter, multiple lines or curves were plotted on top of one another by specifying multiple arguments to the plot function or by using the hold command. The function plot3 and the other 3-D graphics functions offer the same capabilities. For example, the added dimension of plot3 allows multiple 2-D plots to be stacked next to one another along one dimension, rather than directly on top of one another:

```
>> x = linspace(0,3*pi);   % x-axis data
>> z1 = sin(x);            % plot in x-z plane
>> z2 = sin(2*x);
```

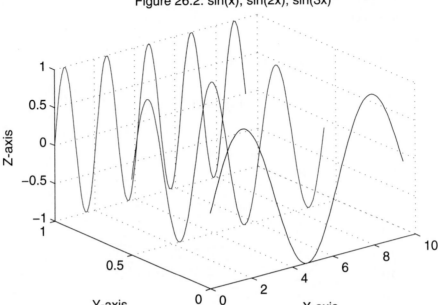

Figure 26.2: sin(x), sin(2x), sin(3x)

```
>> z3 = sin(3*x);
>> y1 = zeros(size(x));   % spread out along y-axes
>> y3 = ones(size(x));    % by giving each curve different y-axis values
>> y2 = y3/2;
>> plot3(x,y1,z1,x,y2,z2,x,y3,z3)
>> grid on
>> xlabel('X-axis'), ylabel('Y-axis'), zlabel('Z-axis')
>> title('Figure 26.2: sin(x), sin(2x), sin(3x)')
>> pause(5)
>> plot3(x,z1,y1,x,z2,y2,x,z3,y3)
>> grid on
>> xlabel('X-axis'), ylabel('Y-axis'), zlabel('Z-axis')
>> title('Figure 26.3: sin(x), sin(2x), sin(3x)')
```

26.2 SCALAR FUNCTIONS OF TWO VARIABLES

As opposed to generating line plots with plot3, it is often desirable to visualize a scalar function of two variables—that is,

$$z = f(x,y)$$

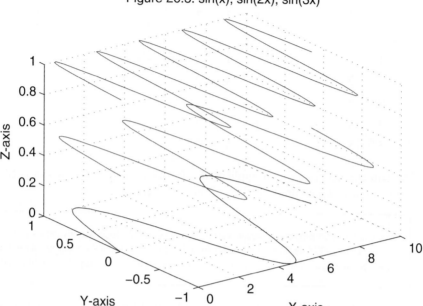

Figure 26.3: sin(x), sin(2x), sin(3x)

Here, each pair of values for x and y produces a value for z. A plot of z as a function of x and y is a surface in three dimensions. To plot this surface in MATLAB, the values for z are stored in a matrix. As described in the section on 2-D interpolation, given that x and y are the independent variables, z is a matrix of the dependent variable and the association of x and y with z is

$$z(i, :) = f(x, y(i)) \qquad \text{and} \qquad z(:, j) = f(x(j), y)$$

That is, the ith row of z is associated with the ith element of y, and the jth column of z is associated with the jth element of x.

When z = f(x,y) can be expressed simply, it is convenient to use array operations to compute all of the values of z in a single statement. To do so requires that we create matrices of all x- and y-values in the proper orientation. This orientation is sometimes called *plaid* by *The Mathworks Inc.* MATLAB provides the function meshgrid to perform this step:

```
>> x = -3:3;    % choose x-axis values
>> y = 1:5;     % y-axis values
>> [X,Y] = meshgrid(x,y)
X =
    -3   -2   -1    0    1    2    3
    -3   -2   -1    0    1    2    3
    -3   -2   -1    0    1    2    3
```

```
-3   -2   -1   0   1   2   3
-3   -2   -1   0   1   2   3
```

Y =

```
1    1    1    1    1    1    1
2    2    2    2    2    2    2
3    3    3    3    3    3    3
4    4    4    4    4    4    4
5    5    5    5    5    5    5
```

As you can see, meshgrid duplicated x for each of the five rows in y. Similarly, it duplicated y as a column for each of the seven columns in x.

An easy way to remember which variable is duplicated which way by meshgrid is to think about 2-D plots. The x-axis varies from left to right, just as the X output of meshgrid does. Similarly, the y-axis varies from bottom to top, just as the Y output of meshgrid does.

Given X and Y, if $z = f(x,y) = (x + y)^2$, then the matrix of data values defining the 3-D surface is given simply as

```
>> Z = (X+Y).^2
```

Z =

```
4    1    0    1    4    9   16
1    0    1    4    9   16   25
0    1    4    9   16   25   36
1    4    9   16   25   36   49
4    9   16   25   36   49   64
```

When a function cannot be expressed simply, you must use For Loops or While Loops to compute the elements of Z. In many cases, it may be possible to compute the elements of Z row wise or column wise. For example, if it is possible to compute Z row wise, the following script file fragment can be helpful:

```
x = ??? % statement defining vector of x axis values
y = ??? % statement defining vector of y axis values

nx = length(x);    % length of x is no. of rows in Z
ny = length(y);    % length of y is no. of columns in Z
Z = zeros(nx,ny); % initialize Z matrix for speed
```

```
for r = 1:nx
  (preliminary commands)
  Z(r,:) = {a function of y and x(r) defining r-th row of Z}
end
```

On the other hand, if Z can be computed column wise, the following script file fragment can be helpful:

```
x = ??? % statement defining vector of x axis values
y = ??? % statement defining vector of y axis values

nx = length(x);     % length of x is no. of rows in Z
ny = length(y);     % length of y is no. of columns in Z
Z = zeros(nx,ny);   % initialize Z matrix for speed

for c = 1:ny
  (preliminary commands)
  Z(:,c) = {a function of y(c) and x defining c-th column of Z}
end
```

Only when the elements of Z must be computed element by element does the computation usually require a nested For Loop, such as in the following script file fragment:

```
x = ??? % statement defining vector of x axis values
y = ??? % statement defining vector of y axis values
nx = length(x);     % length of x is no. of rows in Z
ny = length(y);     % length of y is no. of columns in Z
Z = zeros(nx,ny);   % initialize Z matrix for speed
for r = 1:nx
  for c = 1:ny
    (preliminary commands)
    Z(r,c) = {a function of y(c) and x(r) defining (r,c)-th element}
  end
end
```

26.3 MESH PLOTS

MATLAB defines a ***mesh*** surface by the z-coordinates of points above a rectangular grid in the x–y plane. It forms a mesh plot by joining adjacent points with straight lines. The result looks like a fishing net with knots at the data points. The following example is illustrative:

```
>> [X,Y,Z] = peaks(30);
>> mesh(X,Y,Z)
>> xlabel('X-axis'), ylabel('Y-axis'), zlabel('Z-axis')
>> title('Figure 26.4: Mesh Plot of Peaks')
```

Figure 26.4: Mesh Plot of Peaks

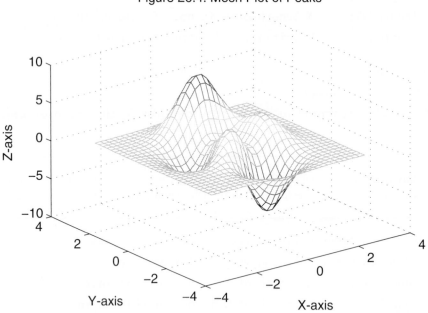

Note on your monitor how the line colors are related to the height of the mesh. In general, mesh accepts optional arguments to control color use in the plot. (This ability to change how MATLAB uses color is discussed in the next chapter.) The use of color is called pseudocolor, since color is used to add a fourth effective dimension to the graph. Note also that the plot was drawn with a grid. ***Most 3-D plots, other than plot3 and a few other exceptions, default to grid on.***

   In addition to these input arguments, mesh and most 3-D plot functions can also be called with a variety of input arguments. The syntax used here is the most specific, in that information is supplied for all three axes. The function mesh(Z) plots the matrix Z versus its row and column indices. The most common variation is to use the vectors that were passed to meshgrid for the *x*- and *y*-axes—for example, mesh(x,y,Z).

As shown in the previous figure, the areas between the mesh lines are opaque rather than transparent. The MATLAB command hidden controls this aspect of mesh plots:

```
>> [X,Y,Z] = sphere(12);
>> subplot(1,2,1)
>> mesh(X,Y,Z), title('Figure 26.5a: Opaque')
>> hidden on
>> axis square off
>> subplot(1,2,2)
>> mesh(X,Y,Z), title('Figure 26.5b: Transparent')
>> hidden off
>> axis square off
```

Figure 26.5a: Opaque   Figure 26.5b: Transparent

The sphere on the left is opaque (the lines are hidden), whereas the one on the right is transparent (the lines are not hidden).

The MATLAB mesh function has two siblings: meshc, which is a mesh plot and underlying contour plot, and meshz, which is a mesh plot that includes a zero plane. The following example uses both of these forms:

```
>> [X,Y,Z] = peaks(30);
>> meshc(X,Y,Z) % mesh plot with underlying contour plot
>> title('Figure 26.6: Mesh Plot with Contours')
>> pause(5)
>> meshz(X,Y,Z) % mesh plot with zero plane
>> title('Figure 26.7: Mesh Plot with Zero Plane')
```

Figure 26.6: Mesh Plot with Contours

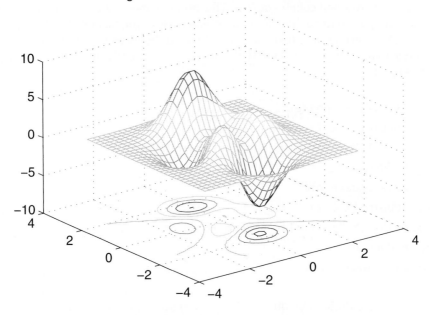

Figure 26.7: Mesh Plot with Zero Plane

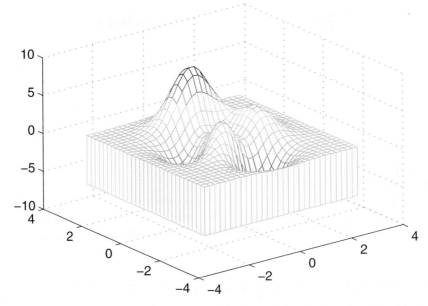

The function `waterfall` is identical to `mesh`, except that the mesh lines appear only in the *x*-direction:

```
>> waterfall(X,Y,Z)
>> hidden off
```

```
>> xlabel('X-axis'), ylabel('Y-axis'), zlabel('Z-axis')
>> title('Figure 26.8: Waterfall Plot')
```

Figure 26.8: Waterfall Plot

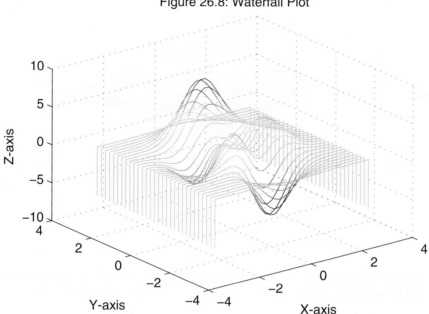

26.4 SURFACE PLOTS

A *surface* plot looks like a mesh plot, except that the spaces between the lines, called *patches*, are filled in. Plots of this type are generated using the surf function:

```
>> [X,Y,Z] = peaks(30);
>> surf(X,Y,Z)
>> xlabel('X-axis'), ylabel('Y-axis'), zlabel('Z-axis')
>> title('Figure 26.9: Surface Plot of Peaks')
```

Note how this plot type is a *dual* of sorts to a mesh plot. Here, the lines are black and the patches have color, whereas in mesh, the patches are the color of the axes and the lines have color. As with mesh, color varies along the z-axis, with each patch or line having constant color. Surface plots default to grid on also.

In a surface plot, one does not think about hidden line removal as in a mesh plot, but rather about different ways to shade the surface. In this surf plot, the shading is *faceted* like a stained-glass window or object, where the black lines are the joints between the

Figure 26.9: Surface Plot of Peaks

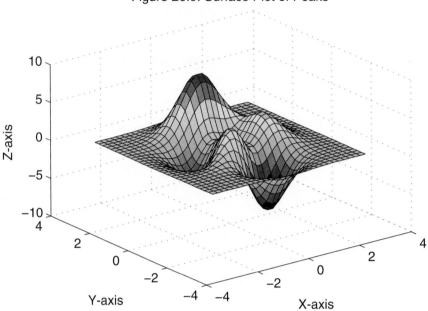

constant-color patches. In addition to faceted shading, MATLAB provides *flat* shading and *interpolated* shading. These are applied by using the function shading:

```
>> [X,Y,Z] = peaks(30);
>> surf(X,Y,Z) % same plot as above
>> shading flat
>> xlabel('X-axis'), ylabel('Y-axis'), zlabel('Z-axis')
>> title('Figure 26.10: Surface Plot with Flat Shading')
>> pause(5)
>> shading interp
>> title('Figure 26.11: Surface Plot with Interpolated Shading')
```

In flat shading, the black lines are removed and each patch retains its single color, whereas in interpolated shading, the lines are removed but each patch is given interpolated shading. That is, the color of each patch is interpolated over its area on the basis of the color values assigned to each of its vertices. Needless to say, interpolated shading requires much more computation than faceted and flat shading. While shading has a significant visual impact on surf plots, it also applies to mesh plots, although in this case the visual impact is relatively minor, since only the lines have color. Shading also affects pcolor and fill plots.

On some computer systems, interpolated shading creates extremely long printing delays or, at worst, printing errors. These problems are not due to the size of the PostScript data file, but rather to the enormous amount of computation required

Figure 26.10: Surface Plot with Flat Shading

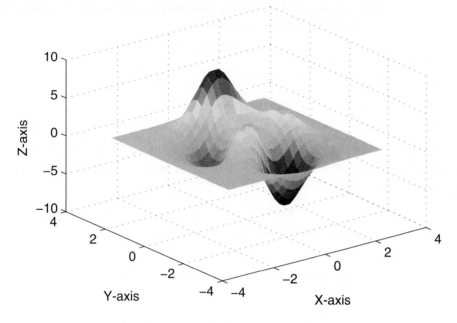

Figure 26.11: Surface Plot with Interpolated Shading

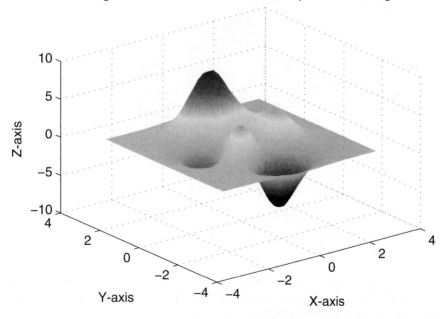

in the printer to generate shading that continually changes over the surface of the plot. Often, the easiest solution to this problem is to use flat shading for printouts.

In some situations, it may be convenient to remove part of a surface so that underlying parts of the surface can be seen. In MATLAB, this is accomplished by setting the data values where holes are desired to the special value NaN. ***Since NaNs have no value, all MATLAB plotting functions simply ignore NaN data points, leaving a hole in the plot where they appear.***

```
>> [X,Y,Z] = peaks(30);
>> x = X(1,:);                  % vector of x axis
>> y = Y(:,1);                  % vector of y axis
>> i = find(y>.8 & y<1.2);      % find y axis indices of hole
>> j = find(x>-.6 & x<.5);      % find x axis indices of hole
>> Z(i,j) = nan;                % set values at hole indices to NaNs
>> surf(X,Y,Z)
>> xlabel('X-axis'), ylabel('Y-axis'), zlabel('Z-axis')
>> title('Figure 26.12: Surface Plot with a Hole')
```

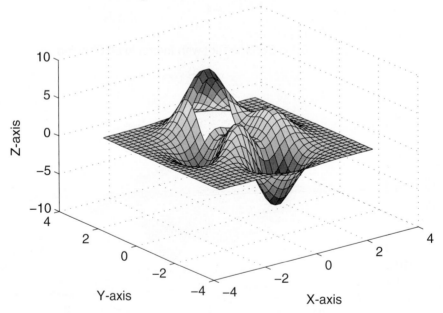

Figure 26.12: Surface Plot with a Hole

The MATLAB surf function also has two siblings: surfc, which is a surface plot and underlying contour plot, and surfl, which is a surface plot with lighting. Both are used in the following code:

```
>> [X,Y,Z] = peaks(30);
>> surfc(X,Y,Z) % surf plot with contour plot
```

```
>> xlabel('X-axis'), ylabel('Y-axis'), zlabel('Z-axis')
>> title('Figure 26.13: Surface Plot with Contours')
>> pause(5)
>> surfl(X,Y,Z)       % surf plot with lighting
>> shading interp     % surfl plots look best with interp shading
>> colormap pink      % they also look better with shades of a single color
>> xlabel('X-axis'), ylabel('Y-axis'), zlabel('Z-axis')
>> title('Figure 26.14: Surface Plot with Lighting')
```

Figure 26.13: Surface Plot with Contours

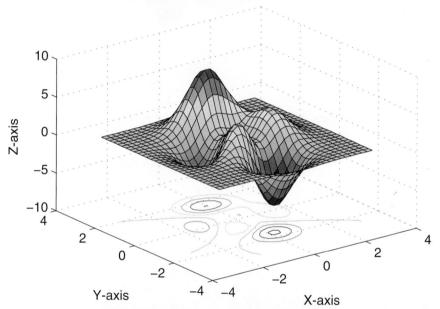

The function surfl makes a number of assumptions regarding the light applied to the surface. It does not use the *light* object. Rather, it simply modifies the color of the surface to give the appearance of lighting. (The next chapter provides more rigorous information regarding light properties.) Also, in the preceding commands, colormap is a MATLAB function for applying a different set of colors to a figure. (This function is discussed in the next chapter as well.)

The surfnorm(X,Y,Z) function computes surface normals for the surface defined by X, Y, and Z, plots the surface, and plots vectors normal to the surface at the data points:

```
>> [X,Y,Z] = peaks(15);
>> surfnorm(X,Y,Z)
>> xlabel('X-axis'), ylabel('Y-axis'), zlabel('Z-axis')
>> title('Figure 26.15: Surface Plot with Normals')
```

Figure 26.14: Surface Plot with Lighting

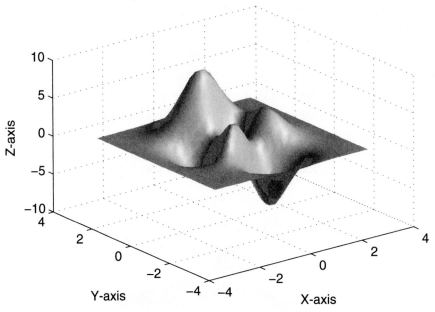

Figure 26.15: Surface Plot with Normals

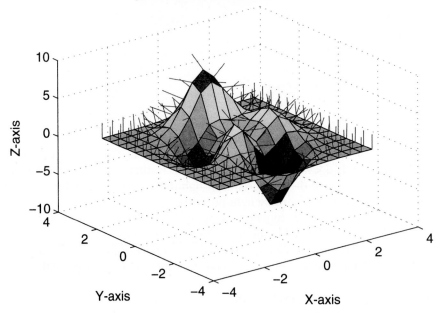

Note that the surface normals are unnormalized and valid at each vertex. The form
$[Nx,Ny,Nz]$ = `surfnorm(X,Y,Z)` computes the 3-D surface normals and returns
their components, but does not plot the surface.

26.5 MESH AND SURFACE PLOTS OF IRREGULAR DATA

Irregular or nonuniformly spaced data can be visualized using the functions trimesh, trisurf, and voronoi:

```
>> x = rand(1,50);
>> y = rand(1,50);
>> z = peaks(x,y*pi);
>> t = delaunay(x,y);
>> trimesh(t,x,y,z)
>> hidden off
>> title('Figure 26.16: Triangular Mesh Plot')
>> pause(5)
>> trisurf(t,x,y,z)
>> title('Figure 26.17: Triangular Surface Plot')
>> pause(5)
>> voronoi(x,y)
>> title('Figure 26.18: Voronoi Plot')
```

(See Chapter 18 for more information on Delaunay triangulation and Voronoi diagrams.)

Figure 26.16: Triangular Mesh Plot

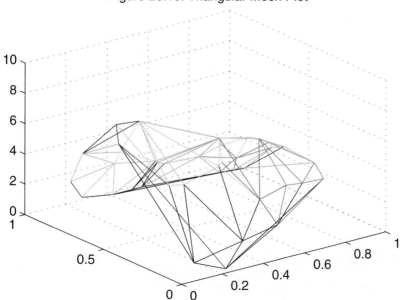

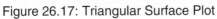

Figure 26.17: Triangular Surface Plot

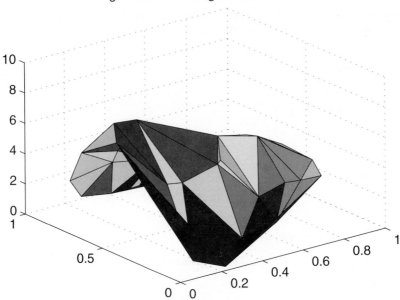

Figure 26.18: Voronoi Plot

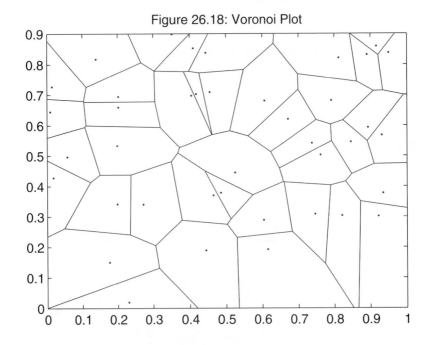

26.6 CHANGING VIEWPOINTS

Note that the default viewpoint of 3-D plots is looking down at the $z = 0$ plane at an angle of 30 degrees and looking up at the $x = 0$ plane at an angle of 37.5 degrees. The angle of orientation with respect to the $z = 0$ plane is called the ***elevation***, and the angle with respect to the $x = 0$ plane is called the ***azimuth***. Thus, the default 3-D viewpoint is an elevation of 30 degrees and an azimuth of –37.5 degrees. The default 2-D viewpoint is an elevation of 90 degrees and an azimuth of 0 degrees. The concepts of azimuth and elevation are described visually in the following figure.

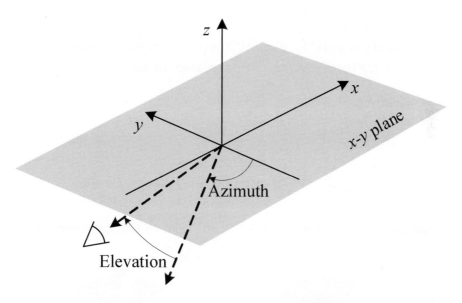

In MATLAB, the function view changes the graphical viewpoint for all types of 2-D and 3-D plots. The forms view(az,el) and view([az,el]) change the viewpoint to the specified azimuth az and elevation el:

```
>> x = -7.5:.5:7.5; y = x;           % create a data set
>> [X,Y] = meshgrid(x,y);
>> R = sqrt(X.^2+Y.^2)+eps;
>> Z = sin(R)./R;
>> subplot(2,2,1)
>> surf(X,Y,Z)
>> view(-37.5,30)
>> xlabel('X-axis'), ylabel('Y-axis'), zlabel('Z-axis')
>> title('Figure 26.19a: Default Az = -37.5, El = 30')
```

```
>> subplot(2,2,2)
>> surf(X,Y,Z)
>> view(-37.5+90,30)
>> xlabel('X-axis'), ylabel('Y-axis'), zlabel('Z-axis')
>> title('Figure 26.19b: Az Rotated to 52.5')

>> subplot(2,2,3)
>> surf(X,Y,Z)
>> view(-37.5,60)
>> xlabel('X-axis'), ylabel('Y-axis'), zlabel('Z-axis')
>> title('Figure 26.19c: El Increased to 60')

>> subplot(2,2,4)
>> surf(X,Y,Z)
>> view(0,90)
>> xlabel('X-axis'), ylabel('Y-axis')
>> title('Figure 26.19d: Az = 0, El = 90')
```

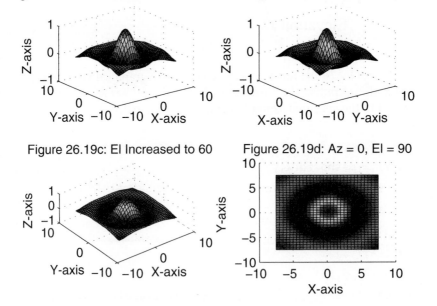

Figure 26.19a: Default Az = –37.5, El = 30 Figure 26.19b: Az Rotated to 52.5

In addition to these forms, view offers additional features that are summarized in its online documentation and help text:

```
>> help view
```
VIEW 3-D graph viewpoint specification.

   VIEW(AZ,EL) and VIEW([AZ,EL]) set the angle of the view from which an
   observer sees the current 3-D plot. AZ is the azimuth or horizontal
   rotation and EL is the vertical elevation (both in degrees). Azimuth
   revolves about the z-axis, with positive values indicating counter-
   clockwise rotation of the viewpoint. Positive values of elevation
   correspond to moving above the object; negative values move below.
   VIEW([X Y Z]) sets the view angle in Cartesian coordinates. The
   magnitude of vector X,Y,Z is ignored.

   Here are some examples:
   AZ = -37.5, EL = 30 is the default 3-D view.
   AZ = 0, EL = 90 is directly overhead and the default 2-D view.
   AZ = EL = 0 looks directly up the first column of the matrix.
   AZ = 180 is behind the matrix.

   VIEW(2) sets the default 2-D view, AZ = 0, EL = 90.
   VIEW(3) sets the default 3-D view, AZ = -37.5, EL = 30.
   [AZ,EL] = VIEW returns the current azimuth and elevation.
   T = VIEW returns the current general 4-by-4 transformation matrix.
   VIEW(AX,...) uses axes AX instead of the current axes.
   See also viewmtx, the axes properties view, Xform.
   Reference page in Help browser
      doc view

        In addition to the view function, the viewpoint can be set interactively with the
   mouse by using the function rotate3d. The function rotate3d on turns on mouse-
   based view rotation, rotate3d off turns it off, and rotate3d with no arguments
   toggles the state. This functionality is also available on the **Tools** menu in a *Figure*
   window, as well as on a button on the *Figure* toolbar. In reality, the **Rotate 3D** menu
   item and toolbar button both call rotate3d to do the requested work.

26.7 CAMERA CONTROL

        The viewpoint control provided by the view function is convenient, but limited in
   capabilities. To provide complete control of a 3-D scene, camera capabilities are needed.
   That is, one must have all of the capabilities available when filming a movie with a

camera. Alternatively, one must have all of the capabilities available in a 3-D computer or console game environment. In this environment, there are two 3-D coordinate systems to manage: one at the camera and one at what the camera is pointed at (i.e., the camera target). The camera functions in MATLAB manage and manipulate the relationships between these two coordinate systems and provide control over the camera lens.

Use of the camera functions in MATLAB is generally not easy for a novice. Therefore, to simplify the use of these functions, most are made available interactively from the **Tools** menu or from the *Camera* toolbar in a *Figure* window. To view the *Camera* toolbar, choose it from the **View** menu in the *Figure* window. By using the interactive camera tools, you avoid dealing with the input and output arguments that the *Command* window functions require. Given the ease with which the interactive tools can be used, the complexity involved in using and describing the camera functions, and the relatively small number of potential users, they are not described in this text. The MATLAB documentation contains a rigorous discussion of these functions and their use. The camera functions available in MATLAB are listed at the end of this chapter.

26.8 CONTOUR PLOTS

Contour plots show lines of constant elevation or height. If you've ever seen a topographical map, you know what a contour plot looks like. In MATLAB, contour plots in 2-D and 3-D are generated by using the contour and contour3 functions, respectively:

```
>> [X,Y,Z] = peaks;
>> subplot(1,2,1)
>> contour(X,Y,Z,20)          % generate 20 2-D contour lines
>> axis square
>> xlabel('X-axis'), ylabel('Y-axis')
>> title('Figure 26.20a: 2-D Contour Plot')
>> subplot(1,2,2)
>> contour3(X,Y,Z,20)         % the same contour plot in 3-D
>> xlabel('X-axis'), ylabel('Y-axis'), zlabel('Z-axis')
>> title('Figure 26.20b: 3-D Contour Plot')
```

The pcolor function maps height to a set of colors and presents the same information as the contour plot, at the same scale:

```
>> subplot(1,2,1)
>> pcolor(X,Y,Z)
>> shading interp   % remove the grid lines
>> axis square
>> title('Figure 26.21a: Pseudocolor Plot')
```

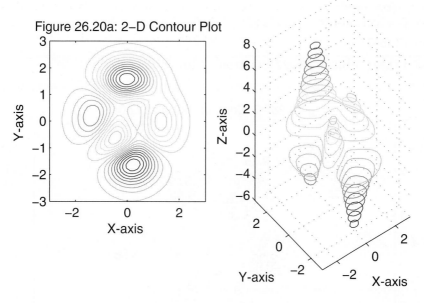

Figure 26.20a: 2–D Contour Plot

Figure 26.20b: 3–D Contour Plot

Combining the idea of a pseudocolor plot with a 2-D contour produces a filled contour plot. In MATLAB, this plot is generated by the function contourf:

```
>> subplot(1,2,2)
>> contourf(X,Y,Z,12)      % filled contour plot with 12 contours
>> axis square
>> xlabel('X-axis'), ylabel('Y-axis')
>> title('Figure 26.21b: Filled Contour Plot')
```

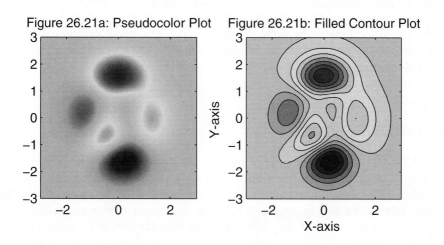

Figure 26.21a: Pseudocolor Plot   Figure 26.21b: Filled Contour Plot

Contour lines can be labeled by using the `clabel` function, which requires a matrix of lines and optional text strings that are returned by `contour`, `contourf`, and `contour3`:

```
>> C = contour(X,Y,Z,12);
>> clabel(C)
>> title('Figure 26.22: Contour Plot With Labels')
```

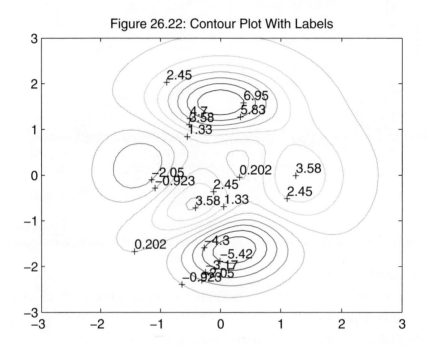

Alternatively, inline labels are generated by using two arguments returned by the contour functions. For example, modifying the preceding MATLAB code to `[C,h] = contour(X,Y,Z,12); clabel(C,h)` produces inline labels that follow the contours. Finally, by using the mouse, you can select which contours to label by providing `'manual'` as the last input argument to `clabel`. Inline or horizontal labels are used, depending on the presence of the second argument h, as illustrated previously.

26.9 SPECIALIZED 3-D PLOTS

MATLAB provides a number of specialized plotting functions in addition to those already discussed. The function `ribbon(Y)` plots the columns of Y as separate ribbons. The function `ribbon(x,Y)` plots x versus the columns of Y. The width of the ribbons can also be specified by using the syntax `ribbon(x,Y,width)`, where the default width is 0.75. The following code uses the first of these forms:

```
>> Z = peaks;
>> ribbon(Z)
>> title('Figure 26.23: Ribbon Plot of Peaks')
```

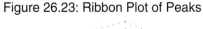

Figure 26.23: Ribbon Plot of Peaks

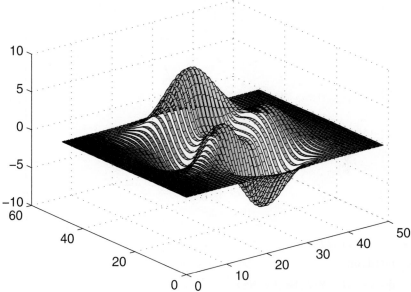

The function `quiver(x,y,dx,dy)` draws directional or velocity vectors `(dx,dy)` at the points `(x,y)`:

```
>> [X,Y,Z] = peaks(16);
>> [DX,DY] = gradient(Z,.5,.5);
>> contour(X,Y,Z,10)
>> hold on
>> quiver(X,Y,DX,DY)
>> hold off
>> title('Figure 26.24: 2-D Quiver Plot')
```

Three-dimensional quiver plots of the form `quiver3(x,y,z,Nx,Ny,Nz)` display the vectors `(Nx,Ny,Nz)` at the points `(x,y,z)`:

```
>> [X,Y,Z] = peaks(20);
>> [Nx,Ny,Nz] = surfnorm(X,Y,Z);
```

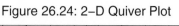

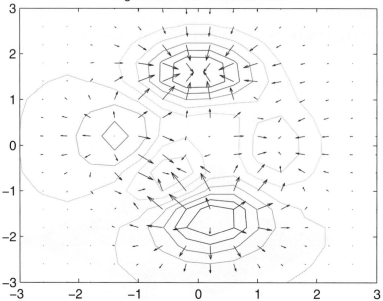

```
>> surf(X,Y,Z)
>> hold on
>> quiver3(X,Y,Z,Nx,Ny,Nz)
>> axis tight
>> hold off
>> title('Figure 26.25: 3-D Quiver Plot')
```

The function fill3, being the 3-D equivalent of fill, draws filled polygons in 3-D space. The form fill3(X,Y,Z,C) uses the arrays X, Y, and Z as the vertices of the polygon; and C specifies the fill color, as in the following example:

```
>> fill3(rand(3,5),rand(3,5),rand(3,5),rand(3,5))
>> grid on
>> title('Figure 26.26: Five Random Filled Triangles')
```

The 3-D equivalent of stem plots discrete sequence data in 3-D space. The form stem3(X,Y,Z,C,'filled') plots the data points in (X,Y,Z) with lines extending to the *x–y* plane. The optional argument C specifies the marker style or color, and the optional 'filled' argument causes the marker to be filled in. The stem3(Z)

Figure 26.25: 3–D Quiver Plot

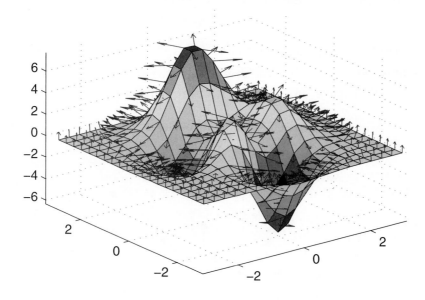

Figure 26.26: Five Random Filled Triangles

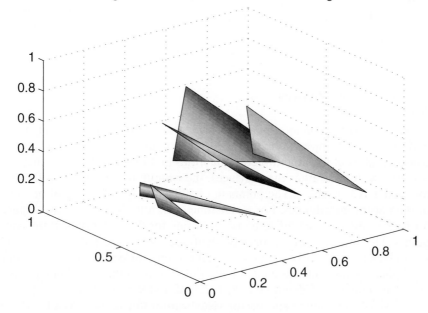

function plots the points in Z and automatically generates X and Y values. The following code uses stem3 with three arguments:

```
>> Z = rand(5);
>> stem3(Z,'ro','filled');
>> grid on
>> title('Figure 26.27: Stem Plot of Random Data')
```

Figure 26.27: Stem Plot of Random Data

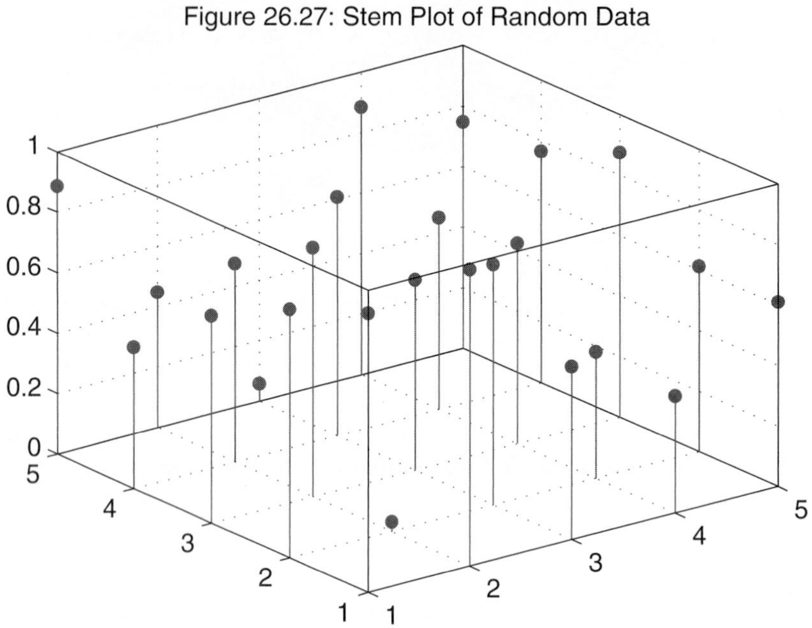

26.10 VOLUME VISUALIZATION

In addition to the common mesh, surface, and contour plots, MATLAB offers a variety of more complex volume and vector visualization functions. These functions construct plots of scalar and vector quantities in 3-D space. Because they generally construct volumes rather than surfaces, their input arguments are 3-D arrays, one for each axis *x*, *y*, and *z*. The points in each 3-D array define a grid of coordinates or data at the coordinates. For scalar functions, four 3-D arrays are required, one for each of the three coordinate axes, plus one for the scalar data at the coordinate points. These arrays are commonly identified as X, Y, Z, and V, respectively. For vector functions, six 3-D arrays are required, one for each of the three coordinate axes, plus one for each axis component of the vector at the coordinate points. These arrays are commonly identified as X, Y, Z, U, V, and W, respectively.

Use of the volume and vector visualization functions in MATLAB requires an understanding of volume and vector terminology. For example, *divergence* and *curl* describe vector processes, and *isosurfaces* and *isocaps* describe visual aspects of volumes. If you are unfamiliar with these terms, using MATLAB volume and visualization functions can be confusing. It is beyond the scope of this text to cover the terminology required to rigorously use these volume and vector visualization functions. However, the structure of the data arrays and the use of several functions are demonstrated in what follows. The MATLAB documentation contains a more thorough introduction and more rigorous explanations and examples.

Consider the construction of a scalar function defined over a volume. First, the volume coordinate axes must be constructed:

```
>> x = linspace(-3,3,13);      % x coordinate points
>> y = 1:20;                   % y coordinate points
>> z = -5:5;                   % z coordinate points
>> [X,Y,Z] = meshgrid(x,y,z);  % meshgrid works here too!
>> size(X)
ans =
      20    13    11
```

Here, X, Y, Z are 3-D arrays defining the grid. The X array contains x duplicated for as many rows as length(y) and as many pages as length(z). Similarly, Y contains y transposed to a column and duplicated for as many columns as length(x) and as many pages as length(z). And Z contains z permuted to a 1-by-1-by-length(z) vector and duplicated for as many rows as length(y) and as many columns as length(x). As described, this is a direct extension of meshgrid to 3-D.

Next, we need to define a function of these data, such as

```
>> V = sqrt(X.^2 + cos(Y).^2 + Z.^2);
```

Now, the 3-D arrays X, Y, Z, and V define a scalar function $v = f(x,y,z)$ defined over a volume. To visualize what this looks like, we can look at slices along planes:

```
>> slice(X,Y,Z,V,[0 3],[5 15],[-3 5])
>> xlabel('X-axis')
>> ylabel('Y-axis')
>> zlabel('Z-axis')
>> title('Figure 26.28: Slice Plot Through a Volume')
```

This plot shows slices on planes defined by $x = 0$, $x = 3$, $y = 5$, $y = 15$, $z = -3$, and $z = 5$, as shown by the last three arguments to the function slice. The color of the plot is mapped to the values in V on the slices.

The slices displayed need not be planes. They can be any surface, as generated by the following code:

```
>> [xs,ys] = meshgrid(x,y);
>> zs = sin(-xs+ys/2); % a surface to use
```

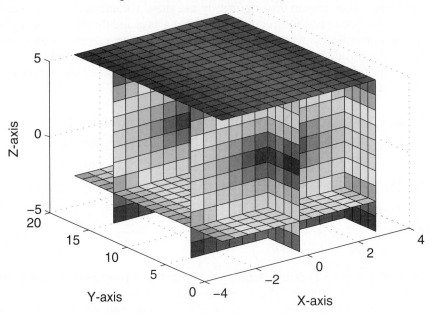

Figure 26.28: Slice Plot Through a Volume

```
>> slice(X,Y,Z,V,xs,ys,zs)
>> xlabel('X-axis')
>> ylabel('Y-axis')
>> zlabel('Z-axis')
>> title('Figure 26.29: Slice Plot Using a Surface')
```

Here, xs, ys, and zs define a surface to slice through the volume.

Going back to the original slice plot, it is possible to add contour lines to selected planes by using the contourslice function:

```
>> slice(X,Y,Z,V,[0 3],[5 15],[-3 5])
>> hold on
>> h = contourslice(X,Y,Z,V,3,[5 15],[]);
>> set(h,'EdgeColor','k','Linewidth',1.5)
>> xlabel('X-axis')
>> ylabel('Y-axis')
>> zlabel('Z-axis')
>> title('Figure 26.30: Slice Plot with Selected Contours')
>> hold off
```

Figure 26.29: Slice Plot Using a Surface

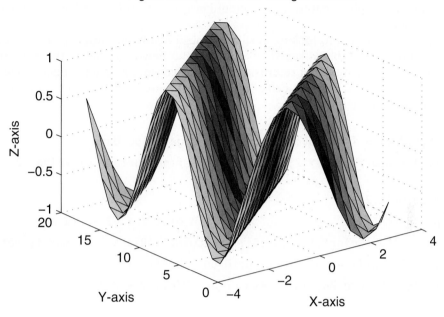

Figure 26.30: Slice Plot with Selected Contours

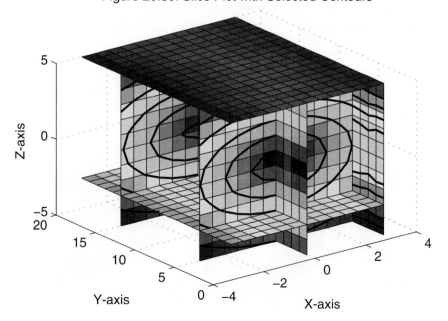

Here, contour lines are added to the $x = 3$, $y = 5$, and $y = 15$ planes. With the use of Handle Graphics features, the contour lines are set to black and their width is set to 1.5 points.

In addition to looking at slices through a volume, surfaces where the scalar volume data V has a specified value can be plotted by using the isosurface function. This function returns triangle vertices in a manner similar to Delaunay triangulation, the results being in the form required by the patch function, which plots the triangles:

```
>> [X,Y,Z,V] = flow(13);              % get flow data
>> fv = isosurface(X,Y,Z,V,-2);       % find surface of value -2

>> subplot(1,2,1)
>> p = patch(fv);                     % plot V = -2 surface
>> set(p,'FaceColor',[.5 .5 .5],'EdgeColor','Black'); % modify patches
>> view(3), axis equal tight, grid on % pretty it up
>> title({'Figure 26.31a:' 'Isosurface Plot, V = 2'})

>> subplot(1,2,2)
>> p = patch(shrinkfaces(fv,.3));     % shrink faces to 30% of original
>> set(p,'Facecolor',[.5 .5 .5],'EdgeColor','Black'); % modify patches
>> view(3), axis equal tight, grid on % pretty it up
>> title({'Figure 26.31b:' 'Shrunken Face Isosurface Plot, V = 2'})
```

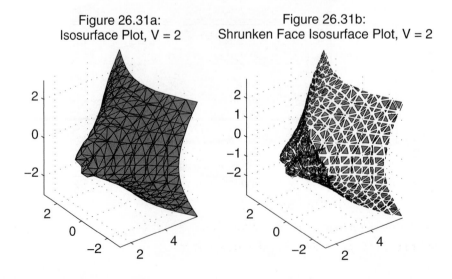

Figure 26.31a:
Isosurface Plot, V = 2

Figure 26.31b:
Shrunken Face Isosurface Plot, V = 2

The preceding plots also demonstrate the use of the function shrinkfaces, which does exactly what its name suggests.

Sometimes volume data contain too many points for efficient display. The functions reducevolume and reducepatch represent two ways to improve the display of an isosurface. The function reducevolume eliminates data before the isosurface is formed, whereas reducepatch seeks to eliminate patches while minimizing distortion in the underlying surface. The following code is illustrative:

```
>> [X,Y,Z,V] = flow;
>> fv = isosurface(X,Y,Z,V,-2);
>> subplot(2,2,1) % Original
>> p = patch(fv);
>> Np = size(get(p,'Faces'),1);
>> set(p,'FaceColor',[.5 .5 .5],'EdgeColor','Black');
>> view(3), axis equal tight, grid on % pretty it up
>> zlabel(sprintf('%d Patches',Np))
>> title('Figure 26.32a: Original')

>> subplot(2,2,2) % Reduce Volume
>> [Xr,Yr,Zr,Vr] = reducevolume(X,Y,Z,V,[3 2 2]);
>> fvr = isosurface(Xr,Yr,Zr,Vr,-2);
>> p = patch(fvr);
>> Np = size(get(p,'Faces'),1);
>> set(p,'FaceColor',[.5 .5 .5],'EdgeColor','Black');
>> view(3), axis equal tight, grid on % pretty it up
>> zlabel(sprintf('%d Patches',Np))
>> title('Figure 26.32b: Reduce Volume')

>> subplot(2,2,3) % Reduce Patch
>> p = patch(fv);
>> set(p,'FaceColor',[.5 .5 .5],'EdgeColor','Black');
>> view(3), axis equal tight, grid on % pretty it up
>> reducepatch(p,.15) % keep 15 percent of the faces
>> Np = size(get(p,'Faces'),1);
>> zlabel(sprintf('%d Patches',Np))
>> title('Figure 26.32c: Reduce Patches')
```

```
>> subplot(2,2,4) % Reduce Volume and Patch
>> p = patch(fvr);
>> set(p,'FaceColor',[.5 .5 .5],'EdgeColor','Black');
>> view(3), axis equal tight, grid on % pretty it up
>> reducepatch(p,.15) % keep 15 percent of the faces
>> Np = size(get(p,'Faces'),1);
>> zlabel(sprintf('%d Patches',Np))
>> title('Figure 26.32d: Reduce Both')
```

Figure 26.32a: Original

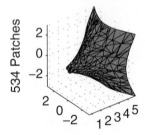

Figure 26.32b: Reduce Volume

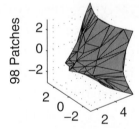

Figure 26.32c: Reduce Patches

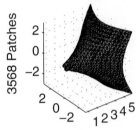

Figure 26.32d: Reduce Both

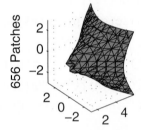

Three-dimensional data can also be smoothed by filtering it with the smooth3 function, as in the following code:

```
>> data = rand(10,10,10);         % random data
>> datas = smooth3(data,'box',3); % smoothed data

>> subplot(1,2,1) % random data
>> p = patch(isosurface(data,.5),...
        'FaceColor','Blue','EdgeColor','none');
>> patch(isocaps(data,.5),...
        'FaceColor', 'interp', 'EdgeColor', 'none');
```

```
>> isonormals(data,p)
>> view(3); axis vis3d tight off
>> camlight; lighting phong
>> title({'Figure 26.33a:' 'Random Data'})

>> subplot(1,2,2) % smoothed random data
>> p = patch(isosurface(datas,.5),...
       'FaceColor','Blue','EdgeColor','none');
>> patch(isocaps(datas,.5), ...
       'FaceColor', 'interp', 'EdgeColor', 'none');
>> isonormals(datas,p)
>> view(3); axis vis3d tight off
>> camlight; lighting phong
>> title({'Figure 26.33b:' 'Smoothed Data'})
```

Figure 26.33a:
Random Data

Figure 26.33b:
Smoothed Data

This example demonstrates the use of the functions isocaps and isonormals. The function isocaps creates the faces on the outer surfaces of the block. The function isonormals modifies properties of the drawn patches, so that lighting works correctly.

26.11 EASY PLOTTING

For those occasions when you don't want to take the time to specify the data points explicitly for a 3-D plot, MATLAB provides the functions ezcontour, ezcontour3, ezcontourf, ezmesh, ezmeshc, ezplot3, ezsurf, and ezsurfc. These functions construct plots like their equivalents without the ez prefix. However, the input arguments are functions defined by function handles, string expressions, or symbolic math objects and, optionally, the axis limits over which the plot is to be generated.

Internally, the functions compute the data and generate the desired plot, as in the following example:

```
>> fstr = ['3*(1-x).^2.*exp(-(x.^2) - (y+1).^2)' ...
   ' - 10*(x/5 - x.^3 - y.^5).*exp(-x.^2-y.^2)' ...
   ' - 1/3*exp(-(x+1).^2 - y.^2)'];

>> subplot(2,2,1)
>> ezmesh(fstr)
>> title('Figure 26.34a: Mesh of peaks(x,y)')
>> subplot(2,2,2)
>> ezsurf(fstr)
>> title('Figure 26.34b: Surf of peaks(x,y)')
>> subplot(2,2,3)
>> ezcontour(fstr)
>> title('Figure 26.34c: Contour of peaks(x,y)')
>> subplot(2,2,4)
>> ezcontourf(fstr)
>> title('Figure 26.34d: Contourf of peaks(x,y)')
```

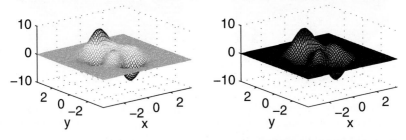

Figure 26.34a: Mesh of peaks(x,y)   Figure 26.34b: Surf of peaks(x,y)

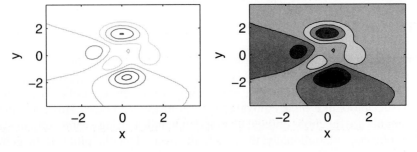

Figure 26.34c: Contour of peaks(x,y) Figure 26.34d: Contourf of peaks(x,y)

26.12 SUMMARY

The following table documents MATLAB functions for 3-D plotting:

Function	Description
plot3	Plots lines and points in 3-D space
mesh	Mesh surface
meshc	Mesh with underlying contour plot
meshz	Mesh with zero plane
surf	Surface plot
surfc	Surface plot with underlying contour plot
surfl	Surface plot with basic lighting
fill3	Filled 3-D polygons
comet3	3-D animated comet-like trajectories
sphere	Generates or plots a sphere
ellipsoid	Generates or plots an ellipsoid
cylinder	Generates or plots a cylinder
shading	Color shading mode
hidden	Mesh hidden line removal
surfnorm	Surface normals
axis	Controls axis scaling and appearance
grid	Grid line visibility
box	Axis box visibility
hold	Holds current plot
subplot	Creates multiple axes in *Figure* window
daspect	Data aspect ratio
pbaspect	Plot box aspect ratio
xlim	*x*-axis limits
ylim	*y*-axis limits
zlim	*z*-axis limits
view	3-D viewpoint specification

Function	Description
viewmtx	View transformation matrix
rotate3d	Interactive axes rotation
campos	Camera position
camtarget	Camera target
camva	Camera view angle
camup	Camera up vector
camproj	Camera projection
camorbit	Camera orbit
campan	Pan camera
camdolly	Dolly camera
camzoom	Zoom camera
camroll	Roll camera
camlookat	Looks at specific object
camlight	Camera lighting creation and placement
title	Plot title
xlabel	x-axis label
ylabel	y-axis label
zlabel	z-axis label
text	Places text on plot
gtext	Places text with mouse
contour	Contour plot
contourf	Filled contour plot
contour3	3-D contour plot
clabel	Contour labeling
pcolor	Pseudocolor plot
voronoi	Voronoi diagram
trimesh	Triangular mesh plot
trisurf	Triangular surface plot
scatter3	3-D scatter plot

stem3	3-D stem plot
waterfall	Waterfall plot
ezmesh	Easy mesh plot of string expression
ezmeshc	Easy mesh plot with contour plot of string expression
ezplot3	Easy 3-D linear plot of string expression
ezsurf	Easy surface plot of string expression
ezsurfc	Easy surface plot with contour plot of string expression
ezcontour	Easy contour plot of string expression
ezcontourf	Easy filled contour plot of string expression
vissuite	Help for visualization suite
isosurface	Isosurface extractor
isonormals	Isosurface normals
isocaps	Isosurface end caps
isocolors	Isosurface and patch colors
contourslice	Contours in slice planes
slice	Volumetric slice plot
streamline	Streamlines from data
stream3	3-D streamlines
stream2	2-D streamlines
quiver3	3-D quiver plot
quiver	2-D quiver plot
divergence	Divergence of a vector field
curl	Curl and angular velocity of a vector field
coneplot	Cone plot
streamtube	Stream tube
streamribbon	Stream ribbon
streamslice	Streamlines in slice planes
streamparticles	Displays stream particles
interpstreamspeed	Interpolates streamline vertices from speed
subvolume	Extracts subset of volume data set

Function	Description
reducevolume	Reduces volume data set
volumebounds	Returns volume and color limits
smooth3	Smoothes 3-D data
reducepatch	Reduces number of patch faces
shrinkfaces	Reduces size of patch faces
rectint	Rectangle intersection area
polyarea	Area of polygon
inpolygon	True for points inside or on a polygonal region

27

Using Color and Light

MATLAB provides a number of tools for displaying information visually in two and three dimensions. For example, the plot of a sine curve presents more information at a glance than a set of data points could. The technique of using plots and graphs to present data sets is known as *data visualization*. In addition to being a powerful computational engine, MATLAB excels in presenting data visually in interesting and informative ways.

Often, however, a simple 2-D or 3-D plot cannot display all of the information you would like to present at one time. Color can provide an additional dimension. Many of the plotting functions discussed in previous chapters accept a *color* argument that can be used to add that additional dimension.

This discussion begins with an investigation of colormaps—how to use them, display them, alter them, and create them. Next, techniques for simulating more than one colormap in a *Figure* window or for using only a portion of a colormap are illustrated. Finally, lighting models are discussed, and examples are presented. As in the preceding chapters, the figures in this chapter do not exhibit color, although they do have color on a computer screen. As a result, if you are not following along in MATLAB, it may take some imagination to understand the concepts covered in this chapter.

27.1 UNDERSTANDING COLORMAPS

MATLAB uses a numerical array with three columns to represent color values. This array is called a *colormap*, and each row in the matrix represents an individual color by using numbers in the range 0 to 1. The numbers in each row indicate the intensity of red, green, and blue that make up a specific color. The following table illustrates the correspondence between numerical values in a colormap and colors:

Red	Green	Blue	Color
1	0	0	Red
0	1	0	Green
0	0	1	Blue
1	1	0	Yellow
1	0	1	Magenta
0	1	1	Cyan
0	0	0	Black
1	1	1	White
0.5	0.5	0.5	Medium gray
0.67	0	1	Violet
1	0.4	0	Orange
0.5	0	0	Dark red
0	0.5	0	Dark green

The first column in a colormap is the intensity of red, the second column is the intensity of green, and the third column is the intensity of blue. Colormap values are restricted to the range from 0 to 1.

A colormap is a sequence of rows containing red–green–blue (RGB) values that vary in some prescribed way from the first row to the last. MATLAB provides a number of predefined colormaps, as shown in the following table:

Colormap Function	Description
hsv	Hue–saturation–value (HSV) colormap; begins and ends with red
jet	Variant of hsv that starts with blue and ends with red
hot	Black to red to yellow to white
cool	Shades of cyan and magenta
summer	Shades of green and yellow
autumn	Shades of red and yellow
winter	Shades of blue and green
spring	Shades of magenta and yellow

white	All white
gray	Linear gray scale
bone	Gray with a tinge of blue
pink	Pastel shades of pink
copper	Linear copper tone
prism	Alternating red, orange, yellow, green, blue, and violet
flag	Alternating red, white, blue, and black
lines	Alternating plot line colors
colorcube	Enhanced color cube
vga	16-color Windows colormap (16-by-3)

By default, each of these colormaps (except vga) generates a 64-by-3 array specifying the RGB descriptions of 64 colors. Each of these functions accepts an argument specifying the number of rows to be generated. For example, hot(m) generates an m-by-3 matrix containing the RGB values of colors ranging from black, through shades of red, orange, and yellow, to white.

27.2 USING COLORMAPS

The statement colormap(M) installs the matrix M as the colormap to be used in the current *Figure* window. For example, colormap(cool) installs a 64-entry version of the cool colormap. The form colormap default installs the default colormap, usually jet.

Line plotting functions, such as plot and plot3, do not use colormaps; they use the colors listed in the plot color and linestyle table. The sequence of colors used by these functions varies, depending on the plotting style you have chosen. Most other plotting functions, such as mesh, surf, contour, fill, pcolor, and their variations, use the current colormap to determine color sequences.

Plotting functions that accept a *color* argument usually accept the argument in one of three forms:

(1) a character string representing one of the colors in the plot color and linestyle table, for example, 'r' or 'red' for red;

(2) a three-entry row vector representing a single RGB value—for example, [.25 .50 .75];

(3) an array. If the color argument is an array, the elements are scaled and used as indices into the current colormap.

27.3 DISPLAYING COLORMAPS

Colormaps can be viewed in a number of ways. One way is to view the elements in a colormap matrix directly:

```
>> hot(8)
ans =
       0.33333            0            0
       0.66667            0            0
             1            0            0
             1      0.33333            0
             1      0.66667            0
             1            1            0
             1            1          0.5
             1            1            1
>> gray(5)
ans =
             0            0            0
          0.25         0.25         0.25
           0.5          0.5          0.5
          0.75         0.75         0.75
             1            1            1
```

This example shows two standard colormaps. The first is an eight-element hot colormap, and the second is a five-element gray colormap. The gray colormap increments all three components equally, thereby producing various shades of gray.

A colormap is best visualized graphically. The pcolor and rgbplot functions are useful in this case, as in the following example:

```
>> n = 21;
>> map = copper(n);
>> colormap(map)
>> subplot(2,1,1)
>> [xx,yy] = meshgrid(0:n,[0 1]);
>> c = [1:n+1;1:n+1];
>> pcolor(xx,yy,c)
>> set(gca,'Yticklabel','')
```

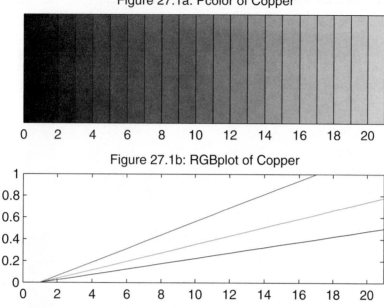

Figure 27.1a: Pcolor of Copper

Figure 27.1b: RGBplot of Copper

```
>> title('Figure 27.1a: Pcolor of Copper')
>> subplot(2,1,2)
>> rgbplot(map)
>> xlim([0,n])
>> title('Figure 27.1b: RGBplot of Copper')
```

These figures show how a copper colormap varies, from its first row (shown on the left) to its last row (shown on the right). The function `rgbplot` simply plots the three columns of the colormap in red, green, and blue, respectively, thereby dissecting the three components visually.

When a 3-D plot is made, the color information in the plot can be displayed as auxiliary information in a colorbar by using the `colorbar` function. For example, consider the following:

```
>> mesh(peaks)
>> axis tight
>> colorbar
>> title ('Figure 27.2: Colorbar Added')
```

In this plot, color is associated with the z-axis, and the colorbar associates z-coordinate values with the colors in the colormap.

Figure 27.2: Colorbar Added

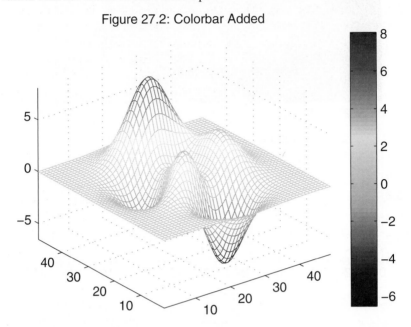

27.4 CREATING AND ALTERING COLORMAPS

The fact that colormaps are arrays means that you can manipulate them exactly like other arrays. The function brighten takes advantage of this feature to adjust a given colormap to increase or decrease the intensity of the colors. The form brighten(beta) brightens (0 < beta ≤ 1) or darkens (-1 ≤ beta < 0) the current colormap. The form brighten(beta) followed by brighten(-beta) restores the original colormap. The command newmap = brighten(beta) creates a brighter or darker version of the current colormap without changing the current map. The command mymap = brighten(cmap,beta) creates an adjusted version of the specified colormap without affecting either the current colormap or the specified colormap cmap.

Colormaps can be created by generating an m-by-3 array mymap and installing it with colormap(mymap). Each value in a colormap matrix must be between 0 and 1. If you try to use a matrix with more or less than three columns, or one containing any values less than 0 or greater than 1, colormap will report an error.

Colormaps can be converted between the red–green–blue standard and the hue–saturation–value standard by using the rgb2hsv and hsv2rgb functions. MATLAB, however, always interprets colormaps as RGB values. Colormaps can be combined as well, as long as the result satisfies the size and value constraints. For example, the colormap called pink is simply

```
pinkmap = sqrt(2/3*gray + 1/3*hot);
```

Again, the result is a valid colormap only if all elements of the m-by-3 matrix are between 0 and 1 inclusive.

Normally, a colormap is scaled to extend from the minimum to the maximum values of your data—that is, the entire colormap is used to render your plot. You may occasionally wish to change the way these colors are used. The caxis function, which stands for *color axis*, allows you to use the entire colormap for a subset of your data range or to use only a portion of the current colormap for your entire data set.

The statement [cmin, cmax] = caxis returns the minimum and maximum data values mapped to the first and last entries of the colormap, respectively. These are normally set to the minimum and maximum values of your data. For example, mesh(peaks) creates a mesh plot of the peaks function and sets caxis to [-6.5466, 8.0752], the minimum and maximum z values. Data points between these values use colors interpolated from the colormap.

The function caxis([cmin,cmax]) uses the entire colormap for data in the range between cmin and cmax. Data points greater than cmax are rendered with the color associated with cmax, and data points less than cmin are rendered with the color associated with cmin. If cmin is less than min(data), or cmax is greater than max(data), the colors associated with cmin or cmax will never be used. Only the portion of the colormap associated with data will be used. The function caxis('manual') fixes the axis scaling at the current range. The function caxis('auto') or the command form caxis auto restores the default values of cmin and cmax. The following example illustrates color axis settings:

```
>> N = 17;
>> data = [1:N+1;1:N+1]';

>> subplot(1,3,1)
>> colormap(jet(N))
>> pcolor(data)
>> set(gca,'XtickLabel','')
>> title('Figure 27.3: Auto Limits')
>> caxis auto        % automatic limits (default)

>> subplot(1,3,2)
>> pcolor(data)
>> axis off
>> title('Extended Limits')
>> caxis([-5,N+5]) % extend the color limits

>> subplot(1,3,3)
>> pcolor(data)
>> axis off
>> title('Restricted Limits')
>> caxis([5,N-5])  % restrict the color limits
```

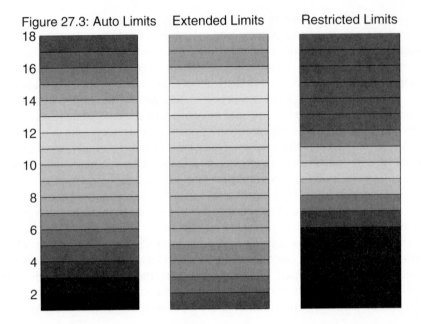

Figure 27.3: Auto Limits   Extended Limits   Restricted Limits

The first plot on the left is the default plot. This plot covers the complete colormap. In the center plot, the color axis is extended, forcing all plotted values to use a subset of the colormap. In the plot on the right, the color axis is restricted, forcing the colormap to cover a region in the center of the plot. The plot extremes simply use the colormap extremes.

27.5 USING COLOR TO DESCRIBE A FOURTH DIMENSION

Surface plots such as mesh and surf vary color along the z-axis, unless a color argument is given—for example, surf(X,Y,Z) is equivalent to surf(X,Y,Z,Z). Applying color to the z-axis produces a colorful plot, but does not provide additional information, since the z-axis already exists. To make better use of color, it is suggested that color be used to describe some property of the data not reflected by the three axes. To do so requires specifying different data for the color argument to 3-D plotting functions.

If the color argument to a plotting function is a vector or a matrix, it is scaled and used as an index into the colormap. This argument can be any real vector or matrix that is the same size as the other arguments. The following code is illustrative:

```
>> x = -7.5:.5:7.5;          % data
>> [X,Y] = meshgrid(x);      % create plaid data
>> R = sqrt(X.^2 + Y.^2)+eps;% create sombrero
>> Z = sin(R)./R;
```

```
>> subplot(2,2,1)
>> surf(X,Y,Z,Z)    % default color order
>> colormap(gray)
>> shading interp
>> axis tight off
>> title('Figure 27.4a: Default, Z')

>> subplot(2,2,2)
>> surf(X,Y,Z,Y) % Y axis color order
>> shading interp
>> axis tight off
>> title('Figure 27.4b: Y axis')

>> subplot(2,2,3)
>> surf(X,Y,Z,X-Y) % diagonal color order
>> shading interp
>> axis tight off
>> title('Figure 27.4c: X - Y')

>> subplot(2,2,4)
>> surf(X,Y,Z,R) % radius color order
```

Figure 27.4a: Default, Z   Figure 27.4b: Y axis

Figure 27.4c: X – Y   Figure 27.4d: Radius

```
>> shading interp
>> axis tight off
>> title('Figure 27.4d: Radius')
```

These subplots demonstrate four simple ways to use color as a fourth dimension. Whatever data are provided as the fourth argument to surf is used to interpolate the colormap. Any function of the first three arguments can be provided, or some completely independent variable works as well. Using the functions del2 and gradient allows one to apply color with respect to curvature and slope, respectively:

```
>> subplot(2,2,1)
>> surf(X,Y,Z,abs(del2(Z)))     % absolute Laplacian
>> colormap(gray)
>> shading interp
>> axis tight off
>> title('Figure 27.5a: |Curvature|')

>> subplot(2,2,2)
>> [dZdx,dZdy] = gradient(Z);  % compute gradient of surface
>> surf(X,Y,Z,abs(dZdx))       % absolute slope in x-direction
>> shading interp
>> axis tight off
>> title('Figure 27.5b: |dZ/dx|')

>> subplot(2,2,3)
>> surf(X,Y,Z,abs(dZdy))       % absolute slope in y-direction
>> shading interp
>> axis tight off
>> title('Figure 27.5c: |dZ/dy|')

>> subplot(2,2,4)
>> dR = sqrt(dZdx.^2 + dZdy.^2);
>> surf(X,Y,Z,abs(dR))         % absolute slope in radius
>> shading interp
>> axis tight off
>> title('Figure 27.5d: |dR|')
```

Figure 27.5a: |Curvature| Figure 27.5b: |dZ/dx|

Figure 27.5c: |dZ/dy| Figure 27.5d: |dR|

Note how color in these subplots provides an additional dimension to the plotted surface. The function de12 is the discrete Laplacian function that applies color based on the curvature of the surface. The function gradient approximates the gradient, or slope, of the surface with respect to the two coordinate directions.

27.6 TRANSPARENCY

Object transparency is another method that can be used to convey additional information in 3-D data visualization. Graphics objects such as surfaces, patches, or images can be made transparent or semi-transparent to show information that would normally be hidden, similar to the effect of the hidden command for mesh plots.

> Transparency functions are only effective on platforms that support OpenGL rendering modes. If OpenGL is not supported, the transparency functions have no effect.

Transparency values (known as ***alpha*** values) are similar to color values in that they are values in the range $0 \leq$ alpha ≤ 1, where 0 means completely transparent and 1 means completely opaque. The alpha(x) function, where x is a scalar, sets the transparency of all surface, patch, and image objects in the current axes to x. The alpha('astring') function, where 'astring' is one of 'clear', 'opaque', 'flat', 'interp', or 'texture' sets appropriate alpha values to the applicable objects. The form alpha(M), where M is a matrix the same size as the color data for the appropriate objects, sets an alpha value for each element in the object's data. The function alpha('alstr'), where 'alstr' is one of 'x', 'y', 'z', 'color', or 'rand', sets the alpha values to be the same as the x data, y data, z data, color data, or random values, respectively. The following code is illustrative:

```
>> subplot(2,2,1)
>> sphere
>> axis square off
>> alpha(0)
>> title({'Figure 27.6a:','Transparent, alpha = 0'})

>> subplot(2,2,2)
>> sphere
>> axis square off
>> alpha(1)
>> title({'Figure 27.6b:','Opaque, alpha = 1'})

>> subplot(2,2,3)
>> sphere
>> axis square off
>> alpha(0.5)
>> title({'Figure 27.6c:','Semi-transparent, alpha = 0.5'})

>> subplot(2,2,4)
>> sphere
```

Figure 27.6a:
Transparent, alpha = 0

Figure 27.6b:
Opaque, alpha = 1

Figure 27.6c:
Semi-transparent, alpha = 0.5

Figure 27.6d:
Graduated, alpha = 'color'

```
>> axis square off
>> alpha('color')
>> title({'Figure 27.6d:','Graduated, alpha = "color"'})
```

The alim function gets or sets the alpha limits similar to the caxis function. The alphamap function gets or sets the alphamap for a figure. The function alphamap(M), where M is an m-by-1 array of alpha values, sets the current figure's alphamap. As with the colormap function, MATLAB supplies a number of predefined alphamaps, including the following examples.

Alphamap Function	Description
alphamap('default')	Sets the alphamap to the default
alphamap('rampup')	Linear alphamap with increasing opacity
alphamap('rampdown')	Linear alphamap with decreasing opacity
alphamap('vup')	Transparent in the center increasing opacity toward each end
alphamap('vdown')	Opaque in the center decreasing opacity toward each end
alphamap('increase')	Increases opacity by 0.1 across the map
alphamap('decrease')	Increases transparency by 0.1 across the map
alphamap('spin')	Rotates the current alphamap
amap = alphamap;	Returns the current alphamap in amap

Finer control is available for alpha and alphamap by setting properties and values. Consult the MATLAB documentation for details.

27.7 LIGHTING MODELS

The graphics functions pcolor, fill, fill3, mesh, and surf discussed in the previous chapters render objects that appear to be well lit from all sides by very diffuse light. This technique emphasizes the characteristics of the objects in the *Figure* window and enhances the user's ability to visualize the data being analyzed. Although the data can be visualized quite clearly, the realism of the scene can be enhanced or diminished by creating different lighting effects.

The shading function selects faceted, flat, or interpolated shading. (Examples of each of these were illustrated in Chapter 26.) Although requiring more computational power and subsequently more time to render, interpolated shading of the objects in a scene can enhance the realism of the scene being rendered.

One or more light sources can be added to simulate the highlights and shadows associated with directional lighting. The function light creates a white light source infinitely far away along the vector [1 0 1]. Once the light source has been created, the lighting function allows you to select from four different lighting models: none (which ignores any light source), flat (the default when a light source is created), phong, and gouraud. Each of these models uses a different algorithm to change the appearance of the object. Flat lighting uses a uniform color for each face of the object. Gouraud lighting interpolates the face colors from the vertices. Phong lighting interpolates the normals at the vertices across each face and calculates the reflectance at each pixel. While colormaps are properties of *Figure* windows, light is a property, or child, of the axes. Therefore, each axis in a *Figure* window can be separately lit, as in the following example:

```
>> subplot(2,2,1)
>> sphere
>> light
>> shading interp
>> axis square off
>> lighting none
>> title('Figure 27.7a: No Lighting')

>> subplot(2,2,2)
>> sphere
>> light
>> shading interp
>> axis square off
>> lighting flat
>> title('Figure 27.7b: Flat Lighting')

>> subplot(2,2,3)
>> sphere
>> light
>> shading interp
>> axis square off
>> lighting gouraud
>> title('Figure 27.7c: Gouraud Lighting')

>> subplot(2,2,4)
>> sphere
>> light
>> shading interp
>> axis square off
>> lighting phong
>> title('Figure 27.7d: Phong Lighting')
```

Figure 27.7a: No Lighting

Figure 27.7b: Flat Lighting

Figure 27.7c: Gouraud Lighting

Figure 27.7d: Phong Lighting

In addition to lighting, the appearance of objects in an axis can be changed by modifying the apparent reflective characteristics, or ***reflectance***, of surfaces. Reflectance is made up of a number of components:

- ***Ambient light***—strength of the uniform directionless light in the figure
- ***Diffuse reflection***—intensity of the soft directionless reflected light
- ***Specular reflection***—intensity of the hard directional reflected light
- ***Specular exponent***—controls the size of the specular "hot spot" or spread
- ***Specular color reflectance***—determines the contribution of the surface color to the reflectance

Some predefined surface-reflectance properties are available by using the `material` function. Options include `shiny`, `dull`, `metal`, and `default` for restoring the default surface-reflectance properties. The form `material([ka kd ks n sc])`, where n and sc are optional, sets the ambient strength, diffuse reflectance, specular reflectance, specular exponent, and specular color reflectance of objects in an axis. The following code illustrates the first four options:

```
>> subplot(2,2,1)
>> sphere
>> colormap(gray)
>> light
>> shading interp
>> axis square off
>> material default
>> title('Figure 27.8a: Default Material')
```

```
>> subplot(2,2,2)
>> sphere
>> light
>> shading interp
>> axis square off
>> material shiny
>> title('Figure 27.8b: Shiny Material')

>> subplot(2,2,3)
>> sphere
>> light
>> shading interp
>> axis square off
>> material dull
>> title('Figure 27.8c: Dull Material')

>> subplot(2,2,4)
>> sphere
>> light
>> shading interp
```

Figure 27.8a: Default Material   Figure 27.8b: Shiny Material

Figure 27.8c: Dull Material   Figure 27.8d: Metal Material

```
>> axis square off
>> material metal
>> title('Figure 27.8d: Metal Material')
```

Note that the function light is fairly limited. It creates white light emanating from an infinite distance away along a given direction. In reality, light is a Handle Graphics object-creation function. The function light offers a variety of properties that can be set. For example, the light color, position, and style can be set, with style meaning that the light can be a point source at a given position or can be infinitely far away along some vector:

```
>> H1 = light('Position',[x,y,z],'Color',[r,g,b],'Style','local');
```

This command creates a light source at position (x,y,z) using light color [r,g,b] and specifies that the position is a location ('local') rather than a vector ('infinite'). It also saves the handle of the *light* object (H1), which can be used to change the properties of the light source at a later time. For example,

```
>> set(H1,'Position',[1 0 1],'Color',[1 1 1],'Style','infinite');
```

sets the light source defined by the handle H1 back to its default characteristics. (For more information see Chapter 30 or consult the MATLAB online documentation.)

27.8 SUMMARY

The following table documents MATLAB functions for color and lighting:

Function	Description
light	Light object-creation function
lighting	Sets lighting mode (flat, gouraud, phong, or none)
lightangle	Positions light object in spherical coordinates
material	Sets material reflectance (default, shiny, dull, or metal)
camlight	Sets light object with respect to camera
brighten	Brightens or darkens colormap
caxis	Sets or gets color axis limits
diffuse	Finds surface diffuse reflectance
specular	Finds surface specular reflectance
surfnorm	Computes surface normals
surfl	Creates surface plot with lighting

Function	Description
colorbar	Creates colorbar
colordef	Defines default color properties
colormap	Sets or gets *Figure* window colormap
colormapeditor	GUI for creating colormaps
hsv2rgb	Converts hue–saturation–value color values to red–green–blue model
rgb2hsv	Converts red–green–blue color values to hue–saturation–value model
rgbplot	Plots colormap
shading	Selects surface shading (flat, faceted, or interp)
spinmap	Spins colormap
whitebg	Changes plot axes background color
graymon	Sets graphics defaults for grayscale monitors
autumn	Colormap with shades of red and yellow
bone	Gray-scale colormap with a tinge of blue
cool	Colormap with shades of cyan and magenta
copper	Colormap with linear copper tone
flag	Colormap with alternating red, white, blue, and black
gray	Colormap with linear gray scale
hot	Colormap with black, red, yellow, and white
hsv	Colormap based on hue, saturation, and value progression
jet	Colormap variant of hsv that starts with blue and ends with red
lines	Colormap based on line colors
prism	Colormap with alternating red, orange, yellow, green, blue, and violet
spring	Colormap with shades of magenta and yellow
summer	Colormap with shades of green and yellow
vga	16-color Windows VGA colormap
winter	Colormap with shades of blue and green
alim	Gets or sets alpha limits
alpha	Gets or sets alpha properties
alphamap	Gets or sets alphamap

28

Images, Movies, and Sound

MATLAB provides commands for displaying several types of images. Images can be created and stored as standard double-precision floating-point numbers (`double`) and, optionally, as 8-bit (`uint8`) or 16-bit (`uint16`) unsigned integers. MATLAB can read and write image files in a number of standard graphics file formats, as well as use `load` and `save` to save image data in MAT-files. MATLAB provides commands for creating and playing animations as movies (sequences of frames). Sound functions are available as well for a number of sound formats.

28.1 IMAGES

Images in MATLAB consist of a data matrix and usually an associated colormap matrix. There are three types of image data matrices, each interpreted differently: indexed images, intensity images, and truecolor, or RGB, images.

An *indexed image* requires a colormap and interprets the image data as indices into the colormap matrix. The colormap matrix is a standard colormap—any m-by-3 array containing valid RGB data. Given an image data array $X(i,j)$ and a colormap array `cmap`, the color of each image pixel P_{ij} is `cmap(X(i,j),:)`. This implies that the data values in X are integers within the range [1 `length(cmap)`]. This image can be displayed by using the following code:

```
>> image(X); colormap(cmap)
```

An *intensity image* scales the image data over a range of intensities. This form is normally used with images to be displayed in gray scale or one of the other monochromatic colormaps, but other colormaps can be used if desired. The image

529

data are not required to be in the range [1 length(cmap)], as is the case with indexed images. The data are scaled over a given range, and the result is used to index into the colormap. For example,

```
>> imagesc(X,[0 1]); colormap(gray)
```

associates the value 0 with the first colormap entry, and the value 1 with the last colormap entry. Values in X between 0 and 1 are scaled and used as indices into the colormap. If the scale is omitted, it defaults to [min(X(:)) max(X(:))].

A *truecolor*, or **RGB, image** is created from an m-by-n-by-3 data array containing valid RGB triples. The row and column dimensions specify the pixel location, and the page of third dimension specifies each color component. For example, pixel P_{ij} is rendered in the color specified by X(i,j,:). A colormap is not required, because the color data are stored within the image data array itself. For example,

```
>> image(X)
```

displays the image, where X is an m-by-n-by-3 truecolor, or RGB, image. The image X can contain uint8, uint16, or double data.

If images are displayed on default axes, the aspect ratio will often be incorrect and the image will be distorted. Issuing

```
>> axis image off
```

sets the axis properties so that the aspect ratio matches the image and the axis labels and ticks are hidden. To force each pixel in the image to occupy one pixel on the display requires setting *figure* and *axes* object properties, as in the following example:

```
>> load clown        % sample image
>> [r,c] = size(X); % pixel dimensions
>> figure('Units','Pixels','Position',[100 100 c r])
>> image(X)
>> set(gca,'Position',[0 0 1 1])
>> colormap(map)
```

Here, the *figure* is set to display exactly the same number of pixels as the image by setting its width and height equal to that of the image. Then, the *axes* position is set to occupy the entire figure in normalized units.

MATLAB installations have a number of sample images in addition to clown.mat used in the preceding example. The demos subdirectory on the MATLAB path contains cape.mat, clown.mat, detail.mat, durer.mat, earth.mat, flujet.mat, gatlin.mat, mandrill.mat, and spine.mat. Each of these files contain images that can be displayed by issuing

```
>> load filename
>> image(X), colormap(map)
>> title(caption)
>> axis image off
```

28.2 IMAGE FORMATS

The image and imagesc commands can display 8- and 16-bit images without first converting them to the double format. However, the range of data values for uint8 is [0 255], as supported in standard graphics file formats, and the range of data values for uint16 is [0 65535].

For indexed uint8 images, image maps the value 0 to the first entry in a 256-entry colormap and the value 255 to the last entry, by automatically supplying the proper offset. Since the normal range of double data for indexed images is [1 length(cmap)], converting between uint8 and double or uint16 and double requires shifting the values by 1. Before MATLAB 7, mathematical operations on uint8 and uint16 arrays were not defined. As a result, to perform mathematical operations on unsigned integers in previous versions of MATLAB requires conversion from integers to the double format. For example, the code

```
>> Xdouble = double(Xuint8) + 1;
>> Xuint8 = uint8(Xdouble - 1);
```

converts the uint8 data in Xuint8 to double and back, taking into account the offset of 1. With the introduction of arithmetic for integer data types in MATLAB 7, it is now possible to perform operations without conversion to double precision. This facilitates easier image manipulation in MATLAB.

For 8-bit intensity and RGB images, the range of values is normally [0 255] rather than [0 1]. To display 8-bit intensity and RGB images, use either of the following commands:

```
>> imagesc(Xuint8,[0 255]); colormap(cmap)
>> image(Xuint8)
```

The 8-bit color data contained in an RGB image is automatically scaled when it is displayed. For example, the color white is normally [1 1 1] when doubles are used. If the same color is stored as 8-bit data, it is represented as [255 255 255]. Conversion of an RGB image between uint8 and double can also be normalized, as in the following code:

```
>> Xdouble = double(Xuint8)/255;
>> Xuint8 = uint8(round(Xdouble*255));
```

Conversion of `uint16` RGB images are normalized the same way, using 65535 rather than 255 as the constant.

The `Xrgb = ind2rgb(Xind,cmap)` function converts an indexed image and associated colormap into an RGB image. The indexed image can be `double`, `uint8`, or `uint16`, while the resulting RGB image is always a `double`. The companion function `rgb2ind` converts an RGB image into an indexed image and associated colormap using uniform quantization, minimum variance quantization, or colormap approximation, depending on usage. See the help text for `rgb2ind` for details. Other image conversion functions available in MATLAB include `imapprox`, `dither`, and `cmunique`. See the help text for complete descriptions and usage of these functions.

The optional *Image Processing Toolbox* available for MATLAB contains many additional functions for manipulating images. This toolbox is valuable if you regularly manipulate images.

28.3 IMAGE FILES

Image data can be saved to files and reloaded into MATLAB using many different file formats. The normal MATLAB `save` and `load` functions support image data in `double`, `uint8`, or `uint16` format in the same way that they support any other MATLAB variable and data type. When saving indexed images or intensity images with nonstandard colormaps, be sure to save the colormap as well as the image data, using, for example, the command

```
>> save myimage.mat X map
```

MATLAB also supports a number of industry-standard image file formats using the `imread` and `imwrite` functions. Information about the contents of a graphics file can be obtained with the `imfinfo` function. The help text for `imread` gives extensive information regarding image read formats and features. Following is a list of the major formats supported.

Type	Description	Extensions
BMP	Windows Bitmap	bmp
CUR	Windows Cursor resources	cur
GIF	Graphics Interchange Format	gif
HDF	Hierarchical Data Format	hdf
ICO	Windows Icon resources	ico
JPEG	Joint Photographic Experts Group	jpg, jpeg

JPEG 2000	JPEG 2000	`j2c, j2k, jp2, jpf, jpx`
PBM	Portable Bitmap	`pbm`
PCX	Windows Paintbrush	`pcx`
PGM	Portable Graymap	`pgm`
PNG	Portable Network Graphics	`png`
PNM	Portable Any Map	`pnm`
PPM	Portable Pixmap	`ppm`
RAS	Sun Raster	`ras`
TIFF	Tagged Image File Format	`tif, tiff`
XWD	X Window Dump	`xwd`

The calling syntax for `imwrite` varies, depending on the image type and file format. The help text for `imwrite`, shown in part as follows, gives extensive information regarding image save formats and features:

```
Table: summary of supported image types
---------------------------------------

BMP        1-bit, 8-bit and 24-bit uncompressed images

GIF        8-bit images

HDF        8-bit raster image datasets, with or without associated
           colormap; 24-bit raster image datasets; uncompressed or
           with RLE or JPEG compression

JPEG       8-bit, 12-bit, and 16-bit Baseline JPEG images

JPEG2000   1-bit, 8-bit, and 16-bit JPEG2000 images

PBM        Any 1-bit PBM image, ASCII (plain) or raw (binary) encoding.

PCX        8-bit images

PGM        Any standard PGM image. ASCII (plain) encoded with
           arbitrary color depth. Raw (binary) encoded with up
           to 16 bits per gray value.
```

PNG	1-bit, 2-bit, 4-bit, 8-bit, and 16-bit grayscale images; 8-bit and 16-bit grayscale images with alpha channels; 1-bit, 2-bit, 4-bit, and 8-bit indexed images; 24-bit and 48-bit truecolor images; 24-bit and 48-bit truecolor images with alpha channels
PNM	Any of PPM/PGM/PBM (see above) chosen automatically.
PPM	Any standard PPM image. ASCII (plain) encoded with arbitrary color depth. Raw (binary) encoded with up to 16 bits per color component.
RAS	Any RAS image, including 1-bit bitmap, 8-bit indexed, 24-bit truecolor and 32-bit truecolor with alpha.
TIFF	Baseline TIFF images, including 1-bit, 8-bit, 16-bit, and 24-bit uncompressed images, images with packbits compression, images with LZW compression, and images with Deflate compression; 8-bit and 24-bit images with JPEG compression; 1-bit images with CCITT 1D, Group 3, and Group 4 compression; CIELAB, ICCLAB, and CMYK images.
XWD	8-bit ZPixmaps

28.4 MOVIES

Animation in MATLAB takes one of two forms. First, if the computations needed to create a sequence of images can be performed quickly enough, *figure* and *axes* properties can be set so that screen rendering occurs sufficiently quickly, making the animation visually smooth. On the other hand, if computations require significant time or the resulting images are complex enough, you must create a movie.

   In MATLAB, the functions getframe and movie provide the tools required to capture and play movies. The getframe command takes a snapshot of the current *figure*, and movie plays back the sequence of frames after they have been captured. The output of getframe is a structure containing all of the information needed by movie. Capturing multiple frames is simply a matter of adding elements to the structure array. The following code is illustrative:

```
% moviemaking example: rotate a 3-D surface plot
[X,Y,Z] = sphere(50);      % create data
surf(X,Y,Z,X)              % plot the sphere
axis vis3d tight off       % fix axes for 3D and turn off axes ticks, etc.
for k = 1:25               % rotate and capture each frame
   view(-37.5+15*(k-1),30) % change the viewpoint for this frame
   m(k) = getframe(gcf);   % add this figure to the frame structure array
end                        % end of loop
movie(gcf,m)               % play the movie in the existing figure window
```

The preceding script file creates a movie by incrementally rotating the sphere and capturing a frame at every increment. Finally, the movie is played in the same figure window. The variable m contains a structure array, with each array element containing a single frame:

```
>> m

m =

1x25 struct array with fields:
      cdata
      colormap

>> m(1)
ans =
   cdata    : [420x560x3 uint8]
   colormap: []
```

The color data holding the image cdata make up a truecolor, or RGB, bitmap image. As a result, the complexity of contents of the *axes* does not influence the bytes required to store a movie. The size of the *axes* in pixels determines the size of the image and therefore the number of bytes required to store a movie. Because the color data is stored in cdata, a colormap is unnecessary and is always empty for RGB images.

Conversion between indexed images and movie frames is possible with the im2frame and frame2im functions. For example,

```
>> [X,cmap] = frame2im(M(n))
```

converts the nth frame of the movie matrix M into an indexed image X and associated colormap cmap. Similarly,

```
>> M(n) = im2frame(X,cmap)
```

converts the indexed image X and colormap cmap into the nth frame of the movie matrix M. Note that im2frame can be used to convert a series of images into a movie in the same way that getframe converts a series of *figures* or *axes* into a movie.

28.5 MOVIE FILES

Movies can be saved to AVI-format (Audio Video Interleave format) files using the movie2avi function. The form movie2avi(m,'spin.avi') creates the file spin.avi in the current directory from the MATLAB movie structure array m using default settings. The help text for movie2avi lists the available parameters:

```
>> help movie2avi
MOVIE2AVI(MOV,FILENAME) Create AVI movie from MATLAB movie
    MOVIE2AVI(MOV,FILENAME) creates an AVI movie from the MATLAB movie MOV.
    MOVIE2AVI(MOV,FILENAME,PARAM,VALUE,PARAM,VALUE...) creates an AVI movie
    from the MATLAB movie MOV using the specified parameter settings.

    Available parameters

    FPS            - The frames per second for the AVI movie. The
                     default is 15 fps.

    COMPRESSION    - A string indicating the compressor to use. On UNIX,
                     this value must be 'None'. Valid values for this param-
                     eter on Windows are 'Indeo3', 'Indeo5', 'Cinepak',
                     'MSVC', or 'None'. To use a custom compressor, the
                     value can be the four character code as specified by
                     the codec documentation. An error will result if the
                     specified custom compressor can not be found. The
                     default is 'Indeo5' on Windows and 'None' on UNIX.

    QUALITY        - A number between 0 and 100. This parameter has no
                     effect on uncompressed movies. Higher quality numbers
                     result in higher video quality and larger file sizes,
                     where lower quality numbers result in lower video
                     quality and smaller file sizes. The default is 75.
```

KEYFRAME   - For compressors that support temporal compression, this is the number of key frames per second. The default is 2 key frames per second.

COLORMAP   - An M-by-3 matrix defining the colormap to be used for indexed AVI movies. M must be no greater than 256 (236 if using Indeo compression). There is no default colormap.

VIDEONAME   - A descriptive name for the video stream. This parameter must be no greater than 64 characters long. The default name is the filename.

See also avifile, aviread, aviinfo, movie.

Reference page in Help browser
    doc movie2avi

The avifile, aviread, and aviinfo functions mentioned in the help text along with the addframe function are depreciated and will be removed in future releases. The newer VideoReader and VideoWriter classes and methods replace all of the functionality of the depreciated functions and also support many additional video formats. The following code saves the movie m into an AVI file:

```
>> vo = VideoWriter('spin2.avi')
```

VideoWriter

General Properties:

                Filename: 'spin2.avi'
                    Path: '/home/bl/matlab/mm'
              FileFormat: 'avi'
                Duration: 0
Video Properties:

           ColorChannels: 3
                  Height:
                   Width:
              FrameCount: 0
               FrameRate: 30

```
    VideoBitsPerPixel:      24
        VideoFormat:        'RGB24'
VideoCompressionMethod:     'Motion JPEG'
            Quality:        75
```

Methods

```
>> open(vo)
>> writeVideo(vo,m)
>> close(vo)
```

Video file formats supported by `VideoWriter` can be listed using the `getProfiles` method:

```
>> VideoWriter.getProfiles()
Summary of installed VideoWriter profiles:

    Name                        Description
------------------- -----------------------------
Archival            Video file compression with JPEG 2000 codec
with lossless mode enabled.
    Motion JPEG 2000 Video file compression with JPEG 2000 codec.
    Motion Jpeg AVI An AVI file with Motion JPEG compression
    Uncompressed AVI An AVI file with uncompressed RGB24 video data
```

Video file formats supported by `VideoReader` vary by platform and by installed software codecs:

```
>> VideoReader.getFileFormats() % Linux platform example
Video File Formats:
    .avi - AVI File
    .mj2 - Motion JPEG2000
    .ogg - Ogg File
    .ogv - Ogg Video

>> VideoReader.getFileFormats() % Windows platform example
Video File Formats:
    .asf - ASF File
    .asx - ASX File
```

```
.avi - AVI File
.mj2 - Motion JPEG2000
.mpg - MPEG-1
.wmv - Windows Media Video
```

The VideoReader function creates a *VideoReader* object which can then be used to get information about a video file and to import data into MATLAB. The following code is illustrative.

```
>> v1 = VideoReader('spin.avi')
Summary of Multimedia Reader Object for 'spin.avi'.
  Video Parameters: 15.00 frames per second, RGB24 560x420.
                    25 total video frames available.

>> get(v1) % display more detail
General Settings:
     Duration = 1.6667
     Name = spin.avi
     Path = /home/bl/matlab/mm
     Tag =
     Type = VideoReader
     UserData = []

Video Settings:
     BitsPerPixel = 24
     FrameRate = 15
     Height = 420
     NumberOfFrames = 25
     VideoFormat = RGB24
     Width = 560

>> f3 = read(v1,3); % read frame 3 of the file into f3
>> size(f3)         % this is a single RGB image
ans =
   420    560    3
>> class(f3)        % f3 contains uint8 data
ans =
uint8

> image(f3)         % render the image
```

Note that while *VideoWriter* objects must be opened and closed, *VideoReader* objects do not. The object-creation functions `VideoWriter` and `VideoReader` return errors if the file cannot be created, or cannot be read, respectively.

28.6 SOUND

Sound manipulation in MATLAB is supported by the versatile *audioplayer* and *audiorecorder* objects and associated methods. The `audioplayer` function creates an *audioplayer* object from supplied data or from an *audiorecorder* object. This *audioplayer* object can be manipulated with methods such as `get`, `set`, `play`, `pause`, `resume`, `stop`, `isplaying`, and `playblocking`. The `audiorecorder` function creates an *audiorecorder* object for recording sound data from a sound card. Similar methods are available including `get`, `set`, `record`, `pause`, `resume`, `stop`, `isrecording`, and `recordblocking`. The `getaudiodata` method creates an array that stores the recorded signal values. The `getplayer` and `play` methods both create *audioplayer* objects. The latter plays the recorded audio as well. Consult the online documentation for detailed usage and additional options.

In addition to these high-level functions, MATLAB supports sound by using a variety of lower-level functions. The function `sound(y,f,b)` sends the signal in vector y to the computer's speaker at sample frequency f. If y is a two-column matrix rather than a vector, the sound is played in stereo on supported platforms. Values in y outside of the range [-1 1] are clipped. If f is omitted, the default sample frequency of 8192 Hz is used. MATLAB plays the sound by using b bits per sample, if possible. Most platforms support b = 8 or b = 16. If b is omitted, b = 8 is used.

The `soundsc` function is the same as `sound`, except that the values in y are *scaled* to the range [-1 1] rather than clipped. This results in a sound that is as loud as possible without clipping. An additional argument is available that permits mapping a range of values in y to the full sound range. The format is `soundsc(y,...,[smin smax])`. If omitted, the default range is [min(y) max(y)].

Two industry-standard sound file formats are supported in MATLAB. NeXT/Sun audio format (*file*.au) files and Microsoft WAVE format (*file*.wav) files can be written and read.

The NeXT/Sun audio sound storage format supports multichannel data for 8-bit mu-law, 8-bit linear, and 16-bit linear formats. The most general form of `auwrite` is `auwrite(y,f,b,'method','filename')`, where y is the sample data, f is the sample rate in hertz, b specifies the number of bits in the encoder, '*method*' is a string specifying the encoding method, and '*filename*' is a string specifying the name of the output file. Each column of y represents a single channel. Any value in y outside of the range [-1 1] is clipped prior to writing the file. The f, b, and '*method*' arguments are optional. If omitted, f = 8000, b = 8, and method = 'mu'. The *method* argument must be either 'linear' or 'mu'. If the file name string contains no extension, '.au' is appended.

Conversion between mu-law and linear formats can be performed by using the `mu2lin` and `lin2mu` functions. (More information about the exact conversion processes involved with these two functions can be found by using the online help.)

Multichannel 8-, 16-, 24-, and 32-bit WAVE format sound files can be created with the wavwrite function. The most general form is wavwrite(y,f,b,'*filename*'), where y is the sample data, f is the sample rate in hertz, b specifies the number of bits in the encoder, and '*filename*' is a string specifying the name of the output file. Each column of y represents a single channel. Any value in y outside of the range [-1 1] is clipped prior to writing the file. The f and b arguments are optional. If omitted, f = 8000 and b = 16. If the file name string contains no extension, '.wav' is appended.

Both auread and wavread have the same syntax and options. The most general form is [y,f,b] = auread('*filename*',n), which loads a sound file specified by the string '*filename*' and returns the sampled data into y. The appropriate extension (.au or .wav) is appended to the file name if no extension is given. Values in y are in the range [-1 1]. If three outputs are requested, as previously illustrated, the sample rate in hertz and the number of bits per sample are returned in f and b, respectively. If n is given, only the first n samples are returned from each channel in the file. If n = [n1 n2], only samples from n1 through n2 are returned from each channel. The form [samples,channels] = wavread('*filename*','size') returns the size of the audio data in the file, rather than the data itself. This form is useful for preallocating storage or estimating resource use.

28.7 SUMMARY

The following table summarizes the image, movie, and sound capabilities in MATLAB:

Function	Description
image	Creates indexed or truecolor (RGB) *image* object
imagesc	Creates intensity *image* object
colormap	Applies colormap to *image*
axis image	Adjusts axis scaling for *image*
uint8	Conversion to unsigned 8-bit integer
uint16	Conversion to unsigned 16-bit integer
double	Conversion to double precision
imread	Reads image file
imwrite	Writes image file
imfinfo	Image file information
imapprox	Approximates indexed image by one with fewer colors
dither	Converts RGB image to indexed image using dithering
cmunique	Eliminates unneeded colors in colormap; converts RGB or intensity image to indexed image

Function	Description
imformats	Manages image file format registry
getframe	Captures movie frame from axis or figure
movie	Plays movie from movie structure
frame2im	Converts movie frame to image
im2frame	Converts image to movie frame
im2java	Converts image to Java image
rgb2ind	Converts RGB image to indexed image
ind2rgb	Converts indexed image to RGB image
avifile	Creates AVI movie file
addframe	Adds frame to AVI movie file
close	Closes AVI movie file
aviread	Reads AVI movie file
aviinfo	Information about an AVI movie file
movie2avi	Converts movie in MATLAB format to AVI format
VideoReader	Multimedia reader object and methods
VideoWriter	Multimedia writer object and methods
audiorecorder	Audio recorder object and methods
audioplayer	Audio player object and methods
audiodevinfo	Audio device information (Windows platform only)
sound	Plays vector as sound
soundsc	Autoscales and plays vector as sound
wavplay	Plays WAVE format sound file (depreciated; use audioplayer instead)
wavrecord	Records sound using audio input device (depreciated; use audiorecorder instead)
wavread	Reads WAVE format sound file
wavwrite	Writes WAVE format sound file
auread	Reads NeXT/Sun sound file
auwrite	Writes NeXT/Sun sound file
lin2mu	Converts linear audio to mu-law
mu2lin	Converts mu-law audio to linear

29

Printing and Exporting Graphics

MATLAB graphics are very effective tools for data visualization and analysis. Therefore, it is often desirable to create hard-copy output or to use these graphics in other applications. MATLAB provides a very flexible system for printing graphics and creating output in many different graphics formats including JPEG, EPS, and TIFF. Most other applications can import one or more of the graphics file formats supported by MATLAB.

Perhaps the most important issue to recognize when printing or exporting graphics is that the *figure* is **rerendered** in the process. That is, what you see on the screen is *not* what you get on a printed page or in an exported file. By default, MATLAB can choose a different renderer (`painters`, `zbuffer`, or `OpenGL`), can change the *axes* tick-mark distribution, and can change the size of the *figure* being printed and exported. Naturally, MATLAB provides the capability to enforce what-you-see-is-what-you-get (WYSIWYG), but this is not the default action taken.

In general, printing and exporting are not simple, because there is an almost uncountable number of possible combinations involving printer drivers, printer protocols, graphics file types, renderers, bit-mapped versus vector graphics descriptions, dots-per-inch selection, color versus black and white, compression, platform limitations, and so on. Most of this complexity is hidden when default output is requested. However, when very specific printed characteristics are required, or when exporting a *figure* for insertion into a word processing or presentation document, the complexities of printing and exporting must be understood. In these cases, it is not uncommon to spend a significant amount of time tweaking output until it matches the characteristics desired.

This chapter introduces the printing and exporting capabilities of MATLAB. As in other areas, MATLAB provides menu items, as well as *Command* window functions, for printing and exporting. The menu approach offers convenience, but limited flexibility, whereas the function approach offers complete flexibility, but requires much more knowledge on the part of the user.

Accessing the online documentation for the function `print` by issuing `>> doc print` provides a great deal of information about printing and exporting.

29.1 PRINTING AND EXPORTING USING MENUS

When graphics are displayed in a *Figure* window, the top of the window contains a menu bar and possibly one or more toolbars. The menu bar has a number of menus, including **File**, **Edit**, **View**, **Insert**, **Tools**, **Desktop**, **Window**, and **Help**. Of these, the **File** menu lists menu items for printing and exporting the current *figure*. In addition, the **Edit** menu on the Windows platform contains menu items for exporting the current *figure* to the system clipboard. The **File** menu includes the menu items **Export Setup**, **Print Preview**, and **Print**. Each of these items offers dialog boxes for setting various aspects of printing and exporting.

The **Export Setup** menu item makes it possible to save the current *figure* in one of many graphics formats. The **Export Setup** dialog box presents an opportunity to set many properties including size and features of the saved *figure*. Customized feature sets (export styles) may be saved and loaded as well. The **Export** button brings up an **Export** dialog, where the user chooses a directory, file name, and file type for the current *figure*.

The **Print Preview** menu item (or the `printpreview` command) opens a window showing how the *figure* will be rendered on the printed page. Options are provided to set many properties of the printed page including page layout, line, text, and color properties, and the renderer used to print the *figure*. Here also, feature sets may be saved and loaded. After any necessary changes are made, a button is available to directly call the **Print** dialog box to print the *figure*.

The **Print** menu item, the **Print** button in the **Print Preview** dialog box, or the `printdlg` command opens a standard **Print** dialog box, from which the printer and the number of copies can be chosen. The option to print to a file is available here as well. The printer, the paper type and source, and other options can be selected. Finally, there is a **Print** button to send the graphic to the printer (or to open a file dialog box.)

On the Windows platform, the **Edit** menu provides export capabilities through the system clipboard. The **Copy Figure** menu item places the current *figure* on the clipboard based on the options set in the preferences dialog box opened by selecting the **Copy Options** menu item. The current *figure* can be copied in either bitmap (BMP) or enhanced meta file (EMF) format. A template can be used to change line widths and font sizes, as well as to set other default options for copy operations. These options apply only when a *figure* is copied to the clipboard by using the **Copy Figure** menu item. They have no effect when printing or exporting a *figure* to a graphics file.

See "How to Print or Export" in the MATLAB Graphics documentation for details on all aspects of printing and exporting using menus and dialog boxes.

29.2 COMMAND LINE PRINTING AND EXPORTING

The function `print` handles all graphics printing and exporting from the *Command* window. This single function offers numerous options that are specified as additional input arguments. The syntax for the command form of `print` is

```
>> print -device -option -option filename
```

where all parameters or arguments are optional. The parameter *-device* specifies the device driver to be used, *-option* specifies one or more options, and *filename* specifies the name of an output file if the output is not sent directly to a printer. Because of command–function duality, `print` can be called as a function as well. For example,

```
>> print('-device','-option','-option','filename')
```

is equivalent to the preceding `print` command statement and is quite useful in scripts and functions.

In general, print options including *-device* can be specified in any order. Recognized print *-option* strings are described in the following table:

-option	Description
-adobecset	Selects PostScript default character set encoding (early PostScript printer drivers and EPS file formats only)
-append	Appends *figure* to existing file (PostScript printer drivers only)
-cmyk	Prints with CMYK colors instead of RGB colors (PostScript printer drivers and EPS file format only)
-device	Printer driver to be used
-dsetup	Displays the **Print Setup** dialog box (Windows only)
-fhandle	Specifies numerical *handle* of *figure* to print or export
-loose	Uses *loose* PostScript bounding box (PostScript, EPS, and Ghostscript only)
-noui	Suppresses printing of *uicontrol* objects
-opengl	Renders using OpenGL algorithm (bitmap format)
-painters	Renders using Painter's algorithm (vector format)
-Pprinter	Specifies name of printer to use (UNIX only)

-*option*	**Description**
-rnumber	Specifies resolution in dots per inch (dpi). Settable for most devices for printing. Settable for built-in MATLAB file formats except EMF and ILL. Not settable for many Ghostscript export formats. Default export resolution is 150 dpi for Z-buffer and OpenGL renderers, and 864 dpi for Painter's renderer.
-swindowtitle	Specifies name of SIMULINK system window to print or export
-v	Verbose. Displays the **Print** dialog box (Windows only)
-zbuffer	Renders using Z-buffer algorithm (bitmap format)

29.3 PRINTERS AND EXPORT FILE FORMATS

The print function supports a number of output devices (printers and file types). Many of these printers are supported through the use of Ghostscript printer drivers, which convert PostScript printer code to native printer code. This conversion process is transparent to the user, but limits printable fonts to those supported in PostScript.

The function print also supports exporting to a file or to the clipboard in a number of graphics formats. Export file formats are also specified by the -*device* option to print. Some of the supported file formats are shown in the following table:

-*device*	**Description**	**Type**	**Generator**
-dbmp16m	24-bit BMP	Bitmap	Ghostscript
-dbmp256	8-bit BMP with fixed colormap	Bitmap	Ghostscript
-dbmp	24-bit BMP	Bitmap	MATLAB
-dbmpmono	Monochrome BMP	Bitmap	Ghostscript
-dmeta	EMF (Windows only)	Vector	MATLAB
-deps	EPS level 1, black and white, including grayscale	Vector	MATLAB
-depsc	EPS level 1, color	Vector	MATLAB
-deps2	EPS level 2, black and white, including grayscale	Vector	MATLAB
-depsc2	EPS level 2, color	Vector	MATLAB
-hdf	HDF, 24-bit	Bitmap	MATLAB
-dill	Adobe Illustrator	Vector	MATLAB

`-djpeg`
 	
`-djpegNN`
	JPEG, 24-bit, quality setting of 75 (uses Z-buffer)
	JPEG, 24-bit, quality setting of NN
		Bitmap	MATLAB

`-dpbm`	PBM plain format	Bitmap	Ghostscript
`-dpbmraw`	PBM raw format	Bitmap	Ghostscript
`-dpcxmono`	PCX, 1-bit	Bitmap	Ghostscript
`-dpcx24b`	PCX, 24-bit color (three 8-bit planes)	Bitmap	Ghostscript
`-dpcx256`	PCX, 8-bit color (256 colors)	Bitmap	Ghostscript
`-dpcx16`	PCX, 16 colors (EGA/VGA)	Bitmap	Ghostscript
`-dpcxgray`	PCX, grayscale	Bitmap	Ghostscript
`-dpdfwrite`	PDF, color	Vector	Ghostscript
`-dpgm`	PGM portable graymap, plain	Bitmap	Ghostscript
`-dpgmraw`	PGM portable graymap, raw	Bitmap	Ghostscript
`-dpng`	PNG, 24-bit	Bitmap	MATLAB
`-dppm`	PPM portable pixmap, plain	Bitmap	Ghostscript
`-dppmraw`	PPM portable pixmap, raw	Bitmap	Ghostscript
`-dsvg`	Scalable Vector graphics	Vector	MATLAB
`-dtiff`	TIFF, 24-bit (rendered using Z-buffer)	Bitmap	MATLAB
`-dtiffnocompression`	TIFF, no compression	Bitmap	MATLAB
`-dtifflzw`	TIFF, LZW compression	Bitmap	MATLAB
`-dtiffpack`	TIFF, Packbits compression	Bitmap	MATLAB
`-dtiff24nc`	TIFF, 24-bit (no compression)	Bitmap	MATLAB
`-tiff`	Adds TIFF preview to EPS formats only—must be used in addition to an EPS device specification	Bitmap	MATLAB

Type `print -d` in the *Command* window for a list of all of the printer drivers supported on your platform.

MATLAB also provides methods for exporting images to graphics files. The functions `getframe`, `imwrite`, `avifile`, and `addframe` provide the capability to create and save image files from *figures*. (See Chapter 28 for more information.)

29.4 POSTSCRIPT SUPPORT

All PostScript devices, as well as devices that use Ghostscript, offer limited font support. This includes both devices that print and those that save images. The following table shows the supported fonts, in addition to common Windows fonts that map into standard PostScript fonts. (Note that fonts not listed in the table are mapped to `Courier`, and that the Windows devices `-dwin` and `-dwinc` use standard Windows printer drivers and therefore support all installed fonts.)

PostScript Font	Windows Equivalent
AvantGarde	
Bookman	
Courier	Courier New
Helvetica	Arial
Helvetica-Narrow	
NewCenturySchlBk	New Century Schoolbook
Palatino	
Symbol	
Times-Roman	Times New Roman
ZapfChancery	
ZapfDingbats	

If your printer supports PostScript, a built-in PostScript driver should be used. Level 1 PostScript is an older specification and is required for some printers. Level 2 or 3 PostScript produces smaller and faster code and should be used if possible.

If you are using a color printer, a color driver should be used. Black-and-white or grayscale printers can use either driver. However, when a color driver is used for a black-and-white printer, the file is larger and colors are dithered, making lines and text less clear in some cases. When colored lines are printed by using a black-and-white driver, they are converted to black. When colored lines are printed by using a color driver on a black-and-white printer, the lines are printed in grayscale. In this case, unless the lines have sufficient width, they often do not have sufficient contrast with the printed page.

As implemented in MATLAB, PostScript supports *surfaces* and *patches* (only with triangular faces) that have interpolated shading. When printed, the corresponding PostScript files contain color information at the *surface* or *patch* vertices, requiring the printer to perform the shading interpolation. Depending on the printer characteristics, this may take an excessive amount of time, possibly leading to a printer error. One way to solve this problem is to use flat shading along with a finer meshed surface. Another alternative that ensures that the printed output matches the screen image is to print by using either the Z-buffer or OpenGL renderer with a sufficiently high resolution. In this case, the output is in bitmap format and may result in a large output file, but no interpolation is required by the printer.

29.5 CHOOSING A RENDERER

A renderer processes graphics data such as arrays of vertices and color data into a format suitable for display, printing, or export. There are two major categories of graphics formats: bitmap (or raster) graphics and vector graphics.

Bitmap graphics formats contain information such as color at each point in a grid. As the divisions between points decrease, and the total number of points increases, the resolution of the resulting graphic increases, and the size of the resulting file increases. Increasing the number of bits used to specify the color of each point in the grid increases the total number of possible colors in the resulting graphic, and also increases the size of the resulting file. Increasing the complexity of the graphic has no effect on the size of the resulting output file.

Vector graphics formats contain instructions for re-creating the graphic by using points, lines, and other geometric objects. Vector graphics are easily resized and usually produce higher-resolution results than bitmap graphics. However, as the number of objects in the graphic increases, the number of instructions required to re-create the graphic increases, and the size of the resulting file increases. At some point, the complexity of the instructions can become too much for an output device to handle, and the graphic simply cannot be output on the specific output device.

MATLAB supports three rendering methods: OpenGL, Z-buffer, and Painter's. Painter's uses vector format, while OpenGL and Z-buffer produce bitmaps. By default, MATLAB automatically selects a renderer based on the characteristics of the *figure* and the printer driver or file format used.

> The renderer MATLAB selects for printing or exporting is not necessarily the same renderer used to display a *figure* on the screen.

MATLAB's default selection can be overridden by the user, and this is often done to make the printed or exported *figure* look the same as it does on the screen or to avoid embedding a bitmap in a vector-format output file, such as PostScript or EPS.

Examples of some situations that require specific renderers are as follows:

- If the *figure* uses RGB color rather than a single color for surface or patch objects, the graphic must be rendered by using a bitmap method to properly capture color.
- HPGL (`-dhpgl`) and Adobe Illustrator (`-dill`) output formats use the Painter's renderer.
- JPEG (`-djpeg`) and TIFF (`-dtiff`) output formats always use Z-buffer rendering.
- Lighting effects cannot be reproduced by using the vector-format Painter's renderer, and so a bitmap method must be used in this case as well.
- The OpenGL renderer is the only method that supports transparency.

The renderer used for printing and exporting can be chosen by using dialog boxes or options to the print command, or by setting Handle Graphics properties. The **Advanced** tab in the **Print Preview** dialog box can be used to select a renderer. The -zbuffer, -opengl, and -painters options to the print command select a specific renderer when printing or exporting that overrides any other selections made.

29.6 HANDLE GRAPHICS PROPERTIES

A number of Handle Graphics properties influence the way graphics are printed or exported. (Handle Graphics are discussed in more detail in Chapter 30.) Many of the options selected from printing and exporting dialog boxes make changes to these properties for the current *figure*. The following table lists the *figure* properties that influence printing and exporting:

'PropertyName'	'PropertyValue' choices, {default}	**Description**
Color	[RGB vector]	Sets *figure* background color
InvertHardcopy	[{on} | off]	Determines whether *figure* background color is printed or exported. When set to on, forces a white *figure* background independent of the Color property.
PaperUnits	[{inches} | centimeters | normalized | points]	Units used to measure the size of a printed or exported *figure (the default units may vary by locale)*
PaperOrientation	[{portrait} | landscape | rotated]	Orientation of *figure* with respect to paper
PaperPosition	[left bottom width height] vector	Position of *figure* on paper or in exported file. width and height determine the size of the *figure* printed or exported.
PaperPositionMode	[auto | {manual}]	Determines whether PaperPosition width, and height are used. When set to auto, the *figure* is printed or exported using the *Figure* window displayed width and height (i.e., output is WYSIWYG)
PaperSize	[width height] vector	Paper size measured in PaperUnits.

PaperType	[{usletter} | uslegal | A0 | A1 | A2 | A3 | A4 | A5 | B0 | B1 | B2 | B3 | B4 | B5 | arch-A | arch-B | arch-C | arch-D | arch-E | A | B | C | D | E | tabloid | <custom>]	Type of paper used. Selecting a PaperType sets the PaperSize accordingly. (Default may vary by locale.)
Renderer	[{painters} | zbuffer | OpenGL | None]	Specifies renderer
RendererMode	[{auto} manual]	Determines how the renderer is chosen. When set to auto, MATLAB chooses the renderer automatically and independently for display, printing, and export. When set to manual, the renderer set by the Renderer property is used for display, printing, and export.

Certain *axes* properties also influence printing and exporting. Particularly important are the *axes* tick mode properties

```
XTickMode [ {auto} | manual ]
YTickMode [ {auto} | manual ]
ZTickMode [ {auto} | manual ]
```

When a *figure* is printed or exported, the resulting graphic is normally rendered at a different size than the *figure* on the screen. Since the width and height are different, MATLAB can rescale the number and placement of tick marks on each axis to reflect the new size. Setting the tick mode properties to 'manual' prevents MATLAB from changing the tick marks on the *axes* when the *figure* is printed or exported.

Some *line* properties can be used to advantage for output as well:

```
Color: [ 3-element RGB vector]
LineStyle: [ {-} | -- | : | -. | none ]
LineWidth: [ scalar]
Marker: [ + | o | * | . | x | square | diamond | v | ^ | > | < | pentagram
        | hexagram | {none} ]
MarkerSize: [ scalar ]
MarkerEdgeColor: [ none | {auto} ] -or- a ColorSpec.
MarkerFaceColor: [ {none} | auto ] -or- a ColorSpec.
```

When colored lines are printed, the result depends on the capabilities of the output device and the printer driver. If a color printer and a color printer driver are used, the result is in color, as expected. If a black-and-white printer driver is used, the result is black and white. If a color printer driver is used with a black-and-white printer, the result is grayscale. This can be a problem, since grayscale is printed by using dithering, which may lead to lines that are not distinct. A different, but related, problem can occur when printing in black and white. When multiple solid lines are printed in black, they lose the distinction that color provides on the screen. In this case, the line properties `'Color'`, `'LineStyle'`, `'LineWidth'`, and `'Marker'` can be used to add distinction to plotted lines.

Finally, it is often advantageous to modify *text* when printing or exporting *figures*. Useful *text* properties include the following:

```
Color: [ 3-element RGB vector]
FontAngle: [ {normal} | italic | oblique ]
FontName: [ font name ]
FontSize: [ scalar ]
FontUnits: [ inches | centimeters | normalized | {points} | pixels ]
FontWeight: [ light | {normal} | demi | bold ]
```

Increasing the size of *text* strings such as titles and *axes* labels when printing or exporting to a smaller-sized graphic can often make the text easier to read. If the font used in the *figure* is not one of the 11 previously listed fonts supported by MATLAB for PostScript output, changing the font to one of these 11 can prevent unwanted font substitution. Sometimes a bold font shows up better in printed output than a normal-weight font. Changes such as these can often improve the appearance of the printed or exported output. Font characteristics are particularly important when exporting graphics for inclusion in presentation software, where fonts must be large and distinct enough to be readable from across a large room.

29.7 SETTING DEFAULTS

MATLAB sets the following factory default options for printing and exporting.

- *Figure* size is 8-by-6 inches (may be different outside of the United States).
- Orientation is portrait.
- *Figure* and *axes* background colors are inverted to white.
- U.S. letter (8.5-by-11 inch) paper is used if available (may be different outside of the United States).
- *Figure* is centered on the paper.
- *Figure* is cropped.
- Output is RGB (not CYMK).

- Tick marks are recalculated.
- MATLAB chooses the renderer.
- Uicontrols print.
- Print device is -dps2 on UNIX and -dwin on the Windows PC platform.

The default print device is set in the $TOOLBOX/local/printopt.m file. Edit this file to change the default print device across MATLAB sessions. If you do not have write access to this file, edit the file by using the command edit printopt.m, make your changes, save the printopt.m file into a local directory, and make sure the local directory is in the MATLABPATH before $TOOLBOX/local. For example, if you use a color PostScript printer for printed output on a UNIX platform, edit printopt.m and add the line dev = '-dpsc2'; to the file at the location specified in the file.

Other option defaults can be changed by setting default properties in the startup.m file. For example, set(0,'DefaultFigurePaperType','A4'); changes the default paper type to A4. The handle 0 addresses the *root* object. The 'DefaultFigurePaperType' property sets the default value for the *figure* property 'PaperType'. Adding the prefix 'Default' to any Handle Graphics property name specifies that the accompanying property value should be used as the default. Default properties for a given object must be set at a higher level in the Handle Graphics hierarchy. For example, default *figure* properties must be set at the *root* object level, and default *axes* properties must be set at the *figure* object or *root* object level. Therefore, defaults are usually set at the *root* level. For example, the code

```
set(0,'DefaultFigurePaperOrientation','landscape');

set(0,'DefaultFigurePaperPosition',[0.25 0.25 10.5 8]);

set(0,'DefaultAxesXGrid','on','DefaultAxesYGrid','on');

set(0,'DefaultAxesLineWidth','1');
```

tells MATLAB to use the landscape mode, fill the page with the plot, print (and display) *x*- and *y*-axis grids, and display and print lines that are 1 point wide, as defaults in all *figures* and *axes*.

Many other options can be set using Handle Graphics properties. Most printing and exporting properties are *figure* and *axes* properties. (For more information on Handle Graphics properties and default values, see Chapter 30.)

29.8 PUBLISHING

MATLAB provides additional options for publishing your formatted results, including graphics, to a number of different formats. Publishing in MATLAB provides the ability to execute M-files and capture the output in a formatted document. Document formats supported by MATLAB include HTML, XML, and LaTeX, as well as formats for Microsoft Office (Word or PowerPoint) and the OpenOffice.org equivalents. The MATLAB Notebook is another option for publishing on PC platforms, but is discouraged in the latest documentation.

Publishing requires that the user define *Cells* (sections of code) in the M-file to be executed. (See the documentation on the MATLAB Editor for additional information on defining Cells.) When an M-file is published, the M-file is executed and the output—including the source code in each Cell, along with any *Command* window output and graphical output—is formatted and published into document files in the selected format. Text markup is also supported. Additions can include an overall title, section titles, descriptive text, and equations and symbols using TeX formatting. Text may be formatted by using bold text, indented or preformatted text, bulleted lists, monospaced text, or even HTML links.

In addition to the **Publish** entries in the **File** menu of the MATLAB Editor, the `publish` function is available in the *Command* window. Publishing preferences can be set in the Editor/Debugger section of the MATLAB **Preferences** menu. Specific instructions and examples of publishing are available in the online documentation.

29.9 SUMMARY

MATLAB provides a very flexible system for printing graphics and creating output in many different graphics formats. Most other applications can import one or more of the graphics file formats supported by MATLAB, but many have limited ability to edit the resulting graphic. The best results are achieved if the *figure* is edited and appropriate options are set before the *figure* is printed or exported. The most widely used, flexible output formats are PostScript, EPS, EMF, and TIFF. MATLAB contains native support for all of these formats and uses Ghostscript to translate them into many others.

MATLAB also provides the ability to publish a document that includes portions of an M-file—along with *Command* window output, graphics output, descriptive text, and comments—in a number of document formats, including HTML, XML, LaTeX, Microsoft Word, and PowerPoint.

30

Handle Graphics

Handle Graphics is the name given to a large collection of low-level graphics features that specify how graphics behave and are displayed in MATLAB. Through interaction with Handle Graphics *objects* and their associated properties, you can gain almost infinite control over the graphical features available in MATLAB. As MATLAB continues to evolve, many Handle Graphics features can be manipulated interactively by using the numerous menus, contextual menus, toolbars, palettes, browsers, and editors available within *Figure* windows. These interactive tools make it possible to customize graphics while requiring minimal knowledge of Handle Graphics. As a result, detailed knowledge of Handle Graphics is required only when noninteractive graphics customization is desired. In this case, Handle Graphics function calls are commonly placed in an M-file.

Because of the diversity and breadth of the graphics features in MATLAB, the MATLAB-supplied documentation on graphics is very extensive. As a result, it is simply not possible to provide a comprehensive survey of Handle Graphics in this text. Doing so would at least double the size of this text.

Given the space limitations and broader goals of the text, this chapter develops a basic understanding of Handle Graphics features that provides a basis for learning the many more specific aspects of graphics in MATLAB.

30.1 OBJECTS

Handle Graphics is based on the idea that every component of a MATLAB graphic is an *object*, that each object has a unique identifier, or *handle*, associated with it, and that each object has *properties* that can be modified as desired. Here, the term *object* is defined as a closely related collection of data and functions that form a unique whole. In MATLAB, a graphics object is a distinct component that can be manipulated individually.

All plotting and graphics in MATLAB create graphics objects. Some produce **composite objects,** and some produce **core objects.** Graphical objects in MATLAB are arranged in a hierarchy of parent objects and associated child objects, as shown in the following diagram:

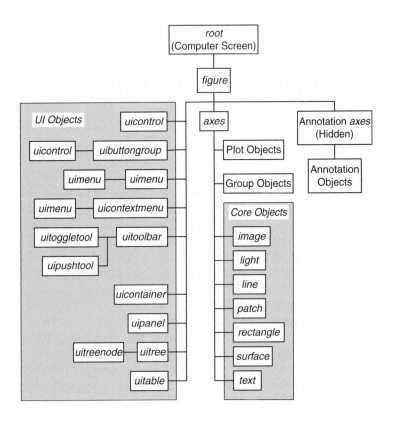

The computer screen itself is the *root* object and the parent of everything else. One or more *figure* objects are children of the *root*. The objects *axes, annotation axes*, and the collection of UI objects are children of *figures*. Plot objects, group objects, and the collection of core objects are children of *axes*. All objects that are children of the *figure* object in the hierarchy, except for the core objects, are called composite objects.

This hierarchy was new to MATLAB 7. In MATLAB 6, the graphics hierarchy did not document the hidden *annotation axes* and did not have **plot** objects or **group** objects. Core objects and some UI objects existed, but were simply not grouped and named. UI objects and core objects behave in the same way in MATLAB 7 as they did in previous MATLAB versions, whereas plot objects and group objects are new in MATLAB 7 and beyond.

30.2 OBJECT HANDLES

In MATLAB, each object has an associated identifier, called a *handle*, which is a double-precision number. Each time an object is created, a unique handle is created for it. The handle of the *root* object, the computer screen, is always zero. The `Hf_fig = figure` command creates a new *figure* and returns its handle in the `Hf_fig` variable. *Figure* handles are normally small integers and are usually displayed in the *Figure* window title bar. Other object handles are typically larger integers or noninteger numbers in full MATLAB precision. All object-creation functions return the handles of the objects they create.

Before MATLAB 7, the high-level graphics functions (discussed in Chapters 25 and 26) returned a column vector of handles to the core objects they created. For example, `Hl = plot(…)` returned handles to all *line* objects created by the `plot` function. Similarly, `Hs = surf(…)` returned a handle to a *surface* object. Now many high-level graphics creation functions return handles to plot objects. That is, `Hls = plot(…)` returns handles to *lineseries* plot objects. In the presence of these changes in the graphics hierarchy, fundamental operations using Handle Graphics remain unchanged. As a result, before these new features are addressed, the fundamental creation and manipulation of Handle Graphics objects through use of their handles is discussed.

It is suggested that variable names used to store handles start with an upper case H, followed by one or more letters identifying the object type, then an underscore, and finally one or more descriptive characters. Thus, `Hf_fig` is a handle to a *figure*, `Ha_ax1` is a handle to an *axes* object, and `Ht_title` is a handle to a *text* object. When an object type is unknown, the letter x is used, as in `Hx_obj`. While handles can be given any name, following this convention makes it easy to spot object handle variables in an M-file.

30.3 OBJECT PROPERTIES

All objects have a set of ***properties*** that define their characteristics. It is by setting these properties that you modify how graphics are displayed. The properties associated with each object type (e.g., *axes*, *line*, and *surface*) are unique, although a number of property names are common to all objects. Object properties include characteristics such as an object's position, color, object type, parent object handle, or child object handles. Each distinct object has properties associated with it that can be changed without affecting other objects of the same type.

Object properties consist of property ***names*** and their associated property ***values***. Property names are character strings. They are typically displayed in mixed case, with the initial letter of each word capitalized. For example, the property name `'LineStyle'` identifies the linestyle for a *line* object. When identifying a property name in an M-file or at the *Command* window prompt, MATLAB recognizes properties regardless of case. In addition, you need use only enough characters to uniquely identify the property name. For example, the position property of an *axes* object can be called `'Position'`, `'position'`, or even `'pos'`.

When an object is created, it is initialized with a full set of default property values that can be changed in either of two ways. The object-creation function can be issued with ('Property-name', Property-value) pairs, or property values can be changed after the object is created. For example,

```
>> Hf_1 = figure('Color','yellow')
```

creates a new *figure* with default properties, except that the background color is set to yellow rather than to the default color.

In addition to the *Figure* window menu and toolbar features in MATLAB, the function inspect provides a GUI for inspection and modification of object properties. To use this function, simply issue inspect(H), where H is the handle of the desired object.

30.4 Get AND Set

The two functions get and set are used to obtain or change Handle Graphics object properties from the *Command* window or an M-file. The function get returns the current value of one or more properties of an object. The most common syntax is get(handle, 'PropertyName'). For example,

```
>> p = get(Hf_1,'Position')
```

returns the position vector of the *figure* having the handle Hf_1. Similarly,

```
>> c = get(Hl_a,'Color')
```

returns the color of an object identified by the handle Hl_a.

The function set changes the values of Handle Graphics object properties and uses the syntax set(handle, 'PropertyName', PropertyValue). For instance,

```
>> set(Hf_1,'Position',p_vect)
```

sets the position of the *figure* having the handle Hf_1 to that specified by the vector p_vect. Likewise,

```
>> set(Hl_a,'Color','r')
```

sets the color of the object having the handle Hl_a to red. In general, the set function can have any number of ('PropertyName', PropertyValue) pairs. For example,

```
>> set(Hl_a,'Color',[1 0 0],'LineWidth',2,'LineStyle','--')
```

changes the color of the *line* having the handle Hl_a to red, its line width to 2 points, and its linestyle to dashed.

In addition to these primary purposes, the get and set functions provide feed-back about object properties. For example, set(handle, 'PropertyName') returns a list of values that can be assigned to the object described by handle. For example,

```
>> set(Hf_1,'Units')
[ inches | centimeters | normalized | points | {pixels} | characters ]
```

shows that there are six allowable character-string values for the 'Units' property of the *figure* referenced by Hf_1 and that 'pixels' is the default value.

If you specify a property that does not have a fixed set of values, MATLAB informs you of that fact:

```
>> set(Hf_1,'Position')
A figure's "Position" property does not have a fixed set of property values.
```

In addition to the set command, Handle Graphics object-creation functions accept multiple pairs of property names and values. For example,

```
>> figure('Color','blue','NumberTitle','off','Name','My Figure')
```

creates a new *figure* with a blue background entitled 'My Figure', rather than the default window title 'Figure 1'.

As an illustration of the preceding concepts, consider the following example:

```
>> Hf_fig = figure  % create a figure
Hf_fig =
     1
>> Hl_light = light % add default light to an axes in the figure
Hl_light =
         174
>> set(Hl_light)
    Position
    Color
    Style: [ {infinite} | local ]

    ButtonDownFcn: string -or- function handle -or- cell array
    Children
    Clipping: [ {on} | off ]
    CreateFcn: string -or- function handle -or- cell array
    DeleteFcn: string -or- function handle -or- cell array
```

```
    BusyAction: [ {queue} | cancel ]
    HandleVisibility: [ {on} | callback | off ]
    HitTest: [ {on} | off ]
    Interruptible: [ {on} | off ]
    Parent
    Selected: [ on | off ]
    SelectionHighlight: [ {on} | off ]
    Tag
    UIContextMenu
    UserData
    Visible: [ {on} | off ]

>> get(Hl_light) % get all properties and names for light
    Position = [1 0 1]
    Color = [1 1 1]
    Style = infinite

    BeingDeleted = off
    ButtonDownFcn =
    Children = []
    Clipping = on
    CreateFcn =
    DeleteFcn =
    BusyAction = queue
    HandleVisibility = on
    HitTest = on
    Interruptible = on
    Parent = [175.002]
    Selected = off
    SelectionHighlight = on
    Tag =
    Type = light
    UIContextMenu = []
    UserData = []
    Visible = on
```

The *light* object was used because it contains the fewest properties of any MATLAB graphics object. A *figure* was created and returned a handle. A *light* object was then created that returned a handle of its own. An *axes* object was also created, since *light* objects are children of *axes*; the *axes* handle is available in the 'Parent' property of the *light*.

Note that the property lists for each object are divided into two groups. The first group lists properties that are unique to the particular object type, and the second group lists properties common to all object types. Note also that the set and get functions return slightly different property lists. The function set lists only properties that can be changed with the set command, while get lists all visible object properties. In the previous example, get listed the 'Type' property, while set did not. This property can be read but not changed; that is, it is a read-only property.

The number of properties associated with each object type is fixed in each release of MATLAB, but the number varies among object types. As already shown, a *light* object lists 3 unique and 18 common properties, or 21 properties in all. On the other hand, an *axes* object lists 103 properties. Clearly, it is beyond the scope of this text to thoroughly describe and illustrate all of the properties of all object types!

As an example of the use of object handles, consider the problem of plotting a line in a nonstandard color. In this case, the line color is specified by using an RGB value of [1 .5 0], a medium orange color:

```
>> x = -2*pi:pi/40:2*pi;   % create data
>> y = sin(x);             % find sine of x
>> Hls_sin = plot(x,y)     % plot sine and save lineseries handle
Hls_sin =
   59.0002
>> set(Hls_sin,'Color',[1 .5 0],'LineWidth',3) % Change color and width
```

Now add a cosine curve in light blue:

```
>> z = cos(x);             % find cosine of x
>> hold on                 % keep sine curve
>> Hls_cos = plot(x,z);    % plot cosine and save lineseries handle
>> set(Hls_cos,'Color',[.75 .75 1]) % color it light blue
>> hold off
```

It's also possible to do the same thing with fewer steps:

```
>> Hls_line = plot(x,y, x,z);      % plot both curves and save handles
>> set(Hls_line(1),'Color',[1 .5 0],'LineWidth', 3)
>> set(Hls_line(2),'Color',[.75 .75 1])
```

How about adding a title and making the font size larger than normal?

```
>> title('Handle Graphics Example') % add a title
>> Ht_text = get(gca,'Title')       % get handle to title
>> set(Ht_text,'FontSize',16)       % customize font size
```

The last example illustrates an interesting point about *axes* objects. Every object has a `'Parent'` property, as well as a `'Children'` property, which contains handles to descendant objects. A *lineseries* object on a set of axes has the handle of the *axes* object in its `'Parent'` property and the empty array in the `'Children'` property. At the same time, the *axes* object has the handle of its *figure* in its `'Parent'` property and the handles of *lineseries* objects in the `'Children'` property. Even though *text* objects created with the `text` and `gtext` commands are children of *axes* and their handles are included in the `'Children'` property, the handles associated with the title string and axis labels are not. These handles are kept in the `'Title'`, `'XLabel'`, `'YLabel'`, and `'ZLabel'` properties of the *axes*. These *text* objects are always created when an *axes* object is created. The `title` command simply sets the `'String'` property of the title *text* object within the current *axes*. Finally, the standard MATLAB functions `title`, `xlabel`, `ylabel`, and `zlabel` return handles and accept property and value arguments. For example, the following command adds a 24-point green title to the current plot and returns the handle of the title *text* object:

```
>> Ht_title = title('This is a title.','FontSize',24,'Color','green')
```

In addition to `set` and `get`, MATLAB provides several other functions for manipulating objects and their properties. Objects can be ***copied*** from one parent to another using the `copyobj` function. For instance,

```
>> Ha_new = copyobj(Ha_ax1,Hf_fig2)
```

makes copies of the *axes* object with handle `Ha_ax1` and all of its children, assigns new handles, and places the objects in the *figure* with handle `Hf_fig2`. A handle to the new *axes* object is returned in `Ha_new`. Any object can be copied into any valid parent object based on the hierarchy described earlier. Either one or both arguments to `copyobj` can be vectors of handles.

Note that any object can be ***moved*** from one parent to another simply by changing the `'Parent'` property value to the handle of another valid parent object. For example,

```
>> figure(1)
>> set(gca,'Parent',2)
```

moves the current *axes* and all of its children from the *figure* having handle 1 to the *figure* having handle 2. Any existing objects in *figure* 2 are not affected, except that they may become obscured by the relocated objects.

Any object and all of its children can be **_deleted_** using the delete(handle) function. Similarly, reset(handle) resets all object properties associated with handle, except for the 'Position' and 'Units' properties, to the defaults for that object type. 'PaperPosition', 'PaperUnits', and 'WindowStyle' properties of *figure* objects are also unaffected. If handle is a column vector of object handles, all referenced objects are affected by set, reset, copyobj, and delete.

The get and set functions return a structure when an output is assigned. Consider the following example:

```
>> lprop = get(Hl_light)
lprop =
          BeingDeleted: 'off'
           BusyAction: 'queue'
         ButtonDownFcn: ''
             Children: [0x1 double]
             Clipping: 'on'
                Color: [1 1 1]
            CreateFcn: ''
            DeleteFcn: ''
     HandleVisibility: 'on'
              HitTest: 'on'
        Interruptible: 'on'
               Parent: 175
             Position: [1 0 1]
             Selected: 'off'
    SelectionHighlight: 'on'
                Style: 'infinite'
                  Tag: ''
                 Type: 'light'
         UIContextMenu: []
             UserData: []
              Visible: 'on'
>> class(lprop) % class of get(Hl_light)
ans =
struct

>> lopt = set(Hl_light)
```

```
lopt =

            BusyAction: {2x1 cell}
         ButtonDownFcn: {}
              Children: {}
              Clipping: {2x1 cell}
                 Color: {}
             CreateFcn: {}
             DeleteFcn: {}
      HandleVisibility: {3x1 cell}
               HitTest: {2x1 cell}
         Interruptible: {2x1 cell}
                Parent: {}
              Position: {}
              Selected: {2x1 cell}
    SelectionHighlight: {2x1 cell}
                 Style: {2x1 cell}
                   Tag: {}
         UIContextMenu: {}
              UserData: {}
               Visible: {2x1 cell}
>> class(lopt) % class of set(Hl_light)
ans =
struct
```

The field names of the resulting structures are the object property-name strings and are assigned alphabetically. Note that even though property names are not case sensitive, these field names are:

```
>> lopt.BusyAction
ans =
    'queue'
    'cancel'
>> lopt.busyaction
??? Reference to non-existent field 'busyaction'.
```

Combinations of property values can be set by using structures as well. For example,

```
>> newprop.Color = [1 0 0];
>> newprop.Position = [-10 0 10];
```

```
>> newprop.Style = 'local';
>> set(Hl_light,newprop)
```

changes the 'Color', 'Position', and 'Style' properties, but has no effect on any other properties of the *light* object. Note that you cannot simply obtain a structure of property values using get and use the same structure to set values:

```
>> light_prop = get(Hl_light);
>> light_prop.Color = [1 0 0];  % change the light color to red
>> set(Hl_light,light_prop);    % reapply the property values
??? Error using ==> set
Attempt to modify a property that is read-only.
Object Name : light
Property Name : 'BeingDeleted'.
```

Since 'BeingDeleted' and 'Type' are the only read-only properties of a *light* object, you can work around the problem by removing these fields from the structure:

```
>> light_prop = rmfield(light_prop,{'BeingDeleted','Type'});
>> set(Hl_light,light_prop)
```

For objects with more read-only properties, all read-only properties must be removed from the structure before using it to set property values.

A cell array can also be used to query a selection of property values. To do so, create a cell array containing the desired property names in the desired order and pass the cell array to get. The result is returned as a cell array as well:

```
>> plist = {'Color','Position','Style'}
plist =
    'Color'      'Position' 'Style'
>> get(Hl_light,plist)
ans =
    [1x3 double]    [1x3 double]    'local'
>> class(ans) % cell array in, cell array out
ans =
cell
```

One more point about the get function should be noted. If H is a vector of handles, get(H, 'Property') returns a cell array rather than a vector. Consider the following example, given a *Figure* window with four subplots:

```
>> Ha = get(gcf,'Children')       % get axes handles
Ha =
         183
         180
         177
         174
>> Ha_kids = get(Ha,'Children') % get handles of axes children
Ha_kids =
         [184]
         [181]
         [178]
         [175]
>> class(Ha_kids)
ans =
     cell
>> Hx = cat(1,Ha_kids{:})          % convert to column vector
Hx =
         184
         181
         178
         175
>> class(Hx)
ans =
     double
```

Now Hx can be used as an argument to Handle Graphics functions expecting a vector of object handles.

30.5 FINDING OBJECTS

As has been shown, Handle Graphics provides access to objects in a *figure* and allows the user to customize graphics by using the get and set commands. The use of these functions requires that you know the handles of the objects to be manipulated. In cases where handles are unknown, MATLAB provides a number of functions for finding object handles. Two of these functions, gcf and gca, were introduced earlier. For example,

```
>> Hf_fig = gcf
```

returns the handle of the current *figure*, and

```
>> Ha_ax = gca
```

returns the handle of the current *axes* in the current *figure*.

In addition to the above, MATLAB includes gco, a function to obtain the handle of the current object. For example,

```
>> Hx_obj = gco
```

returns the handle of the current object in the current *figure*; alternatively,

```
>> Hx_obj = gco(Hf_fig)
```

returns the handle of the current object in the *figure* associated with the handle Hf_fig.

The current object is defined as the last object clicked on with the mouse within a *figure*. This object can be any graphics object except the *root*. When a *figure* is initially created, no current object exists, and gco returns an empty array. The mouse button must be clicked while the pointer is within a *figure* before gco can return an object handle.

Once an object handle has been obtained, the object type can be found by querying an object's 'Type' property, which is a character-string object name such as 'figure', 'axes', or 'text'. The 'Type' property is common to all objects. For example,

```
>> x_type = get(Hx_obj,'Type')
```

is guaranteed to return a valid object string for all objects.

When something other than the 'CurrentFigure', 'CurrentAxes', or 'CurrentObject' is desired, the function get can be used to obtain a vector of handles to the children of an object. For example,

```
>> Hx_kids = get(gcf,'Children');
```

returns a vector containing handles of the children of the current *figure*.

To simplify the process of finding object handles, MATLAB contains the built-in function findobj, which returns the handles of objects with specified property values. The form Hx = findobj(Handles,'flat','PropertyName',PropertyValue) returns the handles of all objects in Handles whose 'PropertyName' property contains the value PropertyValue. Multiple ('PropertyName',PropertyValue) pairs are allowed, and all must match. When Handles is omitted, the *root* object is assumed. When no ('PropertyName',PropertyValue) pairs are given, all objects match and all Handles are returned. When 'flat' is omitted, all objects in Handles **and all descendants of these objects**, including axes titles and labels, are searched. When no objects are found to match the specified criteria, findobj returns an empty array. As an example, finding all green *line* objects is easily accomplished with the following statement:

```
>> Hl_green = findobj(0,'Type','line','Color',[0 1 0]);
```

It is possible to hide the visibility of specific handles by using the 'HandleVisibility' property common to all objects. This property is convenient, because it keeps the user from inadvertently deleting or changing the properties of an object. When an object has its 'HandleVisibility' property set to 'off' or to 'callback', findobj does not return handles to these objects when called from the *Command* window. Hidden handles do not appear in lists of children or as the output of gcf, gca, or gco. However, when the property is set to 'callback', these handles can be found during the execution of a callback.

30.6 SELECTING OBJECTS WITH THE MOUSE

The gco command returns the handle of the current object, which is the last object clicked on with the mouse. When a mouse click is made near the intersection of more than one object, MATLAB uses rules to determine which object becomes the current object. Each object has a selection region associated with it. A mouse click within this region selects the object. For *line* objects, the selection region includes the *line* and all of the area within a 5-pixel distance from the *line*. The selection region of a *surface*, *patch*, or *text* object is the smallest rectangle that contains the object. The selection region of an *axes* object is the *axes* box itself, plus the areas where labels and titles appear. Objects within *axes*, such as *lines* and *surfaces*, are higher in the stacking order, and clicking on them selects the associated object rather than the *axes*. Selecting an area outside of the *axes* selection region selects the *figure* itself.

When a mouse click is within the border of two or more objects, the ***stacking order*** determines which object becomes the current object. The stacking order determines which overlapping object is on *top* of the others. Initially, the stacking order is determined when the object is created, with the newest object at the top of the stack. For example, when you issue two figure commands, two *figures* are created. The second *figure* is drawn on top of the first. The resulting stacking order has *figure* 2 on top of *figure* 1, and the handle returned by gcf is 2. If the figure(1) command is issued or if *figure* 1 is clicked on, the stacking order changes. *Figure* 1 moves to the top of the stack and becomes the current *figure*.

In the preceding example, the stacking order was apparent from the window overlap on the computer screen. However, this is not always the case. When two *lines* are plotted, the second *line* drawn is on top of the first at the points where they intersect. If the first *line* is clicked on with the mouse at some other point, the first *line* becomes the current object, but the stacking order does not change. A click on the intersecting point continues to select the second *line* until the stacking order is explicitly changed.

The stacking order is given by the order in which 'Children' handles appear for a given object. That is, Hx_kids = get(handle,'Children') returns handles of child objects, in the stacking order. The first element in the vector Hx_kids is at the top of the stack, and the last element is at the bottom of the stack. The stacking order can be changed by changing the order of the 'Children' property value of the parent object. For example,

```
>> Hf = get(0,'Children');
>> if length(Hf) > 1
     set(0,'Children',Hf([end 1:end-1]);
  end
```

moves the bottom *figure* to the top of the stack, where it becomes the new current *figure*. The uistack function can be used to change the visual stacking order of children of the same parent.

30.7 POSITION AND UNITS

The 'Position' property of *figure* objects and of most other Handle Graphics objects is a four-element row vector. As shown in the following figure, the values in this vector are [left, bottom, width, height], where [left, bottom] is the position of the lower left corner of the object relative to its parent and [width, height] is the width and height of the object.

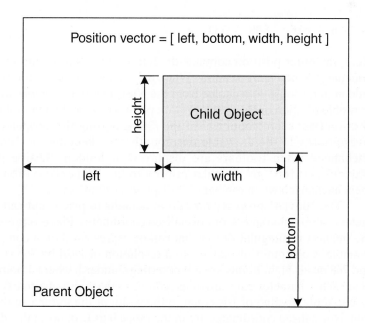

These position vector values are in units specified by the 'Units' property of the object. For example, the code

```
>> get(gcf,'Position')
ans =
   920    620    672    504
>> get(gcf,'units')
ans =
pixels
```

shows that the lower left corner of the current *figure* object is 920 pixels to the right and 620 pixels above the lower left corner of the screen and that the *figure* object is 672 pixels wide and 504 pixels high. ***Note that the* 'Position' *vector for a figure gives the drawable area within the figure object itself and does not include window borders, scroll bars, menus, or the title bar of the Figure window.***
Alternatively, *figures* have an undocumented 'OuterPosition' property that does include the window borders:

```
>> get(gcf,'Position')      % drawable position
ans =
   920    620    672    504
>> get(gcf,'OuterPosition') % outside position
ans =
   916    616    680    531
```

Here, the outer position contains the left, bottom, width, and height of the outer boundary of the *Figure* window. When the *Figure* or *Camera* toolbar is displayed or hidden in response to a choice from the **View** menu in a *Figure* window, either the drawable position or the outer position parameters can be maintained by setting the 'ActivePositionProperty' of the *figure*. Setting this property to 'Position' gives priority to the drawable position, thereby keeping the 'Position' values unchanged when toolbars are displayed or hidden. Setting this property to 'OuterPosition' gives similar priority to the 'OuterPosition' values, leaving them unchanged when toolbars are displayed or hidden.
The 'Units' property for *figures* defaults to pixels, but can be inches, centimeters, points, characters, or normalized coordinates. Pixels represent screen pixels, the smallest rectangular object that can be represented on a computer screen. For example, a computer display set to a resolution of 1024 by 768 is 1024 pixels wide and 768 pixels high. Points are a typesetting standard, where 1 point is equal to 1/72 of an inch. Character units are units relative to the width of a character in the default system font. A value of 1 is equal to the width of the letter x in the default system font. Normalized coordinates are in the range 0 to 1. In normalized coordinates, the

lower left corner of the parent is at [0, 0] and the upper right corner is at [1, 1]. Inches and centimeters are self-explanatory.

To illustrate various 'Units' property values, consider again the preceding example:

```
>> set(gcf,'units','inches')    % INCHES
>> get(gcf,'position')
ans =
        7.9224        5.3362        5.7931        4.3448
>> set(gcf,'units','cent')      % CENTIMETERS
>> get(gcf,'position')
ans =
        20.108        13.544        14.703        11.027
>> set(gcf,'units','normalized')% NORMALIZED
>> get(gcf,'position')
ans =
        0.57438       0.51583       0.42          0.42
>> set(gcf,'units','points')    % POINTS
>> get(gcf,'position')
ans =
        570.41        384.21        417.1         312.83
>> set(gcf,'units','char')      % CHARACTERS
>> get(gcf,'position')
ans =
        153.17        38.688        112           31.5
```

All of these values represent the same *figure* position relative to the computer screen for a particular monitor and screen resolution.

The positions of *axes* objects are also four-element vectors having the same form, [left, bottom, width, height], but specifying the object position relative to the lower left corner of the parent *figure*. ***In general, the 'Position' property of a child is relative to the position of its parent.***

To be more descriptive, the computer screen, or *root* object, position property is not called 'Position', but rather 'ScreenSize'. In this case, [left, bottom] is always [0, 0], and [width, height] are the dimensions of the computer screen in units specified by the 'Units' property of the *root* object. One exception is when the 'Units' property is set to 'Pixels'. In this case, [left, bottom] is always [1, 1], and [width, height] are the dimensions of the computer screen in pixels.

30.8 DEFAULT PROPERTIES

MATLAB assigns default properties to each object as it is created. The built-in defaults are referred to as *factory* defaults. To override these defaults, you must set or get the values using set and get. In cases where you want to change the same properties every time, MATLAB allows you to set your own default properties. It lets you change the default properties for individual objects and for object types at any point in the object hierarchy. When an object is created, MATLAB looks for a default value at the *parent* level. If no default is found, the search continues up the object hierarchy until a default value is found or until it reaches the built-in factory default.

You can set your own default values at any level of the object hierarchy by using a special property-name string consisting of 'Default', followed by the object type and the property name. The handle you use in the set command determines the point in the object parent–child hierarchy at which the default is applied. For example,

```
>> set(0,'DefaultFigureColor',[.5 .5 .5])
```

sets the default background color for all new *figure* objects to medium gray rather than the MATLAB default. This property applies to the *root* object (whose handle is always zero), and so all new *figures* will have a gray background. Other examples include the following:

```
>> set(0,'DefaultAxesFontSize',14)   % larger axes fonts - all figures
>> set(gcf,'DefaultAxesLineWidth',2) % thick axis lines - this figure only
>> set(gcf,'DefaultAxesXColor','y') % yellow X axis lines and labels
>> set(gcf,'DefaultAxesYGrid','on') % Y axis grid lines - this figure
>> set(0,'DefaultAxesBox','on')      % enclose axes - all figures
>> set(gca,'DefaultLineLineStyle',':') % dotted linestyle - these axes only
```

When a default property is changed, only objects created *after* the change is made are affected. Existing objects already have property values assigned and do not change.

When working with existing objects, it is always a good idea to restore them to their original state after they are used. If you change the default properties of objects in an M-file, save the previous settings and restore them when exiting the routine:

```
oldunits = get(0,'DefaultFigureUnits');
set(0,'DefaultFigureUnits','normalized');
    <MATLAB statements>
set(0,'DefaultFigureUnits',oldunits);
```

To customize MATLAB to use user-defined default values at all times, simply include the desired set commands in your startup.m file. For example, the code

```
set(0,'DefaultAxesXGrid','on')
set(0,'DefaultAxesYGrid','on')
set(0,'DefaultAxesZGrid','on')
set(0,'DefaultAxesBox','on')
set(0,'DefaultFigurePaperType','A4')
```

creates all axes with grids, and an enclosing box is turned on and sets the default paper size to A4. Defaults set at the *root* level affect every object in every *Figure* window.

There are three special property value strings that reverse, override, or query user-defined default properties. They are `'remove'`, `'factory'`, and `'default'`. If you've changed a default property, you can reverse the change, thereby resetting it to the original defaults using `'remove'`:

```
>> set(0,'DefaultFigureColor',[.5 .5 .5]) % set a new default
>> set(0,'DefaultFigureColor','remove')   % return to MATLAB defaults
```

To temporarily override a default and use the original MATLAB default value for a particular object, use the special property value `'factory'`:

```
>> set(0,'DefaultFigureColor',[.5 .5 .5]) % set a new user default
>> figure('Color','factory')              % figure using default color
```

The third special property value string is `'default'`. This value forces MATLAB to search up the object hierarchy until it encounters a default value for the desired property. If one is found, MATLAB uses this default value. If the *root* object is reached and no user-defined default is found, the MATLAB factory default value is used. This feature is useful when you want to set an object property to a default property value after it was created with a different property value, as in the following code:

```
>> set(0,'DefaultLineColor','r')    % set default at the root level
>> set(gcf,'DefaultLineColor','g') % current figure level default
>> Hl_rand = plot(rand(1,10));      % plot a line using 'ColorOrder' color
>> set(Hl_rand,'Color','default')  % the line becomes green
>> close(gcf)                       % close the window
>> Hl_rand = plot(rand(1,10));      % plot a line using 'ColorOrder' color again
>> set(Hl_rand,'Color','default')  % the line becomes red
```

Note that the `plot` command does not use *line* object defaults for the line color. If a color argument is not specified, the `plot` command uses the *axes* `'ColorOrder'` property to specify the color of each *line* it generates.

A list of all of the factory defaults can be obtained by issuing

```
>> get(0,'factory')
```

Default properties that have been set at any level in the object hierarchy can be listed by issuing

```
>> get(handle,'default')
```

The *root* object contains default values for a number of color properties and the *figure* position at startup:

```
>> get(0,'default')
ans =
            defaultFigurePosition: [520 678 560 420]
                defaultTextColor: [0 0 0]
               defaultAxesXColor: [0 0 0]
               defaultAxesYColor: [0 0 0]
               defaultAxesZColor: [0 0 0]
           defaultPatchFaceColor: [0 0 0]
           defaultPatchEdgeColor: [0 0 0]
                defaultLineColor: [0 0 0]
       defaultFigureInvertHardcopy: 'on'
               defaultFigureColor: [0.8000 0.8000 0.8000]
                defaultAxesColor: [1 1 1]
            defaultAxesColorOrder: [7x3 double]
            defaultFigureColormap: [64x3 double]
          defaultSurfaceEdgeColor: [0 0 0]
```

Other defaults are not listed until they have been created by the user:

```
>> get(gcf,'default')
ans =
0x0 struct array with fields:
>> set(gcf,'DefaultLineMarkerSize',10)
>> get(gcf,'default')
ans =
    defaultLineMarkerSize: 10
```

30.9 COMMON PROPERTIES

All Handle Graphics objects share the common set of object properties shown in the following table:

Property	Description
BeingDeleted	Sets to 'on' when object is about to be deleted
BusyAction	Controls how MATLAB handles callback interruptions
ButtonDownFcn	Callback code executed when the mouse button is pressed over the object
Children	Handles of visible children
Clipping	Enables or disables clipping of *axes* children
CreateFcn	Callback code executed immediately *after* an object is created
DeleteFcn	Callback code executed immediately *before* an object is deleted
HandleVisibility	Determines whether the object handle is visible in the *Command* window or while executing callbacks
HitTest	Determines whether the object can be selected with the mouse and then become the current object
Interruptible	Determines whether callbacks to this object are interruptible
Parent	Handle of parent object
Selected	Determines whether an object has been selected as the current object
SelectionHighlight	Determines whether a selected object shows visible selection handles or not
Tag	User-defined character string used to identify or *tag* the object. Often useful in association with findobj. For example, findobj(0,'Tag','mytagstring').
Type	Character string identifying object type
UIContextMenu	Handle of contextual menu associated with an object
UserData	Storage of any user-defined variable associated with an object
Visible	Visibility of object

Three of these properties contain *callbacks*:'ButtonDownFcn', 'CreateFcn', and 'DeleteFcn'. Callbacks identify MATLAB code to be executed when the action described by the property occurs. In many cases, the code is a function call. The 'Parent' and 'Children' properties contain handles of other objects in the hierarchy. Objects drawn on the *axes* are clipped at the *axes* limits if `'Clipping'` is `'on'`, which is the default for all axes children except *text* objects. `'Interruptible'` and `'BusyAction'` control the behavior of callbacks if a later callback is triggered while a previous callback is currently executing. `'Type'` is a string specifying the object type. `'Selected'` is `'on'` if this object is the `'CurrentObject'` of the *figure*, and `'SelectionHighlight'` determines whether the object changes appearance when selected. `'HandleVisibility'` specifies whether the object handle is visible, invisible, or visible only to callbacks. The *root* `'ShowHiddenHandles'` property overrides the `'HandleVisibility'` property of all objects, if needed. If `'Visible'` is set to `'off'`, the object disappears from view. It is still there, and the object handle is still valid, but it is not rendered on the computer screen. Setting `'Visible'` to `'on'` restores the object to view. The `'Tag'` and `'UserData'` properties are reserved for the user. The `'Tag'` property is typically used to tag an object with a user-defined character string. For example,

```
>> set(gca,'Tag','My Axes')
```

attaches the string `'My Axes'` to the current *axes* in the current *figure*. This string does not display in the *axes* or in the *figure*, but you can query the `'Tag'` property to identify the object. For instance, when there are numerous *axes*, you can find the handle to the above *axes* object by issuing the command

```
>> Ha_myaxes = findobj(0,'Tag','My Axes');
```

The `'UserData'` property can contain any variable you wish to put into it. A character string, a number, a structure, or even a multidimensional cell array can be stored in any object's `'UserData'` property. No MATLAB function changes, or makes assumptions about, the values contained in these properties.

The properties listed for each object using the `get` and `set` commands are the documented properties. There are also undocumented, or hidden, properties used by MATLAB developers. Some of them can be modified, but others are read-only. Undocumented properties are simply hidden from view. These properties still exist and can be modified. The undocumented *root* property `'HideUndocumented'` controls whether `get` returns all properties or only documented properties. For example,

```
>> set(0,'HideUndocumented','off')
```

makes undocumented object properties visible within MATLAB.

> Since undocumented properties have been purposely left undocumented, one must be cautious when using them. They are sometimes less robust than documented properties and are always subject to change. Undocumented properties may appear, disappear, change functionality, or even become documented in future versions of MATLAB.

30.10 PLOT OBJECTS

As stated earlier, MATLAB includes plot objects and group objects. Plot objects are objects associated with the high-level graphics functions discussed in Chapters 25 and 26. Plot objects were created as a way to group the core objects created by high-level graphics functions, so that the overall properties of the graphics are more easily identified and modified. That is, plot objects have additional properties that do not exist in the core objects used in their creation. These additional properties specify properties specific to the type of graphic created. For example, consider the creation of a bar graph:

```
>> hbs_mm = bar(randn(1,6));
>> title('Figure 30.1 Random Bar Graph')
>> get(hbs_mm,'type')
ans =
hggroup
>> hx_hbs = get(hbs_mm,'children');
>> get(hx_hbs,'type')
ans =
patch
```

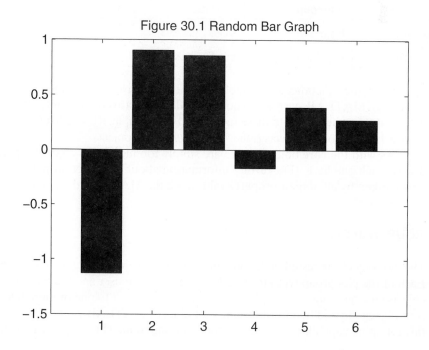

Figure 30.1 Random Bar Graph

This bar graph contains a composite *barseries* plot object whose handle type is designated hggroup. The children of this object are single *patch* objects. Rather than have all of the properties of the *patch* object, the *barseries* object has properties such as 'BaseLine', 'BarLayout', and 'BarWidth', which directly specify aspects of the created bar chart. Changing these properties in turn changes the appropriate properties of the underlying *patch* object. By adding this layer on top of the *patch* object, manipulation of the bar chart is much easier. You do not need to know how to manipulate *patch* object properties to make changes in the bar chart.

The plot objects in MATLAB are shown in the following table (also included is the 'type' property returned when get(H, 'type') is called, where H is the handle of the corresponding plot object):

Plot Object	**Get(H,'type')**	**Created by the Graphics Functions**
areasseries	hggroup	area
barseries	hggroup	bar, bar3, bar3h.
contourgroup	hggroup	contour, contour3, contourf
errorbarseries	hggroup	errorbar
lineseries	line	plot, plot3, semilogx, semilogy, loglog
quivergroup	hggroup	quiver, quiver3
scattergroup	hggroup	scatter, scatter3
stairseries	hggroup	stairs
stemseries	hggroup	stem, stem3
surfaceplot	surface	surf, mesh

Since the functions in this table returned handles to core objects in previous versions of MATLAB, current versions provide backward compatibility when 'v6' is used as the first argument to these functions. For example, H1 = plot('v6', ...) returns handles to *line* objects rather than to *lineseries* objects.

As with the core objects, there are simply too many plot object properties to discuss each one here. (For further information about all object properties and the action taken by all of their property values, see the MATLAB documentation.)

30.11 GROUP OBJECTS

The plot objects discussed in the previous section are an example of grouped objects. Each of the plot group types forms a layer between the user and the underlying core objects created. That is, each plot group is a parent object to one or more underlying core objects. In addition to furnishing plot objects, MATLAB 7 and above provides this grouping capability through the function hggroup. That is, you can group any number of *axes* object children into a group. These children can include any of the core objects, plot objects, or other objects grouped with the use of the hggroup function. As such, a group object has an *axes* object, or another group object, as its parent object.

By grouping Handle Graphics objects and giving them a single group handle, the visibility and selectability of the group are determined globally. That is, setting the 'visible' property of the group sets the individual 'visible' properties of all underlying group children. In addition, it is possible to select all members in a group by clicking on any group member with the mouse. While the plot objects created as part of MATLAB have different or additional property names and property values than their underlying core objects, MATLAB does not provide this capability for groups created by means of the hggroup function. Group objects have all of the common object properties listed in Section 30.9. In addition, they have an 'EraseMode' property that controls the 'EraseMode' property of all core objects in the group.

To illustrate the creation of a group object, consider the following code segment:

```
Ha = newplot; % create new axes object

Hl = line(1:6,[rand(1,6);1:6]);   % create 2 core line objects
Hp = patch([3 5 4],[2 2 5],'w'); % create a core patch object

Hg_mm = hggroup; % create a new group object and return handle

set(Hl,'Parent',Hg_mm)    % place lines in group
set(Hp,'Parent',Hg_mm)    % place patch in group

get(Hg_mm) % look at group properties
    Annotation = [ (1 by 1) hg.Annotation array]
    DisplayName =
    HitTestArea = off

    BeingDeleted = off
    ButtonDownFcn =
    Children = [ (3 by 1) double array]
    Clipping = on
    CreateFcn =
    DeleteFcn =
    BusyAction = queue
    HandleVisibility = on
    HitTest = on
```

```
Interruptible = on
Parent = [174.004]
Selected = off
SelectionHighlight = on
Tag =
Type = hggroup
UIContextMenu = []
UserData = []
Visible = on
```

The children of the group object having handle Hg_mm are the two *line* objects and one *patch* object. The properties of the lines and patch are hidden under the group object, and the group 'type' property is hggroup.

There are many cases where it is desirable to rotate, translate, or scale a group of objects within an axes. While the hggroup function does not provide this capability, the group function hgtransform does. Like the hggroup function, hgtransform creates a group object that becomes the parent of any number of *axes* object children. These children can include any of the core objects, plot objects, or other objects grouped by means of the hgtransform function. As such, a group object has an *axes* object, or another group object, as its parent object. The construction of a hgtransform group object follows that of the hggroup function. Group objects created by using the hgtransform function have all of the common object properties listed in Section 30.9. In addition, they have a 'Matrix' property that specifies a graphical transformation matrix to be applied to all members of the group object. The content of this transformation matrix is constructed following standard graphics transformation practice. The function makehgtform facilitates the construction of this matrix. (For further information, see the MATLAB documentation and other resources on graphics transformations.)

30.12 ANNOTATION *AXES*

The *annotation axes* shown in the Handle Graphics hierarchy at the beginning of this chapter is simply a core object *axes* with its handle visibility turned Off. This object contains all of the annotations added to a *figure*, using the numerous menus, toolbars, palettes, browsers, and editors available within *Figure* windows. The 'units' property of this axes is set to 'normalized', and its position is set to [0 0 1 1], so that it covers the entire drawing area of the *figure*. Lines, arrows, rectangles, text boxes, and so on, can be created by the use of the interactive aspects of *Figure* windows; in addition, they can be created with the annotation function. Because annotations added to a *figure* exist in this separate *axes*, they do not change in response to additions or modifications to any underlying data plots. Therefore, it is important to create annotations after the underlying plots are finalized.

30.13 LINKING OBJECTS

In some situations, it is desirable to link identical properties of a number of objects, so that changing a linked property of any one of the objects makes the same change to all linked objects. To support this capability, MATLAB provides the functions linkprop and linkaxes. The function linkprop provides general object-property linking, whereas linkaxes is an M-file function that uses linkprop to specifically link axis limits on two or more *axes* objects.

To demonstrate the use of these functions, consider the following code:

```
>> H_a1 = subplot(2,1,1);      % create axes object
>> plot(rand(1,30))            % plot some random data
>> H_a2 = subplot(2,1,2);      % create a second axes
>> plot(randn(1,30))           % plot random data on second axes

>> linkaxes([H_a1 H_a2],'xy')  % link x and y axis limits
```

After this code is executed, the axis limits on the two *axes* are linked. Using the pan or zoom features on the *Figure* window toolbar, panning or zooming one *axes* performs the identical function on the other *axes*. Alternatively, issuing an axis command on one *axes* applies the command to all linked *axes* objects.

The previous example can also be implemented with the use of linkprop by executing the following code:

```
>> close                       % close current figure to start over
>> H_a1 = subplot(2,1,1);      % create axes object
>> plot(rand(1,30))            % plot some random data
>> H_a2 = subplot(2,1,2);      % create a second axes
>> plot(randn(1,30))           % plot random data on second axes

>> H_link = linkprop([H_a1 H_a2],{'Xlim','Ylim'})
H_link =

       graphics.linkprop

>> class(H_link)
ans =
graphics.linkprop

>> methods(H_link)
Methods for class graphics.linkprop:
addprop      addtarget      linkprop      removeprop      removetarget
```

The output H_link of the linkprop function is a handle to a *linkprop* object named graphics.linkprop. This object is an object-oriented class created by using MATLAB's object-oriented programming features (as discussed in Chapter 31). As shown, this class has five methods that facilitate manipulation of *linkprop* objects. Once properties are linked via the function linkprop, the returned handle H_link must remain in existence. In the preceding example, H_link exists in the base workspace. If H_link is deleted or destroyed by being overwritten, the linked properties are unlinked. Modifying a *linkprop* object requires passing the associated object handle to one of its methods. For example,

```
>> addprop(H_link,'Zlim')
```

adds z-axis limit linking to the object in H_link. (See the MATLAB documentation for more information on Handle Graphics object linking.)

30.14 NEW PLOTS

When a new graphics object is created by using a low-level command such as line or text, the object appears on the current *axes* in the current *figure* by default. High-level graphics functions like mesh and plot, however, clear the current *axes* and reset most *axes* properties to their defaults before displaying a plot. As discussed earlier, the hold command can be used to change this default behavior.

Both *figures* and *axes* have a 'NextPlot' property used to control how MATLAB reuses existing *figures* and *axes*. The hold, newplot, reset, clf, and cla functions all affect the 'NextPlot' property of *figures* and *axes*. 'NextPlot' has four possible values:

```
>> set(gcf,'NextPlot')
   [ new | {add} | replace | replacechildren ]
>> set(gca,'NextPlot')
   [ new | add | {replace} | replacechildren ]
```

The default setting for *figures* is 'add'; the default for *axes* is 'replace'. When 'NextPlot' is set to 'add', a new plot is added without clearing or resetting the current *figure* or *axes*. When the value is 'replace', a new object causes the *figure* or *axes* to remove all child objects and reset all properties except 'Position' and 'Units' to their defaults before drawing the new plot. When 'NextPlot' is set to 'new' for a *figure*, a new *figure* window is created for the new object. For *axes*, 'new' and 'add' are equivalent. The default settings clear and reset the current *axes* and reuse the current *figure*.

The fourth possible setting for 'NextPlot' is 'replacechildren'. This setting removes all child objects, but does not change the current *figure* or *axes* properties. The command hold on sets both *figure* and *axes* 'NextPlot' properties to 'add'. The command hold off sets the *axes* 'NextPlot' property to 'replace'.

The `newplot` function returns a handle to an *axes* prepared by following the `'NextPlot'` guidelines. This function is meant to be called to create an *axes* that will contain children created by a core object-creation function such as `line` or `patch` as opposed to a high-level graphics function such as `plot` or `surf`. The `newplot` function prepares an *axes* and returns its handle using code whose effect is similar to the following code segment:

```
Hf = gcf;                        % get current figure or create one
next = lower(get(Hf,'NextPlot'));
switch next
  case 'new', Hf = figure;      % create new figure
  case 'replacechildren', clf;  % delete figure children
  case 'replace', clf('reset'); % delete children and reset properties
end
set(Hf,'NextPlot','add');
Ha = gca;                        % get current axes or create one
next = lower(get(Ha,'NextPlot'));
switch next
  case 'replacechildren', cla;  % delete axes children
  case 'replace', cla('reset'); % delete children and reset properties
end
```

If an argument consisting of a column vector of object handles is passed to `newplot`, the listed objects are not deleted and the *figure* and *axis* containing these objects is prepared for use.

30.15 CALLBACKS

All Handle Graphics objects have the properties `'ButtonDownFcn'`, `'CreateFcn'`, and `'DeleteFcn'`. In addition, *figures* have the properties `'CloseRequestFcn'`, `'KeyPressFcn'`, `'WindowButtonDownFcn'`, `'WindowButtonMotionFcn'`, `'WindowScrollWheelFcn'`, and many others. The user interface functions have a property `'CallBack'`. The property names associated with each of these properties are called ***callbacks***. These callbacks identify code to be executed when the specific user actions associated with a property name are taken. In the simplest case, callbacks are character strings that are evaluated by `eval` in the *Command* window workspace. These strings can contain any sequence of valid MATLAB statements. In most cases, callbacks are function calls, often to the same function where the callbacks are defined. The setting of callbacks tells MATLAB to perform certain tasks in response to an action taken by the user. These callbacks form the basis for MATLAB's GUI features.

The simplest callback is the close request callback, which, by default, is not empty:

```
>> get(gcf,'CloseRequestFcn')
ans =
closereq
>> class(ans)
ans =
char
```

By default, when a *Figure* window is closed by clicking the close box in the *figure* title bar, the string 'closereq' is passed to eval. This string is the name of a function in MATLAB that simply deletes the current *figure*. So, by default, clicking the close box deletes the associated *Figure* window. This behavior can be changed simply by replacing the preceding string with another to be evaluated on a close request. Consider the following example:

```
>> set(gcf,'CloseRequestFcn','')
```

This replacement disables closure via the close box. The close request function is an empty string, and so no action is taken. This callback string can be any valid sequence of MATLAB statements. Therefore, the string could prompt the user for confirmation of the close request before actually doing it.

30.16 M-FILE EXAMPLES

There are many, many examples of Handle Graphics usage in MATLAB itself. Almost all of the specialized plotting functions in the specgraph directory (>> doc specgraph) are composed of Handle Graphics functions. Even the M-file function axis is implemented by using Handle Graphics function calls. This section provides further illustrations of Handle Graphics usage.

The function mmis2d returns logical True if an axis is a 2-D view of the *x–y* plane:

```
function [tf,xa,ya] = mmis2d(H)
%MMIS2D True for Axes that are 2D.
% MMIS2D(H) returns True if the axes having handle H displays
% a 2D viewpoint of the X-Y plane where the X- and Y-axes are
```

```
% parallel to the sides of the associated figure window.
%
% [TF,Xa,Ya] = MMIS2D(H) in addition returns the angles of x- and y-axes
%
% e.g., if the x-axis increases from right-to-left Xa = 180
% e.g., if the y-axis increases from left-to-right Ya = 0
% e.g., if the x-axis increases from bottom-to-top Xa = 90

if ~ishandle(H)
      error('H Must be a Handle.')
end
if ~strcmp(get(H,'Type'),'axes')
      error('H Must be a Handle to an Axes Object.')
end
v = get(H,'view');
az = v(1);
el = v(2);
tf = rem(az,90)==0 && abs(el)==90;

if nargout==3
      xdir = strcmp(get(H,'Xdir'),'reverse');
      ydir = strcmp(get(H,'Ydir'),'reverse');
      s = sign(el);

      xa = mod(-s*az - xdir*180,360);
      ya = mod(s*(90-az) - ydir*180,360);
end
```

This function makes use of the function ishandle, which returns logical True for arguments that are valid object handles. It checks to see if the supplied handle is that of an *axes* by getting the 'Type' property. If successful, it then gets the 'View' property to determine the requested output.

The function mmgetpos finds the position of an object in a specific set of units. This function does the right thing, in that it gets the current 'Units' property, sets the units to the 'Units' of the desired output, gets the 'Position' property in the desired units, and then resets the 'Units' property:

```matlab
function p = mmgetpos(H,u,cp)
%MMGETPOS Get Object Position Vector in Specified Units.
% MMGETPOS(H,'Units') returns the position vector associated with the
% graphics object having handle H in the units specified by 'Units'.
% 'Units' is one of: 'pixels', 'normalized', 'points', 'inches', 'cent',
% or 'character'.
% 'Units' equal to 'data' is valid for text objects only.
%
% MMGETPOS does the "right thing", i.e., it: (1) saves the current units,
% (2) sets the units to those requested, (3) gets the position, then
% (4) restores the original units.
%
% MMGETPOS(H,'Units','CurrentPoint') returns the 'CurrentPoint' position
% of the figure having handle H in the units specified by 'Units'.
%
% MMGETPOS(H,'Units','Extent') returns the 'Extent' rectangle of the text
% object having handle H.
%
% 'Uimenu', 'Uicontextmenu', 'image', 'line', 'patch', 'surface',
% 'rectangle' and 'light' objects do NOT have position properties.

if ~ischar(u)
    error('Units Must be a Valid String.')
end
if ~ishandle(H)
        error('H is Not a Valid Handle.')
end
Htype = get(H,'Type');

if nargin==3 && ~isempty(cp) && ischar(cp)

    if strcmp(Htype,'figure') && lower(cp(1))=='c'
```

```
                pname = 'CurrentPoint';
        elseif strcmp(Htype,'text') && lower(cp(1))=='e'
                pname = 'Extent';
        else
                error('Unknown Input Syntax.')
        end

elseif H~=0
        pname = 'Position';

elseif H==0 % root object
        pname = 'ScreenSize';

else
        error('Unknown Input Syntax.')
end

hu = get(H,'units');  % get original units
set(H,'units',u)      % set to desired units

p = get(H,pname);     % get position in desired units
set(H,'units',hu)     % set units back to original units
```

The mmzap function illustrates a technique that is very useful when writing Handle Graphics function M-files. It uses a combination of waitforbuttonpress and gco to get the handle to an object selected using the mouse. The command waitforbuttonpress is a built-in MATLAB function that waits for a mouse click or key press. Part of its help text is as follows:

```
>> help waitforbuttonpress

WAITFORBUTTONPRESS Wait for key/buttonpress over figure.
    T = WAITFORBUTTONPRESS stops program execution until a key or
    mouse button is pressed over a figure window. Returns 0
    when terminated by a mouse buttonpress, or 1 when terminated
    by a keypress. Additional information about the terminating
    event is available from the current figure.
```

After a mouse button is pressed with the mouse pointer over a *figure*, gco returns the handle of the selected object. This handle is then used to manipulate the selected object:

```
function mmzap(arg)
%MMZAP Delete Graphics Object Using Mouse.
% MMZAP waits for a mouse click on an object in
% a figure window and deletes the object.
% MMZAP or MMZAP text erases text objects.
% MMZAP axes erases axes objects.
% MMZAP line erases line objects.
% MMZAP surf erases surface objects.
% MMZAP patch erases patch objects.
%
% Clicking on an object other than the selected type, or striking
% a key on the keyboard aborts the command.

if nargin<1
    arg = 'text';
end
Hf = get(0,'CurrentFigure');

if isempty(Hf)
    error('No Figure Window Exists.')
end
if length(findobj(0,'Type','figure'))==1
        figure(Hf) % bring only figure forward
end
key = waitforbuttonpress;

if key % key on keyboard pressed
        return
        else    % object selected
                object = gco;
                type = get(object,'Type');
```

```
          if strncmpi(type,arg,4)
                delete(object)
          end
end
```

The functions xlim, ylim, and zlim in MATLAB allow you to set and get axis limits for the three plot axes independently. There are no equivalent functions for grid lines. The function grid turns grid lines on and off along all three axes. To add this feature, the function mmgrid allows you to turn grid lines on and off along any individual axis:

```
function mmgrid(varargin)
%MMGRID Individual Axis Grid Lines on the Current Axes.
% V is a single character X, Y, or Z denoting the desired axis.
% MMGRID V toggles the major grid lines on the V-axis.
% MMGRID V ON adds major grid lines to the V-axis.
% MMGRID V ON MINOR adds minor grid lines to the V-axis.
% MMGRID V OFF removes major grid lines from the V-axis.
% MMGRID V OFF MINOR removes minor grid lines from the V-axis.
%
% See also GRID

ni = nargin;
if ni==0
    error('At Least One Input Argument Required.')
end
if ~iscellstr(varargin)
    error('Input Arguments Must be Strings.')
end

Hf = get(0,'CurrentFigure'); % get current figure if it exists
if isempty(Hf) % no figure so do nothing
    return
end
```

```matlab
Ha = get(Hf,'CurrentAxes'); % get current axes if it exists
if isempty(Ha) % no axes so do nothing
    return
end
% parse input and do work
V = varargin{1};

idx = strfind('xXyYzZ',V(1));
if isempty(idx)
    error('Unknown Axis Selected.')
end

VGrid = [upper(V(1)) 'Grid']; % XGrid, YGrid, or ZGrid

if ni==1 % MMGRID V    Toggle Grid
    Gstate = get(Ha,VGrid);
    if strcmpi(Gstate,'on')
       set(Ha,VGrid,'off')
    else
       set(Ha,VGrid,'on')
    end

elseif ni==2 % MMGRID V ON or MMGRID V OFF
    OnOff = varargin{2};
    if strcmpi(OnOff,'on')
       set(Ha,VGrid,'on')
    elseif strcmpi(OnOff,'off')
       set(Ha,VGrid,'off')
    else
       error('Second Argument Must be On or Off.')
    end

elseif ni==3 % MMGRID V ON MINOR or MMGRID V OFF MINOR
    if ~strcmpi(varargin{3},'minor')
       error('Unknown Third Argument.')
```

```
    end
    VGrid = [upper(V(1)) 'MinorGrid'];
    OnOff = varargin{2};
    if strcmpi(OnOff,'on')
        set(Ha,VGrid,'on')
    elseif strcmpi(OnOff,'off')
        set(Ha,VGrid,'off')
    else
        error('Second Argument Must be On or OFF.')
    end
end
```

Rather than use the functions `gcf` and `gca` to get the current *figure* and *axes* respectively, Handle Graphics calls are used to get the `'CurrentFigure'` and `'CurrentAxes'` properties. By doing so, no automatic creation of a *figure* or *axes* is performed. This allows the function to terminate without performing any action, if no *figure* or *axes* exists. The rest of the function simply modifies the `'XGrid'`, `'YGrid'`, `'ZGrid'`, `'XMinorGrid'`, `'YMinorGrid'`, or `'ZMinorGrid'` properties of the current *axes*, as directed by the input arguments.

30.17 SUMMARY

Handle Graphics functions provide the ability to fine-tune the appearance of visual aspects of MATLAB. Each graphics object has a handle associated with it that can be used to manipulate the object. The following table documents pertinent Handle Graphics functions in MATLAB:

Function	Description
get	Gets object properties
set	Sets object properties
gcf	Gets current *figure*
gca	Gets current *axes*
gco	Gets current object
shg	Shows most recent graph window

Function	Description
findobj	Finds visible objects having specified properties
findall	Finds visible and invisible objects having specified properties
findfigs	Returns visible *figure* windows to the computer screen
allchild	Gets visible and invisible children handles for an object
ancestor	Ancestor of graphics object
copyobj	Copies object to new parent
inspect	Opens Handle Graphics property inspector GUI
root	Root computer object, handle = 0
figure	*figure* object creation
axes	*axes* object creation
image	*image* object creation
light	*light* object creation
line	*line* object creation
patch	*patch* object creation
rectangle	*rectangle* object creation
surface	*surface* object creation
text	*text* object creation
axis	Axis scaling and appearance
box	Axes border
grid	Grid lines for 2-D and 3-D plots
linkaxes	Creates a link object for linking axis limits on two or more *axes* objects
linkprop	Creates a link object for linking listed properties of two or more Handle Graphics objects
uibuttongroup	User interface container object for managing 'radiobutton' and 'togglebutton' style *uicontrol* objects
uicontrol	User interface *uicontrol* object creation
uimenu	User interface *uimenu* object creation
uicontextmenu	User interface *uicontextual menu* object creation
uipanel	User interface *uipanel* object creation

`uitoolbar`	User interface *uitoolbar* object creation
`uipushtool`	Momentary contact pushbutton for a *uitoolbar*
`uitoggletool`	On/Off pushbutton for a *uitoolbar*
`uitable`	User interface *uitable* object creation
`areasseries`	Plots object created by the function `area`
`barseries`	Plots object created by the functions bar, bar3, bar3h
`contourgroup`	Plots object created by the functions `contour`, `contour3`, `contourf`
`errorbarseries`	Plots object created by the function errorbar
`lineseries`	Plots object created by the functions `plot`, `plot3`, `semilogx`, `semilogy`, `loglog`
`quivergroup`	Plots object created by the functions `quiver`, `quiver3`
`scattergroup`	Plots object created by the functions `scatter`, `scatter3`
`stairseries`	Plots object created by the function `stairs`
`stemseries`	Plots object created by the functions `stem`, `stem3`
`surfaceplot`	Plots object created by the functions `surf`, `mesh`
`hggroup`	Creates group object
`hgtransform`	Creates group object
`makehgtform`	Creates transformation matrix for `hgtransform` group object
`annotation`	Creates annotation *axes* object
`reset`	Resets object properties to default values
`clf`	Clears current *figure*
`cla`	Clears current *axes*
`ishandle`	True for arguments that are object handles
`ishold`	Returns current hold state
`hgload`	Loads Handle Graphics object hierarchy from file
`hgsave`	Saves Handle Graphics object hierarchy to file
`saveas`	Saves figure using specified format
`delete`	Deletes object
`close`	Closes *figure* using close request function
`refresh`	Refreshes *figure*

Function	Description
refreshdata	Refreshes data in graph when data source is specified
drawnow	Flushes event queue and updates figure window
waitfor	Blocks execution and waits for event or condition
gcbo	Gets current callback object
gcbf	Gets current callback *figure*
closereq	Default *figure* 'CloseRequestFcn' callback
newplot	Creates *axes* with knowledge of 'NextPlot' properties
opengl	Controls OpenGL rendering
propedit	Opens Property Editor tool

31

MATLAB Classes and Object-Oriented Programming

MATLAB has a number of fundamental data types otherwise known as classes. For example, arrays of numbers are commonly double-precision arrays. A variable containing such an array has a class called `double`. Similarly, character strings are another data type or class. Variables containing character strings have a class called `char`. Consider the following example:

```
>> pi          % a simple double
ans =

      3.1416

>> class(pi)
ans =
double

>> s = 'pi'    % a simple string
s =
pi
```

```
>> class(s)

ans =

char
```

Data types or classes in basic MATLAB include double, char, logical, cell, and struct. These data types are the most commonly used classes in MATLAB. In addition, MATLAB includes the lesser-used classes function_handle, map, single, and a variety of integer data types.

For each of these classes, MATLAB defines operations that can be performed. For example, addition is a defined operator for elements of the class double, but is not defined for elements of the class char, or for elements of the class cell:

```
>> x = pi+2
x =
      5.1416


>> y = 'hello' + 'there'
y =
   220    205    209    222    212


>> {'hello' 'there'}+{'sunny' 'day'}
??? Undefined function or method 'plus' for input arguments of type 'cell'.
```

Here, adding two character strings created a numerical array rather than a character string. Rather than report an error, MATLAB chose to convert 'hello' and 'there' to their ASCII numerical equivalents, and then perform element-by-element numerical addition. Even though MATLAB produced a result for this character-string example, it did so only after converting the elements on the right-hand side to the class double. MATLAB does this implicit type or class conversion for convenience, not because addition is defined for character strings. On the other hand, trying to add two cell arrays produces an immediate error.

Starting with version 5, MATLAB added the ability to define new operations for the basic data types and, more importantly, added the ability to create user-defined data types or classes. Creating and using data types is called *object oriented programming* (OOP), in which variables in each data type or class are called *objects*. Operations on objects are defined by methods that encapsulate data and overload operators and functions. The vocabulary of OOP includes terms such as *operator* and *function overloading*, *data encapsulation*, *methods*, *inheritance*, and *aggregation*. These terms and the fundamental principles of object-oriented programming in MATLAB are discussed in this chapter.

31.1 OVERLOADING

Before getting involved in the details of OOP and creating new variable classes, consider the process of overloading standard classes in MATLAB. The techniques used to overload standard classes are identical to those used for user-created classes. Once overloading is understood for standard classes, it is straightforward for user-created classes.

When the MATLAB interpreter encounters an operator such as addition, or a function with one or more input arguments, it considers the data type or class of the arguments to the operator or function and acts according to the rules it has defined internally. For example, addition means to compute the numerical sum of the arguments if the arguments are numerical values or can be converted to numerical values, such as character strings. When the internal rules for an operation or function are redefined, the operator or function is said to be ***overloaded***.

Operator and ***function overloading*** allow a user to redefine what actions MATLAB performs when it encounters an operator or function.

In MATLAB, the redefined rules for interpreting operators and functions are simply function M-files stored in ***class directories*** just off of the MATLAB search path.

The collection of rules or M-files for redefining operators and functions define ***methods***. The files themselves are commonly referred to as method functions.

That is, class directories themselves are not and cannot be on the MATLAB search path, but they are and must be subdirectories of directories that are on the MATLAB search path. To find class subdirectories, MATLAB requires that class directories be named as `@class`, where `class` is the variable class that the M-files in `@class` apply to. In addition, MATLAB supports multiple class directories. That is, there can be multiple `@class` directories for the same data type just off of the MATLAB path. When looking for functions in class directories, MATLAB follows the order given by the MATLAB search path and uses the first matching method function file found.

For example, if a directory @char appears just ***off*** of the MATLAB search path, the M-files contained in this directory can redefine operations and functions on character strings. To illustrate this, consider the function M-file `plus.m`:

```
function s = plus(s1,s2)
% Horizontal Concatenation for char Objects.

if ischar(s1)&ischar(s2)
```

```
        s = cat(2,s1(:).',s2(:).');
    elseif isnumeric(s2)
        s = double(s1)+s2;
    else
        error('Operator + Not Defined.')
    end
```

If this M-file is stored in any @char directory just off of the MATLAB search path, addition of character strings is redefined as horizontal concatenation. For example, repeating the statement y = 'hello' + 'there' made earlier becomes

```
>> y = 'hello' + 'there'
y =
hellothere
```

MATLAB no longer converts the strings on the right-hand side to their ASCII equivalents and adds the numerical results! What MATLAB did was: (1) construe character strings to appear on both sides of the addition symbol +; (2) look down the MATLAB search path for an @char subdirectory; (3) find the one we created and look for a function M-file named plus.m; (4) find the previous plus.m function, pass the two arguments to the addition operator to the function, and let it determine what action to perform; and (5) finally, return the function output as the result of the addition operation.

> To speed operation, MATLAB *caches* class subdirectories at startup. So, for this example to work, you must create the subdirectory and M-file and then restart MATLAB or issue the rehash command to get MATLAB to cache the newly created class subdirectory and associated M-files.

When addition is performed between two different data types, such as char and double, MATLAB considers the precedence and order of the arguments. For variables of equal precedence, MATLAB gives precedence to the leftmost argument to an operator or function, as in the following example:

```
>> z = 2 + 'hello'
z =
   106    103    110    110    113
```

The classes double and char have equal precedence. As a result, MATLAB considers addition to be numerical and applies its internal rules, converting 'hello' to its ASCII equivalent and performing numerical addition. On the other hand, if the order of the preceding operands is reversed, as in the code,

```
>> z = 'hello' + 2
z =
    106    103    110    110    113
```

then MATLAB considers addition to be a char class operation. In this case, the @char/plus.m function is called as plus('hello',2). As written, plus.m identifies this mixed class call with isnumeric(s2) and returns the same result as z = 2 + 'hello'.

As illustrated earlier, a function named plus.m defines addition in a class subdirectory. To support overloading of other operators, MATLAB assigns the function names shown in the following table to operators.

Operator	Function Name	Description
a + b	plus(a,b)	Numerical addition
a – b	minus(a,b)	Numerical subtraction
–a	uminus(a)	Unary minus
+a	uplus(a)	Unary plus
a .* b	times(a,b)	Element-by-element multiplication
a * b	mtimes(a,b)	Matrix multiplication
a ./ b	rdivide(a,b)	Element-by-element right division
a .\ b	ldivide(a,b)	Element-by-element left division
a / b	mrdivide(a,b)	Matrix right division
a \ b	mldivide(a,b)	Matrix left division
a .^ b	power(a,b)	Element-by-element exponentiation
a ^ b	mpower(a,b)	Matrix exponentiation
a < b	lt(a,b)	Less than
a > b	gt(a,b)	Greater than
a <= b	le(a,b)	Less than or equal to
a >= b	ge(a,b)	Greater than or equal to
a ~= b	ne(a,b)	Not equal
a == b	eq(a,b)	Equal
a & b	and(a,b)	Logical AND
a | b	or(a,b)	Logical OR

Operator	Function Name	Description
~a	not(a)	Logical NOT
a:d:b	colon(a,d,b)	Colon operator
a:b	colon(a,b)	
a'	ctranspose(a)	Conjugate transpose
a.'	transpose(a)	Transpose
[a b]	horzcat(a,b)	Horizontal concatenation
[a; b]	vertcat(a,b)	Vertical concatenation
a(s1,s2,...)	subsref(a,s)	Subscripted reference
a(s1,s2,...) = b	subsasgn(a,s,b)	Subscripted assignment
b(a)	subsindex(a)	Subscript index
	display(a)	*Command* window output
end	end(a,k,n)	Subscript interpretation of end

Continuing with the previous char class example, note that subtraction can be overloaded with the following function:

```
function s = minus(s1,s2)
% Subtraction for char Objects.
% Delete occurrences of s2 in s1.

if ischar(s1)&ischar(s2)
    s = strrep(s1,s2,'');
elseif isnumeric(s2)
    s = double(s1)-s2;
else
    error('Operator - Not Defined.')
end
```

As defined, subtraction is interpreted as the deletion of matching substrings:

```
>> z = 'hello' - 'e'
z =
hllo
```

```
>> a = 'hello' - 2
a =
    102    99   106   106   109
```

Again, this mixed-class case returns the MATLAB default action.

When multiple operators appear in a statement, MATLAB adheres to its usual order of precedence rules, working from left to right in an expression, as in the following example:

```
>> a = 'hello' + ' ' + 'there'
a =
hello there


>> a - 'e'
ans =
hllo thr


>> a = 'hello' + ' ' + ('there' - 'e')
a =
hello thr
```

This discussion concludes how MATLAB overloads operators.

Overloading functions follows the same procedure. In this case, the function stored in the class subdirectory has the same name as that of the standard MATLAB function:

```
function s = cat(varargin)
%CAT Concatenate Strings as a Row.

if length(varargin)>1 & ~ischar(varargin{2})
    error('CAT Not Defined for Mixed Classes.')
else
    s = cat(2,varargin{:});
end
```

This function overloads the function `cat` for character strings. It is called only if the first argument to the `cat` function is a character string; that is, it is of class `char`. If the first argument is numerical, the standard `cat` function is called:

```
>> cat('hello','there')  % call overloaded cat
hellothere

>> cat('hello',2)        % call overloaded cat
??? Error using ==> char.cat at 5
CAT Not Defined for Mixed Classes.

>> cat(2,'hello')        % call built in cat
ans =
hello
```

In addition to the operator overloading functions listed in this section, MATLAB provides several OOP utility functions. They include `methods`, `isa`, `class`, `loadobj`, and `saveobj`. The functions `isa` and `class` help to identify the data type or class of a variable or object:

```
>> a = 'hello';
>> class(a) % return class of argument
ans =
char
>> isa(a,'double') % logical class test
ans =
     0
>> isa(a,'char') % logical class test
ans =
     1
```

The function `methods` returns a listing of the methods, or overloading operators and functions associated with a given class:

```
>> methods cell
```

Methods for class cell:

accumarray	newdepfun	setdiff	strtok
cell2struct	permute	setxor	transpose
ctranspose	regexp	sort	union
display	regexpi	strcat	unique
intersect	regexprep	strfind	
ismember	regexptranslate	strjust	
issorted	reshape	strmatch	

This result shows that MATLAB itself has overloading functions that are called when input arguments are cell arrays. These functions extend the functionality of basic MATLAB functions to cell arrays without requiring the basic functions themselves to be rewritten to accept cell array arguments.

Finally, the functions loadobj and saveobj are called (if they exist) whenever the functions load and save are called with user-defined classes, respectively. Adding these functions to a class subdirectory allows you to modify a user-defined variable after a load operation or before a save operation.

31.2 CLASS CREATION

Operator and function overloading are key aspects of OOP. MATLAB's preferred implementation relies on a simple scheme whereby all of the properties and methods associated with a variable class are stored in a class definition file in class subdirectory just off of the MATLAB search path. Although class subdirectories containing individual method M-files as illustrated in the previous section are still supported, a single class definition file has been the preferred style since MATLAB 7.6. This section illustrates the creation of user-defined classes in the preferred style.

A new variable class is created when a class directory @*classname* is created and populated by a class definition M-file. The M-file has the name classname.m, and is used to define the properties and methods of the new variable class. This file is structured as follows:

```
classdef classname % class definition instance
    properties % properties of the class
        property1;
        property2;
        propertyn;
    end % properties

    methods % class methods (function and operator definitions)
        function output = classname(input) % class constructor method
```

```
        statements;
    end % classname

    additional function and operator definitions

  end % methods
end % classdef
```

The following example creates a new rational polynomial variable class. A directory is created just off of the MATLAB path named @mmrp. Then the M-file mmrp.m is created within this class directory structured as shown above. The file begins with a classdef statement, followed by properties of the class, followed by class method functions:

```
classdef mmrp
%MMRP Mastering MATLAB Rational Polynomial Object Class

properties (Access = private) % restrict property access to class methods
    n;
    d;
    v;
end % properties
methods

% insert method functions here

end % methods

end % classdef
```

In this case, the class variable is constructed as a structure with fields n (numerator), d (denominator), and v (variable name). The attribute 'Access = private' indicates that the properties are accessible within the class methods, but not externally.

The first required class method is a function to create variables of the new class from existing classes. This function is called the **constructor** and is named for the class itself. The second required method is the function display and is used to display the new variable in the *Command* window. No variable class is useful without additional method functions, but the constructor and display functions are a minimum requirement.

> In the vocabulary of OOP, the constructor creates an ***instance*** of the class. This instance is an object that utilizes the methods that overload how operators and functions act in the presence of the object.

The constructor method is a standard function call with input arguments containing the data needed to create an output variable of the desired class. For greatest flexibility, the constructor should handle three different sets of input arguments. Just as there are empty strings, arrays, cells, and so on, the constructor should produce an empty variable if no arguments are passed to it. On the other hand, if the constructor is passed a variable of the same class as that created by the constructor, the constructor should simply pass it back as an output argument. Finally, if creation data are provided, a new variable of the desired class should be created. In this last case, the input data can be checked for appropriateness. Inside of the constructor, the data used to create a variable of the desired class are stored in the fields of a structure. Once the structure fields are populated, the new variable is created. For example, the following is the constructor function for a rational polynomial object:

```
function r = mmrp(varargin)
%MMRP Mastering MATLAB Rational Polynomial Object Constructor.
% MMRP(p) creates a polynomial object from the polynomial vector p
% with 'x' as the variable.
% MMRP(p,'s') creates the polynomial object using the letter 's' as
% the variable in the display of p.
% MMRP(n,d) creates a rational polynomial object from the numerator
% polynomial vector n and denominator polynomial d.
% MMRP(n,d,'s') creates the rational polynomial using the letter 's' as
% the variable in the display of p.
%
% All coefficients must be real.

[n,d,v,msg] = local_parse(varargin); % parse input arguments

if isempty(v) % input was mmrp so return it
    r = n;
else
    error(msg) % return error if it exists
```

```
    tol = 100*eps;
    if length(d)==1 & abs(d)>tol % enforce scalar d = 1
        r.n = n/d;
        r.d = 1;
    elseif abs(d(1))>tol % make d monic if possible
        r.n = n/d(1);
        r.d = d/d(1);
    else                    % can't be made monic
        r.n = n;
        r.d = d;
    end
    r.v = v(1);
    r.n = mmpsim(r.n);    % strip leading zero terms
    r.d = mmpsim(r.d);    % see chapter 19 for mmpsim
    r = minreal(r);       % perform pole-zero cancellation
end % if

    function [n,d,v,msg]=local_parse(args); % nested function
        % parse input arguments to mmrp
        statements omitted for simplification
    end % local_parse

end % mmrp function
```

To simplify the function, parsing of the input arguments is done in the nested function local_parse. Details are not shown in the preceding code, but local_parse sets the values of four variables, n, d, v, and msg. The variables n and d are numerical row vectors containing the coefficients of the numerator and denominator of the rational polynomial, respectively. The variable v contains the string variable used to display the polynomial. Last, msg contains an error message if the parsing routines encounter invalid inputs.

     The previous constructor considers all three sets of input arguments. If no input arguments exist, n, d, and v are returned to create an empty rational polynomial. If the input argument is a rational polynomial object, it is simply returned as the output argument. Finally, if data are supplied, a rational polynomial object is created. In the simplest case, the denominator is simply equal to 1 and the display variable is 'x'. The last assignment statement in mmrp passes the created rational polynomial to the overloaded but not shown function `minreal`, which returns a minimal realization of the object by canceling like poles and zeros.

Since the `mmrp` function is contained within the `methods` section of the class definition M-file and will be followed by other functions, the function itself must be terminated by an `end` statement as does the nested function `local_parse`.

Within the constructor method, it is important that the structure fields be created in the same order under all circumstances. Violation of this rule may cause the created object to behave erratically.

Given the `mmrp` constructor method, we have the following associated `display` method:

```
function display(r)
%DISPLAY Command Window Display of Rational Polynomial Objects.

loose = strcmp(get(0,'FormatSpacing'),'loose');
if loose, disp(' '), end
var = inputname(1);
if isempty(var)
    disp('ans =')
else
    disp([var ' ='])
end
nstr = mmp2str(r.n,r.v); % convert polynomial to string
nlen = length(nstr);
if length(r.d)>1 | r.d~=1
    dstr = mmp2str(r.d,r.v);
else
    dstr = [];
end
dlen = length(dstr);
dash = '-';
if loose, disp(' '), end
if dlen % denominator exists
    m = max(nlen,dlen);
    disp('MMRP Rational Polynomial Object:')
    disp([blanks(ceil((m-nlen)/2)) nstr]);
```

```
        disp(dash(ones(1,m)));
        disp([blanks(fix((m-dlen)/2)) dstr]);
else
        disp('MMRP Rational Polynomial Object:')
        disp(nstr);
end
if loose, disp(' '), end
end % display function
```

This `display` function calls the function `mmp2str` to convert a numerical polynomial vector and a desired variable to a character-string representation. Like the `mmpsim` function used in the `mmrp` method, this `mmp2str` function must exist elsewhere on the MATLAB search path, not in the @mmrp directory or as a method in the class definition file. In this case, `mmp2str` is not part of the MATLAB installation. If it existed in the class directory or as a class method, MATLAB would not find it, since the arguments to `mmp2str` are `double` and `char`, respectively, and not of class `mmrp`. As described in the last section, a method function is called only if the leftmost or highest-precedence input argument has a class that matches that of the method. Note that a user-defined class has the highest precedence.

Within a method function, it is possible to act on objects as shown in the last assignment statement in the constructor r = minreal(r), where the variable r on the right-hand side is an object having class mmrp. There are two exceptions to this property. ***The overloading functions subsref and subsasgn are not called when subscripted reference and subscripted assignment appear within a method function.*** This allows the user to more freely access and manipulate a class variable within a method function.

It is also possible for class methods to act on the data contained in an object by simply addressing the structure fields of the object, as shown in numerous places within the `display` method. In this case, the class of the data determines how MATLAB acts.

> Outside of method functions (e.g., in the *Command* window) it is not possible to gain access to the contents of fields of an object, nor is it possible to determine the number of or names of the fields. This property is called data encapsulation.

The following examples demonstrate the creation and display of rational polynomial objects:

```
>> p = mmrp([1 2 3])

p =

MMRP Rational Polynomial Object:
x^2 + 2x^1 + 3
```

```
>> q = mmrp([1 2 3],[4 5 6],'z')
q =
MMRP Rational Polynomial Object:
0.25z^2 + 0.5z^1 + 0.75
----------------------
  z^2 + 1.25z^1 + 1.5

>> r = mmrp(conv([1 2],[1 4]),conv([1 2],[1 3]))
r =
MMRP Rational Polynomial Object:
x^1 + 4
-------
x^1 + 3
```

Rational polynomial objects have little value unless operators and functions are overloaded. In particular, it is convenient to define arithmetic operations on mmrp objects. The following methods define addition, subtraction, multiplication, and division for mmrp objects (since multiplication and division offer multiple methods, they are all overloaded with the associated polynomial manipulation):

```
function r = plus(a,b)
%PLUS Addition for Rational Polynomial Objects.

if isnumeric(a)
    rn = mmpadd(a*b.d,b.n); % see chapter 19 for mmpadd
    rd = b.d;
    rv = b.v;
elseif isnumeric(b)
    rn = mmpadd(b*a.d,a.n);
    rd = a.d;
    rv = a.v;
else % both polynomial objects
    if ~isequal(a.d,b.d)
        rn = mmpadd(conv(a.n,b.d),conv(b.n,a.d));
        rd = conv(a.d,b.d);
```

```
        else
            rn = mmpadd(a.n,b.n);
            rd = b.d;
        end
        if ~strcmp(a.v,b.v)
            warning('Variables Not Identical')
        end
        rv = a.v;
end
r = mmrp(rn,rd,rv); % create new MMRP object from results
end % plus
```

```
function r = uminus(a)
%UMINUS Unary Minus for Rational Polynomial Objects.

r = mmrp(-a.n,a.d,a.v);
end % uminus
```

```
function r = minus(a,b)
%MINUS Subtraction for Rational Polynomial Objects.

r = a+uminus(b); % use plus and uminus to implement minus
end % minus
```

```
function r = times(a,b)
%TIMES Dot Times for Rational Polynomial Objects.

a = mmrp(a); % convert inputs to mmrp if necessary
b = mmrp(b);
rn = conv(a.n,b.n);
rd = conv(a.d,b.d);
```

```
if ~strcmp(a.v,b.v)
    warning('Variables Not Identical')
end
rv = a.v;
r = mmrp(rn,rd,rv); % create new MMRP object from results
end % times
```

```
function r = mtimes(a,b)
%MTIMES Times for Rational Polynomial Objects.

r = a.*b; % simply call times.m
end % mtimes
```

```
function r = rdivide(a,b)
%RDIVIDE Right Dot Division for Rational Polynomial Objects.

a = mmrp(a); % convert inputs to mmrp if necessary
b = mmrp(b);
rn = conv(a.n,b.d);
rd = conv(a.d,b.n);
if ~strcmp(a.v,b.v)
    warning('Variables Not Identical')
end
rv = a.v;
r = mmrp(rn,rd,rv); % create new MMRP object from results
end % rdivide
```

```
function r = mrdivide(a,b)
%MRDIVIDE Right Division for Rational Polynomial Objects.

r = a./b; % simply call rdivide.m
end % mrdivide
```

```
function r = ldivide(a,b)
%LDIVIDE Left Dot Division for Rational Polynomial Objects.

r = b./a; % simply call rdivide.m
end % ldivide
```

```
function r = mldivide(a,b)
%MLDIVIDE Left Division for Rational Polynomial Objects.

r = b./a; % simply call rdivide.m
end % mldivide
```

The preceding method functions are self-explanatory, in that they implement simply polynomial arithmetic. Note that each method is terminated by an end statement. Examples of their use include the following:

```
>> a = mmrp([1 2 3])
a =
MMRP Rational Polynomial Object:
x^2 + 2x^1 + 3

>> b = a + 2                        % addition
b =
MMRP Rational Polynomial Object:
x^2 + 2x^1 + 5

>> a - b                           % subtraction
ans =
MMRP Rational Polynomial Object:
-2

>> a + b                           % addition
ans =
MMRP Rational Polynomial Object:
2x^2 + 4x^1 + 8
```

```
>> 2*b                           % multiplication
ans =
MMRP Rational Polynomial Object:
2x^2 + 4x^1 + 10


>> a * b                         % multiplication
ans =
MMRP Rational Polynomial Object:
x^4 + 4x^3 + 12x^2 + 16x^1 + 15


>> b/2                           % division
ans =
MMRP Rational Polynomial Object:
0.5x^2 + x^1 + 2.5


>> 2/b                           % division
ans =
MMRP Rational Polynomial Object:
      2
--------------
x^2 + 2x^1 + 5


>> c = a/b                       % division
c =
MMRP Rational Polynomial Object:
x^2 + 2x^1 + 3
--------------
x^2 + 2x^1 + 5


>> d = c/(1+c)                   % mixed
d =
MMRP Rational Polynomial Object:
0.5x^2 + x^1 + 1.5
------------------
  x^2 + 2x^1 + 4
```

```
>> (a/b)*(b/a)                    % mixed
ans =
MMRP Rational Polynomial Object:
1
```

Given the polynomial functions available in MATLAB and the ease with which they can be manipulated, there are many functions that can be overloaded. For example, the basic MATLAB functions roots and zeros can be overloaded by the following method M-files:

```
function [z,p] = roots(r)
%ROOTS Find Roots of Rational Polynomial Objects.
% ROOTS(R) returns the roots of the numerator of R.
% [Z,P] = ROOTS(R) returns the zeros and poles of R in
% Z and P respectively.

z = roots(r.n);
if nargout==2
    p = roots(r.d);
end
end % roots
```

```
function z = zeros(r)
%ZEROS Zeros of a Rational Polynomial Object.

z = roots(r.n);
end % zeros
```

The method function roots calls the basic MATLAB function roots, because the arguments within the method are of class double. With the creation of the zeros method, the function zeros has two entirely different meanings, depending on what its arguments are. The beauty of OOP is that functions can have multiple meanings or contexts without having to embed them all in a single M-file. The class of the input arguments dictates which function is called into action.

Mimicking their Handle Graphics usage, it is common to use them to set or get individual class structure fields, as in the following example:

```matlab
function set(r,varargin)
%SET Set Rational Polynomial Object Parameters.
% SET(R,Name,Value,...) sets MMRP object parameters of R
% described by the Name/Value pairs:
%
% Name              Value
% 'Numerator'       Numeric row vector of numerator coefficients
% 'Denominator'     Numeric row vector of denominator coefficients
% 'Variable'        Character Variable used to display polynomial

if rem(nargin,2)~=1
   error('Parameter Name/Values Must Appear in Pairs.')
end
for i = 2:2:nargin-1
   name = varargin{i-1};
   if ~ischar(name), error('Parameter Names Must be Strings.'), end
   name = lower(name(isletter(name)));
   value = varargin{i};
   switch name(1)
   case 'n'
      if ~isnumeric(value) | size(value,1)>1
         error('Numerator Must be a Numeric Row Vector.')
      end
      r.n = value;
   case 'd'
      if ~isnumeric(value) | size(value,1)>1
         error('Denominator Must be a Numeric Row Vector.')
      end
      r.d = value;
   case 'v'
      if ~ischar(value) | length(value)>1
         error('Variable Must be a Single Character.')
      end
      r.v = value;
```

```
    otherwise
        warning('Unknown Parameter Name')
    end
end
vname = inputname(1);
if isempty(vname)
    vname = 'ans';
end
r = mmrp(r.n,r.d,r.v);
assignin('caller',vname,r);
end % set
```

```
function varargout = get(r,varargin)
%GET Get Rational Polynomial Object Parameters.
% GET(R,Name) gets the MMRP object parameter of R described by
% one of the following names:
%
% Name              Description
% 'Numerator'       Numeric row vector of numerator coefficients
% 'Denominator'     Numeric row vector of denominator coefficients
% 'Variable'        Character Variable used to display polynomial
%
% [A,B,...] = get(R,NameA,NameB,...) returns multiple parameters
% in the corresponding output arguments.

if (nargout+(nargout==0))~=nargin-1
    error('No. of Outputs Must Equal No. of Names.')
end
for i = 1:nargin-1
    name = varargin{i};
    if ~ischar(name), error('Parameter Names Must be Strings.'), end
    name = lower(name(isletter(name)));
```

```
    switch name(1)
    case 'n'
      varargout{i} = r.n;
    case 'd'
      varargout{i} = r.d;
    case 'v'
      varargout{i} = r.v;
    otherwise
      warning('Unknown Parameter Name')
    end
end
end % get
```

These functions allow you to modify an `mmrp` object or to get data out of one, as the following example shows:

```
>> c % recall data
c =
MMRP Rational Polynomial Object:
x^2 + 2x^1 + 3
--------------
x^2 + 2x^1 + 5

>> n = get(c,'num') % get numerator vector
n =
               1           2           3

>> set(c,'Numerator',[3 1]) % change numerator
>> c
c =
MMRP Rational Polynomial Object:
    3x^1 + 1
--------------
x^2 + 2x^1 + 5
```

```
>> class(c)   % class and isa know about mmrp objects
ans =
mmrp
>> isa(c,'mmrp')
ans =
      1
```

31.3 SUBSCRIPTS

Because of MATLAB's array orientation, user-defined classes can also make use of subscripts. In particular, V(...), V{...}, and V.field are all supported. In addition, these constructions can appear on either side of an assignment statement. When they appear on the right-hand side of an assignment statement, they are referencing the variable V; and when they appear on the left-hand side, they are assigning data to some part of the variable V. These indexing processes are called subscripted reference and subscripted assignment, respectively. The method functions that control how they are interpreted when applied to an object are subsref and subasgn, respectively. These functions are not as straightforward to understand as other operator and function overloading methods. As a result, this section specifically addresses them. To facilitate this discussion, the mmrp object created in the preceding section is used in the examples.

When dealing with rational polynomials, there are two obvious interpretations for subscripted reference. For a rational polynomial object R, R(x), where x is a data array, can return the results of evaluating R at the points in x. Alternatively, R('v'), where 'v' is a single character, could change the variable used to display the object to the letter provided.

The part of the help text for subsref that describes how to write a subsref method is as follows:

```
B = SUBSREF(A,S) is called for the syntax A(I), A{I}, or A.I
when A is an object. S is a structure array with the fields:
     type -- string containing '()', '{}', or '.' specifying the
             subscript type.
     subs -- Cell array or string containing the actual subscripts.
For instance, the syntax A(1:2,:) invokes SUBSREF(A,S) where S is a
1-by-1 structure with S.type = '()' and S.subs = {1:2,':'}. A colon
used as a subscript is passed as the string ':'.
```

Similarly, the syntax A{1:2} invokes SUBSREF(A,S) where S.type = '{}'
and the syntax A.field invokes SUBSREF(A,S) where S.type = '.' and
S.subs = 'field'.

These simple calls are combined in a straightforward way for
more complicated subscripting expressions. In such cases
length(S) is the number of subscripting levels. For instance,
A(1,2).name(3:5) invokes SUBSREF(A,S) where S is 3-by-1 structure
array with the following values:

S(1).type = '()'	S(2).type = '.'	S(3).type = '()'
S(1).subs = {1,2}	S(2).subs = 'name'	S(3).subs = {3:5}

According to the help text, if R is an mmrp object, R(x) creates S.type = '()' and
S.subs = x, where x contains the values where R is to be evaluated, not indices into
an array. Similarly, R('v') creates S.type = '()' and S.subs = 'v'. All other
possibilities should produce an error.

Using this information leads to the subsref method function:

```
function y = subsref(r,s)
%SUBSREF(R,S) Subscripted Reference for Rational Polynomial Objects.
% R('z') returns a new rational polynomial object having the same numerator
% and denominator, but using the variable 'z'.
%
% R(x) where x is a numerical array, evaluates the rational polynomial R
% at the points in x, returning an array the same size as x.

if length(s)>1
        error('MMRP Objects Support Single Arguments Only.')
end
if strcmp(s.type,'()') % R(x) or R('v')
    arg = s.subs{1};
    argc = class(arg);
    if strcmp(argc,'char')
```

```
        if strcmp(arg(1),':')
            error('MMRP Objects Do Not Support R(:).')
        else
            y = mmrp(r.n,r.d,arg(1)); % change variables
        end
    elseif strcmp(argc,'double')
        if length(r.d)>1
            y = polyval(r.n,arg)./polyval(r.d,arg);
        else
            y = polyval(r.n,arg);
        end
    else
        error('Unknown Subscripts.')
    end
else % R{ } or R.field
    error('Cell and Structure Addressing Not Supported.')
end
end % subsref
```

Examples using this method include the following:

```
>> c % recall data
c =
MMRP Rational Polynomial Object:
    3x^1 + 1
---------------
x^2 + 2x^1 + 5

>> c = c('t') % change variable
c =
MMRP Rational Polynomial Object:
    3t^1 + 1
---------------
t^2 + 2t^1 + 5
```

```
>> x = -2:2
x =

    -2   -1   0   1   2
>> c(x)        % evaluate c(x)
ans =

         -1              -0.5            0.2            0.5          0.53846


>> c{3}         % try cell addressing
??? Error using ==> mmrp.mmrp>mmrp.subsref at 463
Cell and Structure Addressing Not Supported.

>> c.n          % Try field addressing
??? Error using ==> mmrp.mmrp>mmrp.subsref at 463
Cell and Structure Addressing Not Supported.
```

As stated earlier, outside of method functions, the field structure of objects is hidden from view. If it were not, issuing c.n above would have returned the numerator row vector from the object c. To enable this feature, it must be explicitly included in the subsref method:

```
function y = subsref(r,s)
%SUBSREF(R,S) Subscripted Reference for Rational Polynomial Objects.
% R('z') returns a new rational polynomial object having the same numerator
% and denominator, but using the variable 'z'.
%
% R(x) where x is a numerical array, evaluates the rational polynomial R
% at the points in x, returning an array the same size as x.
%
% R.n returns the numerator row vector of R.
% R.d returns the denominator row vector of R.
% R.v returns the variable associated with R.

if length(s)>1
   error('MMRP Objects Support Single Arguments Only.')
end
```

```matlab
if strcmp(s.type,'()') % R( )
   arg = s.subs{1};
   argc = class(arg);
   if strcmp(argc,'char')
      if strcmp(arg(1),':')
         error('MMRP Objects Do Not Support R(:).')
      else
         y = mmrp(r.n,r.d,arg(1));
      end
   elseif strcmp(argc,'double')
      if length(r.d)>1
         y = polyval(r.n,arg)./polyval(r.d,arg);
      else
         y = polyval(r.n,arg);
      end
   else
      error('Unknown Subscripts.')
   end
elseif strcmp(s.type,'.') % R.field
   arg = lower(s.subs);
   switch arg(1)
   case 'n'
      y = r.n;
   case 'd'
      y = r.d;
   case 'v'
      y = r.v;
   otherwise
      error('Unknown Data Requested.')
   end
else % R{ }
   error('Cell Addressing Not Supported.')
end
```

Examples using this method include the following:

```
>> c.n   % return numerator
ans =
     3     1


>> c.v   % return variable
ans =
t

>> c.nadfdf % only first letter is checked in this example method
ans =
     3     1


>> c.t       % not n, d, or v
??? Error using ==> mmrp.mmrp>mmrp.subsref at 472

Unknown Data Requested.

>> c.d(1:2) % we didn't include subaddressing in subsref
??? Error using ==> mmrp.mmrp>mmrp.subsref at 442
MMRP Objects Support Single Arguments Only.
```

As stated earlier, the overloading functions subsref and subsasgn are *not* implicitly called when subscripted reference and subscripted assignment appear within a method function. Overloading the MATLAB polynomial evaluation function polyval demonstrates this fact, as shown by the following method:

```
function y = polyval(r,x)
%POLYVAL Evaluate Rational Polynomial Object.
% POLYVAL(R,X) evaluates the rational polynomial R at the
% values in X.

if isnumeric(x)
  %y = r(x);          % what we'd like to do, but can't
```

```
    S.type = '()';
    S.subs = {x};
    y = subsref(r,S); % must call subsref explicitly
else
    error('Second Input Argument Must be Numeric.')
end
end % polyval
```

Because of how subsref is written for the mmrp object, polynomial evaluation is simply a matter issuing R(x), where R is an mmrp object and x contains the values where R is to be evaluated. Since this matches the expected operation of polyval, simply issuing y = r(x) within polyval should cause MATLAB to call the subsref method to evaluate the rational polynomial. This does not happen, because MATLAB does not call subsref or subsasgn within methods. However, to force this to happen, one can explicitly call subsref with the desired arguments, as previously shown.

When dealing with rational polynomials, there is one obvious interpretation for subscripted assignment. For a rational polynomial object R, R(1,p)= v, where p is a numerical vector identifying variable powers and v is a numerical vector of the same length, the elements of v become the coefficients of the numerator polynomial associated with the powers in p. Likewise, R(2,q) = w changes the denominator coefficients for those powers identified in q.

The part of the help text for subsasgn that describes how to write a subsasgn method is as follows:

```
A = SUBSASGN(A,S,B) is called for the syntax A(I) = B, A{I} = B, or
A.I = B when A is an object. S is a structure array with the fields:
    type -- string containing '()', '{}', or '.' specifying the
            subscript type.
    subs -- Cell array or string containing the actual subscripts.

For instance, the syntax A(1:2,:) = B calls A = SUBSASGN(A,S,B) where
S is a 1-by-1 structure with S.type = '()' and S.subs = {1:2,':'}. A
colon used as a subscript is passed as the string ':'.
```

Similarly, the syntax A{1:2} = B invokes A = SUBSASGN(A,S,B) where
S.type = '{}' and the syntax A.field = B invokes SUBSASGN(A,S,B) where
S.type = '.' and S.subs = 'field'.

These simple calls are combined in a straightforward way for
more complicated subscripting expressions. In such cases
length(S) is the number of subscripting levels. For instance,
A(1,2).name(3:5) = B invokes A = SUBSASGN(A,S,B) where S is 3-by-1
structure array with the following values:

S(1).type = '()'	S(2).type = '.'	S(3).type = '()'
S(1).subs = {1,2}	S(2).subs = 'name'	S(3).subs = {3:5}

On the basis of this help text and the desired subscripted assignment, the following subsasgn method is created:

```
function a = subsasgn(a,s,b)
%SUBSASGN Subscripted assignment for Rational Polynomial Objects.
%
% R(1,p) = C sets the coefficients of the Numerator of R identified
% by the powers in p to the values in the vector C.
%
% R(2,p) = C sets the coefficients of the Denominator of R identified
% by the powers in p to the values in the vector C.
%
% R(1,:) or R(2,:) simply replaces the corresponding polynomial
% data vector.
%
% For example, for the rational polynomial object
%              2x^2 + 3x + 4
% R(x) = --------------------
%            x^3 + 4x^2 + 5x + 6
```

```
%
% R(1,2)=5            changes the coefficient 2x^2 to 5x^2
% R(2,[3 2])=[7 8] changes x^3 + 4x^2 to 7x^3 + 8x^2
% R(1,:)=[1 2 3]    changes the numerator to x^2 + 2x + 3

if length(s)>1
    error('MMRP Objects Support Single Arguments Only.')
end
if strcmp(s.type,'()') % R(1,p) or R(2,p)
   if length(s.subs)~=2
      error('Two Subscripts Required.')
   end
   nd = s.subs{1}; % numerator or denominator
   p = s.subs{2};  % powers to modify
   if ndims(nd)~=2 | length(nd)~=1 | (nd~=1 & nd~=2)
      error('First Subscript Must be 1 or 2.')
end
if isnumeric(p) & ...
   (ndims(p)~=2 | any(p<0) | any(fix(p)~=p))
   error('Second Subscript Must Contain Nonnegative Integers.')
end
if ndims(b)~=2 | length(b)~=prod(size(b))
   error('Right Hand Side Must be a Vector.')
end
b = b(:).';      % make sure b is a row
p = p(:)';       % make sure p is a row
if ischar(p)     & length(p)==1 & strcmp(p,':') % R(1,:) or R(2,:)
   if nd==1      % replace numerator
      r.n = b;
      r.d = a.d;
   else          % replace denominator
      r.n = a.n;
```

```
        r.d = b;
     end
  elseif isnumeric(p) % R(1,p) or R(2,p)
     plen = length(p);
     blen = length(b);
     nlen = length(a.n);
     dlen = length(a.d);
     if plen~=blen
        error('Sizes Do Not Match.')
     end
     if nd==1 % modify numerator
        r.d = a.d;
        rlen = max(max(p)+1,nlen);
        r.n = zeros(1,rlen);
        r.n = mmpadd(r.n,a.n);
        r.n(rlen-p) = b;
     else     % modify denominator
        r.n = a.n;
        rlen = max(max(p)+1,dlen);
        r.d = zeros(1,rlen);
        r.d = mmpadd(r.d,a.d);
        r.d(rlen-p) = b;
     end
  else
     error('Unknown Subscripts.')
  end
else % R{ } or R.field
   error('Cell and Structure Addressing Not Supported.')
end
a = mmrp(r.n,r.d,a.v);
end % subsasgn
```

Examples using this method include the following:

```
>> a = mmrp([3 1],[1 2 5 10]) % create test object
a =
MMRP Rational Polynomial Object:
       3x^1 + 1
----------------------
x^3 + 2x^2 + 5x^1 + 10

>> a(1,:) = [1 2 4]      % replace entire numerator
a =
MMRP Rational Polynomial Object:
    x^2 + 2x^1 + 4
----------------------
x^3 + 2x^2 + 5x^1 + 10

>> a(2,2) = 12           % replace x^2 coef in denominator
a =
MMRP Rational Polynomial Object:
    x^2 + 2x^1 + 4
----------------------
x^3 + 12x^2 + 5x^1 + 10

>> a(1,0) = 0            % replace 4x^0 with 0x^0
a =
MMRP Rational Polynomial Object:
    x^2 + 2x^1
----------------------
x^3 + 12x^2 + 5x^1 + 10

>> a(1,:) = a.d          % subsref and subsasn! (a.n and a.d cancel)
a =
MMRP Rational Polynomial Object:
1
```

31.4 CONVERTER FUNCTIONS

As demonstrated in earlier chapters, the functions double, char, and logical convert their inputs to the data type matching their name. For example, double('hello') converts the character string 'hello' to its numerical ASCII equivalent. Whenever possible, methods for these converter functions should be included in a class. For mmrp objects, double and char have obvious interpretations. The double method should extract the numerator and denominator polynomials, and the char method should create a string representation such as that displayed by display.m, as in the following code:

```
function [n,d] = double(r)
%DOUBLE Convert Rational Polynomial Object to Double.
% DOUBLE(R) returns a matrix with the numerator of R in
% the first row and the denominator in the second row.
% [N,D] = DOUBLE(R) extracts the numerator N and denominator D
% from the rational polynomial object R.

if nargout<=1 & length(r.d)>1
   nlen = length(r.n);
   dlen = length(r.d);
   n = zeros(1,max(nlen,dlen));
   n = [mmpadd(n,r.n);mmpadd(n,r.d)];
elseif nargout<=1
      n = r.n;
else % nargout==2
      n = r.n;
      d = r.d;
end
end % double
```

```
function [n,d,v] = char(r)
%CHAR Convert Rational Polynomial Object to Char.
% CHAR(R) returns a 3 row string array containing R in the
% format used by DISPLAY.M
% [N,D] = CHAR(R) extracts the numerator N and denominator D
```

```
% as character strings from the rational polynomial object R.
% [N,D,V] = CHAR(R) in addition returns the variable V.

if nargout<=1
    nstr = mmp2str(r.n,r.v);
    nlen = length(nstr);
    if length(r.d)>1
        dash = '-';
        dstr = mmp2str(r.d,r.v);
        dlen = length(dstr);
        m = max(nlen,dlen);
        n = char([blanks(ceil((m-nlen)/2)) nstr],...
            dash(ones(1,m)),...
            [blanks(fix((m-dlen)/2)) dstr]);
    else
        n = nstr;
    end
elseif nargout>1
    n = mmp2str(r.n); % converts polynomial to string
    d = mmp2str(r.d);
end
if nargout>2
    v = r.v;
end
end % char
```

31.5 PRECEDENCE, INHERITANCE, AND AGGREGATION

MATLAB automatically gives user-defined classes ***higher precedence*** than the built-in classes. Therefore, operators and functions containing a mixture of built-in classes and a user-defined class always call the methods of the user-defined class. While this default precedence is usually sufficient for simple classes, the presence of multiple user-defined classes requires that some mechanism exist to allow the user to control the precedence of classes with respect to one another.

The `InferiorClasses` attribute to the `classdef` function can be used to specify that one user-defined class has higher precedence than other user-defined classes. For example, the statement

```
classdef (InferiorClasses = {?mmp}) mmrp
```

in a class definition file asserts that the `mmrp` class has higher precedence than the `mmp` class. MATLAB built-in classes (`double`, `char`, `single`, `int16`, `struct`, etc.) are always inferior to user-defined classes and should not be used in this list.

The functions `inferiorto` and `superiorto` provided this capability prior to MATLAB version 7.6. *If you are using them, calls to these functions must appear within the constructor function for a class.* They only apply to the older, non-classdef style class constructors that used the `class` function to create an object.

For large programming projects, it may be convenient to create a hierarchy of object types. In this case, it may be beneficial to let one object type inherit methods from another type. In doing so, fewer methods need to be written, and method modifications are more centralized. In the vocabulary of OOP, an object that inherits the properties of another is called a ***child*** class, and the class it inherits from is called the ***parent*** class. In the simplest case, a child class inherits methods from a single parent class. This is called ***simple*** or ***single inheritance***. It is also possible for a child class to inherit methods from multiple classes, which is called ***multiple inheritance***.

In single inheritance, the child class is given all of the fields of the parent class, plus one or more unique fields of its own. As a result, methods associated with the parent can be directly applied to objects of the child class. Quite naturally, methods of the parent class have no knowledge of the fields unique to the child and therefore cannot use them in any way. Similarly, fields of the parent class cannot be accessed by methods of the child. The child must use the methods it inherited from the parent to gain access to the parent fields. The `lti` object in the *Control Toolbox* is an example of a parent class having child classes `tf`, `zpk`, `ss`, `dss`, and `frd`.

In multiple inheritance, a child class is given all of the fields of all parent classes, plus one or more unique fields of its own. As in single inheritance, the parents and child do not have direct access to each other's fields. With multiple parents, the complexity of determining which parent methods are called under what circumstances is more difficult to describe. (Information and detailed examples regarding inheritance can be found in the MATLAB documentation.)

In the `mmrp` class used as an example earlier, the object fields contained data that were elements of the MATLAB classes `double` and `char`. In reality, there is no reason object fields cannot contain other data types, including user-defined classes. In the vocabulary of OOP, this is called ***containment*** or ***aggregation***. The rules for operator and function overloading do not change. Within method functions of one class, the methods of other classes are called as needed to operate on the fields of the original class.

31.6 HANDLE CLASSES

This chapter detailed the creation of a user-defined *value* class. As shown above, a value class constructor returns an instance that is associated with the variable to which it is assigned. If the variable is assigned to another variable, a copy is created. If the variable is passed to a function that modifies the object, the function must return the modified object.

MATLAB also supports the creation of user-defined *handle* classes. A handle class constructor returns a handle object that is a reference to the object created. If a handle object is reassigned, additional references to the original object are created. If passed to a function that modifies the object, it need not be returned since the changed object is referenced by all of the handles. Handle objects are essentially pointers to objects rather than the objects themselves.

The reference behavior of handles enables these classes to support features like events, listeners, and dynamic properties. User-defined handle classes are all subclasses of the abstract handle class and as such inherit a number of useful methods. The power of handle classes is self-evident. Although a detailed discussion of user-defined handle classes is beyond the scope of this text, the MATLAB documentation includes a wealth of detailed information and examples of user-defined handle classes.

32

Examples, Examples, Examples

It's been said that a picture is worth a thousand words. Likewise, when it comes to software, an example is worth a thousand words. This chapter is devoted to extensive examples. Some of the examples demonstrate *vectorization*, which is the process of writing code that maximizes the use of array operations. Other examples illustrate creating code that maximizes the benefits achieved from MATLAB's *JIT-accelerator* features. Still other examples demonstrate the solution of typical problems. Before considering the examples, it is beneficial to introduce vectorization and JIT-acceleration.

32.1 VECTORIZATION

Vectorization means to write or rewrite code so that scalar operations on array elements are replaced by array operations. For example,

```
>> for i=1:n
       y(i) = sin(2*pi*i/m);
   end
```

is replaced by the *vectorized* code

```
>> i = 1:n;
>> y = sin(2*pi*i/m);
```

The For Loop above represents poor programming practice. Not only is the For Loop unnecessary as seen by its vectorized equivalent but also is very slow since memory is reallocated for the variable y each time through the loop.

Contrary to what some may believe, vectorization does not mean eliminating *all* For Loops. For Loops serve a very useful purpose; after all, MATLAB wouldn't include them as a control flow structure if they didn't. For Loops are a good choice when the loop makes full use of JIT-acceleration features. When that is not possible, For Loops are often a good choice when: (1) the amount of code to be interpreted within the loop is small, especially if this code requires substantial floating-point operations; (2) the code within the loop makes minimum calls to M-file functions; (3) no memory is allocated or reallocated after the first pass through the loop; or (4) using a For Loop eliminates the need to create arrays larger than the computer can access with minimum delay. Obviously, the last case is not only platform-dependent but also dependent on the attributes of an individual platform.

While vectorization leads to efficient MATLAB programming, it does have a downside. That is, vectorized code is often more difficult to read or follow. The above example is clearly an exception to this fact—the vectorized code is much easier to read than the nonvectorized code. The above example represents a simple, easy-to-learn vectorization. Most often, it is the more difficult or less obvious vectorization challenges that lead to code that is more difficult to follow.

Vectorizing code makes use of a small number of MATLAB operators and functions. These operators and functions generally involve the manipulation of indices or the replication of arrays and can be divided into three categories as shown in the tables below. The first two categories are basic internal MATLAB capabilities and are therefore fast. The last category consists of optimized M-file code for implementing common array-manipulation functions.

Operator	Description
:	Colon notation. n:m creates a row array that starts with n and ends with m. n:inc:m creates a row array that starts with n, counts by inc, and ends at or before m. As an array index, : means take all elements. Also, A(:) on the right-hand side of an equal sign means reshape A as a column vector. On the left-hand side of an equal sign, A(:) means fill contents of A with results of the right-hand side without reallocating memory for A.
.'	Nonconjugate transpose. Exchange rows for columns.
[ ]	Brackets. Array concatenation.

Built-in Function	Description
all(x)	True if all elements of x are nonzero.
any(x)	True if any elements of x are nonzero.
cat(Dim,A,B,...)	Concatenates A, B, ... along dimension Dim.

`cumprod(x)`	Cumulative product of elements of vector x.
`cumsum(x)`	Cumulative sum of elements of vector x.
`diff(x)`	Difference between elements in x.
`end`	Last index. Inside array index identifies the last element along given dimension.

`find(x)`
`find(x,n)`
`find(x,n,'first')`
`find(x,n,'last')`
	Find indices where x is nonzero. (This is usually slower than using the logical argument x directly for array addressing.)

`logical(x)`	Converts x to logical data type to enable logical array addressing.
`permute(A,Order)`	Generalized transpose. Rearranges the dimensions of A so that they are in the order specified by the vector Order.
`prod(x)`	Product of elements in x.
`reshape(A,r,c)`	Reshapes array A to be r-by-c.
`sort(x)` `sort(x,'descend')`	Sort array x in ascending or descending order.
`sum(x)`	Sum of elements in x.

M-file Function	**Description**
`arrayfun(fun,S)`	Applies function to each element of array.
`bsxfun(fun,A,B)`	Element-by-element binary operation between A and B.
`ind2sub(Size,idx)`	Converts single indices in idx to array subscripts of an array having dimensions Size.
`ipermute(A,Order)`	Generalized transpose. Inverse of `permute(A,Order)`.
`kron(A,B)`	Kronecker tensor product of A and B.
`meshgrid(x,y)`	Mesh domain generation from vectors x and y.
`repmat(A,r,c)`	Replicates array A creating an r-by-c block array.
`shiftdim(A,n)`	Shifts dimensions of A by integer n.
`squeeze(x)`	Removes singleton dimensions from array A.
`sub2ind(Size,r,c)`	Converts array subscripts r and c of an array having dimensions Size to single indices.

Using the above operators, built-in functions, and M-file functions, vectorization involves the substitution of scalar operations with equivalent array operations when such operations increase execution speed.

32.2 JIT-ACCELERATION

Just-In-Time acceleration, commonly referred to as JIT-acceleration, describes features in the MATLAB interpreter that convert whole sections of code into native instructions in one pass, rather than doing so on a line-by-line basis, which is the conventional way MATLAB code is interpreted and executed. In particular, the JIT-accelerator minimizes the processing overhead involved in executing loops. When the JIT-accelerator can process and execute a whole loop structure at one time, each line of code within the loop is not reinterpreted each time through the loop. As a result, the loop structure is interpreted once and executed as a block. This process speeds the execution of MATLAB code immensely.

Not all loops fit the requirements of the JIT-accelerator. From one MATLAB release to the next, the features and limitations of the JIT-accelerator change, and these changes are often not well documented. In any case, MATLAB code containing loops benefits from the JIT-accelerator if it has the following features and properties:

1. The loop structure is a For Loop.
2. The loop contains only logical data, character-string data, double-precision real data, and less than 64-bit integer data.
3. The loop utilizes arrays that are three-dimensional or less.
4. All variables within a loop are defined prior to loop execution.
5. Memory for all variables within the loop is preallocated outside the loop and the variables maintain constant size, orientation, and data type for all loop iterations.
6. Loop indices are scalar quantities, for example, the index k in for k = 1:N.
7. Only built-in MATLAB functions are called within the loop.
8. Conditional statements using *if-then-else* or *switch-case* constructions involve scalar comparisons.
9. All lines within the block contain no more than one assignment statement.

JIT-acceleration provides the greatest benefit when the arrays addressed within the loop are relatively small. As array sizes increase beyond cpu cache capacities, total execution time increases due to the time consumed by memory transfers, thereby leading to less dramatic improvements in overall execution time.

32.3 THE BIRTHDAY PROBLEM

The classic birthday problem is this: What is the probability p that in a random group of n people at least one pair of them shares the same birthday?

The analytic solution to this problem is well known and is given by,

$$p(n) = 1 - \frac{n!\binom{365}{n}}{365^n}$$

and one approximate solution is given by

$$p_a(n) = 1 - \exp\left(\frac{-n^2}{2 \times 365}\right)$$

To explore this problem, let's plot these two solutions as a function of the group size n. The following code cell and the resulting figure demonstrate the results.

```
%% Birthday Problem Solutions

% create anonymous functions

% analytic solution
p = @(n) 1 - factorial(n)*nchoosek(365,n)/(365^n);

% note that the function factorial implements n!
% and nchoosek implements (365 over n)

% approximate solution
p_a = @(n) 1 - exp(-(n^2)/(2*365));

% the functions operate on scalar inputs,
% so a loop is required to gather data.

% preallocate memory for results (ALWAYS do this!)

N = 50; % maximum number of people
pexact = zeros(1,N);
papprox = zeros(1,N);

for n = 1:N        % JIT acceleration will work here!
    pexact(n) = p(n);
    papprox(n) = p_a(n);
end

nn = 1:N; % x axis for plot
```

```
plot(nn,pexact,nn,papprox)
grid
xlabel('Number of People in Group')
ylabel('Probability of Shared Birthdays')
title('Figure 32.1: Birthday Problem')
```

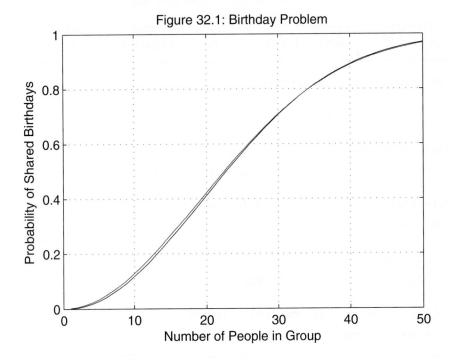

Figure 32.1: Birthday Problem

When n is equal to 8 or greater, the function nchoosek gives a warning stating that the output it provides has reduced accuracy. This warning appears because double-precision arithmetic does not have sufficient integer range to compute these values without loss of precision.

Figure 32.1 shows that there is a 50 percent probability of shared birthdays in a group of 23 people and a 90 percent probability of shared birthdays in a group of 41 people.

To confirm the analytic results, let's run some random trials. That is, start with a group of n people, assign random birthdays to them, then see if there are any shared birthdays. The following code cell demonstrates the basic features of this testing.

```
%% test birthday problem results

n = 23; % number of people in test group
```

```
bdays = randi(365,1,n) % generate n random birthdays

% the number of duplicate values in bdays
% is the number of shared birthdays in the group.

% sort bdays to get duplicates next to each other

sortedbdays = sort(bdays)

% find duplicates by taking differences

diffbdays = diff(sortedbdays)

% shared birthdays occur where values are zero

numshared = sum(diffbdays==0)
```

One run of this code cell produced the following results.

```
bdays =
  Columns 1 through 12
      96   293    11   340   267   179   212    87   168   352   200   191
  Columns 13 through 23
      85   179   228   248   145   135   361    14   324   334   291
sortedbdays =
  Columns 1 through 12
      11    14    85    87    96   135   145   168   179   179   191   200
  Columns 13 through 23
     212   228   248   267   291   293   324   334   340   352   361
diffbdays =
  Columns 1 through 12
       3    71     2     9    39    10    23    11     0    12     9    12
  Columns 13 through 22
      16    20    19    24     2    31    10     6    12     9
numshared =
       1
```

The birthday 179 appears twice in `sortedbdays`, making the ninth value in `diffbdays` equal to zero. Therefore, this group of 23 people has one pair with shared birthdays. Therefore, it appears that the number of zeros in `diffbdays` is equal to the number of shared birthdays.

Depending on the random integers generated by `randi`, there may be any number of shared birthdays. Of particular interest is the case when one or more birthdays are shared by more than two people. In this case, there are more zeros in `diffbdays` than there are shared birthdays. When birthdays are shared by more than two people, there will be consecutive zeros in `diffbdays`. If we want to know exactly the number of shared birthdays, further processing of `diffbdays` would be needed. However, since we are only interested in knowing when there is at least one shared birthday, we only need to know when there is at least one zero in the array `diffbdays`.

To have value, the above random trial must be run many, many times. The following code cell demonstrates an implementation.

```
%% test birthday problem

Ntests = 1000; % number of random trials to run
n = 23; % size of group to test

shared = zeros(1,Ntests); % preallocate result array

for k = 1:Ntests

    bdays = randi(365,1,n);
    % the number of duplicate values in bdays
    % is the number of shared birthdays in the group.

    % sort bdays to get duplicates next to each other
    sortedbdays = sort(bdays);

    % find duplicates by taking differences
    diffbdays = diff(sortedbdays);

    % shared birthdays occur when there are zero values
    shared(k) = any(diffbdays==0);
end
% probability is equal to normalized number of times
% that shared birthdays appear in the trials
probshared = sum(shared)/Ntests
```

Running this code multiple times produces values that are very close to the analytic solutions given earlier. Consider the following example:

```
probshared =
       0.5060
probshared =
       0.4820
probshared =
       0.4990
probshared =
       0.5110
probshared =
       0.5080
probshared =
       0.5240
probshared =
       0.4990
probshared =
       0.5180
```

32.4 UP–DOWN SEQUENCE

Consider the following simple algorithm. Let N be some positive integer. If N is even, divide it by 2. On the other hand, if it is odd, multiply it by 3 and add 1. Repeat these steps until N becomes 1. This algorithm has some interesting properties. It appears to converge to 1 for all numbers N. Some numbers require many iterations to converge. Others such as $N = 2^m$ converge very quickly. While it is interesting to study the sequence of values generated by different values of N, let's just compute the number of iterations required to achieve convergence.

First, let N be a scalar. That is, let's write the algorithm for a single number. The following script M-file implements the above algorithm.

```
% updown1.m
% up-down algorithm

N = 25;     % number to test
count = 0; % iteration count
```

```
while N>1
    if rem(N,2)==0 % even, since division by 2 gives zero remainder
        N = N/2;
        count = count+1;
    else             % odd
        N = (3*N+1)/2;
        count = count+2;
    end
end
count % display iteration count
```

The above code directly implements the algorithm with one exception. When an odd number is multiplied by 3 and has 1 added, the resulting number is automatically even. As a result, the next pass through the algorithm always divides by 2. Since this always occurs, the divide-by-2 step is included and the count is incremented by 2 to reflect the fact that two steps are taken.

Next, consider letting N be an array of numbers, for each of which we wish to find the iteration count. The most direct approach is to use a For Loop, as shown in the following script M-file.

```
% updown2.m
% up-down algorithm

Nums = 25:50;    % numbers to test

for i = 1:length(Nums)
    N = Nums(i);    % number to test
    count = 0;      % iteration count
    while N>1
        if rem(N,2)==0  % even
            N = N/2;
            count = count+1;
        else              % odd
            N = (3*N+1)/2;
            count = count+2;
        end
```

```
    end
    Counts(i) = count;
end
results = [Nums' Counts']
```

Here, the earlier scalar algorithm appears within a For Loop. At the beginning of the loop, the *i*th element of the vector Nums is copied into N. The algorithm then runs to completion, and the iteration count is copied into the *i*th element of Counts. Finally, all results are displayed. This code segment violates a key memory allocation guideline, namely, the variable Counts is reallocated to a larger size at every pass through the For Loop. As a result, the above code does not make use of the JIT-acceleration.

To fix this problem, Counts must be preallocated as shown below.

```
% updown3.m
% up-down algorithm

Nums = 25:50;                 % numbers to test
Counts = zeros(size(Nums));  % preallocate array
N = Nums(1);                  % predefine N data type and dimension
count = 0;                    % predefine count data type and dimension
for i = 1:length(Nums)
    N = Nums(i); % number to test
    count = 0;    % iteration count
    while N>1
        if rem(N,2)==0 % even
            N = N/2;
            count = count+1;
        else      % odd
            N = (3*N+1)/2;
            count = count+2;
        end
    end
    Counts(i) = count;
end
results = [Nums' Counts']
```

Now every time through, the For Loop simply inserts the current `count` into a preexisting location in `Counts`. Preallocation is always a first and most important step in any code optimization process. In addition, since `N` and `count` are used within the For Loop, their data types and dimensions are predefined. With these changes, `updown3.m` makes full use of JIT-acceleration.

For comparison purposes, it is beneficial to develop a vectorized version of this algorithm. In a vectorized version, all input data must be processed simultaneously. This eliminates the For Loop. The function `rem` returns an array the same size as its input, and so all numbers can be tested simultaneously at every iteration of the While Loop. Using this fact leads to the following script M-file.

```matlab
% updown4.m
% up-down algorithm

Nums = 25:50; % numbers to test
N = Nums;                    % duplicate numbers
Counts = zeros(size(N)); % preallocate array

not1 = N>1;                      % True for numbers greater than one

while any(not1)

  odd = rem(N,2)~=0;             % True for odd values

  odd_not1 = odd & not1;     % True for odd values greater than one
  even_not1 = ~odd & not1;   % True for even values greater than one

  N(even_not1) = N(even_not1)/2;            % Process evens
  Counts(even_not1) = Counts(even_not1)+1;

  N(odd_not1) = (3*N(odd_not1)+1)/2;        % Process odds
  Counts(odd_not1) = Counts(odd_not1)+2;

  not1 = N>1;                    % Find remaining numbers not converged
end
results = [Nums' Counts']
```

When considering all elements simultaneously, find a way to operate only on array elements that have not converged to 1. The statement `not1 = N>1;` logically identifies all elements of `N` that haven't converged. Inside the While Loop, the nonconverged odd and even values are identified by the `odd_not1` and `even_not1` logical variables, respectively. Using these variables, the corresponding elements of N are processed and the Counts values are updated.

This vectorized solution makes use of logical array addressing. Addressing using numerical indices could have been done as well by using the find function; for example, the odd indices are given by find(odd_not1). However, use of this function adds more statements to the solution, which makes the code run slower.

Running the profiler on this solution shows that the greatest percentage of time is consumed by the odd = rem(N,2)~=0; statement. There are two reasons for this. First, this statement finds the remainder for all elements of N every time, not just the remainder of nonconverged numbers. For example, if there are 1000 data points in N and only 10 of them are nonconverged, the remainder is computed for 990 elements in N that are not required. Second, the remainder function incurs overhead because it is a function call and because it performs internal error checking. Both of these reasons for the time consumed by the call to the rem function can be addressed. Eliminating the cause of the first reason requires removing the converged values from N as they appear. Implementing this requires substantial complexity. However, the rem function overhead can be minimized by replacing it with its definition, as shown in the code below.

```
% updown5.m
% up-down algorithm

Nums = 25:50;       % numbers to test

N = Nums;                   % duplicate numbers
Counts = zeros(size(N)); % preallocate array

not1 = N>1;                 % True for numbers greater than one

while any(not1)

   odd = (N-2*fix(N/2))~=0;  % True for odd values

   odd_not1 = odd & not1;    % True for odd values greater than one
   even_not1 = ~odd & not1;  % True for even values greater than one

   N(even_not1) = N(even_not1)/2;              % Process evens
   Counts(even_not1) = Counts(even_not1)+1;

   N(odd_not1) = (3*N(odd_not1)+1)/2;          % Process odds
   Counts(odd_not1) = Counts(odd_not1)+2;
```

```
    not1 = N>1;                        % Find remaining numbers
end
results = [Nums' Counts']
```

Running the MATLAB profiler on this example shows that (N-2*fix(N/2))is generally faster than rem(N,2).

As a final test, it is worth considering the use of integer arithmetic. Since N and Counts are always integers, it is possible to cast them into integer arrays. In this case, (N-2*fix(N/2))no longer works as a replacement for rem(N,2). However, odd = 2*(N/2)~=N; is an alternative test for odd integers. The code shown below illustrates this integer implementation.

```
% updown6.m
% up-down algorithm

Nums = 25:50; % numbers to test

N = uint32(Nums);                      % duplicate numbers as uint32
Counts = zeros(size(N),'uint32');% preallocate array as uint32

not1 = N>1;                        % True for numbers greater than one

while any(not1)

    odd = 2*(N/2)~=N;              % True for odd values

    odd_not1 = odd & not1;        % True for odd values greater than one
    even_not1 = ~odd & not1;      % True for even values greater than one

    N(even_not1) = N(even_not1)/2;        % Process evens
    Counts(even_not1) = Counts(even_not1)+1;

    N(odd_not1) = (3*N(odd_not1)+1)/2;    % Process odds
    Counts(odd_not1) = Counts(odd_not1)+2;

    not1 = N>1;                        % Find remaining numbers
end
results = [Nums' Counts']
```

As a comparison, the last four implementations were timed with the profiler using Nums = 1:2049; and with the last line results = [Nums' Counts'] removed. Of these implementations, the fifth and sixth are about equally fast, with the fourth being a little slower, followed by the third which was about one-half as fast as the sixth.

32.5 ALTERNATING SEQUENCE MATRIX

MATLAB has a number of matrix creation functions such as ones, zeros, rand, diag, eye, pascal, and hilb. In this example, let's create a MATLAB function that returns a matrix containing alternating values. That is, if the function is called as plusminus(K,m,n) or plusminus(K,[m n]), where K is a real-valued scalar, and m and n are scalar integers, then the output returned has the form shown below.

$$\begin{bmatrix} K & -K & \cdots \\ -K & K & \cdots \\ \vdots & \vdots & \ddots \end{bmatrix}$$

As with most problems, there are many ways to solve this in MATLAB. The most direct approach is to use nested For loops to set each individual element of the result individually, as shown in the code cell segment and its resulting output below.

```
%% Alternating Sequence For loop approach

K = 1; % chosen value
m = 4; % number of rows
n = 5; % number of columns

out = zeros(m,n); preallocate memory for results

for i = 1:m % loop over rows

    for j = 1:n % loop over the columns

        oddrow = rem(i,2)==1; % true if in odd numbered row
        oddcol = rem(j,2)==1; % true if in odd numbered column

        if (oddrow && ~oddcol) || ~oddrow && oddcol
            out(i,j) = -K;
        else
```

```
        out(i,j) = K;
      end

   end

end

out % display results
```

```
out =
    1  -1   1  -1   1
   -1   1  -1   1  -1
    1  -1   1  -1   1
   -1   1  -1   1  -1
```

Another solution approach uses the fact that $(-1)^{m+n}$, where m and n are the row and column indices, respectively, gives the correct sign for each element of the matrix. Once a matrix of plus and minus ones is computed, the final result is found by multiplying by K. This approach is shown in the code cell segment and its resulting output below.

```
%% Alternating Sequence indices approach

K = 2; % chosen value
m = 4; % number of rows
n = 5; % number of columns

% create m-by-n matrices of row and column indices
[NN,MM] = meshgrid(1:n,1:m);

% sum of column indices gives exponent
out = K*(-1).^rem(MM+NN,2);

out % display results
```

```
out =
    2   -2    2   -2    2
   -2    2   -2    2   -2
    2   -2    2   -2    2
   -2    2   -2    2   -2
```

Yet another solution uses the MATLAB function `repmat` to replicate the basic matrix [K -K; -K K] to the correct final dimensions. The only issue with this approach is when the number of rows or columns are odd. This approach is shown in the code cell segment and its resulting output below.

```
%% Alternating Sequence repmat approach.

K = 3; % chosen value
m = 4; % number of rows
n = 5; % number of columns

% function ceil rounds up to next integer
mrep = ceil(m/2); % number of times to replicate rows
nrep = ceil(n/2); % number of times to replicate columns

out = repmat([K -K;-K K],mrep,nrep); % replicate
out = out(1:m,1:n); % eliminate any extra rows and columns

out % display results
```

```
out =
    3   -3    3   -3    3
   -3    3   -3    3   -3
    3   -3    3   -3    3
   -3    3   -3    3   -3
```

The final approach considered here relies on the outer product of two vectors. That is, if x is a column vector having m rows, and y is a row vector having n columns, then the product of x and y is an m-by-n matrix. Furthermore, if x and y contain alternating +1 and −1 values, this outer product will contain a matrix of plus and minus ones. Multiplying this matrix by K produces the desired final value. This approach is shown in the code cell segment and its resulting output below.

```
%% Alternating Sequence outer product approach

K = 4; % chosen value
m = 4; % number of rows
n = 5; % number of columns

pm = ones(2,ceil(max(m,n)/2)); % two rows of ones
pm(2,:) = -1; % set second row to -1
% resulting pm(:) is the alternating vector [1 -1 1 -1 1...]'

% use pm array as needed to create output variable
out = K*(pm(1:m).' * pm(1:n)); % outer product: column * row = matrix

out % display results
```

```
out =
    4   -4    4   -4    4
   -4    4   -4    4   -4
    4   -4    4   -4    4
   -4    4   -4    4   -4
```

The approach using `repmat` performs no multiplications and does not create numerous large intermediate variables. For that reason, it is marginally faster than the outer product approach. The code below demonstrates how this algorithm can be encapsulated into a MATLAB function. The following code contains needed initial help text and parses and error checks the input arguments before implementing the algorithm.

```
function out = plusminus(K,m,n)
%PLUSMINUS Array of Alternating Plus and Minus K Values.
% PLUSMINUS(K,N) returns an N-by-N array.
%
% PLUSMINUS(K,M,N) or
% PLUSMINUS(K,[M N]) return an M-by-N array.
%
% PLUSMINUS(K,size(A)) returns an array
% the same size as matrix A.
%
```

```
% For example: PLUSMINUS(8,3,5)
% produces the matrix
%                     [ 8 -8  8 -8  8
%                      -8  8 -8  8 -8
%                       8 -8  8 -8  8 ]

% parse and error check inputs

if nargin<2
   error('At least 2 input arguments are required.')
elseif nargin==2 && isscalar(m) % PLUSMINUS(K,N)
   n = m;
elseif nargin==2 && numel(m)==2 % PLUSMINUS(K,[M N])
                                % or PLUSMINUS(K,size(A))
   n = m(2);
   m = m(1);
elseif nargin==2 && numel(m)~=2
   error('Second argument must be a scalar or two element vector.')
end

if numel(m)~=1 || fix(m)~=m || m<1
   error('M must be a positive scalar.')
end
if numel(n)~=1 || fix(n)~=n || n<1
   error('M must be a positive scalar.')
end
if ~isscalar(K) || ~isnumeric(K)
   error('K must be a numeric scalar.')
end

% implement the algorithm
out = repmat([K -K;-K K],ceil(m/2),ceil(n/2));
out = out(1:m,1:n);
```

32.6 VANDERMONDE MATRIX

There are a number of numerical linear algebra problems that require the generation of a Vandermonde matrix. For a vector x, a Vandermonde matrix has the form

$$V = \begin{bmatrix} x_1^m & x_1^{m-1} & \cdots & x_1 & 1 \\ x_2^m & x_2^{m-1} & \cdots & x_2 & 1 \\ \vdots & \vdots & \ddots & \vdots & \vdots \\ x_n^m & x_n^{m-1} & \cdots & x_n & 1 \end{bmatrix}$$

As shown, the columns of V are element-by-element powers of the components of x. Let's consider a variety of approaches to constructing this matrix.

The first approach that comes to mind is the straightforward application of a For Loop, as in the following code cell.

```
%% construct a Vandermonde matrix, approach 1

x = (1:6)';          % column vector for input data
m = 5;               % highest power to compute
V = [];

for i = 1:m+1        % build V column by column
   V = [V x.^(m+1-i)];
end
```

The above approach builds V column by column, starting from an empty matrix. There are a number of weaknesses in this implementation, the most obvious being that memory is reallocated for V each time through the loop. So the first vectorization step is to preallocate V, as shown in the code cell below.

```
%% construct a Vandermonde matrix, approach 2

x = (1:6)';          % column vector for input data
m = 5;               % highest power to compute
n = length(x);       % number of elements in x
V = ones(n,m+1);     % preallocate memory for result

for i = 0:m-1        % build V column by column
   V(:,i+1) = x.^(m-i);
end
```

Here V is initialized as a matrix containing all ones. Then the individual columns of V are assigned within the For Loop. The last column is not assigned in the For Loop since it already contains ones and there is no use in computing x.^0. There are still two problems with the above code. First, the columns of V are explicitly computed without making use of prior columns, and second, the For Loop should be able to be eliminated. The code cell below solves the first problem.

```
%% construct a Vandermonde matrix, approach 3

x = (1:6)';            % column vector for input data
m = 5;                 % highest power to compute
n = length(x);         % number of elements in x
V = ones(n,m+1);       % preallocate memory for result

for i = m:-1:1         % build V column by column
   V(:,i) = x.*V(:,i+1);
end
```

Now the columns of V are assigned starting with the second last column and proceeding backward to the first column. This is done because the *i*th column of V is equal to the $(i + 1)$th column multiplied elementwise by x. This is the implementation found in the MATLAB functions polyfit and vander.

At this point, the above implementation cannot be optimized further without eliminating the For Loop. Eliminating the For Loop requires some ingenuity and a lot of familiarity with the functions in MATLAB. Using the array-manipulation tables found earlier in this chapter, the functions repmat and cumprod offer some promise. The script M-file below demonstrates an approach that uses repmat.

```
%% construct a Vandermonde matrix, approach 4

x = (1:6)';            % column vector for input data
m = 5;                 % highest power to compute
n = length(x);         % number of elements in x

p = m:-1:0;            % column powers

V = repmat(x,1,m+1).^repmat(p,n,1);
```

This implementation uses repmat twice, once to replicate x creating a matrix of m+1 columns each containing x, and the second time to create a matrix containing the powers to be applied to each element of the matrix containing x. Given these two matrices, element-by-element exponentiation is used to create the desired result. As with the second approach, this implementation explicitly computes each column without using information from other columns. The function cumprod solves this problem, as shown in the code cell below.

```matlab
% construct a Vandermonde matrix, approach 5

x = (1:6)';            % column vector for input data
m = 5;                 % highest power to compute
n = length(x);         % number of elements in x

V = ones(n,m+1);       % preallocate memory for result

V(:,2:end) = cumprod(repmat(x,1,m),2);

V = V(:,m+1:-1:1);   % reverse column order
```

Here, the function cumprod is used to compute the columns of V after using repmat to duplicate x. Since cumprod proceeds from left to right, the final result is found by reversing the columns of V. This implementation uses only one M-file function, repmat. Eliminating this function by array addressing should lead to the fastest possible implementation. Doing so leads to the code cell below.

```matlab
% construct a Vandermonde matrix, approach 6

x = (1:6)';            % column vector for input data
m = 5;                 % highest power to compute
n = length(x);         % number of elements in x

V = ones(n,m+1);       % preallocate memory for result

V(:,2:end) = cumprod(x(:,ones(1,m)),2); % avoid call to repmat

V = V(:,m+1:-1:1);   % reverse column order
```

Given these six implementations, the third and sixth are the fastest. The MATLAB functions `polyfit` and `vander` utilize the same algrorithm as the third approach, which is easier to read and requires less memory than the sixth.

32.7 REPEATED VALUE CREATION AND COUNTING

This section considers the following problem. Given a vector x containing data and a vector n of equal length containing nonnegative integers, construct a vector where $x(i)$ is repeated $n(i)$ times for every ith element in the two vectors. For example, $x =$ [3 2 0 5 6] and $n =$ [2 0 3 1 2] would produce the result $y =$ [3 3 0 0 0 5 6 6]. Note that since $n(2)$ is zero, $x(2)$ does not appear in the result y.

In addition to repeated value creation, the inverse problem of identifying and counting repeated values is also of interest. Thus, given the example y above, find the vectors x and n that describe it.

Consider repeated value creation first. A scalar approach is straightforward, as shown in the script M-file below.

```
% repeat1.m
% repeated value creation and counting

x = [3 2 0 5 6]; % data to repeat
n = [2 0 3 1 2]; % repeat counts

y = [];
for i = 1:length(x)

    y = [y repmat(x(i),1,n(i))];
end
```

In this implementation, the result is built using brackets to concatenate the repeated values next to each other. JIT-acceleration does not work with this example because memory is reallocated for y every loop iteration and the M-file function `repmat` appears in the loop. These issues can be resolved by preallocating y and by using indexing, as shown in the following script M-file.

```
% repeat2.m
% repeated value creation and counting

x = [2 1 0 4 5];  % data to repeat
n = [3 0 2 1 3];  % repeat counts
```

```
nz = n==0;                    % locations of zero elements
n(nz) = [];                   % eliminate zero counts

x(nz) = [];                   % eliminate corresponding data
y = zeros(1,sum(n));          % preallocate output array

idx = 1;                      % pointer into y

for i=1:length(x)

   y(idx:idx+n(i)-1) = x(i);  % fill y using scalar expansion
   idx = idx+n(i);            % next pointer location
end
```

As shown above, zero counts are eliminated prior to the For Loop since they do not contribute to the result. In this algorithm, sum(n) is the total number of elements in the result, and the variable idx is used to identify where the next data is to be placed in y.

Once again, it is worthwhile to consider a vectorized solution to this problem. To determine how to proceed, it is beneficial to look at how x is related to y. After eliminating zero counts in the example being considered, x, n, and y are

```
>> x
x =
     2   0   4   5
>> n
n =
     3   2   1   3
>> y
y =
     2   2   2   0   0   4   5   5   5
```

If we can create an index vector idx = [1 1 2 2 2 3 4 4], then x is related to y as

```
>> idx = [1 1 2 2 3 3 3 4];
>> y = x(idx)
y =
     2   2   0   0   4   4   4   5
```

So rather than concentrate on getting the values of x into the correct places in y, if we can generate an index vector for x with the desired values, finding y simplifies to just one statement, y = x(idx). This is a common situation when vectorizing. The indices are often more important than the data itself.

The relationship between n and idx above is straightforward. In addition, since idx looks like a cumulative sum, there's a chance that the cumsum function will be useful. If we can generate an array of ones and zeros at the indices where idx changes value, then idx is indeed a cumulative sum. Consider the example:

```
>> tmp = [1 1 0 1 0 0 1 1]
tmp =
     1   1   0   1   0   0   1   1
>> idx = cumsum(tmp)
idx =
     1   2   2   3   3   3   4   5
```

The nonzero values in tmp are related to n. To discover this relationship, look at its cumulative sum:

```
>> csumm = cumsum(n)
csumm =
     3   5   6   9
```

If the last value in csn is discarded and one is added to the remaining values, the indices of all ones in tmp are known, except for the first which is always one. Therefore, the indices of all ones can be computed. Consider the example:

```
>> tmp2 = [1 csumn(1:end-1)+2]
tmp2 =
     1   5   7   8
```

That's it. The values in tmp2 identify the ones to be placed in tmp. All other values in tmp are 0. Now all that remains is creating tmp. This is easily done, as in the following example.

```
>> tmp = zeros(1,csumn(end)) % preallocate with all zeros
tmp =
     0   0   0   0   0   0   0   0   0
>> tmp(tmp2) = 1             % poke in ones with scalar expansion
tmp =
     1   0   0   0   1   0   1   1   0
```

```
>> idx = cumsum(tmp)          % form the desired cumsum
idx =
      1   1   1   1   2   2   3   4   4
>> y = x(idx)                 % idx does the rest!
y =
      2   2   2   2   0   0   4   5   5
```

Using the above approach, a vectorized implementation of creating repeated values
is shown in the script M-file below.

```
% repeat3.m
% repeated value creation and counting

x = [3 2 0 5 6]; % data to repeat
n = [2 0 3 1 2]; % repeat counts

nz = n==0;       % locations of zero elements
n(nz) = [];      % eliminate zero counts
x(nz) = [];      % eliminate corresponding data
csn = cumsum(n);                % cumulative sum of counts
tmp = zeros(1,csn(end));        % preallocate memory
tmp([1 csn(1:end-1)+1]) = 1;    % poke in ones
idx = cumsum(tmp);              % index vector
y = x(idx);                     % let array indexing do the work
```

Let's move on to the inverse of the above algorithm, namely, repeated value
identification and counting. That is, starting with y, find x and n (except for any zero
count values, of course). Again, the most obvious solution is nonvectorized and uses
a For Loop, as shown in the script M-file below.

```
% repeat4.m
% repeated value creation and counting
% inverse operation

y = [3 3 0 0 0 5 6 6]; % data to examine

x = y(1);               % beginning data
```

```
n = 1;                      % beginning count
idx = 1;                    % index value
for i = 2:length(y)
   if y(i)==x(idx) % value matches current x

       n(idx) = n(idx)+1; % increment current count

   else % new value found

      idx = idx+1;         % increment index
      x(idx) = y(i);       % poke in new x
      n(idx) = 1;          % start new count
   end
end
```

Here, a simple If-Else-End construction is used to decide if a particular element of y is a member of the current repeated value. If it is, the count is incremented. If it's not, a new repeat value is created. Though it is not obvious, each time the Else section is executed, memory is reallocated for x and n. As a result, the above solution does not gain benefit from JIT-acceleration.

The next step is to use preallocation, as shown in the script M-file below.

```
% repeat5.m
% repeated value creation and counting
% inverse operation

y = [2 2 0 0 0 4 5]; % data to examine

x = zeros(size(y));     % preallocate results
n = zeros(size(y));

x(1) = y(1);                % beginning data
n(1) = 1;                   % beginning count
idx = 1;                    % index value

for i = 2:length(y)
```

```
    if y(i)==x(idx) % value matches current x

        n(idx) = n(idx)+1; % increment current count

    else % new value found
        idx = idx+1;        % increment index
        x(idx) = y(i);      % poke in new x
        n(idx) = 1;         % start new count
    end
end
nz = (n==0); % find elements not used
x(nz) = []; % delete excess allocations
n(nz) = [];
```

Since the length of x and n are unknown, but they cannot be any longer than y, x and n are preallocated to have the same size as y. At the end, the excess memory allocations are discarded.

Eliminating the For Loop to vectorize this algorithm requires study of an example. Consider the y vector that was the result of applying the creation algorithm to the original x and n data:

```
>> y
y =
     2    2    0    0    0    4    5
```

Because of the structure of this vector, the MATLAB function diff must be useful:

```
>> diff(y)
ans =
     0   -2    0    0    4    1
```

If this vector is shifted one element to the right, the nonzero elements line up with places in y that represent new repeated values. In addition, the first element is always a new repeated value. So, using logical operations, the repeated values can be identified, as in the following example:

```
>> y
y =
     2    2    0    0    0    4    5
```

```
>> tmp = [1 diff(y)]~=0
tmp =
    1    0    1    0    0    1    1
```

Given the logical variable tmp, the vector x is found by logical addressing:

```
>> x = y(tmp)
x =
    2    0    4    5
```

Finding the repeat counts n associated with each value in x takes further study. The repeat counts are equal to the distance between the ones in tmp. Therefore, finding the indices associated with tmp is useful:

```
>> find(tmp)
ans =
    1    3    6    7
>> diff(ans)
ans =
    2    3    1
```

The difference in indices gives the repeat count for all but the last repeated element. The function diff misses this count because there is no marker identifying the end of the array, which marks the end of the last repeated value. Appending one non-zero to tmp solves this problem:

```
>> find([tmp 1])
ans =
    1    3    6    7    8
>> n = diff(ans)
n =
    2    3    1    1
```

The inverse algorithm is now known. Implementing it leads to the script file below.

```
% repeat6.m
% repeated value creation and counting
% inverse operation
```

```
y = [3 3 0 0 0 5 6 6]; % data to examine

tmp = ([1 diff(y)]~=0);
x = y(tmp);
n = diff(find([tmp 1]));
```

The above implementation demonstrates the compactness of vectorization. This solution requires just three lines of code, compared to the 17 used in repeat5.m. This implementation also demonstrates the difficulty often encountered in reading vectorized code. These three lines are essentially meaningless unless one executes and views the results of each line with a simple example.

Before concluding, it is important to note that the above implementations fail if y contains any Inf or NaN elements. The function diff returns NaN for differences between Inf elements, as well as for differences containing NaN elements. Because of the utility of creating and counting repeated values, it is encapsulated in the function mmrepeat shown below.

```
function [y,m] = mmrepeat(x,n)
%MMREPEAT Repeat or Count Repeated Values in a Vector. (MM)
% MMREPEAT(X,N) returns a vector formed from X where X(i) is repeated
% N(i) times. If N is a scalar it is applied to all elements of X.
% N must contain nonnegative integers. N must be a scalar or have the same
% length as X.
%
% For example, MMREPEAT([1 2 3 4],[2 3 1 0]) returns the vector
% [1 1 2 2 2 3] (extra spaces added for clarity)
%
% [X,N] = MMREPEAT(Y) counts the consecutive repeated values in Y returning
% the values in X and the counts in N. Y = MMREPEAT(X,N) and [X,N] = MMREPEAT(Y)
% are inverses of each other if N contains no zeros and X contains unique
% elements.

if nargin==2 % MMREPEAT(X,N) MMREPEAT(X,N) MMREPEAT(X,N) MMREPEAT(X,N)
   nlen = length(n);
   if ~isvector(x)
      error('X Must be a Vector.')
```

```
else
   [r,c] = size(x);
end
if any(n<0) || any(fix(n)~=n)
   error('N Must Contain NonNegative Integers.')
end
if ~isvector(n) || (nlen>1 && nlen~=numel(x))
   error('N Must be a Scalar or Vector the Same Size as X.')
end
x = reshape(x,1,xlen); % make x a row vector

if nlen==1 % scalar n case, repeat all elements the same amount
   if n==0 % quick exit for special case
      y = [];
      return
   end
   y = x (ones(1,n),:); % duplicate x to make n rows each containing x
   y = y(:);            % stack each column into a single column
   if r==1              % input was a row so return a row
      y = reshape(y,1,[]);
   end
else % vector n case
   iz = n~=0;           % take out elements to be repeated zero times
   x = x(iz);
   n = n(iz);
   csn = cumsum(n);
   y = zeros(1,csn(end)); % preallocate temp/output variable
   y(csn(1:end-1)+1) = 1; % mark indices where values increment
   y(1) = 1;              % poke in first index
   y = x(cumsum(y));      % use cumsum to set indices
   if c==1                % input was a column so return a column
      y = reshape(y,[],1);
   end
end
```

```
elseif nargin==1 % MMREPEAT(Y) MMREPEAT(Y) MMREPEAT(Y) MMREPEAT(Y)
   xlen = length(x);
   if ~isvector(x)
      error('Y Must be a Vector.')
   else
      c = size(x,2);
   end
   x = reshape(x,1,[]); % make x a row vector
   xnan = isnan(x);
   xinf = isinf(x);
   if any(xnan|xinf) % handle case with exceptions
      ntmp = sum(rand(1,4))*sqrt(realmax); % replacement for nan's
      itmp = 1/ntmp;                       % replacement for inf's
      x(xnan) = ntmp;
      x(xinf) = itmp.*sign(x(xinf));
      y = [1 diff(x)]~=0;           % places where distinct values begin
      m = diff([find(y) xlen+1]); % counts
      x(xnan) = nan;               % poke nan's and inf's back in
      x(xinf) = inf*x(xinf);       % get correct sign in inf terms
   else % x contains only finite numbers
      y = [1 diff(x)]~=0;          % places where distinct values begin
      m = diff([find(y) xlen+1]); % counts
   end
   y = x(y); % the unique values
   if c==1 % input was a column so return a column
      y = reshape(y,[],1);
   end
else
   error('Incorrect Number of Input Arguments.')
end
```

32.8 DIFFERENTIAL SUMS

This section considers computing the differential sum between elements in an array. That is, if $x = [\,3\,1\,2\,6\,3\,1\,{-}1\,]$, then the differential sum of x is $y = [\,4\,3\,8\,9\,4\,0\,]$. This is the dual or complement of the MATLAB function diff. Just as with diff, we wish to create a function that works along any dimension of its input, no matter how many dimensions the input has. Before generalizing to the n-D case, let's consider the vector and matrix cases. The vector case is straightforward:

```
>> x = [1 2 3 4 5 6 -1]
x =
      1    2    3    4    5    6    -1
>> x(1:end-1)
ans =
      1    2    3    4    5    6
>> x(2:end)
ans =
      2    3    4    5    6    -1
>> y = x(1:end-1) + x(2:end)
y =
      3    5    7    9    11    5
```

Simple array addressing is all that is required, and the above approach works if x is either a row or a column vector. When the input is a matrix, the default action is to perform a differential sum down the columns, which is along the row dimension, as in the example below:

```
>> x = magic(4) % 2-D data
x =
     16     5     9     4
      2    11     7    14
      3    10     6    15
     13     8    12     1
>> y = x(1:end-1,:) + x(2:end,:)
y =
     18    16    16    18
      5    21    13    29
     16    18    18    16
```

It is also desirable to be able to specify an operation across the columns, which is along the column dimension:

```
>> y = x(:,1:end-1) + x(:,2:end)
y =
        21      14      13
        13      18      21
        13      16      21
        21      20      13
```

Here, the row and column indices are reversed from the preceding row operation. This operation along the column dimension can be performed by the preceding row operation if x is transposed first, and then transposed again after the operation. Consider the following example.

```
>> tmp = x'; % transpose data
>> y = tmp(1:end-1,:) + tmp(2:end,:);
>> y = y'    % transpose result
y =
        21      14      13
        13      18      21
        13      16      21
        21      20      13
```

This can also be accomplished with the *n*-D functions permute and ipermute, which are generalizations of the transpose operator, as in the following example:

```
>> tmp = permute(x,[2 1])
tmp =
        16       2       3      13
         5      11      10       8
         9       7       6      12
         4      14      15       1
>> y = tmp(1:end-1,:) + tmp(2:end,:);
>> y = ipermute(y,[2 1])
y =
        21      14      13
        13      18      21
        13      16      21
        21      20      13
```

Before extending this to the *n*-D case, consider the 3-D case since it is relatively easy to visualize:

```
>> x = cat(3,hankel([3 1 6 -1]),pascal(4))
x(:,:,1) =
        3     1     6    -1
        1     6    -1     0
        6    -1     0     0
       -1     0     0     0
x(:,:,2) =
        1     1     1     1
        1     2     3     4
        1     3     6    10
        1     4    10    20
>> y = x(1:end-1,:,:) + x(2:end,:,:) % diff sum along row dimension
y(:,:,1) =
        4     7     5    -1
        7     5    -1     0
        5    -1     0     0
y(:,:,2) =
        2     3     4     5
        2     5     9    14
        2     7    16    30
```

Note that the same process occurs here but that added colons are required to reach both pages of x. If the 1:end-1 and 2:end indices are moved to other dimensions, the differential sum moves to that dimension:

```
>> y = x(:,:,1:end-1) + x(:,:,2:end)
y =
        4     2     7     0
        2     8     2     4
        7     2     6    10
        0     4    10    20
```

This is the differential sum between the two pages of x.

The above example points to one way to generalize this algorithm to *n*-dimensions. The indices into x on the right-hand side are comma-separated lists.

Therefore, if we create cell arrays containing the desired indices, the differential sum can be computed using comma–separated list syntax:

```
>> y = x(1:end-1,:,:) + x(2:end,:,:) % duplicate this case
y(:,:,1) =
        4       7       5      -1
        7       5      -1       0
        5      -1       0       0
y(:,:,2) =
        2       3       4       5
        2       5       9      14
        2       7      16      30

>> c1 = {(1:3) ':' ':'} % first set of indices
c1 =
    [1x3 double]      ':'     ':'
>> c2 = {(2:4) ':' ':'} % second set of indices
c2 =
    [1x3 double]      ':'     ':'

>> y = x(c1{:}) + x(c2{:}) % use comma separated list syntax
y(:,:,1) =
        4       7       5      -1
        7       5      -1       0
        5      -1       0       0
y(:,:,2) =
        2       3       4       5
        2       5       9      14
        2       7      16      30
```

The above example demonstrates the power of comma–separated list syntax. In this example, the keyword end could not be used because it has meaning only when used directly as an index to a variable. As a result, c1 and c2 contain the actual numerical indices.

The following script file generalizes the above implementation for computing differential sums.

```
% diffsum1.m
% compute differential sum along a given dimension

x = cat(3,hankel([3 1 6 -1]),pascal(4)) % data to test
dim = 1 % dimension to work along

xsiz = size(x)
xdim = ndims(x)

tmp = repmat({':'},1,xdim)    % cells of ':'
c1 = tmp;
c1{dim} = 1:xsiz(dim)-1       % poke in 1:end-1
c2 = tmp;
c2(dim) = {2:xsiz(dim)}       % poke in 2:end

y = x(c1{:}) + x(c2{:})       % comma separated list syntax
```

With no semicolons at the end of the statements, diffsum1 produces the following output.

```
x(:,:,1) =
        3       1       6      -1
        1       6      -1       0
        6      -1       0       0
       -1       0       0       0
x(:,:,2) =
        1       1       1       1
        1       2       3       4
        1       3       6      10
        1       4      10      20
dim =
        1
xsiz =
        4       4       2
xdim =
        3
```

```
tmp =
    ':'    ':'    ':'
c1 =
    [1x3 double]    ':'    ':'
c2 =
    [1x3 double]    ':'    ':'
y(:,:,1) =
        4    7    5    -1
        7    5    -1    0
        5    -1    0    0
y(:,:,2) =
        2    3    4    5
        2    5    9    14
        2    7    16    30
```

Here, x is the input data and dim is the dimension chosen for computing the differential sum. Using information about the size and dimensions of x, repmat can produce a cell array for addressing all elements in all dimensions of x. Then indices are inserted into the proper cells to create c1 and c2, as shown earlier. Finally, comma–separated list syntax is used to generate the final result.

This algorithm is not the only way to compute differential sums for an arbitrary *n*-D array. The functions permute and ipermute can be used to transpose x so that the desired dimension for computing the sum is the row dimension. Applying this procedure to the 3-D example earlier gives the following script M-file.

```
% diffsum2.m
% compute differential sum along a given dimension

x = cat(3,hankel([3 1 6 -1]),pascal(4)) % data to test
dim = 3 % dimension to work along

xsiz = size(x);
n = xsiz(dim);                    % size along desired dim
xdim = ndims(x);                  % # of dimensions

perm = [dim:xdim 1:dim-1]         % put dim first
x = permute(x,perm)               % permute so dim is row dimension
```

```
x = reshape(x,n,[])              % reshape into a 2D array

y = x(1:n-1,:) + x(2:n,:)        % Differential sum along row dimension

xsiz(dim) = n-1                  % new size of dim dimension
y = reshape(y,xsiz(perm))        % put result back in original form
y = ipermute(y,perm)             % inverse permute dimensions
```

Here, the variable perm forms a permutation vector for transposing x so that the differential sum is computed along the row dimension. After permuting, x is reshaped into a 2-D array. For 3-D x, this means block-stacking the pages of x as additional columns. Then the differential sum is computed, and the array result is reshaped and inverse-permuted to its original shape using the fact that its size along the chosen dimension has decreased by 1. With no semicolons at the end of the statements, diffsum2 produces the following output.

```
x(:,:,1) =
        3       1       6      -1
        1       6      -1       0
        6      -1       0       0
       -1       0       0       0
x(:,:,2) =
        1       1       1       1
        1       2       3       4
        1       3       6      10
        1       4      10      20
dim =
        3
perm =
        3       1       2
x(:,:,1) =
        3       1       6      -1
        1       1       1       1
x(:,:,2) =
        1       6      -1       0
        1       2       3       4
```

```
x(:,:,3) =
    6    -1     0     0
    1     3     6    10
x(:,:,4) =
   -1     0     0     0
    1     4    10    20
x =
  Columns 1 through 12
    3     1     6    -1     1     6    -1     0     6    -1     0     0
    1     1     1     1     1     2     3     4     1     3     6    10
  Columns 13 through 16
   -1     0     0     0
    1     4    10    20
y =
  Columns 1 through 12
    4     2     7     0     2     8     2     4     7     2     6    10
  Columns 13 through 16
    0     4    10    20
xsiz =
    4     4     1
y(:,:,1) =
    4     2     7     0
y(:,:,2) =
    2     8     2     4
y(:,:,3) =
    7     2     6    10
y(:,:,4) =
    0     4    10    20
y =
    4     2     7     0
    2     8     2     4
    7     2     6    10
    0     4    10    20
```

Of these two approaches to generalizing functions, the *n*-D case, the approach taken in diffsum2, appears more often in M-File functions distributed with MATLAB. There is no significant difference in the speed of the approaches.

32.9 STRUCTURE MANIPULATION

Structures are a convenient data structure in MATLAB. They allow one to group associated data into a single variable and use descriptive field names to identify different data contained within the structure. Given the utility and convenience of structures, it is often convenient to pack a group of variables into a single structure, and then later extract them back out. To illustrate this, first consider the process of gathering variables and storing them as fields within a single structure, with field names matching variable names. Performing this task in the *Command* window is straightforward; one just assigns fields to like-named variables, as in the following example:

```
>> a = speye(2) % test data
a =
      (1,1)      1
      (2,2)      1
>> b = 'Word'
b =
Word
>> c = cell(3)
c =
      []        []
      []        []
      []        []
>> x.a = a; % store variables in a structure
>> x.b = b;
>> x.c = c
x =
      a: [2x2 double]
      b: 'Word'
      c: {3x3 cell}
```

The inverse of this process is also straightforward. For example, using the structure y as an example leads to the following statements.

```
>> a = x.a
a =
      (1,1)      1
      (2,2)      1
```

```
>> b = x.b
b =
Word
>> c = x.c
c =
        []        []        []
        []        []        []
        []        []        []
```

Here, the field names of the structure become the variable names.

This process of packing and unpacking variables is encapsulated in the M-file function mmv2struct, as shown below.

```
function varargout = mmv2struct(varargin)
%MMV2STRUCT Pack/Unpack Variables to/from a Scalar Structure.
% MMV2STRUCT(X,Y,Z,...) returns a structure having fields X,Y,Z,...
% containing the corresponding data stored in X,Y,Z,...
% Inputs that are not variables are stored in fields named ansN
% where N is an integer identifying the Nth unnamed input.
%
% MMV2STRUCT(S)assigns the contents of the fields of the scalar structure
% S to variables in the calling workspace having names equal to the
% corresponding field names.
%
% [A,B,C,...] = MMV2STRUCT(S) assigns the contents of the fields of the
% scalar structure S to the variables A,B,C,... rather than overwriting
% variables in the caller. If there are fewer output variables than
% there are fields in S, the remaining fields are not extracted. Variables
% are assigned in the order given by fieldnames(S).

if nargin==0
    error('Input Arguments Required.')

elseif nargin==1 % Unpack Unpack Unpack Unpack Unpack Unpack Unpack Unpack
```

```
    argin = varargin{1};
    if ~isstruct(argin) || length(argin)~=1
        error('Single Input Must be a Scalar Structure.')
    end
    names = fieldnames(argin);

    if nargout==0 % assign in caller

        for i = 1:length(names)
            assignin('caller',names{i},argin.(names{i}))
        end
    else % deal fields into variables in caller

        varargout = cell(1,nargout); % preallocate output
        for i = 1:nargout
            varargout{i} = argin.(names{i});
        end
    end
else % Pack Pack Pack Pack Pack Pack Pack Pack Pack Pack Pack Pack Pack Pack

    args = cell(2,nargin);
    num = 1;

    for i = 1:nargin % build cells for call to struct

        args(:,i) = {inputname(i); varargin{i}};

        if isempty(args{1,i})
            args{1,i} = sprintf('ans%d',num);
            num = num + 1;
        end
    end
    varargout{1} = struct(args{:}); % create struct using comma-separated list
end
```

Consider the code that unpacks a structure into variables. When called with no output arguments as mmv2struct(structvar), the function uses a For Loop to repeatedly call the function assignin to assign values to variables in the workspace that called mmv2struct. During each For Loop iteration, variables having names that match the corresponding field names of the structure are created. The statement fragment arg.(names{i}) addresses the field name string contained in names{i} of the structure argin, which is equal to structvar. The content of arg.(names{i}) is assigned to a variable named names{i} in the workspace of the calling program.

The code that packs variables into a structure creates a cell array containing the arguments required for the struct function, for example, struct('field1', values1, 'field2', values2,...). For convenience, the arguments to the struct function are created with 'field1', 'field2', and so on, as a first row in the variable args, and values1, values2, and so on, as a second row in the variable args. Then args{:} in the struct(arg{:}) statement creates a single row of arguments as needed by the struct function using comma–separated list syntax.

32.10 INVERSE INTERPOLATION

In Chapter 18, Data Interpolation, 1-D interpolation was demonstrated using the function interp1(x,y,xi). While this function offers a variety of beneficial features, it assumes that there is one y value for each value in xi. If there is not, the function terminates with an error.

In some situations, it is necessary to interpolate in the reverse direction, that is, to inverse interpolate data, whereby given a value of y, the problem is to find all values of x where $y = f(x)$. To illustrate this situation, consider the following example code and associated plot.

```
>> x = (1:10).';     % sample data
>> y = cos(pi*x);

>> yo = -0.2;        % inverse interpolation point
>> xol = [x(1); x(end)];
>> yol = [yo; yo];

>> plot(x,y,xol,yol)
>> xlabel X
>> ylabel Y
>> set(gca,'Ytick',[-1 yo 0 1])
>> title('Figure 32.2: Inverse Interpolation')
```

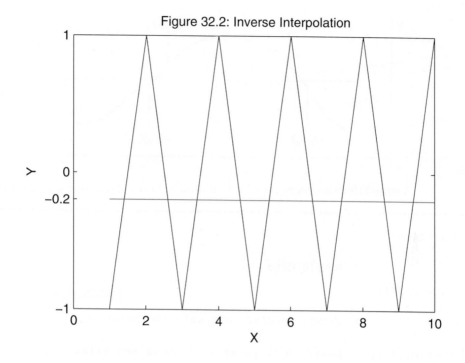

Figure 32.2: Inverse Interpolation

As shown by the triangle waveform in the figure, it is straightforward to use `interp1` to find y values associated with any set of x values since for each x value there is one and only one value of y. On the other hand, in the inverse direction there are numerous values of x for each value of y between −1 and 1. For example, for the horizontal line drawn yo = -0.2, there are nine corresponding values of y. The function `interp1` cannot be used to find those points:

```
>> interp1(y,x,yo)
??? Error using ==> interp1
The data abscissae should be distinct.
```

Since `interp1` does not work in this case, let's consider the development of an M-file function that performs this task.

To solve this problem, we must find all the data points that stagger the desired point. As shown in the figure below, in some cases a data point will appear below the interpolating point y_o and the next point will be above it. In other cases, a data point will appear above y_o and the next point will be below it. Once the pairs of data points (x_k, y_k) and (x_{k+1}, y_{k+1}) are found, it is simply a matter of linearly interpolating between the data point pairs to find each x_o that corresponds with y_o.

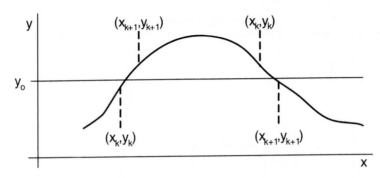

A straightforward, nonvectorized solution is shown in the code segment below.

```
% invinterp1

x = (1:10).';     % sample data
y = cos(pi*x);

yo = -0.2;        % chosen inverse interpolation point

if yo<min(y) || yo>max(y) % quick exit if no values exist
   xo = [];
else              % search for the desired points

   n = length(y);
   xo = nan(size(y)); % preallocate space for found points
   alpha = 0;

   for k = 1:n-1 % look through all data pairs

      if ( y(k)<yo && y(k+1)>=yo ) ||... % below then above
         ( y(k)>yo && y(k+1)<=yo )       % above then below

        alpha = (yo-y(k))/(y(k+1)-y(k));   % distance between x(k+1) and x(k)
        xo(k) = alpha*(x(k+1)-x(k)) + x(k); % linearly interpolate using alpha

      end
   end
   xo = xo(~isnan(xo));      % get rid of unneeded preallocated space
   yo = repmat(yo,size(xo)); % duplicate yo to match xo points found
end
```

Note that this code preallocates the interpolation points xo with an array of NaNs. An array of zeros or ones is not appropriate since zero and one are potential valid points in xo. In some sense, this preallocation is wasteful because xo typically has many fewer points than the input data. In any case, the unused points are eliminated in the second last statement and yo is replicated to be the same size as xo in the last statement. This replication makes it simple to plot the interpolated points, as in plot(xo,yo,'o').

It appears that this function can be vectorized since logical comparisons can be done on an array basis. To investigate this possibility, consider the example code below.

```
>> x = (1:10);    % sample data
>> y = cos(pi*x);

>> yo = -0.2;    % chosen inverse interpolation point

>> below = y<yo   % True where below yo
below =
     1   0   1   0   1   0   1   0   1   0

>> above = y>=yo  % True where at or above yo
above =
     0   1   0   1   0   1   0   1   0   1
```

Because of the choice of x and y in the above example, alternating elements are below and alternating elements are above. The value yo is bracketed by consecutive data points when below(k) and above(k+1) are True or when above(k) and below(k+1) are True. By shifting the below and above arrays by one, these tests are accomplished as shown below.

```
>> n = length(y);
>> below(1:n-1) & above(2:n) % below(k) and above(k+1)
ans =
     1   0   1   0   1   0   1   0   1

>> above(1:n-1) & below(2:n) % above(k) and below(k+1)
ans =
     1   0   1   0   1   0   1   0   1
```

Combining these gives the k points. The points immediately after these are the (k+1) points. These are found by the code

```
>> kth = (below(1:n-1)&above(2:n))|(above(1:n-1)&below(2:n)); % True at k points
>> kp1 = [false kth];                                         % True at k+1 points
```

These two logical arrays address the points in y that require interpolation. Using the same interpolation algorithm as used earlier, a vectorized inverse interpolation solution is shown in the code segment below.

```
% invinterp2

x = (1:10).';              % sample data
y = cos(pi*x);

yo = -0.2;                 % chosen inverse interpolation point

n = length(y);

if yo<min(y) || yo>max(y) % quick exit if no values exist
   xo = [];
else                       % find the desired points

   below = y<yo;           % True where below yo
   above = y>=yo;          % True where at or above yo

kth = (below(1:n-1)&above(2:n))|(above(1:n-1)&below(2:n)); % True at k points
kp1 = [false; kth];                                        % True at k+1 points

alpha = (yo - y(kth))./(y(kp1)-y(kth));% distance between x(k+1) and x(k)
xo = alpha.*(x(kp1)-x(kth)) + x(kth);  % linearly interpolate using alpha

yo = repmat(yo,size(xo)); % duplicate yo to match xo points found
end
```

Searching with a For Loop is replaced with logical comparisons and logical manipulation. Again, the find function has been avoided since it requires additional computations, which would slow the implementation down. All linear interpolations are performed with array mathematics. Here, the size of xo is determined as the code executes, so there is no need to preallocate it. The downside of this vectorized case is that it creates a number of potentially large arrays.

Testing both of these algorithms with the profiler shows that they have essentially identical performance. Since neither implementation dominates, the following code illustrates M-file function creation implementing the vectorized solution.

```
function [xo,yo] = mminvinterp(x,y,yo)
%MMINVINTERP 1-D Inverse Interpolation.
% [Xo,Yo] = MMINVINTERP(X,Y,Yo) linearly interpolates the vector Y to find
% the scalar value Yo and returns all corresponding values Xo interpolated
% from the X vector. Xo is empty if no crossings are found. For
% convenience, the output Yo is simply the scalar input Yo replicated so
% that size(Xo) = size(Yo).
% If Y maps uniquely into X, use INTERP1(Y,X,Yo) instead.
%
% See also INTERP1.

if nargin~=3
   error('Three Input Arguments Required.')
end
n = numel(y);
if ~isequal(n,numel(x))
   error('X and Y Must have the Same Number of Elements.')
end
if ~isscalar(yo)
   error('Yo Must be a Scalar.')
end

x = x(:); % stretch input vectors into column vectors
y = y(:);

if yo<min(y) || yo>max(y)  % quick exit if no values exist
   xo = [];
   yo = [];
else                       % find the desired points

   below = y<yo;           % True where below yo
   above = y>=yo;          % True where at or above yo
```

```
kth = (below(1:n-1)&above(2:n)) | (above(1:n-1)&below(2:n)); % point k
kp1 = [false; kth];                                    % point k+1

alpha = (yo - y(kth)) ./ (y(kp1)-y(kth));% distance between x(k+1) and x(k)
xo = alpha.*(x(kp1)-x(kth)) + x(kth);  % linearly interpolate using alpha

yo = repmat(yo,size(xo)); % duplicate yo to match xo points found
end
```

The above function is useful for finding the intersection of two plotted curves. For example, consider the problem of finding the intersection points of the two curves created by the code below.

```
>> x = linspace(0,10);
>> y = sin(x);
>> z = 2*cos(2*x);
>> plot(x,y,x,z)
>> xlabel X
>> ylabel Y
>> title 'Figure 32.3: Intersection of Two Curves'
```

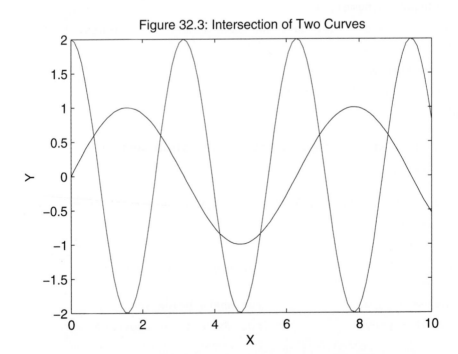

Figure 32.3: Intersection of Two Curves

The intersection of these two curves is given by the zero crossings of their difference. Given the above data, these points are found using `mminvinterp`:

```
>> xo = mminvinterp(x,y-z,0); % find zero crossings of difference

>> yo = interp1(x,y,xo);      % find corresponding y values

>> plot(x,y,x,z,xo,yo,'o')    % regenerate plot showing intersection points
>> xlabel X
>> ylabel Y
>> title 'Figure 32.4: Intersection Points'
```

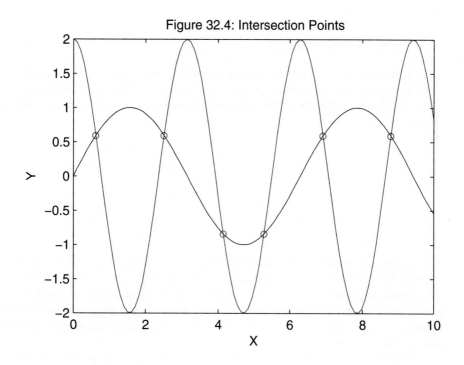

Figure 32.4: Intersection Points

32.11 POLYNOMIAL CURVE FITTING

As discussed in Chapter 19, polynomial curve fitting is performed using the function `polyfit`. Because polynomial curve fitting is a basic numerical analysis topic, it is worth exploring more fully. Consider a general polynomial

$$y = p_1 x^n + p_2 x^{n-1} + \ldots + p_n x + p_{n+1}$$

Here, there are $n+1$ coefficients for an nth order polynomial. For convenience, the coefficients have subscripts that are numbered in increasing order as the power of x decreases. Written in this way, the polynomial can be written as the matrix product as

$$y = [x^n \quad x^{n-1} \quad \ldots \quad x \quad 1] \cdot \begin{bmatrix} p_1 \\ p_2 \\ \ldots \\ p_n \\ p_{n+1} \end{bmatrix}$$

In this format, the polynomial coefficients are grouped into a single column vector.

Common polynomial curve fitting uses this form to compute the polynomial coefficients given a set of data points $\{xi, yi\}$ for $i = 1,2,\ldots N$. Substituting each of these data points into the above relationship and grouping the results leads to the matrix equation

$$\begin{bmatrix} x_1^n & x_1^{n-1} & \ldots & x_1 & 1 \\ x_2^n & x_2^{n-1} & \ldots & x_2 & 1 \\ \vdots & \vdots & \ldots & \vdots & \vdots \\ x_N^n & x_N^{n-1} & \ldots & x_N & 1 \end{bmatrix} \cdot \begin{bmatrix} p_1 \\ p_2 \\ \ldots \\ p_n \\ p_{n+1} \end{bmatrix} = \begin{bmatrix} y_1 \\ y_2 \\ \vdots \\ y_N \end{bmatrix}$$

As written, the matrix on the left is a Vandermonde matrix as discussed earlier in Section 32.6. If $N = n+1$, the Vandermonde matrix is square. Furthermore, if the data points are distinct, the matrix has full rank and the unique polynomial vector $p = [p_1 p_2 \ldots p_n p_{n+1}]'$ is found in MATLAB by using the backslash or left-division operator as p = V\y, where V is the Vandermonde matrix and y is the right-hand-side vector.

On the other hand, when $N \geq n+1$ and the data points are distinct, the Vandermonde matrix has more rows than columns, and no exact solution exists. In this case, p = V\y in MATLAB computes the polynomial coefficient vector that minimizes the least squared error in the set of equations. The following code segment demonstrates an example with and without the use of `polyfit`.

```
% polyfit1.m
% find polynomial coefficients without polyfit

x = [0 .1 .2 .3 .4 .5 .6 .7 .8 .9 1]'; % column vector data
y = [-.447 1.978 3.28 6.16 7.08 7.34 7.66 9.56 9.48 9.30 11.2]';

n = 2; % desired polynomial order

pm = polyfit(x,y,n) % MATLAB polyfit result

% create Vandermonde matrix using code from vander3.m

m = length(x);     % number of elements in x
V = ones(m,n+1);   % preallocate memory for result

for i = n:-1:1     % build V column by column
   V(:,i) = x.*V(:,i+1);
end

p = V\y; % find least squares solution
p = p' % convert to row vector to match MATLAB's convention for polynomials
```

Running this code shows that both polynomial vectors are equal.

```
>> polyfit1
pm =
        -9.8108 20.129 -0.031671
p =
        -9.8108 20.129 -0.031671
```

From this basic understanding, it is possible to consider potential numerical problems. In particular, because the Vandermonde matrix contains elements ranging from 1 to x^n, it can suffer accuracy problems if the x data points differ a great deal from the number one. For example, if the desired polynomial order is increased to 4 and the x data is scaled by 10^4, the Vandermonde matrix becomes

```
>> x = 1e4*[0 .1 .2 .3 .4 .5 .6 .7 .8 .9 1]'; % scale x data by 1e4
>> n = 4;                                     % change order to 4
```

```
>> m = length(x);   % number of elements in x
>> V = ones(m,n+1); % preallocate memory for result

>> for i = n:-1:1   % build V column by column
   V(:,i) = x.*V(:,i+1);
end
>> V
V =
```

0	0	0	0	1
1e+012	1e+009	1e+006	1000	1
1.6e+013	8e+009	4e+006	2000	1
8.1e+013	2.7e+010	9e+006	3000	1
2.56e+014	6.4e+010	1.6e+007	4000	1
6.25e+014	1.25e+011	2.5e+007	5000	1
1.296e+015	2.16e+011	3.6e+007	6000	1
2.401e+015	3.43e+011	4.9e+007	7000	1
4.096e+015	5.12e+011	6.4e+007	8000	1
6.561e+015	7.29e+011	8.1e+007	9000	1
1e+016	1e+012	1e+008	10000	1

Now the values in V vary in value from 1 to 10^{16}. This disparity causes trouble for the function polyfit, as shown below.

```
>> p = polyfit(x,y,n)
  Warning: Polynomial is badly conditioned. Add points with distinct X
           values, reduce the degree of the polynomial, or try centering
           and scaling as described in HELP POLYFIT.
pm =
   2.2057e-015 -2.8038e-011 -6.3531e-008 2.3810e-003 -4.5159e-001
```

Since we have reached the limits of double-precision mathematics, to eliminate this problem we must somehow scale the data so that the Vandermonde matrix does not exhibit this disparity in values. Of the numerous ways to scale the data, use of the mean and standard deviation often leads to good results. That is, instead of fitting a polynomial to the data in x, the fit is done with respect to a new independent variable z given by

$$z = \frac{x - x_m}{s}$$

where x_m and s are the mean and standard deviation, respectively, of the x data. Subtracting the mean shifts the data to the origin, and dividing by the standard deviation reduces the spread in the data values. Applying this to the data that was scaled by 10^4 gives

```
>> xm = mean(x)
>> s = std(x)
>> z = (x - xm)/s;

>> m = length(z);    % number of elements in x
>> V = ones(m,n+1); % preallocate memory for result
>> for i = n:-1:1    % build V column by column
   V(:,i) = z.*V(:,i+1);
end
V
V =
```

5.1653	-3.4263	2.2727	-1.5076	1
2.1157	-1.7542	1.4545	-1.206	1
0.66942	-0.74007	0.81818	-0.90453	1
0.13223	-0.21928	0.36364	-0.60302	1
0.0082645	-0.02741	0.090909	-0.30151	1
0	0	0	0	1
0.0082645	0.02741	0.090909	0.30151	1
0.13223	0.21928	0.36364	0.60302	1
0.66942	0.74007	0.81818	0.90453	1
2.1157	1.7542	1.4545	1.206	1
5.1653	3.4263	2.2727	1.5076	1

Now the Vandermonde matrix is numerically well conditioned and the polynomial curve fit proceeds without difficulty. For this data, the resulting polynomial is

```
>> p = (V\y)';
p =
     0.26689     0.58649    -1.6858      2.4732      7.7391
```

Comparing these polynomial coefficients to those computed before scaling shows that these are much better scaled as well.

While the above result works well, it changes the original problem to

$$y = p_1 z^n + p_2 z^{n-1} + \ldots + p_n z + p_{n+1}$$

That is, the polynomial is now a function of the variable z. Because of this, evaluation of the original polynomial requires a two-step process. First, using the values of x_m and s used to find z, x data points for polynomial evaluation must be converted to their z data equivalents using $z = (x - x_m)/s$. Then, the polynomial can be evaluated:

```
>> xi = 1e4*[.25 .35 .45];  % sample data for polynomial evaluation
>> zi = (xi-xm)/s;          % convert to z data

>> yi = polyval(p,zi)       % evaluate using polyval
yi =
         4.752        6.2326        7.326
```

Rather than performing this step manually when data scaling is used, the MATLAB functions polyfit and polyval implement this data scaling through the use of an additional variable. For example, the above results are duplicated by the following code.

```
>> [p,Es,mu] = polyfit(x,y,n);
>> p
p =
      0.26689       0.58649       -1.6858        2.4732        7.7391
>> yi = polyval(p,xi,Es,mu)
yi =
         4.752        6.2326        7.326
```

The optional polyfit output variable mu contains the mean and the standard deviation of the data used to compute the polynomial. Providing this variable to polyfit directs it to perform the conversion to z before evaluating the polynomial. As shown here, the polynomial p and the interpolated points yi are equal to those computed using the Vandermonde matrix.

The structure variable Es that has not been discussed here is used to compute error estimates in the solution. See the documentation for further information regarding this variable.

Beyond consideration of data scaling, it is sometimes important to perform weighted polynomial curve fitting. That is, in some situations, there may be more confidence in some of the data points than others. When this is true, the curve fitting procedure should take this confidence into account and return a polynomial that reflects the weight given to each data point.

While polyfit does not provide this capability, it is easy to implement in a number of ways. Perhaps the simplest way is to weigh the data before the Vandermonde matrix is formed. For example, consider the case of a third-order polynomial being fit to five data points. Assuming that the confidence in the third data point is α times that of all other data points, the matrix equation to be solved becomes

$$\begin{bmatrix} x_1^3 & x_1^2 & x_1 & 1 \\ x_2^3 & x_2^2 & x_2 & 1 \\ \alpha \cdot x_3^3 & \alpha \cdot x_3^2 & \alpha \cdot x_3 & \alpha \\ x_4^3 & x_4^2 & x_4 & 1 \\ x_5^3 & x_5^2 & x_5 & 1 \end{bmatrix} \cdot \begin{bmatrix} p_1 \\ p_2 \\ p_3 \\ p_4 \end{bmatrix} = \begin{bmatrix} y_1 \\ y_2 \\ \alpha \cdot y_3 \\ y_4 \\ y_5 \end{bmatrix}$$

By multiplying the third row of the matrix equation by α, the error in this equation is weighted by α. The process of minimizing the least squared error forces the error at this data point to decrease relative to the others. As α increases, the error at the data point decreases.

In general, it is possible to give a weight to each data point, not just one of them as shown above. Implementing this approach in MATLAB is shown in the code segment example below.

```
% polyfit2.m
% find weighted polynomial coefficients

x = [0 .1 .2 .3 .4 .5 .6 .7 .8 .9 1]';
y = [-.447 1.978 3.28 6.16 7.08 7.34 7.66 9.56 9.48 9.30 11.2]';

n = 3;

pm = polyfit(x,y,n)   % MATLAB polyfit result

% create Vandermonde matrix using code from vander3.m

m = length(x);      % number of elements in x
V = ones(m,n+1);    % preallocate memory for result

for i = n:-1:1      % build V column by column
   V(:,i) = x.*V(:,i+1);
end

w = ones(size(x));   % default weights of one
w(4) = 2;            % weigh 4th point by 2
w(7) = 10;           % weigh 7th point by 10

V = V.*repmat(w,1,n+1);    % multiply rows by weights
y = y.*w;                  % multiply y by weights

p = (V\y)' % find polynomial
```

Running this code shows that the unweighted and weighted polynomials are different.

```
>> polyfit2
pm =
        16.076      -33.924      29.325      -0.6104
p =
        28.576      -50.761      34.104      -0.67441
```

An alternative to the above approach uses the MATLAB function lscov, which specifically computes weighted least squares solutions. Repeating the previous example using this function is straightforward, as shown in the following code segment and its associated output.

```
% polyfit3
% find weighted polynomial coefficients using lscov

x = [0 .1 .2 .3 .4 .5 .6 .7 .8 .9 1]';
y = [-.447 1.978 3.28 6.16 7.08 7.34 7.66 9.56 9.48 9.30 11.2]';

n = 3;

pm = polyfit(x,y,n)   % MATLAB polyfit result

% create Vandermonde matrix using code from vander3.m

m = length(x);    % number of elements in x
V = ones(m,n+1); % preallocate memory for result

for i = n:-1:1    % build V column by column
   V(:,i) = x.*V(:,i+1);
end

w = ones(size(x)); % default weights of one
w(4) = 2^2;        % here weights are the square of those used earlier
w(7) = 10^2;

p = lscov(V,y,w)'
```

```
>> polyfit3

pm =

        16.076          -33.924          29.325          -0.6104

p =

        28.576          -50.761          34.104          -0.67441
```

To see how these polynomials differ, they can be plotted over the range of the data
as shown below.

```
% mm3205.m
% find weighted polynomial coefficients using lscov

x = [0 .1 .2 .3 .4 .5 .6 .7 .8 .9 1]';
y = [-.447 1.978 3.28 6.16 7.08 7.34 7.66 9.56 9.48 9.30 11.2]';
n = 3;
pm = polyfit(x,y,n); % MATLAB polyfit result

% create Vandermonde matrix using code from vander3.m

m = length(x);        % number of elements in x
V = ones(m,n+1);      % preallocate memory for result

for i = n:-1:1        % build V column by column
    V(:,i) = x.*V(:,i+1);
end

w = ones(size(x)); % default weights of one
w(4) = 2^2;
w(7) = 10^2;

p = lscov(V,y,w)';

xi = linspace(0,1,100);
ym = polyval(pm,xi);
yw = polyval(p,xi);
```

```
pt = false(size(x)); % logical array pointing to weighted points
pt([4 7]) = true;

plot(x(~pt),y(~pt),'x',x(pt),y(pt),'o',xi,ym,'--',xi,yw)
xlabel('x'), ylabel('y = f(x)')
title('Figure 32.5: Weighted Curve Fitting')
```

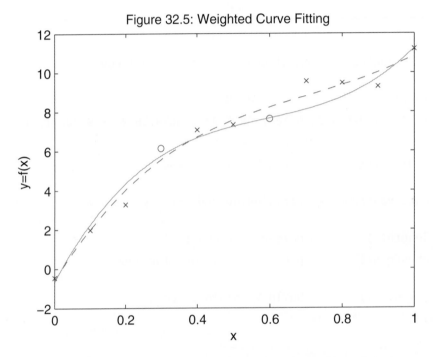

Figure 32.5: Weighted Curve Fitting

The weighted points are marked with circles in the figure, while the unweighted points are marked with an x. Clearly, the weighted polynomial is closer to the fourth and seventh data points than the original polyfit results. In addition, the seventh data point is closer than the fourth because its weight is much higher.

32.12 NONLINEAR CURVE FITTING

The polynomial curve fitting discussed in the previous section is popular in part because the problem can be written in matrix form and solved using linear least squares techniques. Perhaps most important or most convenient is that no initial guess is required. The optimum solution is found without searching an n-dimensional solution space. In the more general case, where the unknown parameters do not appear linearly, that is, nonlinear curve fitting, the solution space must be searched starting with an initial guess. For example, consider fitting data to the function

$$f(t) = a + be^{\alpha t} + ce^{\beta t}$$

If α and β are known constants and a, b, and c are the unknowns to be found, this function can be written as

$$f(t) = [1 \quad e^{\alpha t} \quad e^{\beta t}] \cdot \begin{bmatrix} a \\ b \\ c \end{bmatrix}$$

Substituting a given a set of data points, $\{t_i, y_i = f(t_i)\}$ for $i = 1,2,\ldots N$, into this expression and gathering the results in matrix format results in the equation

$$\begin{bmatrix} 1 & e^{\alpha t_1} & e^{\beta t_1} \\ 1 & e^{\alpha t_2} & e^{\beta t_2} \\ \vdots & \vdots & \vdots \\ 1 & e^{\alpha t_N} & e^{\beta t_N} \end{bmatrix} \cdot \begin{bmatrix} a \\ b \\ c \end{bmatrix} = \begin{bmatrix} y_1 \\ y_2 \\ \vdots \\ y_N \end{bmatrix}$$

Written in this way, a, b, and c appear linearly, just as the polynomial coefficients did in the previous section. As a result, the backslash operation in MATLAB produces the least squares solution for the unknown variables. That is, if E is the N-by-3 matrix on the left, p is the vector $[a\ b\ c]'$, and y is the right-hand-side vector, p = E\y gives the least squares solution for p.

When α and β are not known, the linear least squares solution does not apply because α and β cannot be separated as a, b, and c, as in the preceding equation. In this case, it is necessary to use a minimization algorithm such as fminsearch in MATLAB. There are two ways to set up this problem for solution with fminsearch. The first is to write $f(t)$ as

$$f(t) = x_3 + x_4 e^{x_1 t} + x_5 e^{x_2 t}$$

where the unknown variables have been combined into a vector $x = [x_1\ x_2\ x_3\ x_4\ x_5]'$. Given this form, and a set of data points $\{t_i, y_i\}$ for $i = 1,2,\ldots N$, the least squares solution minimizes the norm of error between the data points y_i and the function evaluated at t_i, $f(t_i)$. That is, if y is a vector containing the y_i data points and if f is a vector containing the above function evaluated at the time points t_i, then the least squares solution is given by

$$\min_x \|y - f\|$$

To use fminsearch, this norm must be expressed as a function. One way to do so is to define it using the anonymous function

```
fitfun = @(x,td,yd) norm(x(3)+x(4)*exp(x(1)*td)+x(5)*exp(x(2)*td)-yd);
```

where td and yd are vectors containing the data points to be fit. Since the function fminsearch expects to call a function of just one variable, that is, the variable it manipulates is x, another anonymous function must be created where td and yd are incorporated into function rather than be arguments to the function. One way of doing so is to use `fitfun`:

```
fitfun1 = @(x) fitfun(x,td,yd);
```

The following code segment uses this approach by creating some data, making an initial guess, and calling `fminsearch` to find a solution.

```
%% testfitfun1
% script file to test nonlinear least squares problem

% create test data
x1 = -2; % alpha
x2 = -5; % beta
x3 = 10;
x4 = -4;
x5 = -6;

xstar = [x1;x2;x3;x4;x5]; % gather optimum solution

td = linspace(0,4,30)';
yd = x3 + x4*exp(x1*td) + x5*exp(x2*td);

% create an initial guess
x0 = zeros(5,1); % not a good one, but a common first guess

% create anonymous function to minimize

fitfun = @(x,td,yd) norm(x(3)+x(4)*exp(x(1)*td)+x(5)*exp(x(2)*td)-yd);

% create anonymous function that incorporates the numerical data
% given rather than pass it as an argument
fitfun1 = @(x) fitfun(x,td,yd);
options = []; % take default options for fminsearch

x = fminsearch(fitfun1,x0,options);
```

```
% compare fminsearch solution to optimum solution
output = [x xstar]

% compute error norm at returned solution
enorm = fitfun1(x)
```

Running this code produces the following result:

```
Exiting: Maximum number of function evaluations has been exceeded
         - increase MaxFunEvals option.
         Current function value: 3.234558

output =
          0.17368      -2
          -1.5111      -5
          13.972       10
          -2.2059      -4
          -9.4466      -6
enorm =
          3.2346
```

For this initial guess, the algorithm has not converged. Increasing the number of function evaluations and algorithm iterations permitted to 2000 leads to

```
>> options = optimset('MaxFunEvals',2e3,'MaxIter',2e3);

>> x = fminsearch(fitfun1,x0,options);

output =
          -0.78361      -2
          -3.7185       -5
          10.061        10
          -0.80533      -4
          -9.1907       -6
enorm =
          0.23727
```

The algorithm now converges to a solution, but not to the values $x^* = [-2\ -5\ 10\ -4\ -6]$ used to create the data. A plot of the data, the actual function, and the fitted solution provide further information as computed and displayed below.

```matlab
%% testfitfun2
% script file to test nonlinear least squares problem

% create test data
x1 = -2; % alpha
x2 = -5; % beta
x3 = 10;
x4 = -4;
x5 = -6;

td = linspace(0,4,30)';
yd = x3 + x4*exp(x1*td) + x5*exp(x2*td);

% create an initial guess
x0 = zeros(5,1); % not a good one, but a common first guess

% create anonymous function to minimize

fitfun = @(x,td,yd) norm(x(3)+x(4)*exp(x(1)*td)+x(5)*exp(x(2)*td)-yd);

% create anonymous function that incorporates the numerical data
% given rather than pass it as an argument
fitfun1 = @(x) fitfun(x,td,yd);

options = optimset('MaxFunEvals',2e3,'MaxIter',2e3);
x = fminsearch(fitfun1,x0,options);

% pick time points to evaluate at
ti = linspace(0,4);

% evaluate true function
actual = x3 + x4*exp(x1*ti) + x5*exp(x2*ti);
```

```
% evaluate fitted solution
fitted = x(3) + x(4)*exp(x(1)*ti) + x(5)*exp(x(2)*ti);

subplot(2,1,1)
plot(td,yd,'o',ti,actual,ti,fitted)
xlabel t
title 'Figure 32.6: Nonlinear Curve Fit'

subplot(2,1,2)
plot(ti,actual-fitted)
xlabel t
ylabel Error
```

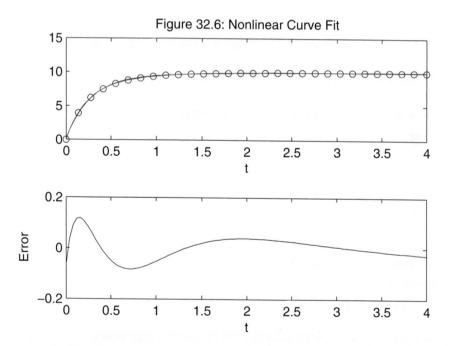

Figure 32.6: Nonlinear Curve Fit

The fitted solution in the upper plot looks good visually. However, the lower plot shows that there is significant error. If the overall goal of this problem is to minimize the error between the actual coefficients represented by the data and the results returned by fminsearch, then this algorithm essentially failed even though the fitted solution plot looks good visually.

At this point, one must decide whether this solution is good enough. Certainly one can try setting tighter convergence criteria. For example,

```
>> options=optimset('TolX',1e-10,'TolFun',1e-10,'MaxFunEvals',2e3,'MaxIter',2e3);

>> x = fminsearch(fitfun1,x0,options)
x =
      -0.78359
      -3.7185
      10.061
      -0.80533
      -9.1907
>> enorm = fitfun1(x)
enorm =
      0.23727
```

In this case, the solution has not changed. One can also try better initial guesses, as the example shows:

```
>> x0 = [-1 -4 8 -3 -7]'; % much closer initial guess

>> options = []; % default options

>> x = fminsearch(fitfun1,x0,options)
x =
            -2
            -5
            10
            -4
            -6
>> enorm = fitfun1(x)
enorm =
      3.2652e-006
```

In this case, fminsearch returns coefficients that very closely match those used to create the data. This example illustrates an ambiguity inherent in most nonlinear optimization algorithms. That is, other than having the error norm be zero, there is no set way of knowing when a solution is the best that can be expected.

When a nonlinear curve fitting problem contains a mixture of linear and nonlinear terms, as is true in this example, it is possible to use linear least squares to compute the linear terms and use a function such as fminsearch to compute the nonlinear terms. In almost all cases, this leads to better results.

To understand how this is done, consider the problem formulation stated earlier where α and β were known constants.

$$\begin{bmatrix} 1 & e^{\alpha t_1} & e^{\beta t_1} \\ 1 & e^{\alpha t_2} & e^{\beta t_2} \\ \vdots & \vdots & \vdots \\ 1 & e^{\alpha t_N} & e^{\beta t_N} \end{bmatrix} \cdot \begin{bmatrix} a \\ b \\ c \end{bmatrix} = \begin{bmatrix} y_1 \\ y_2 \\ \vdots \\ y_N \end{bmatrix}$$

In this case, α and β become the variables manipulated by fminsearch. Within the function that evaluates the error norm, the linear least squares problem is solved using the current estimates for α and β. That is, if $x1 = \alpha$, $x2 = \beta$, and $p = [a\ b\ c]'$, then the function M-file to be evaluated can be written as follows:

```
function [enorm,p] = fitfun2(x,td,yd)
%ENORM Norm of fit to example nonlinear function
% f(t) = p(1)+p(2)*exp(x(1)*t)+p(3)*exp(x(2)*t)
%
% ENORM(X,Td,Yd) returns norm(Yd-f(Td))
%
% [e,p]=ENORM(...) returns the linear least squares
                % parameter vector p
% solve linear least squares problem given x input supplied by fminsearch

E = [ones(size(td)) exp(x(1)*td) exp(x(2)*td)];
p = E\yd; % least squares solution for p = [a b c]'

% use p vector to compute error norm
f = p(1) + p(2)*exp(x(1)*td) + p(3)*exp(x(2)*td);

enorm = norm(f-yd);
```

Because of the multiple steps involved, the above example cannot be written as an anonymous function, but must be written as an M-file.

Now there are only two parameters for fminsearch to manipulate. The following code segment tests this alternative approach.

```
%% testfitfun3
% script file to test nonlinear least squares problem
```

```matlab
% create test data
x1 = -2; % alpha
x2 = -5; % beta
p1 = 10;
p2 = -4;
p3 = -6;

xstar = [x1;x2];      % ideal values
pstar = [p1;p2;p3];

td = linspace(0,4,30)';
yd = p1 + p2*exp(x1*td) + p3*exp(x2*td);

% create an initial guess, only two needed!
x0 = zeros(2,1); % not a good one, but a common first guess

% create anonymous function calls the fitfun2.m function
fitfun1 = @(x) fitfun2(x,td,yd);

options = []; % take default options for fminsearch

x = fminsearch(fitfun1,x0,options);

% compute error norm and get p from solution
[enorm,p] = fitfun1(x);

% compare results
enorm
pout = [p pstar]
xout = [x xstar]
```

Running this code produces the following output.

```
enorm =
    7.1427e-006
```

```
pout =
        10        10
        -4        -4
        -6        -6
xout =
        -2        -2
        -5        -5
```

Although the inner least squares solution reports a warning on several iterations, fminsearch finds a minimum very close to the actual values without difficulty. The incorporation of an internal linear least squares algorithm within an outer nonlinear problem reduces dimension of the parameter space to be searched. This almost always improves the speed and convergence of the nonlinear algorithm. In the above example, the problem was reduced from five dimensions to two.

Finally, note how the parameter p was included as a second output argument to fitfun2. The minimization algorithm fminsearch ignores this argument, so it does not influence the search process. However, after the algorithm terminates, calling the fitfun2 again through the anonymous function fitfun1 with two output arguments returns the values of the linear parameters at the solution point. Calling fitfun2 after fminsearch finishes execution guarantees that the parameter vector p is evaluated at the solution returned by fminsearch.

32.13 CIRCLE FITTING

There are numerous applications where data describing points on a circle are known. From these data points, the coordinates of the circle center and the radius are usually desired. Since there are three unknowns, namely the coordinates of the center given by x_c and y_c, and the radius given by r, three distinct data points uniquely describe a circle provided the data points don't fall on a straight line. When there are more than three distinct data points, no unique circle exists. However, it is possible to find some circle that best fits the data. What is meant by best fit depends on the algorithm chosen. In this section, let's consider a least squared equation error approach.

A circle is commonly described by the equation

$$(x - x_c)^2 + (y - y_c)^2 = r^2$$

Expanding this expression and rearranging the result leads to the equation

$$(2x)x_c + (2y)y_c + (1)z = x^2 + y^2$$

where the new variable z is

$$z = r^2 - x_c^2 - y_c^2$$

from which the circle radius can be found as

$$r = \sqrt{z + x_c^2 + y_c^2}$$

The preceding expanded and rearranged expression is linear with respect to x_c, y_c, and z. For a set of data points x_i and y_i, for $i = 1, 2, 3,...N$, this rearranged equation can be written in matrix form as

$$\begin{bmatrix} 2x_1 & 2y_1 & 1 \\ 2x_2 & 2y_2 & 1 \\ \vdots & \vdots & \vdots \\ 2x_N & 2y_N & 1 \end{bmatrix} \begin{bmatrix} x_c \\ y_c \\ z \end{bmatrix} = \begin{bmatrix} x_1^2 + y_1^2 \\ x_2^2 + y_2^2 \\ \vdots \\ x_N^2 + y_N^2 \end{bmatrix}$$

When there are three distinct data points not on a straight line, there is a unique solution to these three equations in three unknowns. That solution gives x_c, y_c, and z. From the equation for the radius r, the radius is easily computed. When there are more than three distinct data points, the least squares solution gives a least squares solution for x_c, y_c, and z.

As with the Vandermonde matrix used in polynomial fitting, the N-by-3 matrix can have poor numerical properties when the data points are very large or widely separated. This is likely to occur when a small diameter circle is a long way from the origin or when the radius is very large. As a result, the robustness of the solution to the above equation can be improved by appropriate scaling of the data. Depending on how well distributed the data is around the circle, different scaling techniques are appropriate. In the function shown below that implements the above algorithm, the x- and y-data are scaled by their mean values first to center the data at the origin. Then, all the data is scaled by the distance from the origin to the farthest data point.

```
function [xc,yc,r] = circlefit(x,y)
%CIRCLEFIT Fit circle to data.
% [Xc,Yc,R] = CIRCLEFIT(X,Y) finds the center of a circle (Xc,Yc) and its
% radius R that best fits the data points contained in X and Y in a least
% squares sense. X and Y must be arrays containing at least three distinct
% points in the x-y plane. The data pair (X(i),Y(i)) is the coordinates of
% the i-th data point. If three data points are given, the circle will fit
% the three points.
%
% [Xc,Yc,R] = CIRCLEFIT(XY) where XY is a two column array assumes that
% X = XY(:,1) and Y = XY(:,2).
%
% P = CIRCLEFIT(...) alternatively ouput P = [Xc, Yc, R];
%
% Requires Single or Double real-valued data
```

```
% Algorithm:
% Circle Equation (X - Xc)^2 + (Y - Yc)^2 = R^2
% Multiply out and rearrange as
% (2X)*Xc + (2Y)*Yc + (1)*Z = X^2 + Y^2 where Z = R^2 - Xc^2 - Yc^2
%
% Solve for the vector [Xc;Yc;Z] by writing out the above for each data
% point and form a set of equations. Compute R = sqrt(Z + Xc^2 + Yc^2).

if nargin==1 % CIRCLEFIT(XY)
   y = x(:,2);
   x = x(:,1);
end
x = x(:); % make sure data is in columns
y = y(:);
nx = numel(x);

if ~isfloat(x) || ~isfloat(y) || ~isreal(x) || ~isreal(y)
   error('CIRCLEFIT:IncorrectInput',...
         'X and Y Must Contain Real Valued Floating Point Data.')
end
if nx~=numel(y)
   error('CIRCLEFIT:IncorrectInput',...
         'X and Y Must Contain the Same Number of Elements.')
end
if nx<3
   error('CIRCLEFIT:IncorrectInput',...
         'X and Y Must Contain at Least Three Elements.')
end

% shift and scale data for better conditioned results
xm = mean(x); % compute mean of x data
x = x-xm;     % subtract mean from x data

ym = mean(y); % compute mean of y data
y = y-ym;     % subtract mean from y data
```

```
s = max(max(hypot(x,y)),eps); % find point farthest from origin
x = x/s; % scale all data by farthest point
y = y/s;

% form matrix equation to be solved and solve it
xyz = [2*x 2*y ones(nx,1)]\hypot(x,y).^2; % it only takes one line

% unscale data and form output
p = [xm ym 0]+[xyz(1) xyz(2) sqrt(xyz(3)+hypot(xyz(1),xyz(2)).^2)]*s;

if nargout==3      % [Xc,Yc,R] = CIRCLEFIT(...)
    xc = p(1);
    yc = p(2);
    r = p(3);
elseif nargout==1 % P = CIRCLEFIT(...)
    xc = p;
else
    error('CIRCLEFIT:IncorrectOutput',...
          '1 or 3 Output Arguments Required.')
end
```

The code segment and figure below demonstrate use of the function circlefit.

```
%% mm3207
% Ideal data
Xc = 10;
Yc = -4;
R = 5;

% generate data for fitting
theta = 2*pi*rand(20,1);  % random angle around circle
x = Xc + R*cos(theta);    % x data
y = Yc + R*sin(theta);    % y data
```

```
% add a little noise to the data
x = x + randn(size(x))/8;
y = y + randn(size(y))/8;
p = circlefit(x,y); % get p = [Xc Yc R]

% generate data on fitted circle
theta = linspace(0,2*pi);
xf = p(1) + p(3)*cos(theta);
yf = p(2) + p(3)*sin(theta);

% plot original data and fitted circle

plot(x,y,'o',xf,yf,'k')
title('Figure 32.7: Data and Fitted Circle')
text(Xc,Yc,...
     sprintf('X_c = %.2f, Y_c = %.2f, R = %.2f',p),...
     'HorizontalAlignment','center')
```

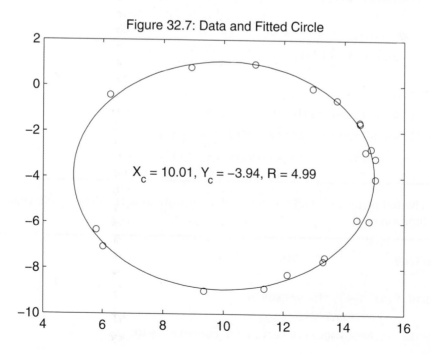

Figure 32.7: Data and Fitted Circle

32.14 LAMINAR FLUID FLOW IN A CIRCULAR PIPE

When fluid flows through a pipe, there is friction between the fluid and the pipe walls that slows the fluid flow as one approaches the pipe's inner surface. At the center of the pipe, the fluid velocity is maximum and it is zero on the pipe's inner surface. The following equation models this behavior.

$$v(r) = v_{max} \left(1 - \frac{r}{r_o} \right)^{1/n}$$

In this equation, $v(r)$ is the fluid flow velocity as a function of radius, v_{max} is the maximum velocity, r_0 is the pipe radius, and n is a variable that depends on the properties of the fluid and its interaction with the pipe walls. The code cell and figure below demonstrate this function for n equal to 7.

```
%% mm3208

% basic fluid flow analysis

n = 7; % chose n value

% generate normalized radii across the pipe diameter
rn = linspace(-1,1);

% compute normalized flow velocity
vn = (1-abs(rn)).^(1/n);

plot(rn,vn)
xlabel('Normalized radius, r / r_o')
ylabel('Normalized velocity, v(r) / v_{max}')
title('Figure 32.8: Flow Analysis Example')
```

The following code cell and figure demonstrate how the flow profile changes as a function of n.

```
%% mm3209

% fluid flow analysis versus n

% create an anonymous function to compute velocity
```

```
vnf = @(rn,n) (1-abs(rn)).^(1./n);

% plot multiple solutions

plot(rn,vnf(rn,3),'k-',... % n = 3 case
    rn,vnf(rn,5),'k--',...% n = 5 case
    rn,vnf(rn,7),'k-.',...% n = 7 case
    rn,vnf(rn,9),'k:') % n = 9 case

xlabel('Normalized radius, r / r_o')
ylabel('Normalized velocity, v(r) / v_{max}')
title('Figure 32.9: Flow Analysis versus \it{n}')
legend('n = 3','n = 5','n = 7','n = 9','Location','SouthEast')
```

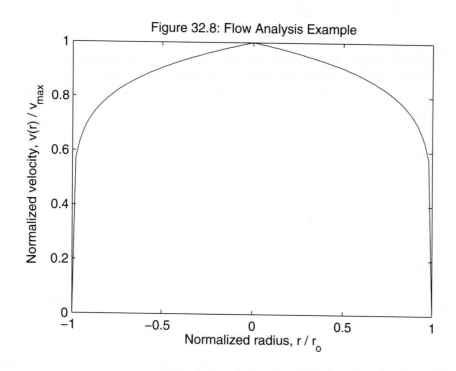

Figure 32.8: Flow Analysis Example

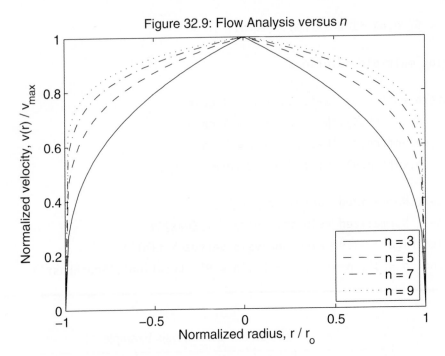

Figure 32.9: Flow Analysis versus *n*

As an alternative to the above line plot, a mesh surface plot can provide a better visual display, as demonstrated in the following code cell and figure.

```
%% mm3210

% mesh plot alternative

% create an anonymous function to compute velocity
vnf = @(rn,n) (1-abs(rn)).^(1./n);

n = 2:10; % define a range of values

[nn,rnn] = meshgrid(n,rn);    % generate data for all n values

vnn = vnf(rnn,nn);            % evaluate function at all data points at once

mh = mesh(nn,rnn,vnn);        % create mesh plot

set(mh,'MeshStyle','column') % get rid of unwanted lines along n axis
set(mh,'FaceColor','none')   % make surface transparent
set(mh,'CData',nn)           % vary color with n
```

```
ylabel('Normalized radius')
xlabel('n')
title('Figure 32.10: Flow Analysis versus \it{n}')
```

Figure 32.10: Flow Analysis versus *n*

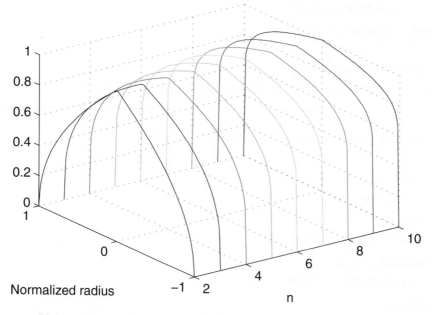

Normalized radius

n

Using the flow velocity equation, we can pose and answer a number of questions. For example, the average flow velocity is given by

$$v_{avg} = \frac{2v_{max}}{r_o^2} \int_0^{r_o} r\left(1 - \frac{r}{r_o}\right)^{1/n} dr$$

Written in terms of normalized radius $r_n = r/r_0$ and normalized velocity, this expression becomes

$$v_{navg} = \frac{v_{avg}}{v_{max}} = 2 \int_0^1 r_n(1 - r_n)^{1/n} dr$$

The following code section computes this integral using a number of MATLAB algorithms for *n* equal to 5.

```
%% compute average flow rate

n = 5;

% anonymous function for normalized average velocity integrand
```

```
% n in function declaration takes the value defined above
vint = @(rn) 2*rn.*(1-abs(rn)).^(1./n);

format long
vquad = quad(vint,0,1)
vquadl = quadl(vint,0,1)
vquadgk = quadgk(vint,0,1)

rn = linspace(0,1,1000);
vtrapz = trapz(rn,vint(rn)) % try simple trapezoidal rule
```

Running the above code cell produces the following output.

```
vquad =
    0.757567393217381
vquadl =
    0.757575561210144
vquadgk =
    0.757575755675408
vtrapz =
    0.757399742733815
```

All four results agree with each to three significant digits, which is usually sufficient for engineering calculations.

One can also find the value of n that achieves a specified normalized average velocity, as demonstrated in the following code cell and its output.

```
%% find n that produces a specific average velocity

% anonymous function for normalized average velocity integrand
% n is now included as an argument

vint2 = @(rn,n) 2*rn.*(1-abs(rn)).^(1./n);

avgvalue = 0.8; % chosen normalized average velocity
```

```
% create an anonymous function for zero finding
fzerofun = @(n) quad( @(rn)vint2(rn,n) ,0,1) - avgvalue;

nstar = fzero(fzerofun,[2,10]) % call fzero to find desired n

fzerofun(nstar) % check result (should be very close to 0)
```

```
nstar =
    6.3169
ans =
    0
```

It is also possible to find the radius where the actual velocity equals the average velocity, as shown in the following code cell and its output. In this case, rn is the argument for fzero.

```
%% find radius where velocity equals average velocity

n = 5; % chosen value
vnf = @(rn,n) (1-abs(rn)).^(1./n); % velocity profile
vint = @(rn) 2*rn.*(1-abs(rn)).^(1./n); % integrand

avgvalue2 = quad(vint,0,1); % normalized average velocity for n = 5

% create an anonymous function for zero finding
fzerofun2 = @(rn) vnf(rn,n)-avgvalue2;

rstar = fzero(fzerofun2,[0,1]) % call fzero to find desired r

vnf(rstar,n) - avgvalue2 % check value (should be very close to 0)
```

```
rstar =
    0.75048
ans =
    0
```

32.15 PROJECTILE MOTION

A common physics problem considers the influence of gravity on the motion of a projectile when there is no wind resistance or other losses or dynamic effects. If the projectile is launched from the origin of a coordinate at a specified velocity and angle with respect to the horizontal axis, the motion can be described by the equations

$$y(t) = v_{yo}t - \frac{g}{2}t^2$$

$$x(t) = v_{xo}t$$

where $y(t)$ describes the height, $x(t)$ describes the distance, g is the gravitational constant, for example, $g = 9.81$ on earth, and the initial y- and x-direction velocities are

$$v_{yo} = V_o \sin(\theta)$$

and

$$v_{xo} = V_o \cos(\theta)$$

in which V_o is the initial velocity along the launched direction and θ is the launch angle.

Given these equations, trajectories can be plotted as a function of initial velocity for a fixed angle. The following code cell and its output demonstrate these trajectories.

```
%% mm3211

g = 9.81; % gravity constant

% anonymous function for height, y(t)
height = @(t,Vo,theta) Vo*sin(theta)*t -(g/2)*t.^2;

% anonymous function for distance, x(t)
distance = @(t,Vo,theta) Vo*cos(theta)*t;

% create time axis
t = linspace(0,2.5);
M = length(t);

% constant angle
theta = pi/4;
```

```
% a variety of velocities
Vo = [10 15 20];
N = length(Vo);

Y = zeros(M,N); % preallocate arrays
X = zeros(M,N);

for k = 1:N      % fill in columns with data
   Y(:,k) = height(t,Vo(k),theta);
   X(:,k) = distance(t,Vo(k),theta);
end

plot(X,Y)
xlabel('Distance, m')
ylabel('Height, m')
legend('V_o = 10','V_o = 15','V_o = 20',...
   'Location','SouthEast')
title('Figure 32.11: Example Trajectories')
```

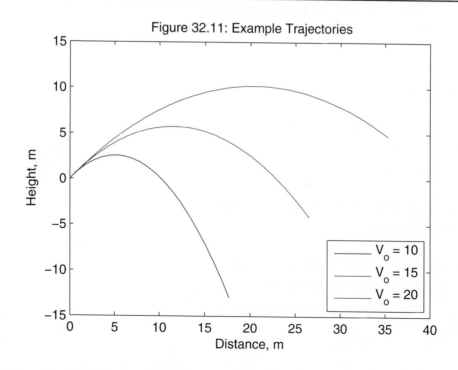

Figure 32.11: Example Trajectories

The trouble with this figure is that the trajectories end at a specified time, not a specified height or distance. In many cases, the trajectory makes the most sense when it ends when the height returns to zero. This can be accomplished in a number of ways numerically. One way is to compute data past the point where the height returns to zero, then simply not plot data beyond that point. This is easily done, as shown in the following code cell and its output.

```
%% mm3212

g = 9.81; % gravity constant

% anonymous function for height, y(t)
height = @(t,Vo,theta) Vo*sin(theta)*t -(g/2)*t.^2;

% anonymous function for distance, x(t)
distance = @(t,Vo,theta) Vo*cos(theta)*t;

% create time axis
% t = linspace(0,6,400); % old time axis

% increase time resolution and time length
t = linspace(0,6,400); % new time axis
M = length(t);

% constant angle
theta = pi/4;

% a variety of velocities
Vo = [10 15 20];
N = length(Vo);

Y = zeros(M,N); % preallocate arrays
X = zeros(M,N);

for k = 1:N % fill in columns with data
    Y(:,k) = height(t,Vo(k),theta);
    X(:,k) = distance(t,Vo(k),theta);
end
```

```
% eliminate data below y = 0
tf = Y<0;    % true where height is less than zero.
Y(tf) = nan; % NaNs in the data do not plot
X(tf) = nan; % NaNs

plot(X,Y)
xlabel('Distance, m')
ylabel('Height, m')
legend('V_o = 10','V_o = 15','V_o = 20',...
   'Location','NorthEast')
title('Figure 32.12: Example Trajectories')
```

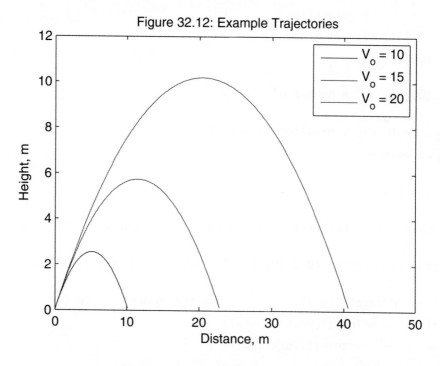

Figure 32.12: Example Trajectories

As shown in this figure, the trajectories do not end exactly at zero height. When this is unacceptable, some other technique must be used. In this case, it is possible to analytically determine the time t_o, when a trajectory returns to zero height, that is, $y(t_o) = 0$. Rather than solve this problem analytically, we can let MATLAB search for the zero crossing, as shown in the following code cell and its output.

```matlab
%% mm3213

g = 9.81; % gravity constant

% anonymous function for height, y(t)
height = @(t,Vo,theta) Vo*sin(theta)*t -(g/2)*t.^2;

% anonymous function for distance, x(t)
distance = @(t,Vo,theta) Vo*cos(theta)*t;

% constant angle
theta = pi/4;

% a variety of velocities
Vo = [10 15 20];
N = length(Vo);

M = 100;          % number of time points

Y = zeros(M,N); % preallocate arrays
X = zeros(M,N);

for k = 1:N

    tofun = @(t) height(t,Vo(k),theta); % look where height is zero

    to = fzero(tofun,[0.1,10]); % let fzero find zero crossing time

    t = linspace(0,to,M);       % just the right time vector
    Y(:,k) = height(t,Vo(k),theta);
    X(:,k) = distance(t,Vo(k),theta);
end

plot(X,Y)
xlabel('Distance, m')
ylabel('Height, m')
```

```
title('Figure 32.13: Example Trajectories')
legend('V_o = 10','V_o = 15','V_o = 20',...
    'Location','NorthEast')
xmm = xlim; % get x axis limits
line(xmm,[0 0],... % put in horizontal axis
      'Color','k',...
      'LineStyle',':')
```

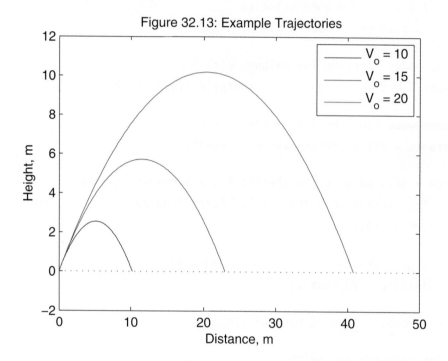

Figure 32.13: Example Trajectories

Now the trajectories end at zero height as identified by the horizontal axis line.

Projectile motion can be used to describe a basketball shot. As a result, we can find the initial velocities and angles that lead to a perfect swish through the center of the basket. In particular, let's closely model a free throw where the distance to the front of the rim is 4.0 meters from the shooter, the distance to the center of the basket is 4.25 meters, the distance to the back of the rim is 4.5 meters, the net height of the rim is 1.0 meter up from where the ball is launched, and the center of the ball must clear the front edge of the rim by 0.125 meters so that the ball swishes through the basket. This geometry and one potential solution is shown in the following code cell and its output.

```
%% mm3214

g = 9.81;        % gravity constant
d2f = 4;         % distance to front of rim
d2c = 4.25;      % distance to center of basket
d2b = 4.5;       % distance to back of rim
h2r = 1;         % height to rim
hmin = 1.125;    % minimum ball height at front of rim

Vo = 7.251;       % good velocity
theta = 0.28*pi; % good angle

% anonymous function for height, y(t)
height = @(t,Vo,theta) Vo*sin(theta)*t -(g/2)*t.^2;

% anonymous function for distance, x(t)
distance = @(t,Vo,theta) Vo*cos(theta)*t;

tofun = @(t) height(t,Vo,theta); % where height is zero
to = fzero(tofun,[0.1,10]); % let fzero find zero crossing time
t = linspace(0,to);

y = height(t,Vo,theta);    % find trajectory
x = distance(t,Vo,theta);

tf = x<=d2b; % stop at back of rim

% plot geometry and swish
plot([0 d2b d2b],[0 0 2.5*h2r],'k',... % border
   [d2f d2b],[h2r h2r],'k',... % rim
   d2c,h2r,'xk',... % center of hoop
   d2f,hmin,'ko',...
   x(tf),y(tf),'-.k')

xlabel('Distance, m')
ylabel('Height, m')
```

```
title('Figure 32.14: Basketball Swish')

text(d2c,h2r,'Rim',...
    'VerticalAlignment','Top',...
    'HorizontalAlignment','Center')
```

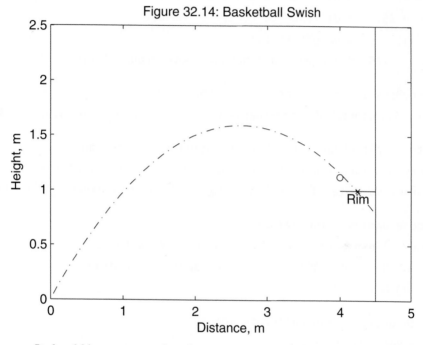

Figure 32.14: Basketball Swish

It should be apparent that there is a unique relationship between initial velocity and initial angle that leads to a swish. That is, for a given initial angle θ, there is
one initial velocity V_o that produces a perfect swish. While this relationship can be
determined analytically, let's find it numerically using MATLAB. Doing so requires
that the distance relationship, $x(t) = v_{xo}t = V_o\cos(\theta)t$ be rewritten to solve for the
swish time, that is, the time it takes the ball to travel to the center of the basket,

$$t_s = \frac{d_{2c}}{V_o\cos(\theta)}$$

where the distance to the center of the basket, d_{2c}, is equal to 4.25 in this example.
Substituting this into the height equation, $y(t) = V_o\sin(\theta)t - (g/2)t^2$, gives the
ball height at the time when the ball is over the center of the basket. The resulting
equation can be written as

$$h_{2r} = d_{2c}\tan(\theta) - \frac{g}{2}\left(\frac{d_{2c}}{V_o\cos(\theta)}\right)^2$$

When h_{2r} is equal to the rim height and the ball clears the front of the rim, the shot is a perfect swish. The following code cell and its output demonstrate the perfect swish solution.

```
%% mm3215

g = 9.81;      % gravity constant
d2f = 4;       % distance to front of rim
d2c = 4.25;    % distance to center of basket
h2r = 1;       % height to rim
hmin = 1.125;  % minimum ball height above front of rim

% anonymous function that is zero for a swish
swish = @(Vo,theta) d2c*tan(theta)-(g/2)*(d2c./(Vo*cos(theta))).^2 - h2r;

% with simple changes to above function we can get an
% anonymous function that gives ball height at front of rim
habove = @(Vo,theta) d2f*tan(theta)-(g/2)*(d2f./(Vo*cos(theta))).^2 - hmin;

% choose angles, find velocities
thetad = linspace(35,75); % angles in degrees to test
theta = thetad*pi/180;    % convert angles to radians
N = length(theta);

Vo = zeros(1,N); % preallocate memory
pass = Vo;

for k = 1:N % gather data

    Vo(k) = fzero(@(Vo) swish(Vo,theta(k)),10); % find swish velocity
    pass(k) = habove(Vo(k),theta(k)); % above front of rim?
end
plot(thetad(pass<0),Vo(pass<0),'k:',... % non-passing solutions
     thetad(pass>0),Vo(pass>0),'k')      % passing solutions

xlabel('Initial Angle, degrees')
ylabel('Initial Velocity, m/s')
title('Figure 32.15: Swish Solution Space')
```

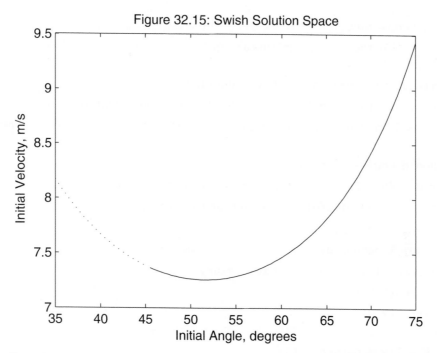

Figure 32.15: Swish Solution Space

Based on the figure, the initial angle must be greater than about 45 degrees for the ball to make it over the front of the rim. The following code cell and its output plot several solution trajectories.

```
%% mm3216

g = 9.81;      % gravity constant
d2f = 4;       % distance to front of rim
d2c = 4.25;    % distance to center of basket
d2b = 4.5;     % distance to back of rim
h2r = 1;       % height to rim
hmin = 1.125;% minimum ball height above front of rim

% anonymous function for height, y(t)
height = @(t,Vo,theta) Vo*sin(theta)*t -(g/2)*t.^2;

% anonymous function for distance, x(t)
distance = @(t,Vo,theta) Vo*cos(theta)*t;
```

```
% anonymous function that is zero for a swish
swish = @(Vo,theta) d2c*tan(theta)-(g/2)*(d2c./(Vo*cos(theta))).^2 - h2r;

% with simple changes to above function we can get an
% anonymous function that gives ball height at front of rim
habove = @(Vo,theta) d2f*tan(theta)-(g/2)*(d2f./(Vo*cos(theta))).^2 - hmin;

% choose angles, find velocities
thetad = 45:5:75;        % angles in degrees to test
theta = thetad*pi/180;  % convert angles to radians
N = length(theta);
M = 100; % number of time points to compute
Vo = zeros(1,N); % preallocate memory
X = zeros(M,N);
Y = zeros(M,N);

for k = 1:N % gather data
   Vo(k) = fzero(@(Vo) swish(Vo,theta(k)),10); % find swish velocity

   tofun = @(t) height(t,Vo(k),theta(k)); % where height is zero
   to = fzero(tofun,[0.1,10]); % let fzero find zero crossing time
   t = linspace(0,to,M);

   Y(:,k) = height(t,Vo(k),theta(k));    % find trajectory
   X(:,k) = distance(t,Vo(k),theta(k));
end

% don't plot past back of rim
Y(X>d2b) = nan;
X(X>d2b) = nan;

% plot geometry and swishes
plot([0 d2b d2b],[0 0 4.5*h2r],'k',... % border
     [d2f d2b],[h2r h2r],'k',... % rim
     d2c,h2r,'xk',... % center of hoop
```

```
    d2f,hmin,'ko',...
    X,Y,'-.k')

xlabel('Distance, m')
ylabel('Height, m')
title('Figure 32.16: Swish Trajectories')
```

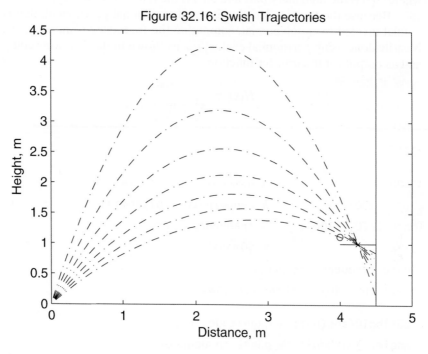

Figure 32.16: Swish Trajectories

32.16 BODE PLOTS

The frequency response of a system provides valuable insight into system operation and performance capabilities. Because frequency response is characterized by a complex-valued function of frequency, there are a number of ways to interpret or visualize this data. Of these, a Bode plot is perhaps the most useful. A Bode plot is composed of two subplots. One is a plot of the magnitude of the frequency response in decibels versus frequency, and the other is a plot of the phase of the frequency response in degrees versus frequency.

Given a transfer function,

$$H(s) = \frac{b_m s^m + b_{m-1} s^{m-1} + \ldots + b_1 s + b_0}{s^n + a_{n-1} s^{n-1} + \ldots + a_1 s + a_0}$$

the frequency response is given by

$$H(\omega) = H(s)|_{s=j\omega}$$

The magnitude of $H(\omega)$ in decibels is given by

$$H(\omega)|_{dB} = 20 \log_{10}(|H(\omega)|)$$

and its phase in degrees is given by

$$\measuredangle(H(\omega) = \arctan\left(\frac{\Im(H(\omega))}{\Re(H(\omega))}\right)$$

where $\Im(\cdot)$ is the imaginary part and $\Re(\cdot)$ is the real part of its argument.

Because the transfer function is simply a rational polynomial, that is, a polynomial divided by a polynomial, computing the frequency response in MATLAB is easily done using polynomial evaluation, as shown in the following code segment and its output for the transfer function

$$H(s) = \frac{100}{s^2 + 6s + 100}$$

```
%% mm3217

num = 100;                 % numerator polynomial
den = [1 6 100];           % denominator polynomial
w = logspace(0,2,100);     % frequency vector
jw = 1i*w;                 % jOmega
% evaluate frequency response
h = polyval(num,jw)./polyval(den,jw);

hdb = 20*log10(abs(h)); % magnitude in dB
hph = angle(h)*180/pi;  % phase in degrees

subplot(2,1,1)
semilogx(w,hdb) % magnitude plot
title('Figure 32.17: First Example Bode Plot')
ylabel('|H(\omega)|, dB')

subplot(2,1,2)
semilogx(w,hph) % phase plot
xlabel('Frequency, rad/s')
ylabel('\angle(H), \circ')
```

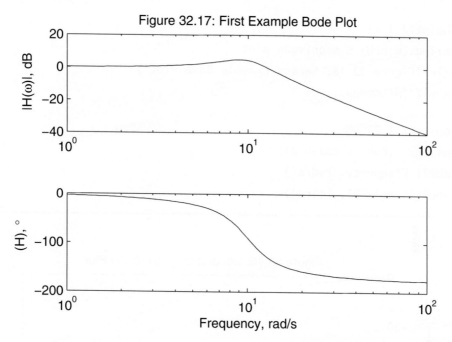

Figure 32.17: First Example Bode Plot

The use of the function polyval makes computation of the frequency response simple. However, it does have a problem when the difference between the number of poles and zeros is greater than two, as shown in the following code segment and its output for the transfer function

$$H(s) = \frac{100(s + 5)}{s^4 + 14s^3 + 168s^2 + 530s + 375}$$

```
%% mm3218
num = 100*[1 5];
den = [1 14 168 530 375];

w = logspace(0,2,100);      % omega axis points
jw = 1i* w;                 % j*Omega

% evaluate frequency response
h = polyval(num,jw)./polyval(den,jw);

hdb = 20*log10(abs(h)); % magnitude in dB
hph = angle(h)*180/pi;  % phase in degrees
```

```
subplot(2,1,1)
semilogx(w,hdb) % magnitude plot
title('Figure 32.18: Second Example Bode Plot')
ylabel('|H(\omega)|, dB')

subplot(2,1,2)
semilogx(w,hph) % phase plot
xlabel('Frequency, rad/s')
ylabel('\angle(H), \circ')
```

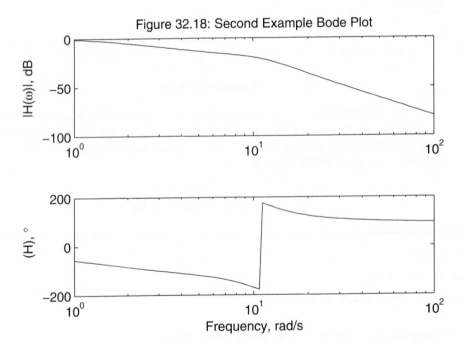

Figure 32.18: Second Example Bode Plot

The discontinuity in phase just past 10 rad/s in the phase plot occurs because the angle of a complex number can only be resolved into its principle range from $-180°$ to $+180°$ by the function angle. In reality, the phase should continue its monotonic decrease beyond 10 rad/s. To fix this, $360°$ needs to be subtracted from all the phase data at and beyond the discontinuity. The MATLAB function unwrap accepts the phase information in radians and returns a revised unwrapped phase vector, as shown in the following code segment and its output.

```
%% mm3219
num = 100*[1 5];
den = [1 14 168 530 375];

w = logspace(0,2,100);      % omega axis points
jw = 1i* w;                 % j*Omega

% evaluate frequency response
h = polyval(num,jw)./polyval(den,jw);

hdb = 20*log10(abs(h)); % magnitude in dB
hph = unwrap(angle(h))*180/pi;   % phase in degrees, unwrapped

subplot(2,1,1)
semilogx(w,hdb) % magnitude plot
title('Figure 32.19: Third Example Bode Plot')
```

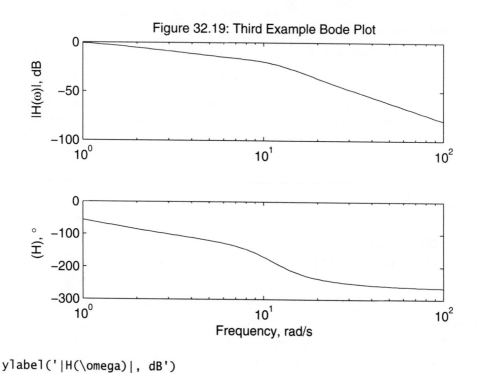

Figure 32.19: Third Example Bode Plot

```
ylabel('|H(\omega)|, dB')
```

```
subplot(2,1,2)
semilogx(w,hph) % phase plot
xlabel('Frequency, rad/s')
ylabel('\angle(H), \circ')
```

The MATLAB function unwrap knows what to do because it finds the discontinuities in the phase data and adds or subtracts the correct multiple of 2π radians as needed to unwrap the phase. If the frequency range over which the frequency response data is computed does not jump over discontinuities, then unwrap is unsuccessful.

To eliminate this phase ambiguity requires that the frequency response be computed differently. Rather than evaluate the numerator and denominator polynomials in their entirety using polyval, the magnitude and phase can be computed term by term when the transfer function is written in the zero-pole-gain form

$$H(s) = k \frac{(s - z_1)(s - z_2)\ldots(s - z_m)}{(s - p_1)(s - p_2)\ldots(s - p_n)}$$

In this case, the magnitude of the frequency response can be written as

$$|H(\omega)| = |k| \frac{|(j\omega - z_1)||(j\omega - z_2)|\ldots|(j\omega - z_m)|}{|(j\omega - p_1)||(j\omega - p_2)|\ldots|(j\omega - p_n)|}$$

and the phase can be written as

$$\sphericalangle(H(\omega)) = \sphericalangle(k) + \sphericalangle(j\omega - z_1) + \sphericalangle(j\omega - z_2)\ldots + \sphericalangle(j\omega - z_m)$$
$$- \sphericalangle(j\omega - p_1) - \sphericalangle(j\omega - p_2)\ldots - \sphericalangle(j\omega - p_n)$$

The MATLAB function tf2zp converts the original rational polynomial transfer function form into this zero-pole-gain form. Given the above frequency response magnitude and phase expressions, the frequency response can be computed as shown in the following code segment and its output.

```
%% mm3220

num = 100*[1 5];
den = [1 14 168 530 375];

w = logspace(0,2,100);      % omega axis points
jw = 1i* w;                 % j*Omega
```

```
[Z,P,K] = tf2zp(num,den); % convert to zero-pole-gain form

HdB = abs(K)+zeros(size(w));      % magnitude of gain term
Hph = zeros(size(w)); % phase is zero for positive gain term
if K<0
   Hph(:) = -pi;           % phase is -pi for negative gain term
end

for k = 1:length(Z)    % add contribution from each zero
   HdB = HdB .* abs(jw-Z(k));
   Hph = Hph + angle(jw-Z(k));
end
for k = 1:length(P)    % add contribution from each pole
   HdB = HdB ./ abs(jw-P(k));
   Hph = Hph - angle(jw-P(k));
end
HdB = 20*log10(HdB);   % convert magnitude to dB
Hph = Hph*180/pi;      % convert phase to degrees

subplot(2,1,1)
semilogx(w,HdB)          % magnitude plot
title('Figure 32.20: Fourth Example Bode Plot')
ylabel('|H(\omega)|, dB')

subplot(2,1,2)
semilogx(w,Hph)          % phase plot
xlabel('Frequency, rad/s')
ylabel('\angle(H), \circ')
```

The following figure matches the previous third example Bode plot without calling the function unwrap to eliminate discontinuities in the phase data.

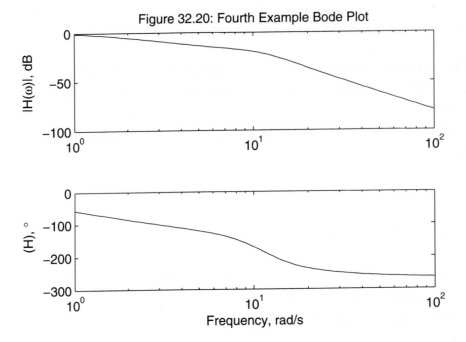

Figure 32.20: Fourth Example Bode Plot

The content of this section can easily be incorporated into a useful function such as that shown below. This implementation allows the user to simply create a Bode plot given the numerator and denominator polynomials, returns frequency response data, or returns a function handle for computing the frequency response data of a given transfer function.

```
function varargout = mmbode(varargin)
%MMBODE Bode Frequency Response.
% MMBODE(Num,Den) computes the frequency response of the transfer function
% having the numerator polynomial coefficients in row vector Num and the
% denominator polynomical coefficients in the row vector Den.
%
% FunH = MMBODE(Num,Den) returns a function handle with the following
% properties:
% [HdB,Hph] = FunH(w) returns in HdB the magnitude in dB and in Hph returns
% the phase in degrees at the frequency points contained in array w.
% FunH('Num') returns the numerator polynomial contained in FunH.
% FunH('Den') returns the denominator polynomial contained in FunH.
%
```

```
% [HdB,Hph] = MMBODE(Num,Den,w) returns in HdB the magnitude in dB and in
% Hph returns the phase in degrees at the frequency points contained in
% array w.
% [HdB,Hph,w] = MMBODE(Num,Den) returns in HdB the magnitude in dB and in
% Hph returns the phase in degrees at program selected radian frequency
% points, which are returned in vector w.
% MMBODE(Num,Den) and MMBODE(Num,Den,w) with no output arguments creates
% Bode magnitude and Bode phase plots.

% check and parse inputs
if nargin<2 || nargin>3
   error('Two or three input arguments required.')
end

Num = varargin{1};    % numerator coefficients
Den = varargin{2};    % denominator coefficients
if ~isnumeric(Num) || ~isvector(Num) ||...
   ~isnumeric(Den) || ~isvector(Den)
   error('Num and Den must be numeric vectors')
end
% convert transfer function to zero/pole/gain form
[Z,P,K] = tf2zp(Num(:)',Den(:)');

if nargin==3
   w = varargin{3};
   if ~isnumeric(w)
      error('w must be a numeric array')
   end
end
% create function handle to compute response
funh = @(w) boderesp(w,Z,P,K); % boderesp is a subfunction

if nargin==2 && nargout==1 % function handle only requested
   varargout{1} = funh;
   return
end
```

```matlab
if nargin==2 % no omega data points given, so find them
   zp = abs(real([Z;P]));  % magnitudes of zero and pole real parts
   zpm = min(zp);          % smallest zero or pole real part
   zpx = max(zp);          % largest zero or pole real part
   if zpm<1e-6             % pole or zero at origin
      lmin = -1;           % start at 10^-1
   else
      lmin = floor(log10(zpm));   % go below smallest
   end
   lmax = ceil(log10(zpx))+1;     % go a decade above largest
   lmax = max(lmin+1,lmax);       % go at least one decade
   lmax = min(lmin+5,lmax);       % but not more than 5 decades
   w = logspace(lmin,lmax,max(50,25*(lmax-lmin))); % good w axis?
end

% compute response at points in w by calling function handle
[HdB,Hph] = funh(w);

% figure out what to return
if nargout==0           % create plot only
   subplot(2,1,1)
   semilogx(w,HdB)      % magnitude plot
   ylabel('|H(\omega)|, dB')
   subplot(2,1,2)
   semilogx(w,Hph)      % phase plot
   xlabel('Frequency, rad/s')
   ylabel('\angle(H), \circ')
elseif nargin==3        % [HdB,Hph] = MMBODE(Num,Den,w)
   varargout{1} = HdB;
   varargout{2} = Hph;
elseif nargin==2        % [HdB,Hph,w] = MMBODE(Num,Den)
   varargout{1} = HdB;
   varargout{2} = Hph;
   varargout{3} = w;
end
```

```
%--------------------------------------------------------------
function [HdB,Hph] = boderesp(w,Z,P,K)
% compute frequency response at points in w given the
% zero/pole/gain form of the transfer function
% handle character string input requests first
if ~isempty(w) && ischar(w)
    if strncmpi(w,'Num',1)        % FunH('Num')
        HdB = K*real(poly(Z));
    elseif strncmpi(w,'Den',1)  % FunH('Den')
        HdB = real(poly(P));
    else
        error('Unknown input')
    end
    return
elseif isempty(w) || ~isnumeric(w)
    error('Numeric input expected')
end
% compute frequency response magnitude and phase
jw = 1i * w;                     % jOmega points
HdB = abs(K)+zeros(size(w));     % magnitude of gain term
Hph = zeros(size(w));            % phase of gain term
if K<0
    Hph(:) = -pi;                % phase is -pi for negative gain
end

for k = 1:length(Z)              % add contribution from each zero
    HdB = HdB .* abs(jw-Z(k));
    Hph = Hph + angle(jw-Z(k));
end
for k = 1:length(P)              % add contribution from each pole
    HdB = HdB ./ abs(jw-P(k));
    Hph = Hph - angle(jw-P(k));
end
HdB = 20*log10(abs(HdB));        % convert magnitude to dB
Hph = Hph*180/pi;                % convert phase to degrees
```

32.17 INVERSE LAPLACE TRANSFORM

Finding the inverse Laplace transform is a common academic task in engineering. Analytically, the inverse is computed by performing a partial-fraction expansion, finding coefficients, and writing the corresponding time function with the help of a Laplace transform table. For simple Laplace transforms, this task is readily accomplished by hand. However, as the complexity of the Laplace transform expression increases, this becomes a tedious task. Moreover, the resulting time function expression provides no visual feedback about the shape or characteristics of the result. To help resolve these issues, let's use MATLAB.

Partial-fraction expansion of rational polynomial functions such as

$$F(s) = \frac{b_m s^m + b_{m-1} s^{m-1} + \ldots + b_1 s + b_0}{s^n + a_{n-1} s^{n-1} + \ldots + a_1 s + a_0}$$

make use of the Laplace transform pair relationship

$$\frac{R}{(s + a)^m} \Leftrightarrow \frac{R}{(m - 1)!} t^{m-1} e^{-at} u_s(t)$$

where $u_s(t)$ is the unit step function. This relationship holds for all real and complex values of the variable a. When a partial-fraction expansion is performed by hand, other Laplace transform pairs are more convenient when a is complex, but that does not diminish the validity of the above transform pair. Since MATLAB works seamlessly with complex numbers, the above Laplace transform pair is all that is required.

Finding the inverse Laplace transform of $F(s)$ requires that it be written in partial-fraction expansion form as

$$F(s) = \frac{R_1}{s + p_1} + \frac{R_2}{s + p_2} + \ldots + \frac{R_n}{s + p_n}$$

when all the poles $-p_i$ are distinct. When a pole is repeated, or has a multiplicity greater than one, additional terms are required, such as

$$\frac{R_m}{(s + p_k)^m} + \frac{R_{m-1}}{(s + p_k)^{m-1}} + \ldots + \frac{R_1}{(s + p_k)}$$

when a pole $-p_k$ has multiplicity m.

The partial-fraction expansion containing terms as shown above holds as long as $F(s)$ is strictly proper, that is, $F(s)$ has more poles than zeros. When this is not true, the resulting time function contains impulse functions and its derivatives at time equal to zero that cannot be evaluated numerically. As a result, in MATLAB these terms are simply discarded.

The MATLAB function residue computes the partial-fraction expansion, as shown in the following code cell and its output.

```
%% use of residue

num = 100
den = poly([0;-2;-4]) % s(s+2)(s+4)
[R,P,K] = residue(num,den)
```

```
num =
      100
den =
      1    6    8    0
R =
      12.5000
     -25.0000
      12.5000
P =
      -4
      -2
       0
K =
      []
```

These results describe the partial-fraction expansion as

$$F(s) = \frac{100}{s^3 + 6s^2 + 8s} = \frac{12.5}{s+4} - \frac{25}{s+2} + \frac{12.5}{s}$$

The output K from `residue` contains polynomial coefficients that characterize impluse functions and its derivatives. Since this $F(s)$ is strictly proper, K is an empty array as shown. If K is not empty, it is discarded as described earlier.

When a pole has multiplicity greater than one, the function `residue` returns all coefficients as shown in the code cell and its output below.

```
%% multiplicities and residue

num = 100
den = poly([0;-2;-4;-4]) % s(s+2)(s+4)^2
[R,P] = residue(num,den) % simply don't ask for K
```

num =

     100

den =

     1   10   32   32   0

R =

     9.3750

     12.5000

     -12.5000

     3.1250

P =

     -4.0000

     -4.0000

     -2.0000

     0

In this case, the output of residue describes the partial-fraction expansion as

$$F(s) = \frac{100}{s^4 + 10s^3 + 32s^2 + 32s} = \frac{9.375}{s + 4} + \frac{12.5}{(s + 4)^2} - \frac{12.5}{s + 2} + \frac{3.125}{s}$$

With this understanding of the output of the function residue, it is possible to perform inverse Laplace transforms in MATLAB and evaluate the resulting time function. This possibility comes with the inherent fact that performing a partial-fraction expansion is an ill-posed numerical problem. When done analytically, that is, by hand, partial-fraction expansion is fine. However, when done numerically, even when using double-precision arithmetic as MATLAB does, the results may produce significant errors. That being said, as long as one is solving academic problems, for example, homework problems, this approach generally works very well.

The solution to this problem is best accomplished by a MATLAB function that accepts the numerator and denominator polynomials and returns a function handle that evaluates the time function at user-chosen time points. In that way, the function handle internally contains all the information necessary to evaluate the resulting time function. This function handle can point to a subfunction or a nested function. In the earlier Bode plot section, a subfunction was used. Therefore, a nested function will be used here for illustrative purposes. In addition to evaluating the inverse transform, it is beneficial to provide the capability of returning the underlying numerator and denominator polynomial vectors. The following function implements these features.

```
function fun = mminvlap(B,A)
%MMINVLAP Time Function from Laplace Transform.
% FUN = INVLAPFUN(B,A) returns a function handle for evaluating the time
```

```
% function FUN(t) associated with the Laplace Transform B(s)/A(s), where
% B and A are the respective row vectors containing the real-valued
% polynomial coefficients.
%
% FUN(t) evaluates the inverse Laplace transform evaluated at the time
% points in the array t. FUN(t) = 0 for t<0.
%
% S = FUN('ba') returns a structure S with fields containing the numerator
% and denominator polynomial vectors. That is, S.B contains the numerator
% polynomial vector and S.A returns the denominator polynomial vector.
%
% Alternatively, S = FUN('rpk') returns a structure S with fields
% containing the outputs from RESIDUE. That is, S.R contains R, S.P
% contains P, and S.K contains K.
%
% This function uses the partial fraction expansion returned by the
% function RESIDUE. As a result, the accuracy of this function is limited
% by the accuracy of the function RESIDUE in MATLAB.

% parse inputs
if nargin~=2
    error('Two input arguments expected')
end
if ~isnumeric(B) || ~isvector(B) || ~isreal(B) ||...
    ~isnumeric(A) || ~isvector(A) || ~isreal(A)
    error('Inputs must be real-valued numeric Vectors')
end
% perform partial fraction expansion
[R,P,K] = residue(B,A);

% create anonymous function handle to time function
fun = @(t) invlaplace(t);

% define invlaplace as a "nested function" so it has access to all the
```

```
% variables in the above mminvlap function, namely R,P,K, and B and A.
% Nested Function -----------------------------------------------------
   function y = invlaplace(t)
      if ~isempty(t) && ischar(t)
         if strncmpi(t,'ba',1) % FUN('ba')
            y.B = B;
            y.A = A;
         elseif strncmpi(t,'rpk',1) % FUN('rpk')
            y.R = R;
            y.P = P;
            y.K = K;
         end
         return
      elseif isempty(t) || ~isnumeric(t)
         error('Numeric Input Array Expected.')
      end
      % time points provided, compute time response
      y = zeros(size(t)); % preallocate memory for result
      t(t<0) = nan;       % throw out negative time points
      tol = 0.001;        % tolerance for repeated roots
      k = 1;              % loop through all roots

      while k <= length(P)

         y = y + R(k)*exp(P(k).*t); % basic term that always exists

         m = sum(abs(P(k)-P) < tol*max(abs(P(k)),1)); % root multiplicity

         if m > 1                % multiplicity, add extra terms as needed
            for n = k+1:k+m-1 % loop through all powers
               y = y + (R(n)/prod(1:n-k)) * t.^(n-k) .* exp(P(k).*t);
            end
         end
         k = k+m;               % next root to consider
      end
```

```
        y(isnan(t)) = 0;   % output is zero for negative time
        % when complex conjugate roots exist, residual complex values may
        % exist. Strip them away.
        y = real(y);
    end % End of Nested Function
end % End of Outer Function
```

The following code cell and its output demonstrate use of this function.

```
%% mm3221

num = 32;
den = poly([0;-2;-4;-4]); % s(s+2)(s+4)^2

myfun = mminvlap(num,den); % get inverse Laplace function handle

myfun('ba') % get B and A polynomial vectors

myfun('rpk')% get R, P, and K

t = linspace(0,4);  % time points
y = myfun(t);        % evaluate inverse Laplace function

plot(t,y) % plot response
xlabel('time, s')
ylabel('Response')
title('Figure 32.21: Inverse Laplace Example')
```

```
    ans =
        B: 32
        A: [1  10  32  32  0]
    ans =
        R: [4x1 double]
        P: [4x1 double]
        K: []
```

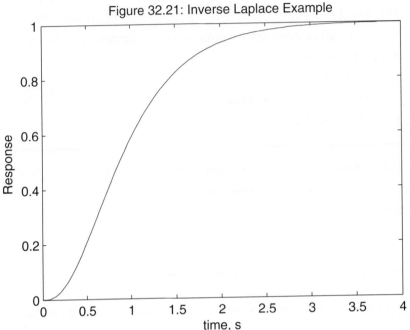

Figure 32.21: Inverse Laplace Example

32.18 PICTURE-IN-A-PICTURE ZOOM

The final example in this chapter demonstrates use of the Handle Graphics features in MATLAB. As illustrated in the figure below, this example implements a picture-in-a-picture zoom. That is, the user calls the function, then goes to the current axes, and drags a selection rectangle using the mouse. Once created, a dotted outline of the selection rectangle remains and a new, but smaller, axes is created showing the graphical contents inside the drawn selection rectangle. This function allows the user to zoom into a portion of an axes without hiding the original plot. Though not shown, the smaller axes can be selected, dragged, and resized.

There are a number of steps involved in the creation of this function. Rather than implement this function as one large function, it is convenient to create supporting functions first that could be subfunctions of the primary functions or simply functions on the MATLAB search path. The first function getn simplifies the assignment of the output of the get function into individual variables.

```
function varargout = getn(H,varargin)
%GETN Get Multiple Object Properties.
% [Prop1,Prop2,...] = GETN(H,PName1,PName2,...) returns the requested
% properties of the scalar handle H in the corresponding output arguments.
%
```

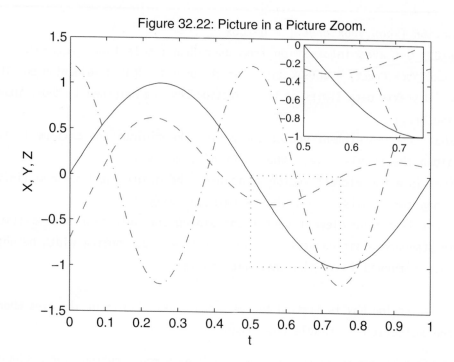

Figure 32.22: Picture in a Picture Zoom.

```
% For example, [Xlim,Ylim,Xlabel] = GETN(gca,'Xlim','Ylim','Xlabel')
% returns the requested axes properties in like-named output variables.
%
% This simplifies the construct
% [Xlim,Ylim,Xlabel] = deal(get(gca,{'Xlim','Ylim','Xlabel'}))

if ~isscalar(H) || ~ishandle(H)
   error('Scalar Object Handle Required.')
end
varargout = get(H,varargin);
```

This function is not absolutely needed, but it is convenient and demonstrates use of varargin and varargout.

The next function to consider is implementation of the selection rectangle and capturing its x- and y-axis limits. As shown below, this requires use of the MATLAB functions rbbox, waitforbuttonpress, and the CurrentPoint properties of the *axes* and *figure* objects.

```
function [xbox,ybox,prect] = getbox
%GETBOX Get axes information from user-drawn selection rectangle.
% [Xbox,Ybox,Prect] = GETBOX waits for the user to drag a selection box The
% x and y axis data limits of the selection box are returned in Xbox, Ybox,
% and Prect.
% Xbox is a two element vector containing the minimum and maximum limits
% along the x axis, i.e., Xbox = [min(x) max(x)]
% Ybox is a two element vector containing the minimum and maximum limits
% along the y axis, i.e., Ybox = [min(y) max(y)]
% Prect is a four element vector containing the selection box position
% in standard position vector format, Prect = [left bottom width height]
% Data returned is in the axis data units.
%
% The selection box is limited to the x and y axis limits of the axes where
% the selection rectangle was drawn.

% waitforbuttonpress waits until user presses a mouse button over a figure
% waitforbuttonpress return False if that happens. Alternatively, it
% returns True if the user presses a key on the keyboard.

if waitforbuttonpress % Returns True if a key is pressed, abort
    xlim = [];
    ylim = [];
    prect = [];
    return
end
% Function only gets here if user presses a mouse button in a figure

Hf = gcf;        % get current figure where button was pressed
Ha = gca(Hf);    % get current axes where button was pressed

AxesPt = get(Ha,'CurrentPoint'); % get first axes data point clicked
FigPt = get(Hf,'CurrentPoint');   % get first figure point clicked

% call the function rbbox, i.e., rubberband box, to create the selection
% rectangle. This function needs to know where to start from. It does not
% automatically start at the mouse click unless told to do so.
```

```
% drag selection rectangle starting at first figure point
rbbox([FigPt 0 0],FigPt) % function returns as soon as mouse button is up

% get point on opposite corner of selection rectangle; add to first point
AxesPt = [AxesPt; get(Ha,'CurrentPoint')];

% get axis limits of axes where selection rectangle was drawn
[Xlim,Ylim] = getn(Ha,'Xlim','Ylim');

% convert AxesPt data into usable output vectors.
xbox = [min(AxesPt(:,1)) max(AxesPt(:,1))]; % x axis limits of selection
xbox = [max(xbox(1),Xlim(1)) min(xbox(2),Xlim(2))]; % limit to axes size

ybox = [min(AxesPt(:,2)) max(AxesPt(:,2))];
ybox = [max(ybox(1),Ylim(1)) min(ybox(2),Ylim(2))]; % limit to axes size

prect = [xbox(1) ybox(1) diff(xbox) diff(ybox)]; % position rectangle
```

The comments contained in `getbox` describe its operation. The `CurrentPoint` property of an *axes* object returns a 2-by-3 matrix containing the *x*, *y*, and *z* data points of the back and front of the current point in 3-D space. Since this is a 2-D axes, only the *x* and *y* data points in the first two columns are relevant. Manipulation of this data produces the desired output variables.

Using `getn` and `getbox`, the function `mmzoom` shown below implements the picture-in-a-picture zoom.

```
function [Hza,Hzr] = mmzoom(arg)
%MMZOOM Picture in a Picture Zoom.
% MMZOOM creates a new axes containing the data inside a box formed by a
% click and drag with the mouse in the current axes. The new zoomed axes
% is placed in the upper right of the current axes, but can be moved with
% the mouse. Clicking in the figure border disables dragging.
%
% Previous axes created by MMZOOM are deleted if MMZOOM is called again.
%
```

```
% [Hza,Hzr] = MMZOOM returns handles to the created axes and rectangle
% marking the zoomed area respectively.
%
% MMZOOM DRAG enables dragging of a zoomed axes.
% MMZOOM RESET disables dragging of a zoomed axes.
% MMZOOM OFF removes the zoomed axes and rectangle marking the zoomed area.

if nargin == 0
   arg = [];
end
if isempty(arg) % zoom zoom zoom zoom zoom zoom zoom zoom zoom zoom zoom

   Hzoom = findobj(0,'Tag','MMZOOM'); % find previous zoomed axes
   if ~isempty(Hzoom) % delete prior zoomed axes if it exists
      delete(Hzoom)
   end

   [xlim,ylim,prect] = getbox; % get selection box for zoom

   if ~isempty(prect) % act only if rectangle exists
      Haxes = gca; % handle of axes where selection box was drawn
      Hzr = rectangle('Position',prect,... % place rectangle object
         'Linestyle',':',...                % to mark selection box
         'Tag','MMZOOM');

      Hfig = gcf; % handle of Figure where selection box was drawn
      Hzoom = copyobj(Haxes,Hfig); % copy original axes and its children

      OldUnits = get(Haxes,'Units');    % get position vector of original
      set(Haxes,'Units','normalized') % axes in normalized units
      Pvect = get(Haxes,'Position');
      set(Haxes,'Units',OldUnits)

      % scale and shift zoomed axes relative to original axes
      alpha = 1/3;   % position scaling for zoomed axes
      beta = 98/100; % position shift for zoomed axes
```

```
% compute position vector for zoomed axes
Zwidth = alpha*Pvect(3);                        % zoomed axes width
Zheight = alpha*Pvect(4);                       % zoomed axes height
Zleft = Pvect(1)+beta*Pvect(3)-Zwidth;     % zoomed axes left
Zbottom = Pvect(2)+beta*Pvect(4)-Zheight; % zoomed axes bottom

% modify zoomed axes as required
set(Hzoom,'units','Normalized',...             % make units normalized
    'Position',[Zleft Zbottom Zwidth Zheight],...% axes position
    'Xlim',xlim,'Ylim',ylim,...                % axis data limits
    'Box','on',...                             % axis box on
    'Xgrid','off','Ygrid','off',...            % grid lines off
    'FontUnits','points',...
    'FontSize',8,...                           % shrink font size
    'ButtonDownFcn',@selectmoveresize,...  % enable drag
    'Tag','MMZOOM',...                         % tag zoomed axes
    'UserData',Haxes)                          % store original axes

[Htx,Hty,Htt] = getn(Hzoom,'Xlabel','Ylabel','Title');
set([Htx,Hty,Htt],'String','')  % delete labels on zoomed axes

set(Haxes,'DeleteFcn',...           % delete both axes together
    'delete(findobj(0,''Type'',''axes'',''Tag'',''MMZOOM''))')

% place zoomed axes at top of object stack
Hchild = findobj(Hfig,'type','axes'); % get all axes in figure
Hchild(Hchild==Hzoom) = [];         % remove zoomed axes from list

set(Hfig,'Children',[Hzoom;Hchild],...% put zoom axes at top of stack
    'CurrentAxes',Haxes,...                 % make original axes current
    'ButtonDownFcn','mmzoom reset')% enable reset

if nargout>=1 % provide output only if requested
    Hza = Hzoom;
end
end
```

```
elseif strncmpi(arg,'d',1) % drag zoom axes drag zoom axes drag zoom axes

    Hzoom = findobj(0,'Type','axes','Tag','MMZOOM');
    if ~isempty(Hzoom)
        set(Hzoom,'ButtonDownFcn',@selectmoveresize)
    end

elseif strncmpi(arg,'r',1) % reset reset reset reset reset reset reset

    Hzoom = findobj(0,'Type','axes','Tag','MMZOOM');
    if ~isempty(Hzoom)
        [Hfig,Haxes] = getn(Hzoom,'Parent','UserData');
        set(Hzoom,'ButtonDownFcn','','Selected','off')% turn off selection
        set(Hfig,'CurrentAxes',Haxes)                 % make Haxes current
    end

elseif strncmpi(arg,'o',1) % off off off off off off off off off off off

    Hzoom = findobj(0,'Tag','MMZOOM');
    if ~isempty(Hzoom)
        delete(Hzoom)
    end
else
    error('Unknown Input Argument.')
end
```

The structure of mmzoom uses a *switchyard* approach that evaluates code based on the input argument. When no input argument is used, mmzoom creates the zoomed *axes*. The function copyobj is used to make a copy of the original *axes* and all its children. This new *axes* is then modified to create the zoomed *axes*. Finally, the zoomed *axes* is placed on top of the original axes, which is made the current *axes*. The ButtonDownFcn callback of the zoomed axes executes the function selectmoveresize, which performs exactly what its name implies. The ButtonDownFcn callback of the *figure* calls mmzoom('reset'), which deselects the zoomed axes and deletes its callback. Since this function is fairly simple, the function findobj is used to find the zoomed *axes* and the *rectangle* marking the selection box. Doing so makes it difficult to allow multiple simultaneous instances of this function to exist. As a result, only one is permitted. If multiple simultaneous instances were desired, function handle callbacks with added arguments would be required.

Appendix A

MATLAB Release Information

These appendices document the new and revised functionality and functions available in individual releases of MATLAB, starting with MATLAB 5. This material makes it possible to identify what functionality and functions appeared in which version of MATLAB. With worldwide MATLAB users running a variety of MATLAB releases, and with the proliferation and ease of file sharing, it is important to be able to identify the backward and forward compatibility of MATLAB M-files. The information contained in these appendices facilitates this process by documenting MATLAB release information that is otherwise not centrally available.

The material presented is limited to the base MATLAB product only and is based on information contained in the release notes for each release. Because the release notes are not always comprehensive, some information may be missing.

Furthermore, two distinct classes of information have been omitted because (1) the topics are outside the scope of this text, and (2) they constitute too much information to be included here: Handle Graphics objects and their properties, and changes relating to the MATLAB API. Specialized texts are available for those interested in building graphical user interfaces and/or interfacing MATLAB with external C, FORTRAN, and/or Java programs or with Windows-specific applications. Therefore, functions, objects, and properties specific to Handle Graphics or the MATLAB API (e.g., MX-/MEX-functions, ActiveX/DDE/COM objects and associated functions) are not included in these appendices.

Each MATLAB release cited here identifies the public release date, the functionality and functions introduced in the release, as well as the revised functionality and functions that became effective with the release. Obsolete functions are also identified. Changed functionality and functions have (Change) in their description. When appropriate, a reference to the chapter number covering the cited functionality is included.

MATLAB 5.0, Release 8 (December 1996)

Functionality	Description	Chapter
Camera viewing model	Axes now have camera properties to set viewing properties	26
Cell arrays	A container variable class identified by a name, with contents enclosed with { and }	8
Character-string storage	(Change) Each character occupies 2 bytes of storage rather than 8	9
Dimension specification in data analysis functions	(Change) Data analysis functions sum, prod, cumprod, and cumsum now accept a last input argument that specifies the dimension along which the function operates	17
Empty arrays	(Change) Empty arrays may have some non-zero dimensions	5, 6
end as last array element	Last element along a dimension can be addressed by using the keyword end (e.g., A(1,2:end))	5
Light object	New graphics object	27, 30
Marker style enhancements	(Change) New line markers are available, and markers can be specified independently from linestyles	25
Multidimensional arrays	Arrays can have any number of dimensions (e.g., A(i,j,k,...))	6
Object-oriented programming	User-defined variable classes with function and operator overloading, data encapsulation, methods, inheritance, and so on	31
Private functions	Function M-files in private subdirectories just off of the MATLAB path	12
Pseudocode creation	Ability to precompile M-files into an encrypted format	12
Scalar expansion assignment	(Change) A(...) = x performs scalar expansion to fill all addressed elements of A with the scalar x	5
Single-byte data type for images	Image object supports 8-bit integer data, using the uint8 data type	7, 28
Structures	A container variable class identified by a variable name and fields, with fields denoted by a preceding period (e.g., varname.field1)	8
Subfunctions	Function M-files can contain multiple functions, with those appearing after the first or primary function being subfunctions	12

Functionality	Description	Chapter
Switch-case statements	Alternative control flow construction	11
TeX support	(Change)Text objects can contain TeX formatting commands	25
Truecolor	Images in Truecolor are supported	28
Variable length function input and output argument lists	Cell arrays `varargin` and `varargout` supported in function input and output argument lists	12
Z-buffer support	Z-buffer graphics rendering is supported	29

Math Function	Description
`airy`	Airy functions
`besselh`	Bessel functions of the third kind (Hankel)
`condeig`	Condition number with respect to eigenvalues
`condest`	1-norm matrix condition number estimate
`dblquad`	Numerical 2-D integration
`mod`	Modulus
`normest`	2-norm estimate

n-D Function	Description
`cat`	Concatenates arrays
`flipdim`	Flips array along specified dimension
`ndgrid`	Generates arrays for *n*-D functions
`ndims`	Number of array dimensions
`permute, ipermute`	Permute and inverse permute dimensions of *n*-D array
`reshape`	Changes the shape of an array
`shiftdim`	Shifts dimensions
`squeeze`	Eliminates singleton dimensions
`sub2ind, ind2sub`	Single index from subscripts and subscripts from single linear index

Cell Array and Structure Function	Description
`cell`	Creates cell array
`cell2struct`	Converts cell to structure
`celldisp`	Displays cell structure

Cell Array and Structure Function	Description
cellplot	Graphically displays cell structure
fieldnames	Field names of structure
getfield	Gets field from structure
num2cell	Converts matrix to cell array
rmfield	Removes field from structure
setfield	Sets field in structure
struct	Creates structure array
struct2cell	Converts structure to cell array

Character Function	Description
char	Converts to string array
mat2str	Converts matrix to string
strcat	String concatenation
strmatch	Finds matches for a string
strncmp	Compares first n characters
strvcat	Vertical string concatenation

Logical Function	Description
iscell	True for cell array
isequal	True if arrays are equal
isfinite	True for finite elements
islogical	True for logical arrays
isnumeric	True for numeric array
isstruct	True for structure
logical	Converts (casts) to logical array

M-file Function	Description
assignin	Assigns variable in a specific workspace
evalin	Evaluates expression in a specific workspace
inmem	Functions in memory
inputname	Input argument name

`mexext`	MEX-file extension
`mfilename`	Name of the currently running M-file
`pcode`	Creates pseudocode
`profile`	Profiles M-file execution
`varargin, varargout`	Passes or returns a variable number of function arguments
`warning`	Displays warning message

File and Directory Function	Description
`addpath`	Appends directory(s) to the MATLAB search path
`edit`	Edits M-file (launch MATLAB Editor/Debugger)
`editpath`	Modifies the MATLAB search path
`fullfile`	Builds full filename from parts

Set, Bit, and Base Function	Description
`base2dec`	Converts base to decimal
`bin2dec`	Converts binary to decimal
`bitand`	Bitwise AND
`bitcmp`	Compares bits
`bitget`	Gets bit
`bitmax`	Maximum floating-point integer
`bitor`	Bitwise OR
`bitset`	Sets bit
`bitshift`	Bitwise shift
`bitxor`	Bitwise XOR
`dec2base`	Converts decimal to base
`dec2bin`	Converts decimal to binary
`intersect`	Set intersection
`ismember`	Detects members of a set
`setdiff`	Set difference
`setxor`	Set exclusive OR
`union`	Union of two sets
`unique`	Unique elements of a vector

Time and Date Function	Description
calendar	Produces monthly calendar
datenum	Serial date number
datestr	Creates date string
datetick	Creates date-formatted tick labels
datevec	Date components
eomday	End of month
now	Current date and time
weekday	Day of week

Matrix Function	Description
cholinc	Incomplete Cholesky factorization
gallery	More than 50 test matrices
luinc	Incomplete LU factorization
repmat	Replicates and tiles an array
sprand	Random uniformly distributed sparse arrays

Sparse Matrix Function	Description
bicg	Biconjugate gradients method
bicgstab	Biconjugate gradients stabilized method
cgs	Conjugate gradients squared method
eigs	Finds several eigenvalues and eigenvectors
gmres	Generalized minimal residual method
pcg	Preconditioned conjugate gradients method
qmr	Quasi-minimal residual method
svds	A few singular values

Data Analysis Function	Description
convhull	Convex hull
cumtrapz	Cumulative trapezoidal numerical integration

delaunay	Delaunay triangularization
dsearch	Searches for nearest point
factor	Prime factors
inpolygon	Detects points inside a polygonal region
isprime	True for prime numbers
nchoosek	All possible combinations of n elements taken k at a time
perms	All possible permutations
polyarea	Area of a polygon
primes	Generates a list of prime numbers
sortrows	Sorts rows in ascending order
sum([]) = 0, prod([]) = 1, max([]) = [], min([]) = []	(Change) Defined output for empty inputs
tsearch	Searches for enclosing Delaunay triangle
voronoi	Voronoi diagram

Interpolation Function	Description
griddata	(Change) Uses triangle-based interpolation
interp3	3-D data interpolation
interpn	n-D data interpolation
ndgrid	Generates arrays for n-D functions and interpolation

ODE Function	Description
ode45, ode23, ode113, ode23s, ode15s	Solves differential equations by using various methods
odefile	Problem definition file for ODE solvers
odeget	Extracts options from ODE options structure
odeset	Creates or edits options structure for ODE solvers

Plot Function	Description
area	Filled area plot
bar3	3-D bar chart
bar3h	3-D horizontal bar chart
barh	2-D horizontal bar chart
box	Turns On or Off axes box
contourf	Filled contour plot
pie	Pie chart
pie3	3-D pie chart
plotyy	Plot with *y*-axis labels on both the left and right
quiver3	3-D quiver plot
ribbon	Draws lines as 3-D strips
stem	(Change) Stem tips can be filled or unfilled
stem3	3-D stem plot
trimesh	Triangular mesh plot
trisurf	Triangular surface plot

Color and Light Function	Description
autumn	Colormap of red and yellow
colorcube	Colormap of regularly spaced colors
colordef	Selects figure color scheme
lines	Colormap that follows axes colororder property
spring	Colormap of magenta and yellow
summer	Colormap of green and yellow
winter	Colormap of blue and green

Image Function	Description
imread	Reads image data
imwrite	Writes image data

Graphics Function	Description
dragrect	Drags rectangle
inputdlg	Displays input dialog

msgbox	Displays message dialog
questdlg	Displays question dialog
rbbox	Rubberband box
selectmoveresize	Interactively selects, moves, or resizes objects
uiresume	Resumes suspended M-file execution
uiwait	Blocks M-file execution
waitfor	Blocks execution until condition is satisfied

MATLAB 5.1, Release 9 (June 1997)

Functionality	Description	Chapter
find function returns empty	(Change) find(...) returns [] if no indices where argument is True are found	5
Multibyte characters	(Change) Two byte characters are supported	9

Function	Description
ismember	Now calls the MEX function ismemc to maximize speed
pagedlg	Open Page Layout dialog
printdlg	Open Print dialog
scatter	2-D scatter plot
scatter3	3-D scatter plot

MATLAB 5.2, Release 10 (March 1998)

Functionality	Description	Chapter
HDF file support	Files in HDF format can be read and written to	13
M-file locking	M-files can now be locked so that clear does not purge that M-file from memory	12
OpenGL support	OpenGL graphics rendering is supported	29
Persistent variables	Variables can be declared persistent so that they persist from one call to a function to the next	12
S = load(...)	(Change) Loads file contents to structure	13
Support for cell arrays of strings	The functions intersect, ismember, lower, setdiff, setxor, sort, union, unique, and upper now handle inputs that are cell arrays of strings	8, 9, 14

Functionality	Description	Chapter
Toggle buttons	New *Uicontrol* object	30
Tooltips	UI controls have a `tooltip` string property	30
try/catch blocks	Error handling using `try/catch` blocks	11
Uicontextmenu	New *Uimenu* object	30

Mathematical, FFT, or ODE Function	Description
`cholinc`	Sparse incomplete Cholesky and Cholesky-infinity factorization
`cholupdate`	Rank 1 update to Cholesky factorization
`ifftshift`	Inverse FFT shift
`ode23t`	Solves moderately stiff differential equations
`ode23tb`	Solves stiff differential equations by using crude error tolerances

Graphics or 3-D Visualization Function	Description
`camdolly`	Translates camera position and target
`camlight`	Creates or moves light object in camera coordinate system
`camorbit`	Rotates camera position around camera target
`campan`	Rotates camera target around camera position
`campos`	Sets or gets camera position and position mode
`camproj`	Sets or gets camera projection type
`camroll`	Rotates camera about camera viewing axis
`camtarget`	Sets or gets camera target and camera target mode
`camup`	Sets or gets camera up vector and up vector mode
`camva`	Sets or gets camera view angle and view angle mode
`camzoom`	Zooms camera in or out
`daspect`	Sets or gets data aspect ratio and aspect ratio mode
`lightangle`	Creates or moves light object in spherical coordinates
`pbaspect`	Sets or gets plot box aspect ratio and plot box aspect mode
`xlim`	Sets or gets *x*-axis limits and limit mode
`ylim`	Sets or gets *y*-axis limits and limit mode
`zlim`	Sets or gets *z*-axis limits and limit mode

M-File Function	Description
lastwarn	Returns last warning string
mislocked	True if M-file cannot be cleared
mlock	Prevents M-file clearing
munlock	Permits M-file clearing
persistent	Declares a variable persistent

Other Function	Description
strcmpi	Compares strings, ignoring case
strjust	(Change) Now does right, left, and center justification
strncmpi	Compares first n characters in strings ignoring case

MATLAB 5.3, Release 11 (March 1999)

Functionality	Description	Chapter
Double buffering	*Figure* windows now support double buffering to reduce flicker	30
Execute function when quitting	When MATLAB is quit, MATLAB executes the function finish	4
Integer data types	Support for int8, uint8, int16, uint16, int32, and uint32 added	7
Mass matrix support	All functions in the ODE suite support the use of a mass matrix	24
Optimization parameters	Options for optimization functions fminbnd, fminsearch, and lsqnonneg are set using an options structure variable rather than an options vector	22
Portable Network Graphics	Support for images in PNG format	28
Rectangle object	New Handle Graphics object for drawing rectangle, ovals, and circle	30
Single-precision data	Storage support for single added	7

Matrix Function	Description
blkdiag	Creates block diagonal matrix
complex	Creates complex array from real and imaginary parts
sum	(Change) Can now be used with all integer data types

Graphics or 3-D Visualization Function	Description
coneplot	Velocity vector cone plot in a 3-D vector field
contourslice	Draws contours in volume slices
ezcontour	Easy contour plotter
ezcontourf	Easy filled contour plotter
ezmesh	Easy mesh plotter
ezmeshc	Easy mesh with contour plotter
ezplot	Easy 2-D line plotter
ezplot3	Easy 3-D line plotter
ezpolar	Easy polar plotter
ezsurf	Easy surface plotter
ezsurfc	Easy surface with contour plotter
findfigs	Finds all visible *Figure* windows
hist	(Change) Now calls MEX function histc
histc	Histogram binning given bin edges
isocaps	Computes isosurface end-cap geometry
isonormals	Computes normals at isosurface vertices
isosurface	Extracts isosurface data from volume data
lsqnonneg	(Change) New name and calling sequence for function nnls
reducepatch	Reduces number of patch faces
reducevolume	Reduces number of volume data elements
shrinkfaces	Reduces size of patch faces
smooth3	Smoothes 3-D data
stream2	Computes 2-D streamlines
stream3	Computes 3-D streamlines
streamline	Draws streamlines
subvolume	Extracts subset of volume data set
surf2patch	Converts surface data to patch data

Optimization Function	Description
fminbnd	(Change) New name and calling sequence for function fmin
fminsearch	(Change) New name and calling sequence for function fmins
optimget, optimset	Get, set, and modify optimization options

Other Function	Description
cellfun	Performs common operations on cell arrays
datenum, datestr, datevec	(Change) Now accepts pivotyear argument
evalc	String evaluation with output converted to a string
int8, uint8, int16, uint16, int32, uint32	Conversion to integer data types
pause	(Change) Fractional second argument is accepted
single	Conversion to single-precision data type
str2double	Converts character string to double-precision value
texlabel	Creates the TeX format from a character string

MATLAB 6.0, Release 12 (November 2000)

Functionality	Description	Chapter
Fast Fourier Transform	(Change) MATLAB now uses the MIT FFTW library for computing Fast Fourier Transforms	21
Function handles	New data type for capturing function information for evaluation	12
Java	MATLAB now supports an interface with the Java language	
MATLAB desktop	User interface organized and named the MATLAB desktop, which is implemented in the Java language	3
Matrix math	(Change) MATLAB now uses LAPACK with BLAS for matrix computations	16
Operator precedence	(Change) Logical AND now has higher precedence than logical OR	10
Transparency	(Change) Surfaces, patches, and images now support transparency	27, 28

Logical Function	Description
iskeyword	Generates or tests if argument is a MATLAB keyword
isvarname	True if input string is a valid variable name

M-file Function	Description
check_syntactic_warnings	Runs syntax check on M-files
func2str	Constructs function name string from function handle
functions	Displays information about a function handle

M-file Function	Description
nargoutchk	Validates number of output arguments
rehash	Refreshes function and file system caches
str2func	Constructs function handle from function name string

Sparse Matrix Function	Description
colamd	Computes approximate column minimum degree permutation
lsqr	LSQR implementation of Conjugate Gradients on Normal Equations
minres	Solves system of equations by using Minimal Residual Method
symamd	Symmetric approximate minimum degree permutation
symmlq	Solves system of equations by using symmetric LQ method

Interpolation Function	Description
convhulln	n-D convex hull
delaunay3	3-D Delaunay tessellation
delaunayn	n-D Delaunay tessellation
desearchn	n-D nearest point search
griddata3	3-D data gridding
griddatan	n-D data gridding
interp1	(Change) 'cubic' option now calls pchip
pchip	Piecewise cubic Hermite interpolating polynomial interpolation
voronoin	n-D Voronoi diagram

Optimization or Integration Function	Description
dblquad	(Change) Can now pass extra arguments to integrand function
fzero	(Change) Calling sequence changed
quadl	Numerically computes integral by using Lobatto quadrature; replaces quad8

Differential Equation Function	Description
bvp4c	Solves two-point boundary value problem by collocation
bvpget	Gets BVP option from option structure

bvpinit	Forms initial guess for bvp4c
bvpset	Creates or changes BVP option structure
bvpval	Evaluates solution from bvp4c
ode*	(Change) ODE solvers can now solve problems without using an ODE file
pdepe	Solves partial differential equations in one dimension
pdeval	Evaluates solution computed by pdepe

3-D Visualization Function	Description
coneplot	Creates 3-D coneplot
curl	Computes curl and angular velocity perpendicular to flow
divergence	Computes divergence of vector field
interpstreamspeed	Interpolates streamline vertices from speed
isocolors	Computes colors of isosurface vertices
isosurface	Extracts isosurface
streamparticles	Draws stream particles
streamribbon	Draws stream ribbons
streamslice	Draws streamlines
streamtube	Draws stream tubes
volumebounds	Gets coordinate and color limits for volume data

Graphics Function	Description
alim	Sets or gets axes alpha limits
alpha	Sets or gets transparency properties
alphamap	Specifies figure alphamap

Java Function	Description
import	Adds to the current Java packages import list
isjava	True for Java Object
javaArray	Creates Java array
javaMethod	Invokes Java method
javaObject	Constructs Java object
methodsview	Displays information on all methods implemented by a Java or MATLAB class

Obsolete Function	Description
errortrap	Replaced by try-catch block
flops	Removed. Floating-point operation count no longer relevant
fmin	Replaced by fminbnd
fmins	Replaced by fminsearch
foptions	Replaced by optimget and optimset
interp4, interp5, interp6	Replaced by interp2
isdir	Replaced with exist
isieee	Obsolete. All MATLAB platforms now use IEEE arithmetic
isstr	Replaced by ischar
meshdom	Replaced by meshgrid
nnls	Replaced by lsnonneg
quad8	Replaced by quadl
saxis	Removed. No longer used
setstr	Replaced by char
str2mat	Replaced by char
table1	Replaced by interp1
table2	Replaced by interp2

Other Function	Description
beep	Makes computer sound a beep
continue	Skips rest of For Loop or While Loop
genpath	Generates path string that includes all directories below a specified directory
numel	Number of elements in an array
polyfit, polyval	(Change) Now support centering and scaling of data
sort	(Change) Now works on data types other than double precision
std	(Change) std([]) now returns NaN rather than empty

MATLAB 6.1, Release 12.1 (June 2001)

Functionality	Description	Chapter
Transparent legend	Can now make axes legend box transparent	25

Mathematics or Data Analysis Function	Description
bvpval	(Obsolete) Replaced by deval
convhull	(Change) [K,a] = convhull(x,y) now returns the area a of the convex hull; in addition, it now ignores the third input argument
convhull, delaunay, griddata, voronoi	(Change) Now make use of Qhull
convhulln	(Change) [K,v] = convhulln(x,y) now returns the volume v of the convex hull
delaunay	(Change) Now ignores third input argument
deval	Evaluates ODE solution; also replaces obsolete bvpval function
erfcinv	Inverse complementary error function
interp1	(Change) Now enables data extrapolation
ode*	(Change) Now optionally returns solution structure for use by deval
polyeig	(Change) Can now return only the eigenvalues
ppval	(Change) Now supports ppval(xx,pp) to permit use of ppval with function functions
quad	(Change) Function sampling bug fixed
svd	(Change) Can now return only the first two outputs, U and S

Graphics Function	Description
histc	(Change) Bug fixed
tetramesh	Tetrahedron mesh plot for use with delaunayn
triplot	2-D triangular plot for use with delaunay

External Interface Function	Description
audioplayer	Creates audio object to play audio data on Windows platforms
audiorecorder	Creates audio object to record audio data on Windows platforms
cdfinfo	Gets information about a CDF file
cdfread	Reads CDF file
fitsinfo	Gets information about a FITS file
fitsread	Reads FITS file
hdfinfo	Gets information about a HDF file
hdfread	Reads HDF file

Other Function	Description
datenum, datestr	(Change) Now accept date vector as an input argument
numel	(Change) numel(A,varargin) returns the number of subscripted elements in A(varargin{:})
reshape	(Change) reshape(A,...,[],...) now calculates the size required for the empty dimension
sortrows	(Change) Now calls MEX function sortrowsc to maximize speed. With cell array of strings input, now calls MEX function sortcellchar to maximize speed
strfind	Search for occurrence of second string argument in first string argument

MATLAB 6.5, Release 13 (August 2002)

Functionality	Description	Chapter
BLAS for Pentium 4	Specific BLAS provided for the Pentium 4 processor	
Delay differential equations	Support for solving delay differential equations provided	24
Dynamic structure field names	Need for getfield and setfield eliminated; var.(fstr), where fstr is a character string, addresses the field identified by fstr in the variable var	8
Empty comparisons	[]==[] and []==scalar now return empty ([]) result to coincide with other operators	10
Formatted error and warning strings	Functions error and warning now accept data formatting inputs similar to sprintf	12
Integer array subscripts	Array subscripts must be real positive integer values or logical values (e.g., x(1.3) reports an error)	5
JIT-acceleration	For Loops with specific properties now execute at maximum speed, thereby avoiding the need to vectorize code under some circumstances	11
Logical class	Logical arrays are now a separate MATLAB class that use one byte of storage per array element	10
Logical operators for short-circuiting	New operators && and || short circuit comparisons when an early exit is possible	10

Logical True and False	MATLAB now supports direct creation of arrays containing logical True and logical False by using the functions `true` and `false`	10
Maximum name length for variables and functions	Variable names and function names can now have up to 63 characters	2
Message identifiers	Error and warning messages can now contain identifier tags to make the message uniquely identifiable	12
Platform support	First release to support Macintosh PowerPC (PPC) platform	
Regular expressions	MATLAB now supports regular expressions	9
Relational operators and 64 bit-integer arrays	(MATLAB 6.5.1) All relational operators now support `int64` and `uint64` data types	7
Scheduled execution	New timer object permits scheduled execution of MATLAB code	4
Sparse class	Sparse is now an attribute of the underlying variable class, as opposed to a distinct or separate variable class. In MATLAB 6.5, logical and double data types can be sparse	16
Text object properties	Text objects now have properties for their background box	30
UMFPACK support	When appropriate, sparse matrix solutions now use UMFPACK library functions	16
Warning control	Individual warnings can be suppressed	12

Data Analysis or Mathematical Function	Description
`corrcoef`	(Change) Three new syntaxes
`dde23`	Solves delay differential equations with constant delays
`ddeget`	Gets properties from DDE options structure
`ddeset`	Creates or modifies properties in DDE options structure
`deval`	(Change) Now accepts output from `dde23`
`lu`	(Change) Now uses UMFPACK for sparse matrices
`psi`	Evaluates Digamma function
`qrdelete, qrinsert`	(Change) Can now insert or delete rows as well as columns
`triplequad`	Evaluates triple integral

Logical Function	Description
false	Creates array of logical False
isequal	(Change) When used to compare structures, input argument field creation order no longer has an impact on equality
isequalwithequalnans	True if arrays are equal with NaNs considered equal
ismember	(Change) [tf, idx] = ismember(...) now returns indices idx of located members; in this case, MEX function ismemc2 is called to maximize speed
issorted	True if array is sorted
true	Creates array of logical True

External Interface Function	Description
audiodevinfo	Gets information about installed audio devices on Windows platforms
cdfepoch	Converts MATLAB date number or string to CDF format
cdfwrite	Writes data to CDF file
copyfile	(Change) Now also copies directories
fileattrib	Sets or gets file attributes
imformats	Eases the task of adding read and write support for new file formats
movefile	(Change) Now also renames a file or directory
multibandread	Supports reading data from raw files
multibandwrite	Supports writing data to raw files
namelengthmax	Returns maximum variable and function name lengths
pcode	(Change) Internal format of P-files change
perl	Calls Perl script using platform executable
rmdir	Removes directory and, optionally, contents as well
sendmail	Sends e-mail
urlread, urlwrite	Reads and writes content using URL
winopen	On Windows platforms, opens a file in its appropriate application
xmlread, xmlwrite	Reads and writes XML document
xslt	Transforms XML document using XSLT engine
zip, unzip	Compresses and uncompresses files and directories

Other Function	Description
cell2mat	Combines cell array of matrices into one matrix; previously part of the *Neural Networks Toolbox*
colormapeditor	Interactive colormap editor
gallery	(Change) New test matrices available
int64, uint64	Creates signed or unsigned 64-bit integer array
lasterror	Returns last error message and related information
mat2cell	Breaks matrix up into a cell array of matrices; previously part of the *Neural Networks Toolbox*
orderfields	Order fields of a structure
profview	Produces graphical profile report
regexp	Matches regular expression
regexpi	Matches regular expression, ignoring case
regexprep	Replaces string, using regular expression
rethrow	Reissues error
timer	Creates and controls timer objects to schedule execution of MATLAB code
ver	(Change) Now returns more detailed information, and hostid information is no longer provided

MATLAB 7.0, Release 14 (June 2004)

Functionality	Description	Chapter
Annotation layer	*Figure* windows now have an annotation layer where annotations appear. Annotations include rectangles, ellipses, arrows, double arrows, text arrows, text boxes, lines, colorbars, and legends	30
Anonymous functions	Single line function specification as a replacement for in-line functions	12
Block comments	Now supports block comments using %{ and %} syntax	4
Case sensitivity	Function and directory names are now case sensitive; in prior versions, they were case sensitive only on UNIX platforms	4, 12, 13
Code cells	Editor permits creation, execution, and so on, on sections of code called code cells	4
Code checking	M-file code can be assessed for potential problems and improvement by using the functions mlint or mlintrpt	12

Functionality	Description	Chapter
Desktop	Native MATLAB HTML reader/web browser introduced. Drag and drop supported between Desktop tools. Tooltips and tool preferences enhanced	3
FIG files	(Change) The format of FIG files has changed	29
File FTP support	Access files using FTP commands in MATLAB	13
Freeform date and time	Date and time functions now support user-specified date and time specifications	15
Function precedence	Built-in and M-file functions now share the same calling precedence; in prior versions, built-in functions had higher precedence	12
Generic dynamic linked libraries	MATLAB supports interaction with Generic DLLs on the Windows platform	
Group objects	Graphics objects can now be grouped or linked to each other	30
Java 1.4	MATLAB now uses Java version 1.4	
Nested functions	M-file functions can be nested, allowing shared workspaces among the primary and all nested functions	12
Nondouble arithmetic	Mathematical operations on integer and single-precision data are now supported; many built-in functions support these data types as well	7
Panels and button groups	New user interface container objects	30
Performance	Significant overall performance improvements	
Plot objects	High-level and specialized plotting functions now create plot objects that have properties specific to the type of plot created. In the past, these functions returned handles to the core graphics objects used to create the graphics	30
Regular expressions	Expanded regular expression capabilities	9
Results publishing	M-files and figures can be published to HTML, XML, LaTeX, Word, and PowerPoint	29
TeX support	Text objects now optionally support complete TeX capabilities	25, 30
Toolbars	New user interface object	30
Unicode character storage	MATLAB now encodes character strings in Unicode format	9

Mathematical Function	Description
`acosd, acotd, acscd, asecd, asind, atand`	Inverse trigonometric functions returning angles in degrees
`balance`	(Change) Now returns different outputs and offers balancing without permuting rows and columns
`bvp4c`	(Change) Can now solve multipoint boundary value problems
`convhulln, delaunayn, voronoin`	(Change) Now support user-settable options
`cosd, cotd, cscd, secd, sind, tand`	Trigonometric functions with arguments in degrees
`decic`	Computes consistent initial conditions for `odeo15i`
`deval`	(Change) Now optionally returns derivative at points as well
`eps`	(Change) Now accepts arguments to specify single- or double-precision values and to return `eps` relative to any value, not just 1
`expm1`	Computes $\exp(x)-1$ accurately
`eye, ones, zeros`	(Change) Now accept a final argument specifying numeric data type of result
`fftw`	Tunes or sets options in FFTW library for FFT computations
`fminbnd, fminsearch, fzero`	(Change) Now support calling an output function at each iteration
`funm`	(Change) The optional second output is now an exit flag rather than a (sometimes inaccurate) error estimate
`gammainc`	Can now specify the tail of the incomplete gamma function for nonnegative input
`int8, int16, int32, int64, uint8, uint16, uint32, uint64`	(Change) Now round noninteger inputs rather than truncating
`interp1`	(Change) Now optionally returns a pp-form for evaluation with `ppval`
`interp1, ppval, spline`	(Change) Now support multidimensional arrays for Y
`linsolve`	Solves $Ax = y$, given specific structure of A
`log1p`	Computes $\log(1+x)$ accurately

Mathematical Function	Description
lscov	Can now specify either Cholesky or orthogonal decomposition
ltitr	(Change, previously undocumented) Linear time-invariant time response kernel
mimofr	(Change, previously undocumented) Linear time-invariant frequency response
nthroot	n^{th} real root
ode15i	Ordinary differential equation solver for implicit equations
odextend	Extends solution of ordinary differential equations
ordqz	Reorders QZ factorization
ordschur	Reorders Shur factorization
poleig	Can now return a vector of condition numbers for the eigenvalues
pwch	Piecewise cubic Hermite interpolation
quadv	Vectorized quad function
sort	(Change) Now supports an optional last argument that specifies the sort direction
svd	(Change) Adds support for economy decomposition on matrices having few rows and many columns using svd(A,'econ'). (Documented in MATLAB 7.1.)

Logical Function	Description
iscom	True for COM/ActiveX objects
isevent	True if event of object
isfloat	True for floating-point data
ishghandle	True for Handle Graphics Object handle
isinteger	True for integer data
isinterface	True for COM interface
ispuma	True for computers running Mac OS X 10.1.x
isscalar	True if argument is a scalar
isstrprop	True for string elements matching a variety of specifications
isvector	True if argument is a row or column vector

M-file Function	Description
auditcontents	Audits Contents.m file for a given directory
dbstack	Now supports nested functions
dbstatus	Display breakpoints function now supports anonymous and nested functions as well as a new -completenames argument
dbstop	Now supports nested and anonymous functions
deleteconfirm	Confirms the deletion of a file with a dialog box
deprpt	Scans a file or directory for dependencies
diff2asv	Compares file to autosaved version, if one exists
diffrpt	Visual directory browser
dofixrpt	Scans a file or directory for all TODO, FIXME, or NOTE messages
function	(Change) Function definition line no longer requires commas separating output values
helprpt	Scans a file or directory for help
makecontentsfile	Makes a new Contents.m file
makemcode	Makes M-file for regenerating object and its children
mlint, mlintrpt	Examine M-file or a directory of M-files for potential problems and make suggestions for possible improvements
path2rc	(Obsolete) Replaced by savepath
profile	No longer supports the -detail flag's builtin option
profreport	(Obsolete) Replaced by profsave
publish	Runs a script M-file and saves the results
recycle	Determines if deleted files go to Recycle Bin
restoredefaultpath	Restores default MATLAB path
savepath	Saves current MATLAB path; replaces path2rc function

Obsolete Function	Description
colmmd	(Removed) Use colamd instead
fmin, fmins, icubic, interp4, interp5, interp6, meshdom, nnls, saxis	(Removed) No longer supported
quad8	(Removed) Use quadl instead
symmmd	(Removed) Use symamd instead
terminal	(Removed) No longer supported

Graphics Function	Description
addsubplot	Adds subplot to figure in given location
ancestor	Gets ancestor of graphics object
annotation	Adds *annotation* object
axes	(Change) `ActivePositionProperty`, `OuterPosition`, `TightInset` properties added
axescheck	Processes leading *axes* object from input list
axis	(Change) Now accepts *axes* handle as first argument
commandhistory	Opens *Command History* window or selects it
commandwindow	Opens *Command* window or selects it
datacursormode	Interactively creates data cursors on plot
exportsetupdlg	Shows figure export style dialog
figure	(Change) `DockControls` property added, `KeyPressFcn` property modified
figureheaderdlg	Shows figure header dialog
figurepalette	Shows or hides the palette for a figure
getpixelposition	Gets position of object in pixels
hasbehavior	Sets or gets behaviors of Handle Graphics objects
hgexport	Exports a figure
hggroup	Creates a Handle Graphics group object
hgtransform	Creates graphics transformation object
hold	(Change) Now supports `all` option, which holds the plot so that subsequent plots do not reset the color and linestyle order
linkaxes	Synchronizes limits of specified *axes* objects
linkprop	Maintains same value for corresponding Handle Graphics properties
makehgtransform	Creates graphical transformation matrix
pan	Interactively pans the plot view
plotbrowser	Shows or hides the plot browser for a figure
plottools	Shows or hides the plot-editing tools for a figure
printdlg	Prints dialog box
printpreview	Displays preview of figure to be printed
propertyeditor	Shows or hides the property editor for a figure
refreshdata	Refreshes data in plot
setpixelposition	Sets position of object in pixels

showplottool	Shows or hides one of the plot-editing components for a figure
title	(Change) Now accepts *axes* handle as first argument
uibuttongroup	Creates *buttongroup* object
uicontainer	Creates *container* object
uicontrol	(Change) uicontrol(h) now transfers focus to the *uicontrol* having handle h. Multiline 'edit' style *uicontrol* objects now have a vertical scroll bar. *uicontrol* objects now have a 'KeyPressFcn' callback
uigetfile	(Change) Now permits selection of multiple files
uipanel	Creates *uipanel* container object
uipushtool	Creates pushbutton in *uitoolbar* object
uitable	Creates a *uitable* object
uitoggletool	Creates togglebutton in *uitoolbar* object
uitoolbar	Creates *uitoolbar* object
uitree	Creates *uitree* object
uitreenode	Creates a *node* object in a *uitree* component
uiwait	(Change) uiwait(handle,t) now times out after time t has elapsed
xlabel, ylabel, zlabel	(Change) Now accept *axes* handle as first argument

External Interface Function	Description
calllib, libfunctions, libfunctionsview, libisloaded, libpointer, libstruct, loadlibrary unloadlibrary	Generic DLL Interface functions
callsoapservice	Sends a SOAP message off to an endpoint
eventlisteners	Lists all events that are registered
ftp	Creates *ftp* object
instrfindall	Finds all serial port objects with specified property values
javaaddpath	Adds directories to the dynamic Java path
javaclasspath	Gets and sets Java path
javacomponent	Creates a Java AWT Component and puts in a figure
javarmpath	Removes directory from dynamic Java path

External Interface Function	Description
registerevent	Registers events for a specified control at runtime
unregisterallevents	Unregisters all events for a specified control at runtime
unregisterevent	Unregisters events for a specified control at runtime

Other Function	Description
accumarray	Constructs array with accumulation; that is, if any element is specified more than once, later elements add to the current value rather than overwrite it
addtodate	Modifies a particular field of a date number
audiorecorder, audioplayer	(Change) Now supported on both UNIX and Windows platforms
bin2dec	Now ignores any spaces in the input string
cast	Casts a variable to a different data type
datatipinfo	Produces short description of a variable
datenum, datevec, datestr	Can now specify output formats and local date options
deal	No longer needed in many cases, for example, [a,b,c,d] = C{:} is equivalent to [a,b,c,d] = deal(C{:})
dlmwrite	New input arguments and output options
docsearch	Searches HTML documentation in the Help browser
find	(Change) Now supports optional arguments specifying an upper limit on the number of indices returned, and whether the search begins at the start or end of the array
fixquote	Double up single quotes in a string
genvarname	Generates variable name from candidate name
hd5info, hd5read, hdf5write	HDF5 file information, read, and write
hex2num, num2hex	Converts number to and from IEEE hexadecimal format
imread	Added ability to read a specified portion of a TIFF image
inmem	Now reports path information
intmax, intmin	Maximum and minimum integer values given integer data type
intwarning	Controls state of the integer data type warnings
mmcompinfo	Multimedia compressor information
mmfileinfo	Gets information about a multimedia file (PC only)

`recycle`	Sets option to move deleted files to a recycle folder
`save`	(Change) Now supports compressing MAT-files
`strfind, strtok`	(Change) Now support cell array of strings as input
`strtrim`	Removes leading and trailing white space from a string
`textscan`	Reads text file into a cell array; has more features than `textread`
`timerfindall`	Finds all timer objects with specified property values
`weekday`	New output options
`xlsfinfo`	Output format change
`xlsread`	Date import enhanced
`xlswrite`	Writes Matrix to Excel spreadsheet

MATLAB 7.0.1, Release 14 Service Pack 1 (September 2004)

Functionality	Description	Chapter
Character set conversion	The `native2unicode` and `unicode2native` functions convert between Unicode and a native character set	9
Desktop tools	Enhancements for desktop tools including additional keyboard shortcuts	3
FIG-files	FIG file format change	28
Library path	Now uses the `librarypath.txt` file to locate native Java method libraries. PATH and LD_LIBRARY_PATH environment variables are no longer used	4
Math libraries	Intel/AMD BLAS libraries used; FDLIBM is now version 5.3	
Persistent variables	Multiple declarations of persistent variables are no longer supported	12
Single-precision data	More functions now accept single-precision data inputs in addition to the usual double-precision inputs	7
Web services	Expanded support for web services such as SOAP and WDSL. Integrated MATLAB web browser only used for files smaller than 1.5MB; system default browser used for larger files	

Mathematical Function	Description
`ordeig`	Returns the vector of eigenvalues of a quasitriangular matrix or matrix pair

M-file Function	Description
dbstatus	(Change) Display breakpoints function now supports anonymous and nested functions as well as a new -completenames argument
depfun	(Change) Now supports a number of new options
nargin, nargout	(Change) Now accept either a function name or function handle as an input argument

Graphics Function	Description
opengl	Allows switching between hardware and software-based OpenGL® rendering

External Interface Function	Description
fwrite	Now supports saving uint64 and int64 values on all platforms

Other Function	Description
clear	(Change) The clear mex command no longer clears M-functions in addition to MEX-functions
funm	(Change) Second output argument changed from error estimate to result flag
mat2str	(Change) Now converts non double data types
native2unicode	Converts a character string from a native character set to Unicode
regexprep	(Change) Now supports the use of character representations (like \t or \n) in replacement strings
unicode2native	Converts a character string from Unicode to a native character set

MATLAB 7.0.4, Release 14 Service Pack 2 (March 2005)

Functionality	Description	Chapter
BLAS libraries	Vendor BLAS libraries have changed for the Macintosh and 64-bit Linux platforms	
Installation folder	Folder names with spaces in the installation path are now supported on Windows platforms	13
JVM updated	Version 1.5 of the Java Virtual Machine is now installed	
Memory-mapped files	Memory-mapped files are now supported. Memory-mapping speeds up and simplifies access to file contents and permits access to data from MATLAB and other applications	13

Preferences	Various interface preferences have been added or enhanced	3
Publishing results	Cell publishing image file types and extensions have changed	29
Subfunction help	Use `help functionname>subfunctionname` to get help for a subfunction	12

Mathematical Function	**Description**
`max, min`	(Change) Now return an error when input arguments are complex integers

Graphics and GUI Function	**Description**
`imwrite`	(Change) Now supports GIF format

External Interface Function	**Description**
`gzip, gunzip`	Compress or uncompress files to/from gzip format or uncompress from a URL
`tar, untar`	Archive or extract files to/from a tar file or extract from a URL
`unzip`	(Change) Argument can be a file or a URL
`xlsread`	(Change) Now operates on function handles; date formats changed

Other Function	**Description**
`format`	(Change) Two new display formats are introduced: `short eng` and `long eng`
`textscan`	(Change) Now reads data from strings in addition to files

MATLAB 7.1, Release 14 Service Pack 3 (September 2005)

Functionality	**Description**	**Chapter**
Built-in functions	Built-in functions no longer use files with the `.bi` extension. All `.bi` files have been removed	
Demos	Demos now run in the *Command* window; variables are available in the workspace. The `playshow` helper function is replaced by the `echodemo` function	

Functionality	Description	Chapter
Editor/Debugger	Various enhancements including split screen views, highlighting, commenting, and indenting options. Mathworks introduces a new bug reporting system on their website	3
Find Files	More filtering options are now available in the Find Files tool	3
JIT-accelerator	Now supported on Macintosh platforms	11
LAPACK location	The location of the LAPACK libraries have been relocated on Windows platforms	
Notebook	No longer supports Word 97 on the Windows platform	
Plot tools	Now available on the Macintosh platform	3
Preferences	A number of changes were made to preference options including the name of the preferences directory/folder	3, 4

Mathematical Function	Description
accumarray	(Change) Now supports additional data types and classes
hypot	Square root of the sum of squares
mode	Finds most frequent values in a sample. mode is now a reserved word
odeset	(Change) Nonnegativity constraints can be imposed on computed solutions
rand	(Change) Now supports the Mersenne twister algorithm
svd	(Change) Now supports economy decomposition (introduced but undocumented in v7.0)
timeseries, tscollection	New functions for time series analysis. Must be manually enabled on Linux 64-bit platforms

M-file Function	Description
mlint	(Change) New -notok option; hyperlink to Editor/Debugger in output
swapbytes	Swap byte-ordering
typecast	Converts data types without changing the underlying data

Obsolete Function	Description
clruprop	(Obsolete) Use rmappdata instead
ctlpanel	(Obsolete) Use guide instead
extent	(Obsolete) Use get(h, 'extent') instead

`figflag`	(Obsolete) Use `findobj` and/or `figure` instead
`getuprop`	(Obsolete) Use `getappdata` instead
`hthelp`	(Obsolete) Use `web` instead
`layout`	(Obsolete) No replacement
`matq2ws`, `matqdlg`, `matqparse`, `matqueue`, `ws2matq`	(Obsolete) No replacement
`menuedit`	(Obsolete) Use `guide` instead
`menulabel`	(Obsolete) Use `'&'` and the `'Accelerator'` property instead
`playshow`	(Obsolete) Replaced by the `echodemo` helper function
`setuprop`	(Obsolete) Use `setappdata` instead
`wizard`	(Obsolete) No replacement

Date and Time Function	Description
`datenum`, `datestr`, `datevec`	(Change) Now support milliseconds using Free Form Date Format Specifiers
`datestr`	(Change) Seconds field truncates instead of rounding

External Interface Function	Description
`exifread`	Reads EXIF information from JPEG and TIFF image files
`fread`	(Change) Added a precision argument (in MATLAB 7.1 and earlier)
`mexext`	(Change) Now returns `mexw32` on 32-bit Windows platforms

Other Function	Description
`arrayfun`	Applies a function to each element of an array
`cellfun`	(Change) Applies a function to each cell of a cell array
`error`	(Change) Now saves stack information accessible by `lasterror`
`isfield`	(Change) Now supports cell array input
`lasterror`	(Change) Now returns stack information from the last error
`rethrow`	(Change) Now accepts stack information as input
`structfun`	Applies a function to each field of a structure
`who`, `whos`	(Change) Now displays information for nested functions separately

MATLAB 7.2, Release R2006a (March 2006)

Functionality	Description	Chapter
Desktop	The MATLAB desktop has been reorganized	3
Error logs	Error logs are now generated upon a segmentation fault. The user is prompted to e-mail the log to the Mathworks, Inc	
Installation	The default installation directory structure has changed on Windows platforms	
Java Virtual Machine	The JVM software has been updated on 64-bit Linux platforms	
Libraries	CHOLMOD for sparse matrices, BLAS libraries updated, GCC on Linux platforms must be version 3.4 or later	
Preferences	Some preferences have been reorganized and renamed	3
Profiling	A number of profiling enhancements have been made	12
Regular expressions	Expanded support for regular expressions	9
Time series tools	Full support of 64-bit Linux; importing data from text files (.csv, .dat, .txt) now supported	17

Mathematical Function	Description
accumarray	(Change) Now accepts a cell vector as the subs input
condest	(Change) Efficiency improvements
ddesd	New solver for delay differential equations that have general delays
expm	(Change) Efficiency improvements
gallery	(Change) New optional classname input argument: either 'single' or 'double'
idivide	Integer division similar to A./B; fractional quotients are rounded to integers using a specified rounding mode

Graphics and GUI Function	Description
inspect	The Property Inspector has a new interface but no new functionality

External Interface Function	Description
fopen	(Change) Now supports an argument specifying a character encoding scheme

`fread, fwrite`	(Change) Now use character encoding specified by `fopen`; calls to `unicode2native` and `native2unicode` are no longer needed or appropriate
`xlsread, xlsinfo`	Now support Excel files in formats other than XLS

Other Function	Description
`dbstop`	(Change) Now supports stopping just before a non M-file function (such as a built-in function)
`gzip, gunzip, tar, zip`	Partial paths and wildcards (e.g., `'~'` and `'*'`) are accepted in filename arguments
`issorted`	Now works on cell arrays of strings
`mexect`	MEX-files on 64-bit Windows now use the `.mexw64` extension
`mlint`	(Change) The `mlint` functionality has been built into the Editor/Debugger. Some of the message text has changed as well
`profile`	(Change) New –nohistory option added
`publish`	(Change) New `catchError` option available. Italic text now supported
`regexp, regexpi`	(Change) Calling these functions with `'tokenExtents'` and `'once'` options now returns a double array
`regexptranslate`	Returns a regular expression for a literal string containing wildcard or metacharacters
`setenv`	Sets an environment variable in the underlying operating system
`toolboxdir`	Returns the absolute path to the specified toolbox

MATLAB 7.3, Release R2006b (September 2006)

Functionality	Description	Chapter
Desktop tools	Enhancements to Plotting Tools, Editor/Debugger, and Figure window user interface. File Comparisons tool introduced	3
Help	Enhanced searching in the Help browser	
Import Wizard	Can generate M-code to import similar files	
Installation	A new utility has been added to the Help browser to support file extension associations on Windows platforms	
Libraries	New versions of FFTW, AMD, COLAMD, CHOLMOD, and UMFPACK libraries	
MAT-files	Format change permits saving files larger than 2 GB. HDF5-based MAT-file support	13

Functionality	Description	Chapter
Performance	Improved performance on 64-bit systems and accessing cell arrays	8
Printing	Enhancements to the Printing user interface	29
Redirection	When run in −nodesktop mode, MATLAB now conforms to standard Unix redirection: errors are logged to stderr	
Sparse arrays	Changes to internal storage of sparse arrays on 64-bit systems	16
Time series tools	Enhancements to time series tools and objects	17
User interface	The user interface on Linux and Solaris platforms has changed; functionality was added but none removed	3

Mathematical Function	Description
amd	Interface to the amd algorithm; similar to symamd but typically faster
bvpxtend	Generates a guess structure for extending a boundary value problem solution
fftw	(Change) Default planner method is now estimate. New syntax for importing/exporting wisdom databases. Wisdom exported by earlier versions of FFTW (prior to 3.1.1) cannot be imported
ldl	Full ldl factorization and solving for Hermitian matrices
lu	(Change) Additional output helps improve numerical stability of sparse lu factorization
lu, luinc, ldl, chol, spparms, symbfact	Enhancements to improve control, performance, and memory usage, as well as access to upper and lower triangular factors and lower symbolic factor
max, min	(Change) Now use the phase angle in the case of equal magnitude complex input. No longer return warning messages for inputs of different data types

Obsolete Function	Description
axlimdlg	(Removed) No replacement
bessela	(Removed) No replacement
beta	No longer accepts three inputs
betacore	(Obsolete) Generates a warning message but still works
bvpval	(Removed) No replacement
edtext	(Obsolete) Sets the Editing property of the *text* object

eigs	The eigs options structure no longer accepts the opts.cheb and opts.stagtol fields
gamma	No longer accepts two inputs
ge, gt, le, lt	No longer accept complex integer inputs
menubar	(Obsolete) Sets the MenuBar property of the *figure* to none
pagedlg	(Obsolete) Use pagesetupdlg instead
propedit	The -v6 option is no longer supported. The Version 6 Property Editor has been removed
quad8	(Removed) No replacement
save	No longer accepts -compress, -uncompress, -unicode, or -nounicode options. Compression and Unicode are the defaults. Use the -v6 argument to disable compression and Unicode
sort	No longer accepts complex integer inputs
table1, table2	(Removed) No replacement
umtoggle	(Obsolete) Sets the Checked property of the uimenu object

External Interface Function	Description
fprintf, fwrite	(Change) Writing to stdin (fwrite(0,...)) now generates an error
mwIndex	New platform-independent index type declaration
mwSize	New platform-independent size type declaration for values such as array dimensions and number of elements
notebook	(Change) The -setup switch no longer accepts arguments
save	(Change) Now supports -v7.3 option to save MAT-files in uncompressed HDF5 format permitting file sizes greater than 2 GB. Compressed HDF5 format will be the default MAT-file format in a future version

Other Function	Description
dbstop	(Change) The ability to restore breakpoints was added
mlint	(Change) Enhanced preferences and message text changes
mlintrpt	(Change) New option applies preferences from a settings file
parfor, classdef	New reserved words
regexp, regexpi	(Change) New ?%cmd operator to help with debugging regular expressions

Other Function	Description
unique	(Change) New 'first' and 'last' options
whos	(Change) Modifications to the output format

MATLAB 7.4, Release R2007a (March 2007)

Functionality	Description	Chapter
Desktop	The stand-alone editor is depreciated and will be removed. Use the MATLAB Editor/Debugger instead. Enhancements to a number of MATLAB desktop windows and preferences	3
Ghostscript printing	Ghostscript printing software has been upgraded; stand-alone Ghostscript executable is no longer included	29
JVM updated	The Java Virtual Machine supplied with MATLAB on Windows platforms has been updated to version 1.5.0_07	
Libraries	BLAS libraries updated	
M-lint enhanced	M-lint now has the ability to suggest corrections or enhancements to your code in the Editor/Debugger	3, 12
Multithreaded computation support added	Enable multithreaded computation support for multicore processors or multiprocessor systems using preferences. It is disabled by default	
Performance	Improved performance on 64-bit platforms. Faster scalar indexing and assignment of cell array data into variables. More efficient matrix multiplication of sparse matrices	
Platform support	First release to support Intel Macintosh platform	
Publishing	Added publishing options include publishing M-files and adding in-line links with text, graphics, and/or HTML markup	29
Startup directory	The default startup directory on Microsoft Windows platforms has changed from $MATLAB\Work to My Documents\MATLAB, or Documents\MATLAB on the Vista platform.	

Mathematical Function	Description
bsxfun	Applies element-by-element binary operation to two full arrays with singleton expansion enabled
ilu	Performs the sparse incomplete LU factorization
mode	(Change) Returns NaN for an empty array input
rand	(Change) Now uses the Mersenne twister algorithm as default rather than Subtract-with-Borrow

M-File Function	Description
`assert`	Generates an error when a specified condition is violated
`inputParser`	New OOP class available for parsing and validating inputs to functions
`ismac`	Returns `true` on OSX platforms
`verLessThan`	Compares specified toolbox version with currently running version

Obsolete Function	Description
`cshelp`	(Depreciated) No replacement available
`figflag`	(Depreciated) No replacement available
`getstatus`	(Depreciated) No replacement available
`hidegui`	(Obsolete) Set the figure `HandleVisibility` property instead
`ispuma`	(Depreciated) OS X 10.1 is no longer supported on the Mac platform
`menulabel`	(Depreciated) No replacement available
`popupstr`	(Depreciated) No replacement available
`print`	(Change) The `-dln03` option has been removed
`setstatus`	(Depreciated) No replacement available
`uigettoolbar`	(Depreciated) No replacement yet available

External Interface Function	Description
`publish`	(Change) New `maxOutputLines` field added; default value is `Inf`
`save`	(Fix) The `-rexgrep` argument is no longer taken to be a file name if no file name is supplied

Other Function	Description
`builddocsearchdb`	Creates a search database for your own HTML help files for the *Help browser*
`dir`	(Change) Now returns additional numeric date (`datenum`) field
`mat2str`	(Fix) Now returns correct values for character array input
`mexext`	(Change) New MEX-file extensions for Intel Mac (`mexmaci`) and Solaris 64-bit (`mexs64`) platforms
`sprintf, fprintf, warning, error, assert`	(Change) Additional numeric positional arguments available for formatted string format specifiers
`textscan`	(Change) New `CollectOutput` switch to return like values in the same cell array

MATLAB 7.5, Release R2007b (September 2007)

Functionality	Description	Chapter
Array size	The maximum array size has increased from 2^{31} elements to $2^{48}-1$ elements on 64-bit platforms given sufficient memory	5
Desktop enhancements	Desktop user interface enhanced; buttons and tools redesigned, links and targets added. Help browser and Editor/Debugger enhanced	3
Error messages	Text of some error messages changed	
GUIDE	Enhancements including new toolbar editor and icon editors	
Libraries	LAPACK and BLAS libraries upgraded; symbol libraries separated out	
Line endings	Text files supplied with MATLAB on the Windows platform no longer include a carriage return and line feed at the end of each line; Notepad does not recognize line endings	
P-code format	Internal P-code format change. P-code files created in version 7.5 are not recognized in earlier releases	12
Platform support	Last release to support PowerPC (PPC) Macintosh and 32-bit SPARC platforms; 64-bit UltraSPARC support continues	
Publishing	Notebook supports Word of Office 2007 on Windows platforms. Editor/Debugger menu items added for enhanced cell markup	

Mathematical Function	Description
bvp5c	Solves boundary value problems for ODEs, most useful for small error tolerances
dmperm	(Change) New output arguments for the indices of the Dulmage–Mendelsohn coarse decomposition
ldl	(Change) Now supplies factorization and solving for an additional output argument, the scaling matrix, when the input matrix is real, sparse, and symmetric
maxNumCompThreads	Gets/sets the maximum number of complex threads
MException	New OOP class available for exception handling in function M-files
ones, zeros, rand, true, false, eye, ..	(Change) Matrix generating functions no longer accept complex dimension arguments
quadgk	Numerically evaluates the integral, adaptive Gauss–Kronrod quadrature

M-File Function	**Description**
catch	(Change) New optional MException class argument
pcode	(Change) Internal format of P-files change
validateattributes	Checks the validity (e.g., numeric, nonempty) of an input array
validatestring	Checks the validity (e.g., character, nonempty) of text string input

Obsolete Function	**Description**
axlimdlg	(Removed) No replacement
cbedit	(Removed) Use guide instead
clruprop	(Removed) Use rmappdata instead
ctlpanel	(Removed) Use guide instead
edtext	(Removed) Set the editing property of the *text* object
extent	(Removed) Set the extent property of the *text* object
finite	(Depreciated) Use isfinite instead
freeserial	(Obsolete) Use fclose instead
getuprop	(Removed) Use getappdata instead
hthelp	(Removed) Use web instead
layout	(Removed) No replacement
matq2ws	(Removed) No replacement
matqdlg	(Removed) No replacement
matqparse	(Removed) No replacement
matqueue	(Removed) No replacement
menubar	(Removed) Set the menubar property of the *figure* to none
menuedit	(Removed) Use guide instead
pagedlg	(Removed) Use guide instead
setuprop	(Removed) Use guide instead
umtoggle	(Removed) Set the Checked property of the *uimenu* object
wizard	(Removed) Use guide instead
ws2matq	(Removed) No replacement

Graphics Function	Description
annotation	(Change) New FitBoxToText property permits textboxes to automatically resize to fit the enclosed text
area, bar, barh, colorbar, contour, contourf, errorbar, legend, loglog, mesh, plot, plot3, quiver, quiver3, scatter, scatter3, semilogx, semilogy, stairs, stem, stem3, subplot, surf	(Change) The v6 option for high-level plotting functions is obsolete
drawnow	(Change) Can now selectively update UI components

External Interface Function	Description
avifile	(Change) Now generates uncompressed AVI files on Windows platforms unless the Indio5 codec is installed
aviread	(Change) No longer reads compressed AVI files on Windows platforms unless the Indio5 codec is installed
hdfread	(Change) Now generates an error rather than a warning on a failed I/O operation
imread	(Change) Improved support for TIFF files. Improved performance in some instances
mmfileinfo	(Change) Now reads files on the MATLAB path as well as the current directory. New Path field added to the mmfileinfo output struct
mmreader	Video file reader for Windows platforms
movie2avi	(Change) Now generates uncompressed AVI files on Windows platforms unless the Indio5 codec is installed
tempname	(Change) Now generates a longer and more unique filename

Other Function	Description
colon (:)	(Change) Colon operations on characters iterating a For Loop (for x = 'a':'c') now returns character data type rather than double
mlint	(Change) New cyc option to determine McCabe (cyclomatic) complexity of M-files
rexgrep	(Change) New 'split' option to split an input string into sections

MATLAB 7.6, Release R2008a (March 2008)

Functionality	Description	Chapter
Desktop tools	Array Editor renamed and enhanced, Help browser reorganized, stand-alone Editor removed, Editor/Debugger enhancements, Data Brushing and Data Linking tools introduced	3
Desktop UI	New Desktop UI customization, license management, and check for updates enhancements. Default startup directory on all platforms is now *userhome*/Documents/MATLAB	3
JIT support	JIT support extended to statements executed in the *Command w*indow and in cell mode in the Editor/Debugger	11
JVM updated	The JVM is updated to Version 6 on Solaris platforms	
Libraries	BLAS, MKL, SPL, LAPACK libraries upgraded. FFTW may be overridden	
Multithreaded support	Multithreaded support added to elementwise math functions. Multithreading now enabled by default	5
Object-oriented programming	Enhancements to MATLAB object-oriented programming include a new `classdef` keyword, a new `handle` class for reference behavior, support for events and listeners, support for packages of classes and functions, and the availability of metaclasses	31
Publishing	Multiple configurations supported, additional cell nesting options, new code evaluation, and output publishing options	29

Mathematical Function	Description
`funm`	(Change) New algorithms
`ldivide`	(Change) Multithreaded support added
`ldl`	(Change) New algorithms
`log, log2`	(Change) Multithreaded support added
`logm`	(Change) New algorithms
`rdivide`	(Change) Multithreaded support added
`rem`	(Change) Multithreaded support added

M-File Function	Description
`onCleanup`	Specifies task(s) to perform just before exiting the current function

Obsolete Function	Description
bessela	(Obsolete) Use besselj instead
beta	No longer supports three input arguments. Use betainc with three input arguments instead
betacore	(Obsolete) Use betainc instead
colmmd	(Obsolete) Use colamd instead
flops	(Obsolete) No replacement
hidegui	(Depreciated) Set the *figure's* handlevisibility property
meditor	(Removed) Stand-alone editor removed; use the Editor/Debugger instead
quad8	(Obsolete) Use quadl instead
symmmd	(Obsolete) Use symamd instead
table1	(Obsolete) Use interp1 or interp1g instead
table2	(Obsolete) Use interp2 instead

Graphics Function	Description
brush	Turns data brushing on or off and selects a color for brushing graphs
get, set	(Change) Do not use get or set to manage properties of a Java object; this usage will generate an error in future releases
linkdata	Turns data linking on or off for a figure
uigetfile, uiputfile	(Change) Now supports meta directories such as '.', '..', and '/'
uitable	Creates a new graphic table component

External Interface Function	Description
imfinfo	(Change) Now returns Exif data for JPEG or TIFF files. New DigitalCamera and GPSInfo fields
imwrite	(Change) New RowsPerStrip parameter to specify number of image rows per strip written to TIFF files. Default <= 8KB
publish	(Change) New codetoEvaluate option; stopOnError option removed
save	(Change) Data items over 2 GB stored in a MAT-file using the -v7.3 option are now compressed

snapnow	Includes a snapshot of output in a published document
userpath	Views/sets/clears the user directory from the top of the MATLAB search path. Default is "$HOME/Documents/MATLAB" or "My Documents\MATLAB" or "Documents\MATLAB" depending on platform

Other Function	**Description**
clearvars	Clears variables from the workspace; supports exceptions
dbclear, dbstop	(Change) New -completenames option permits clearing or setting breakpoints for M-files not on the search path
edit	(Change) Now accepts file path information to enable editing files not in the current directory
memory	(Change) Displays memory usage and availability information on Windows platforms. Formerly provided help text describing how to free additional memory for MATLAB
mlint	(Change) New -config option to override preference settings

MATLAB 7.7, Release R2008b (October 2008)

Functionality	**Description**	**Chapter**
Desktop	New Desktop default layout. New Function Browser tool and popup function hints. Help browser, Current Directory browser, File and Directory comparisons tool enhancements	3
Editor/ Debugger	Now supports templates for new function and class M-files. M-lint and code folding enhancements	3
JVM updated	The Java Virtual Machine has been updated to JVM version 6 update 4.	
Libraries	Intel Math Kernel Libraries updated	
Map object introduced	New *map* object implements associative arrays with key lookup	31
Publishing	In-line LaTeX math symbols supported within text	29
Startup changes	Mac version now uses standard Mac App bundle packaging. Windows version has a new -shield startup option for memory allocation and protection. The -memmgr and -check_malloc startup options are now depreciated	

Mathematical Function	Description
bvp5c	(Change) Now supports multipoint boundary value problems
lsqnonneg	(Change) More efficient; now accepts sparse matrices as input and maintains sparsity throughout
randi	(New) Returns random integers from a uniform discrete distribution
randn	(Change) New longer period random number algorithm is the default

M-File Function	Description
pause	(Change) New query argument to return pause state

Obsolete Function	Description
betacore	(Removed) Use betainc instead
colmmd	(Removed) Use colamd instead
flops	(Removed) No replacement—no longer practical to count flops
print, saveas	(Depreciated) Certain printer and graphics format option strings are depreciated and will be removed in future versions
symmmd	(Removed) Use symamd instead

Graphics Function	Description
hist	(Change) Changes to the data tips in the histogram display

External Interface Function	Description
dir	(Change) Empty matrices now returned when appropriate
fopen	(Change) No longer supports 'vaxd', 'vaxg', or 'cray' ('d', 'g', or 'c') formats
publish	(Change) New figureSnapMethod option to specify figure details be included in the snapshot. Now supports in-line LaTeX math symbols

Other Function	Description
addtodate	(Change) Now supports hours, minutes, seconds, and milliseconds in addition to years, months, and days
getReport	(Change) Several new options available

tic, toc	(Change) Now support multiple consecutive timings
what	(Change) New package information available

MATLAB 7.8, Release R2009a (March 2009)

Functionality	**Description**	**Chapter**
Desktop tools	Enhancements to Current Directory browser, File and Directory Comparisons tool, extended M-lint messages in the Editor/Debugger. Integrated text editor and block indenting options removed from Editor/Debugger	3
Libraries	New computational geometry tools and library (CGAL). ACML and HDF5 libraries upgraded. LAPACK and BLAS support 64-bit integers for matrix dimensions	
Multithreading	New -singleCompThread startup flag limits MATLAB to a single computational thread. Default is multithreading if supported. More functions support multithreading	
Profiling	Profile summary report now includes information on profiling overhead when available	12
Publishing	New options to snapshot the figure or the entire figure window. Dynamic links to files on the MATLAB path are now supported	29
Serial port	Now supported on all platforms	
Timer object format	The format in which MATLAB saves timer objects has changed	15

Mathematical Function	**Description**
betaincinv	Implements the inverse incomplete beta function
bicgstabl	Implements the stabilized biconjugate gradients method for solving systems of linear equations
conv2, convn	(Change) Now accept the shape parameter as input
fft, fft2, fftn, ifft, ifft2, ifftn	(Change) These functions are now multithreaded
gammaincinv	Implements the inverse incomplete gamma function
prod, sum, max, min	(Change) These functions are now multithreaded
quad2d	Provides additional quadrature functionality for nonrectangular areas of integration
tfqmr	Implements a transpose-free quasi-minimal-residual method for solving systems of linear equations

M-File Function	Description
isempty	(Change) Now supports `containers.Map` objects
str2func	(Change) Can now convert an anonymous function definition to a function handle
TriRep, DelaunayTri, TriScatteredInterp	New OOP classes provide enhanced computational geometry tools
validateattributes	(Change) Now supports checking size and range of input values

Obsolete Function	Description
finite	(Obsolete) Function removed. Use `isfinite` instead
nextpow2	Due to change to element-by-element calculation in a future release

Graphics Function	Description
cmpermute	Rearranges colors in colormap
cmunique	Eliminates unneeded colors in colormap of indexed image
dither	Converts image using dithering
immapprox	Approximates indexed image by one with fewer colors
rgb2ind	Converts RGB image to indexed image

External Interface Function	Description
mmreader	(Change) Now supported on Linux platforms
serial	(Change) Now supported on all platforms
web	(Change) Use `-browser` parameter to specify which system browser to use (default is Firefox.) Any `docopt.m` browser specification is now ignored
xlsread, xlswrite, importdata	(Change) Now support XLS, XLSX, XLSB, and XLSM formats on Windows platforms with an appropriate version of Excel installed

Other Function	Description
docsearch	(Change) Now accepts multiple arguments including wildcards without function syntax
maxNumCompThreads	(Change) Ability to adjust the maximum number of computational threads has been removed; multithreading support is either on or off

| publish | (Change) New `figureSnapMethod` options are `entireGUIWindow` (default) and `entireFigureWindow` which include the title bar and all other window decorations in the snapshot |

MATLAB 7.9, Release R2009b (September 2009)

Functionality	Description	Chapter
Desktop tools	Enhancements to the Help browser, Plotting tool, Current Directory browser, Variable Editor, File and Folder Comparison tool, and the Editor/Debugger. New File Exchange desktop tool	3
Multithreading	Support for many more functions including `bsxfun`, `sort`, `filter`, gamma and `gammaln`, `mldivide` and `qr` for sparse matrices, and `erf`	
Performance	Performance improvements for basic math, binary and relational operations, exponential functions, and indexing for sparse matrices. Significant performance improvements for `conv2`	
Platform support	Last release to support UltraSPARC 64-bit platform	
Preferences	Changes to keyboard bindings and shortcut preferences and desktop tools font preferences	
Publishing	Now supports publishing to PDF files	29
Startup options	The `-memmgr` and `-check_malloc` startup options are obsolete and are ignored	

Mathematical Function	Description
`conv2`	(Change) Significant performance improvements
`convhull`, `delaunay`, `delaunay3`, `griddata`, `griddata3`, `voronoi`	(Change) No longer support the QHULL or QHULL options arguments
`fft`, `fft2`, `fftn`, `ifft`, `ifft2`, `ifftn`	(Change) Now support larger input arrays (size in one dimension greater than $2^{31}-1$) on 64-bit platforms
`mldivide`	(Change) Now supports complex regular sparse input matrices. Performance improvements for sparse rectangular matrix input
`qr`	(Change) Now supports complex sparse input matrices with a third output argument containing a fill-reducing permutation for sparse matrix input

M-File Function	Description
~	(Change) New usage: to specify unused input arguments or unused outputs

Obsolete Function	Description
delaunay3	(Depreciated) Use DelaunayTri method instead
dsearch	(Depreciated) Use DelaunayTri/nearestNeighbor methods instead
erfcore	(Obsolete) Use erf, erfc, erfcx, erfinv, or erfcinv instead
griddata3	(Depreciated) Use TriScatteredInterp method instead
lasterr, lasterror	(Depreciated) Use MException objects instead
maxNumCompThreads	(Depreciated) To be discontinued with no replacement planned
rethrow	(Change) The rethrow(lasterror) usage is depreciated. Use the rethrow(MException) or rethrow(MException.last) instead
tsearch	(Depreciated) Use DelaunayTri/pointLocation methods instead

Graphics Function	Description
print	(Change) The -adobecset option and -dill device are depreciated and will be removed
view	(Change) No longer supports 4-by-4 transformation matrices as inputs. Use view([az el]) instead

External Interface Function	Description
imread, imwrite	(Change) Now support rewriting portions of a TIFF file rather than replacing the entire file. New Tiff object allows access to functions in the LibTIFF library
mmreader	(Change) Now supports Motion JPEG 2000 (.mj2) files on all platforms except Solaris

Other Function	Description
gallery	(Change) New integerdata, normaldata, and uniformdata options for gallery suite of test matrices
isequal	(Change) Now ignores class: isequal(double(3), single(3)); now returns True
numel	(Fix) Now returns consistent results for built-in classes and subclasses

MATLAB 7.9.1, Release R2009bSP1 (Fall 2009)

Functionality	Description	Chapter
Bug fixes	A number of bug fixes were incorporated in this release	
MCR version change	The MATLAB Compiler Runtime (MCR) version number changed	

MATLAB 7.10, Release R2010a (March 2010)

Functionality	Description	Chapter
Desktop tools	Ability to zip and unzip files and folders in the Current Folder browser. Enhancements for the Editor/Debugger, Plot Selector, Help Browser, and File and Folder Comparison tool	3, 13
Keyboard shortcuts	Keyboard shortcuts are now organized into sets for easier set management	
Libraries and OOP	CDFlib package added to enable low-level access to CDF files. CDF, HDF5, HDF4, HDF-EOS2, and PNG file format libraries upgraded. Tiff class enhancements	13
Multithreading	Additional math functions support multithreading	5
Performance	Multicore support and performance improvements for over 50 functions including sparse matrix indexing and functions	
Platform support	Last release to support 32-bit Intel Macintosh; 64-bit support continues	
Time series objects	Now accept duplicate times in adjacent positions; time vectors must be nondecreasing	17

Mathematical Function	Description
`bvp4c, bvp5c`	(Change) Significant performance improvements for sparse problems
`delaunay, convhull, griddata, voronoi, delaunay3, griddata3`	(Change) These computational geometry functions no longer use the `options` arguments.
`fft, conv2, int8, int16, int32, int64, uint8, uint16, uint32, uint64`	(Change) Multithreading extended to `fft` for long vectors, the two-input form of `conv2`, and integer conversion and arithmetic

Mathematical Function	Description
lsqnonneg	(Change) No longer uses the optional input x0 as a starting point
nextpow2	(Change) Now returns a vector the same size as the input
qr	(Change) The upper triangular matrix produced by the factorization routine in qr always contains real, nonnegative diagonal elements
sortrows, mrdivide, convn, hisrtc	(Change) Significant performance improvements
spones, spfun, sprand, sprandn, sprandsym, cat, horzcat, vertcat	(Change) Better error checking for sparse matrix functions

M-File Function	Description
containers.Map	(Change) New syntax available to construct Map object containers

Obsolete Function	Description
aviinfo	To be removed. Use mmreader and get instead
aviread	To be removed. Use mmreader and read instead
delaunay3	Depreciated. Use DelaunayTri instead
docopt	Removed. Preferred web browser now set in Preferences
dsearch	Depreciated. Use DelaunayTri/nearestNeighbor instead
erfcore	Removed. Replace with erf, erfc, erfcx, erfinv, or erfcinv.
exifread	To be removed. Use imfinfo instead
fileparts	(Change) Fourth output (versn) is always empty
griddata3	Depreciated. Use TriScatteredInterp instead
intwarning	To be removed. Use warning instead
intwarning	Depreciated and will be removed in a future release
isstr	To be removed. Use ischar instead
mmreader. isPlatformSupported	(Change) Always returns True. Use mmreader.getFileFormats instead
setstr	To be removed. Use char instead
str2mat	To be removed. Use char instead
strread	To be removed. Use textscan instead

strvcat	To be removed. Use `char` instead
textread	To be removed. Use `textscan` instead
tsearch	Depreciated. Use `DelaunayTri`/`pointLocation` instead
wk1finfo	To be removed. Use `xlsinfo` instead
wk1read	To be removed. Use `xlsread` instead
wk1write	To be removed. Use `xlswrite` instead

Graphics Function	**Description**
`mmreader.getFileFormats`	Returns a platform-specific list of video file formats `mmreader` supports

External Interface Function	**Description**
`imwrite`	Added support for JPEG 2000 format files

Other Function	**Description**
`migrateistimefirst`	Time series object `isTimeFirst` property behavior is changing in a future release. `migrateistimefirst` is a temporary function to help migrate this property to conform to the newer usage; it will be removed in future releases
`unzip`	Now preserves original write attribute of all extracted files

MATLAB 7.11, Release R2010b (September 2010)

Functionality	**Description**	**Chapter**
Desktop tools	Ability to manage ZIP files as folders in the Current Folder browser and File and Folder Comparison tool. Additional enhancements for the Editor/Debugger, Plot Selector, and File and Folder Comparison tool	3, 13
Graphic objects	Support for Motion JPEG and uncompressed AVI files greater than 2 GB. Support for NetCDF and HDF5 data storage	28
Libraries and OOP	Support for enumerated data types with sets of named values. Enumeration class templates supplied. Arrays of time series objects are now supported. A number of new HDF4, HDF5, HDF-EOS, and NetCDF functions support high- and low-level access to their respective library routines	

Functionality	Description	Chapter
Mathematics	Support for 64-bit integer arithmetic in core functions	5
Performance	Significant performance improvements for the three-output form of svd, sparse column assignment, and all degree-based trigonometric functions	
Toolboxes	*Spline Toolbox* merged into *Curve Fitting Toolbox*. RF Blockset renamed as SimRF	

Mathematical Function	Description
arrayfun	(Change) Now accepts arrays of objects as input
convhull	(Change) Now supports 3-D input. The simplify option allows removal of noncontributing vertices
delaunay	(Change) Now supports 3-D input in either multiple vector or multicolumn matrix format
DelaunayTri/ NearestNeighbor	(Change) The two-output form returns the corresponding Eculidean distances between the query points and their nearest neighbors
plus (+), minus (-), uminus (-), times (.*), rdivide (./), rem, mod, ldivide (.\), bitcmp, power (.^), sign, any, all, sum, diff, colon (:), accumarray, bsxfun	(Change) Now support int64 and uint64 classes natively
sind, cosd, tand, cotd, secd, cscd, asind, acosd, acotd, asecd, acscd	(Change) Significant performance improvements
svd	(Change) Significant performance improvement for the three-output form of svd

Logical Function	Description
isa, islogical	(Change) Now return consistent results for objects with base class of logical
isequal, isequalwithequalnans	(Fix) Now return correct values when inputs are arrays of objects containing NaN values
isrow, iscolumn, ismatrix	Provide basic information about inputs

M-File Function	Description
`hdf, hdf5, netcdf`	A number of new functions support MATLAB interface to the HDF4, HDF5, and NetCDF libraries. Use `help hdf`, `help dhf5`, and `help netcdf` for overviews of the available functionality

Obsolete Function	Description
`bessel`	(Obsolete) Use `besselj` instead
`erfcore`	(Removed) Use `erf`, `erfc`, `erfcx`, `erfinv`, or `erfcinv` instead
`fileparts`	(Change) Returns an empty fourth argument (file versions not supported)
`intwarning`	(Removed) No replacement
`mmreader`	To be removed. Use `VideoReader` instead
`print, printdmfile`	(Change) The `-dmfile` option to `print` and the `printdmfile` function will be removed in a future release
`saveas`	(Change) The `mmat` format option is depreciated and will be removed in a future release
`wavplay`	To be removed. Use `audioplayer` and `play` instead
`wavrecord`	To be removed. Use `audioplayer` and `record` instead
`wk1finfo`	To be removed. Use `xlsinfo` instead
`wk1read`	To be removed. Use `xlsread` instead
`wk1write`	To be removed. Use `xlswrite` instead

Graphics Function	Description
`movie`	(Change) No longer a built-in function; the syntax is unchanged
`uitab, uitabgroup`	(Change) Changes are being made to these undocumented functions
`VideoReader`	Replacement for `mmreader`; both return identical VideoReader objects
`VideoWriter`	Improvement over `avifile` supports files larger than 2 GB

External Interface Function	Description
`imread, imwrite`	(Change) Now support n-channel J2C JPEG 2000 files

MATLAB 7.11.1, Release R2010bSP1 (March 2011)

Functionality	Description	Chapter
Bug fixes	There were a couple of minor bug fixes in MATLAB; this release contained mostly bug fixes for Simulink	

MATLAB 7.12, Release R2011a (April 2011)

Functionality	Description	Chapter
Desktop	Menus now conform to the Mac standard (top of screen vs. top of window) on the Macintosh platform. Plot Catalog tool enhanced	3
File management	Editor/Debugger reflects changes in directory or file names. MAT-file comparisons support viewing and manipulating variables. Folder comparisons support filters. Text file comparisons support difference-only viewing	3, 13
Help browser	Support requests may be submitted directly from the Help browser. Mathworks website help and documentation enhancements and URL changes	
Libraries	HDF4 and HDF EOS functions grouped into three packages. Two new functions added to CDFLIB package	
Performance	Performance enhancements to matrix transpose, element-wise single-precision functions, sparse matrix indexed assignment, many linear algebra functions, and convolution for large matrices and long vectors using conv and conv2	
Preferences	Changes to default preference for deleting autosaved files. Color preference changes for nonlocal variable display	
Publishing	New option to include in-line LaTeX equations in comment lines when publishing code	29
Toolboxes	Some toolboxes reorganized and renamed	

Mathematical Function	Description
gammainc	(Change) New 'scaledlower' or 'scaledupper' argument returns scaled versions of the incomplete gamma function
ichol	New incomplete Cholesky factorization function; preferred replacement for the cholinc function

qr	Reverts to pre-v7.10 behavior; the diagonal of R may contain complex or negative elements
rng	New function to control the random number generator. Use rng rather than the `'seed'`, `'state'`, or `'twister'` inputs to the rand or randn functions

Obsolete Function	**Description**
bessel	(Removed) Use besselj instead
cholinc	(Obsolete) Use ichol instead
hdf5info	(Depreciated) Use h5info instead
hdf5read	(Depreciated) Use h5read instead
hdf5write	(Depreciated) Use h5write instead
intwarning	(Removed) No replacement
luinc	(Obsolete) Use ilu instead
pagesetupdlg	(Depreciated) Use printpreview instead
RandStream object	(Changes) RandnAlg property replaced by NormalTransform property, setDefaultStream method replaced by setGlobalStream method, getDefaultStream method replaced by getGlobalStream method

External Interface Function	**Description**
audioplayer, audiorecorder	(Change) Now support device selection on all platforms
fitsread	(Change) New options to support data subsetting
h5create	(New) Creates HDF5 data set
h5disp	(New) Displays the contents of an HDF5 file
h5info	(New) Returns information about an HDF5 file
h5read	(New) Reads data from an HDF5 data set
h5readatt	(New) Reads an attribute from an HDF5 group or data set
h5write	(New) Writes to an HDF5 data set

External Interface Function	Description
h5writeatt	(New) Writes an attribute to an HDF5 group or data set
nccreate	(New) Creates variable in a NetCDF file
ncdisp	(New) Displays contents of a NetCDF file
ncinfo	(New) Returns information about a NetCDF file
ncread	(New) Reads data and attributes of a NetCDF file
ncreadatt	(New) Reads an attribute from a NetCDF file
ncwrite	(New) Writes data to a NetCDF file
ncwriteatt	(New) Writes an attribute to a NetCDF file
ncwriteschema	(New) Adds NetCDF schema definitions to a NetCDF file
VideoWriter	(Change) Now supports Motion JPEG 2000 files

Appendix B

MATLAB Function Information

Each MATLAB function cited here was (1) introduced in MATLAB version 5.0 (Release 8) or later, (2) had significant changes in one or more of these releases, or (3) was depreciated, made obsolete, or was removed. Each entry identifies the function name, a description of the function introduced, changed, or obsoleted, as well as the MATLAB version in which the introduction or change occurred. Changed functions have (Change) in their description. Obsolete functions are also identified.

This is not a comprehensive list of functions in every version of MATLAB. The current version alone contains over 1600 functions and methods. A full list of functions available in the current release organized by category and alphabetically, along with hyperlinks to the help text for each function, may be found in the documentation distributed with MATLAB.

Function	Description	Version
~	(Change) New usage: to specify unused input arguments or unused outputs	7.9
accumarray	Constructs array with accumulation; that is, if any element is specified more than once, later elements add to the current value rather than overwrite it	7.0
accumarray	(Change) Now supports additional data types and classes	7.1
accumarray	(Change) Now accepts a cell vector as the subs input	7.2
accumarray	(Change) Now supports int64 and uint64 classes natively	7.11

Function	Description	Release
acosd	Inverse trigonometric functions returning angles in degrees	7.0
acosd	(Change) Significant performance improvements	7.11
acotd	Inverse trigonometric functions returning angles in degrees	7.0
acotd	(Change) Significant performance improvements	7.11
acscd	Inverse trigonometric functions returning angles in degrees	7.0
acscd	(Change) Significant performance improvements	7.11
addpath	Appends directory(s) to the MATLAB search path	5.0
addsubplot	Adds subplot to figure in given location (undocumented)	7.0
addtodate	Modifies a particular field of a date number	7.0
addtodate	(Change) Now supports hours, minutes, seconds, and milliseconds in addition to years, months, and days	7.7
airy	Airy functions	5.0
alim	Sets or gets axes alpha limits	6.0
all	(Change) Now supports int64 and uint64 classes natively	7.11
alpha	Sets or gets transparency properties	6.0
alphamap	Specifies figure alphamap	6.0
amd	Interface to the amd algorithm; similar to symamd but typically faster	7.3
ancestor	Gets ancestor of graphics object	7.0
annotation	Adds *annotation* object	7.0
annotation	(Change) New FitBoxToText property permits textboxes to automatically resize to fit the enclosed text	7.5
any	(Change) Now supports int64 and uint64 classes natively	7.11
area	Filled area plot	5.0
area	(Change) The v6 option for high-level plotting functions is obsolete	7.5
arrayfun	Applies a function to each element of an array	7.1
arrayfun	(Change) Now accepts arrays of objects as input	7.11
asecd	Inverse trigonometric functions returning angles in degrees	7.0
asecd	(Change) Significant performance improvements	7.11
asind	Inverse trigonometric functions returning angles in degrees	7.0
asind	(Change) Significant performance improvements	7.11
assert	(Change) Additional numeric positional arguments available for formatted string format specifiers	7.4

`assert`	Generates an error when a specified condition is violated	7.4
`assignin`	Assigns variable in a specific workspace	5.0
`atand`	Inverse trigonometric functions returning angles in degrees	7.0
`atand`	(Change) Significant performance improvements	7.11
`audiodevinfo`	Gets information about installed audio devices on Windows platforms	6.5
`audioplayer`	Creates audio object to play audio data on Windows platforms	6.1
`audioplayer`	(Change) Now supported on both UNIX and Windows platforms	7.0
`audioplayer`	(Change) Now supports device selection on all platforms	7.12
`audiorecorder`	Creates audio object to record audio data on Windows platforms	6.1
`audiorecorder`	(Change) Now supported on both UNIX and Windows platforms	7.0
`audiorecorder`	(Change) Now supports device selection on all platforms	7.12
`auditcontents`	Audits `Contents.m` file for a given directory	7.0
`autumn`	Colormap of red and yellow	5.0
`avifile`	(Change) Now generates uncompressed AVI files on Windows platforms unless the Indio5 codec is installed	7.5
`aviinfo`	To be removed; use `mmreader` and `get` instead	7.10
`aviread`	(Change) No longer reads compressed AVI files on Windows platforms unless the Indio5 codec is installed	7.5
`aviread`	To be removed; use `mmreader` and `read` instead	7.10
`axes`	(Change) `ActivePositionProperty`, `OuterPosition`, `TightInset` properties added	7.0
`axescheck`	Processes leading *axes* object from input list	7.0
`axis`	(Change) Now accepts *axes* handle as first argument	7.0
`axlimdlg`	(Removed) No replacement	7.3
`axlimdlg`	(Removed) No replacement	7.5
`balance`	(Change) Now returns different outputs and offers balancing without permuting rows and columns	7.0
`bar`	(Change) The `v6` option for high-level plotting functions is obsolete	7.5
`bar3`	3-D bar chart	5.0
`bar3h`	3-D horizontal bar chart	5.0

Function	Description	Release
barh	2-D horizontal bar chart	5.0
barh	(Change) The v6 option for high-level plotting functions is obsolete	7.5
base2dec	Converts base to decimal	5.0
beep	Makes computer sound a beep	6.0
bessel	(Obsolete) Use besselj instead	7.11
bessel	(Removed) Use besselj instead	7.12
bessela	(Removed) No replacement	7.3
bessela	(Obsolete) Use besselj instead	7.6
besselh	Bessel functions of the third kind (Hankel)	5.0
beta	No longer accepts three inputs	7.3
beta	No longer supports three input arguments. Use betainc with three input arguments instead	7.6
betacore	(Obsolete) Generates a warning message but still works	7.3
betacore	(Obsolete) Use betainc instead	7.6
betacore	(Removed) Use betainc instead	7.7
betaincinv	Implements the inverse incomplete beta function	7.8
bicg	Biconjugate gradients method	5.0
bicgstab	Biconjugate gradients stabilized method	5.0
bicgstabl	Implements the stabilized biconjugate gradients method for solving systems of linear equations	7.8
bin2dec	Converts binary to decimal	5.0
bin2dec	Now ignores any spaces in the input string	7.0
bitand	Bitwise AND	5.0
bitcmp	Compare bits	5.0
bitcmp	(Change) Now supports int64 and uint64 classes natively	7.11
bitget	Gets bit	5.0
bitmax	Maximum floating-point integer	5.0
bitor	Bitwise OR	5.0
bitset	Sets bit	5.0
bitshift	Bitwise shift	5.0
bitxor	Bitwise XOR	5.0
blkdiag	Creates block diagonal matrix	5.3

box	Turns On or Off axes box	5.0
brush	Turns data brushing on or off and selects a color for brushing graphs	7.6
bsxfun	Applies element-by-element binary operation to two full arrays with singleton expansion enabled	7.4
bsxfun	(Change) Now supports int64 and uint64 classes natively	7.11
builddocsearchdb	Creates a search database for your own HTML help files for the *Help browser*	7.4
bvp4c	Solves two-point boundary value problem by collocation	6.0
bvp4c	(Change) Can now solve multipoint boundary value problems	7.0
bvp4c	(Change) Significant performance improvements for sparse problems	7.10
bvp5c	Solves boundary value problems for ODEs, most useful for small error tolerances	7.5
bvp5c	(Change) Now supports multipoint boundary value problems	7.7
bvp5c	(Change) Significant performance improvements for sparse problems	7.10
bvpget	Gets BVP option from option structure	6.0
bvpinit	Forms initial guess for bvp4c	6.0
bvpset	Creates or changes BVP option structure	6.0
bvpval	Evaluates solution from bvp4c	6.0
bvpval	(Obsolete) Replaced by deval	6.1
bvpval	(Removed) No replacement	7.3
bvpxtend	Generates a guess structure for extending a boundary value problem solution	7.3
calendar	Produces monthly calendar	5.0
calllib	Generic DLL Interface function	7.0
callsoapservice	Sends a SOAP message off to an endpoint	7.0
camdolly	Translates camera position and target	5.2
camlight	Creates or moves light object in camera coordinate system	5.2
camorbit	Rotates camera position around camera target	5.2
campan	Rotates camera target around camera position	5.2
campos	Sets or gets camera position and position mode	5.2
camproj	Sets or gets camera projection type	5.2

Function	Description	Release
camroll	Rotates camera about camera viewing axis	5.2
camtarget	Sets or gets camera target and camera target mode	5.2
camup	Sets or gets camera up vector and up vector mode	5.2
camva	Sets or gets camera view angle and view angle mode	5.2
camzoom	Zooms camera in or out	5.2
cast	Casts a variable to a different data type	7.0
cat	Concatenates arrays	5.0
cat	(Change) Better error checking for sparse matrix functions	7.10
catch	(Change) New optional MException class argument	7.5
cbedit	(Removed) Use guide instead	7.5
cdfepoch	Converts MATLAB date number or string to CDF format	6.5
cdfinfo	Gets information about a CDF file	6.1
cdfread	Reads CDF file	6.1
cdfwrite	Writes data to CDF file	6.5
cell	Creates cell array	5.0
cell2mat	Combines cell array of matrices into one matrix; previously part of the *Neural Networks Toolbox*	6.5
cell2struct	Converts cell to structure	5.0
celldisp	Displays cell structure	5.0
cellfun	Performs common operations on cell arrays	5.3
cellfun	(Change) Applies a function to each cell of a cell array	7.1
cellplot	Graphically displays cell structure	5.0
cgs	Conjugate gradients squared method	5.0
char	Converts to string array	5.0
check_syntactic_warnings	Runs syntax check on M-files	6.0
chol	Enhancements to improve control, performance, and memory usage, as well as access to upper and lower triangular factors and lower symbolic factor	7.3
cholinc	Incomplete Cholesky factorization	5.0
cholinc	Sparse incomplete Cholesky and Cholesky-infinity factorization	5.2
cholinc	(Obsolete) Use ichol instead	7.12

`cholupdate`	Rank 1 update to Cholesky factorization	5.2
`classdef`	New reserved word	7.3
`clear`	(Change) The `clear mex` command no longer clears M-functions in addition to MEX-functions	7.0.1
`clearvars`	Clears variables from the workspace; supports exceptions	7.6
`clruprop`	(Obsolete) Use `rmappdata` instead	7.1
`clruprop`	(Removed) Use `rmappdata` instead	7.5
`cmpermute`	Rearranges colors in colormap	7.8
`cmunique`	Eliminates unneeded colors in colormap of indexed image	7.8
`colamd`	Computes approximate column minimum degree permutation	6.0
`colmmd`	(Depreciated) Use `colamd` instead	7.0
`colmmd`	(Obsolete) Use `colamd` instead	7.6
`colmmd`	(Removed) Use `colamd` instead	7.7
`colon (:)`	(Change) Colon operations on characters iterating a For Loop (`for x = 'a':'c'`) now returns character data type rather than double	7.5
`colon (:)`	(Change) Now supports `int64` and `uint64` classes natively	7.11
`colorbar`	(Change) The `v6` option for high-level plotting functions is obsolete	7.5
`colorcube`	Colormap of regularly spaced colors	5.0
`colordef`	Selects figure color scheme	5.0
`colormapeditor`	Interactive colormap editor	6.5
`commandhistory`	Opens *Command History* window or selects it	7.0
`commandwindow`	Opens *Command* window or selects it	7.0
`complex`	Creates complex array from real and imaginary parts	5.3
`condeig`	Condition number with respect to eigenvalues	5.0
`condest`	1-norm matrix condition number estimate	5.0
`condest`	(Change) Efficiency improvements	7.2
`coneplot`	Velocity vector cone plot in a 3-D vector field	5.3
`coneplot`	Creates 3-D coneplot	6.0
`containers.Map`	(Change) New syntax available to construct Map object containers	7.10

Function	Description	Release
continue	Skips rest of For Loop or While Loop	6.0
contourf	(Change) The v6 option for high-level plotting functions is obsolete	7.5
contourslice	Draws contours in volume slices	5.3
conv2	(Change) Now accepts the shape parameter as input	7.8
conv2	(Change) Significant performance improvements	7.9
conv2	(Change) Multithreading extended to the two-input form of conv2	7.10
convhull	Convex hull	5.0
convhull	(Change) Now makes use of Qhull	6.1
convhull	(Change) [K,a] = convhull(x,y) now returns the area a of the convex hull; in addition, it now ignores the third input argument	6.1
convhull	(Change) No longer supports the QHULL or QHULL options arguments	7.9
convhull	(Change) No longer uses the options argument	7.10
convhull	(Change) Now supports 3-D input. The simplify option allows removal of noncontributing vertices	7.11
convhulln	n-D convex hull	6.0
convhulln	(Change) [K,v] = convhulln(x,y) now returns the volume v of the convex hull	6.1
convhulln	(Change) Now supports user-settable options	7.0
convn	(Change) Now accepts the shape parameter as input	7.8
convn	(Change) Significant performance improvements	7.10
copyfile	(Change) Now also copies directories	6.5
corrcoef	(Change) Three new syntaxes	6.5
cosd	Trigonometric function with arguments in degrees	7.0
cosd	(Change) Significant performance improvements	7.11
cotd	Trigonometric function with arguments in degrees	7.0
cotd	(Change) Significant performance improvements	7.11
contour	(Change) The v6 option for high-level plotting functions is obsolete	7.5
contourf	Filled contour plot	5.0
cscd	Trigonometric function with arguments in degrees	7.0

`cscd`	(Change) Significant performance improvements	7.11
`cshelp`	(Depreciated) No replacement available	7.4
`ctlpanel`	(Obsolete) Use `guide` instead	7.1
`ctlpanel`	(Removed) Use `guide` instead	7.5
`cumtrapz`	Cumulative trapezoidal numerical integration	5.0
`curl`	Computes curl and angular velocity perpendicular to flow	6.0
`daspect`	Sets or gets data aspect ratio and aspect ratio mode	5.2
`datacursormode`	Interactively creates data cursors on plot	7.0
`datatipinfo`	Produces short description of a variable	7.0
`datenum`	Serial date number	5.0
`datenum`	(Change) Now accepts `pivotyear` argument	5.3
`datenum`	(Change) Now accepts date vector as an input argument	6.1
`datenum`	Can now specify output formats and local date options	7.0
`datenum`	(Change) Now supports milliseconds using Free Form Date Format Specifiers	7.1
`datestr`	Creates date string	5.0
`datestr`	(Change) Now accepts `pivotyear` argument	5.3
`datestr`	(Change) Now accepts date vector as an input argument	6.1
`datestr`	Can now specify output formats and local date options	7.0
`datestr`	(Change) Seconds field truncates instead of rounding	7.1
`datestr`	(Change) Now supports milliseconds using Free Form Date Format Specifiers	7.1
`datetick`	Creates date-formatted tick labels	5.0
`datevec`	Date components	5.0
`datevec`	(Change) Now accepts `pivotyear` argument	5.3
`datevec`	Can now specify output formats and local date options	7.0
`datevec`	(Change) Now supports milliseconds using Free Form Date Format Specifiers	7.1
`dbclear`	(Change) New `-completenames` option permits clearing or setting breakpoints for M-files not on the search path	7.6
`dblquad`	Numerical 2-D integration	5.0
`dblquad`	(Change) Can now pass extra arguments to integrand function	6.0
`dbstack`	Now supports nested functions	7.0

Function	Description	Release
`dbstatus`	Display breakpoints function now supports anonymous and nested functions as well as a new -completenames argument	7.0
`dbstatus`	(Change) Display breakpoints function now supports anonymous and nested functions as well as a new-completenames argument	7.0.1
`dbstop`	Now supports nested and anonymous functions	7.0
`dbstop`	(Change) Now supports stopping just before a non M-file function (such as a built-in function)	7.2
`dbstop`	(Change) The ability to restore breakpoints was added	7.3
`dbstop`	(Change) New -completenames option permits clearing or setting breakpoints for M-files not on the search path	7.6
`dde23`	Solves delay differential equations with constant delays	6.5
`ddeget`	Gets properties from DDE options structure	6.5
`ddeset`	Creates or modifies properties in DDE options structure	6.5
`ddesd`	New solver for delay differential equations that have general delays	7.2
`deal`	No longer needed in many cases, for example, [a,b,c,d] = C{:} is equivalent to [a,b,c,d] = deal(C{:})	7.0
`dec2base`	Converts decimal to base	5.0
`dec2bin`	Converts decimal to binary	5.0
`decic`	Computes consistent initial conditions for `ode15i`	7.0
`delaunay`	Delaunay triangularization	5.0
`delaunay`	(Change) Now ignores third input argument	6.1
`delaunay`	(Change) Now makes use of Qhull	6.1
`delaunay`	(Change) No longer supports the QHULL or QHULL options arguments	7.9
`delaunay`	(Change) No longer uses the `options` argument	7.10
`delaunay`	(Change) Now supports 3-D input in either multiple vector or multicolumn matrix format	7.11
`delaunay3`	3-D Delaunay tessellation	6.0
`delaunay3`	(Depreciated) Use `DelaunayTri` method instead	7.9
`delaunay3`	(Change) No longer supports the QHULL or QHULL options arguments	7.9
`delaunay3`	(Change) No longer uses the `options` argument	7.10

`delaunay3`	Depreciated. Use `DelaunayTri` instead	7.10
`delaunayn`	*n*-D Delaunay tessellation	6.0
`delaunayn`	(Change) Now supports user-settable options	7.0
`DelaunayTri`	New OOP class provides enhanced computational geometry tools	7.8
`deleteconfirm`	Confirms the deletion of a file with a dialog box	7.0
`depfun`	(Change) Now supports a number of new options	7.0.1
`deprpt`	Scans a file or directory for dependencies	7.0
`desearchn`	*n*-D nearest point search	6.0
`deval`	Evaluates ODE solution; also replaces obsolete `bvpval` function	6.1
`deval`	(Change) Now accepts output from `dde23`	6.5
`deval`	(Change) Now optionally returns derivative at points as well	7.0
`diff`	(Change) Now supports `int64` and `uint64` classes natively	7.11
`diff2asv`	Compares file to autosaved version, if one exists	7.0
`diffrpt`	Visual directory browser	7.0
`dir`	(Change) Now returns additional numeric date (`datenum`) field	7.4
`dir`	(Change) Empty matrices now returned when appropriate	7.7
`dither`	Converts image using dithering	7.8
`divergence`	Computes divergence of vector field	6.0
`dlmwrite`	New input arguments and output options	7.0
`dmperm`	(Change) New output arguments for the indices of the Dulmage–Mendelsohn coarse decomposition	7.5
`docopt`	Removed. Preferred web browser now set in Preferences	7.10
`docsearch`	Searches HTML documentation in the Help browser	7.0
`docsearch`	(Change) Now accepts multiple arguments including wildcards without function syntax	7.8
`dofixrpt`	Scans a file or directory for all TODO, FIXME, or NOTE messages	7.0
`dragrect`	Drags rectangle	5.0
`drawnow`	(Change) Can now selectively update UI components	7.5
`dsearch`	Searches for nearest point	5.0
`dsearch`	Depreciated. Use `DelaunayTri/nearestNeighbor` instead	7.9

Function	Description	Release
dsearch	(Depreciated) Use `DelaunayTri/nearestNeighbor` methods instead	7.10
edit	Edits M-file (launch MATLAB Editor/Debugger)	5.0
edit	(Change) Now accepts file path information to enable editing files not in the current directory	7.6
editpath	Modifies the MATLAB search path	5.0
edtext	(Obsolete) Set the `Editing` property of the *text* object	7.3
edtext	(Removed) Set the `editing` property of the *text* object	7.5
eigs	Finds several eigenvalues and eigenvectors	5.0
eigs	The `eigs` options structure no longer accepts the `opts.cheb` and `opts.stagtol` fields	7.3
eomday	End of month	5.0
eps	(Change) Now accepts arguments to specify single- or double-precision values and to return `eps` relative to any value, not just 1	7.0
erfcinv	Inverse complementary error function	6.1
erfcore	(Obsolete) Use `erf`, `erfc`, `erfcx`, `erfinv`, or `erfcinv` instead	7.9
erfcore	(Removed) Replace with `erf`, `erfc`, `erfcx`, `erfinv`, or `erfcinv`	7.10
erfcore	(Removed) Use `erf`, `erfc`, `erfcx`, `erfinv`, or `erfcinv` instead	7.11
error	(Change) Now saves stack information accessible by `lasterror`	7.1
error	(Change) Additional numeric positional arguments available for formatted string format specifiers	7.4
errorbar	(Change) The v6 option for high-level plotting functions is obsolete	7.5
errortrap	Replaced by try-catch block	6.0
evalc	String evaluation with output converted to a string	5.3
evalin	Evaluates expression in a specific workspace	5.0
eventlisteners	Lists all events that are registered	7.0
exifread	Reads EXIF information from JPEG and TIFF image files	7.1
exifread	To be removed. Use `imfinfo` instead	7.10
expm	(Change) Efficiency improvements	7.2

expm1	Computes $\exp(x)-1$ accurately	7.0
exportsetupdlg	Shows figure export style dialog	7.0
extent	(Obsolete) Use get(h, 'extent') instead	7.1
extent	(Removed) Set the extent property of the *text* object	7.5
eye	(Change) Now accepts a final argument specifying numeric data type of result	7.0
eye	(Change) No longer accepts complex dimension arguments	7.5
ezcontour	Easy contour plotter	5.3
ezcontourf	Easy filled contour plotter	5.3
ezmesh	Easy mesh plotter	5.3
ezmeshc	Easy mesh with contour plotter	5.3
ezplot	Easy 2-D line plotter	5.3
ezplot3	Easy 3-D line plotter	5.3
ezpolar	Easy polar plotter	5.3
ezsurf	Easy surface plotter	5.3
ezsurfc	Easy surface with contour plotter	5.3
factor	Prime factors	5.0
false	Creates array of logical False	6.5
false	(Change) No longer accepts complex dimension arguments	7.5
fft	(Change) FFT functions are now multithreaded	7.8
fft	(Change) Now supports larger input arrays (size in one dimension greater than $2^{31}-1$) on 64-bit platforms	7.9
fft	(Change) Multithreading extended to fft for long vectors	7.10
fft2	(Change) FFT functions are now multithreaded	7.8
fft2	(Change) Now supports larger input arrays (size in one dimension greater than $2^{31}-1$) on 64-bit platforms	7.9
fftn	(Change) FFT functions are now multithreaded	7.8
fftn	(Change) Now supports larger input arrays (size in one dimension greater than $2^{31}-1$) on 64-bit platforms	7.9
fftw	Tunes or sets options in FFTW library for FFT computations	7.0
fftw	(Change) Default planner method is now estimate. New syntax for importing/exporting wisdom databases. Wisdom exported by earlier versions of FFTW (prior to 3.1.1) cannot be imported	7.3

Function	Description	Release
fieldnames	Field names of structure	5.0
figflag	(Obsolete) Use findobj and/or figure instead	7.1
figflag	(Depreciated) No replacement available	7.4
figure	(Change) DockControls property added, KeyPressFcn property modified	7.0
figureheaderdlg	Shows figure header dialog	7.0
figurepalette	Shows or hides the palette for a figure	7.0
fileattrib	Sets or gets file attributes	6.5
fileparts	(Change) Fourth output (versn) is always empty	7.10
fileparts	(Change) Returns an empty fourth argument (file versions not supported)	7.11
find	(Change) Now supports optional arguments specifying an upper limit on the number of indices returned, and whether the search begins at the start or end of the array	7.0
findfigs	Finds all visible *Figure* windows	5.3
finite	(Depreciated) Use isfinite instead	7.5
finite	(Removed) Use isfinite instead	7.8
fitsinfo	Gets information about a FITS file	6.1
fitsread	Reads FITS file	6.1
fitsread	(Change) New options to support data subsetting	7.12
fixquote	Double up single quotes in a string	7.0
flipdim	Flips array along specified dimension	5.0
flops	(Depreciated) Floating-point operation count no longer relevant	6.0
flops	(Obsolete) No replacement	7.6
flops	(Removed) No replacement—no longer practical to count flops	7.7
fmin	Replaced by fminbnd	6.0
fmin	(Removed) No longer supported	7.0
fminbnd	(Change) New name and calling sequence for function fmin	5.3
fminbnd	(Change) Now supports calling an output function at each iteration	7.0
fmins	Replaced by fminsearch	6.0
fmins	(Removed) No longer supported	7.0

`fminsearch`	(Change) New name and calling sequence for function `fmins`	5.3
`fminsearch`	(Change) Now supports calling an output function at each iteration	7.0
`fopen`	(Change) Now supports an argument specifying a character encoding scheme	7.2
`fopen`	(Change) No longer supports `'vaxd'`, `'vaxg'`, or `'cray'` (`'d'`, `'g'`, or `'c'`) formats	7.7
`foptions`	Replaced by `optimget` and `optimset`	6.0
`format`	(Change) Two new display formats are introduced: `short eng` and `long eng`	7.0.4
`fprintf`	(Change) Writing to `stdin` (`fprintf(0,...)`) now generates an error	7.3
`fprintf`	(Change) Additional numeric positional arguments available for formatted string format specifiers	7.4
`fread`	(Change) Added a precision argument (in MATLAB 7.1 and earlier)	7.1
`fread`	(Change) Now uses character encoding specified by `fopen`; calls to `unicode2native` and `native2unicode` are no longer needed or appropriate	7.2
`freeserial`	(Obsolete) Use `fclose` instead	7.5
`ftp`	Creates *ftp* object	7.0
`fullfile`	Builds full filename from parts	5.0
`func2str`	Constructs function name string from function handle	6.0
`function`	(Change) Function definition line no longer requires commas separating output values	7.0
`functions`	Displays information about a function handle	6.0
`funm`	(Change) The optional second output is now an exit flag rather than a (sometimes inaccurate) error estimate	7.0
`funm`	(Change) Second output argument changed from error estimate to result flag	7.0.1
`funm`	(Change) New algorithms	7.6
`fwrite`	Now supports saving `uint64` and `int64` values on all platforms	7.0.1
`fwrite`	(Change) Now uses character encoding specified by `fopen`; calls to `unicode2native` and `native2unicode` are no longer needed or appropriate	7.2

Function	Description	Release
fwrite	(Change) Writing to stdin (fwrite(0,...)) now generates an error	7.3
fzero	(Change) Calling sequence changed	6.0
fzero	(Change) Now supports calling an output function at each iteration	7.0
gallery	More than 50 test matrices	5.0
gallery	(Change) New test matrices available	6.5
gallery	(Change) New optional classname input argument: either 'single' or 'double'	7.2
gallery	(Change) New integerdata, normaldata, and uniformdata options for gallery suite of test matrices	7.9
gamma	No longer accepts two inputs	7.3
gammainc	Can now specify the tail of the incomplete gamma function for nonnegative input	7.0
gammainc	(Change) New 'scaledlower' or 'scaledupper' argument returns scaled versions of the incomplete gamma function	7.12
gammaincinv	Implements the inverse incomplete gamma function	7.8
ge	No longer accepts complex integer inputs	7.3
genpath	Generates path string that includes all directories below a specified directory	6.0
genvarname	Generates variable name from candidate name	7.0
get	(Change) Do not use get or set to manage properties of a Java object; this usage will generate an error in future releases	7.6
getfield	Gets field from structure	5.0
getpixelposition	Gets position of object in pixels	7.0
getReport	(Change) Several new options available	7.7
getstatus	(Depreciated) No replacement available	7.4
getuprop	(Obsolete) Use getappdata instead	7.1
getuprop	(Removed) Use getappdata instead	7.5
gmres	Generalized minimal residual method	5.0
griddata	(Change) Uses triangle-based interpolation	5.0
griddata	(Change) Now makes use of Qhull	6.1

griddata	(Change) No longer supports the QHULL or QHULL options arguments	7.9
griddata	(Change) No longer uses the `options` argument	7.10
griddata3	3-D data gridding	6.0
griddata3	(Depreciated) Use `TriScatteredInterp` method instead	7.9
griddata3	(Change) No longer supports the QHULL or QHULL options arguments	7.9
griddata3	(Change) No longer uses the `options` argument	7.10
griddata3	Depreciated. Use `TriScatteredInterp` instead	7.10
griddatan	*n*-D data gridding	6.0
gt	No longer accepts complex integer inputs	7.3
gunzip	Uncompresses files from gzip format or from a URL	7.0.4
gunzip	Partial paths and wildcards (e.g., '~' and '*') are accepted in filename arguments	7.2
gzip	Compresses files to gzip format	7.0.4
gzip	Partial paths and wildcards (e.g., '~' and '*') are accepted in filename arguments	7.2
h5create	(New) Creates HDF5 data set	7.12
h5disp	(New) Displays the contents of an HDF5 file	7.12
h5info	(New) Returns information about an HDF5 file	7.12
h5read	(New) Reads data from an HDF5 data set	7.12
h5readatt	(New) Reads an attribute from an HDF5 group or data set	7.12
h5write	(New) Writes to an HDF5 data set	7.12
h5writeatt	(New) Writes an attribute to an HDF5 group or data set	7.12
hasbehavior	Sets or gets behaviors of Handle Graphics objects	7.0
hd5info	Returns HDF5 file information	7.0
hd5read	Reads HDF5 file	7.0
hdf	A number of new functions support MATLAB interface to the HDF4 libraries. Use `help hdf` for overviews of the available functionality	7.11
hdf5	A number of new functions support MATLAB interface to the HDF5 libraries. Use `help dhf5` for overviews of the available functionality	7.11
hdf5info	(Depreciated) Use `h5info` instead	7.12
hdf5read	(Depreciated) Use `h5read` instead	7.12

Function	Description	Release
hdf5write	Writes HDF5 file	7.0
hdf5write	(Depreciated) Use h5write instead	7.12
hdfinfo	Gets information about a HDF file	6.1
hdfread	Reads HDF file	6.1
hdfread	(Change) Now generates an error rather than a warning on a failed I/O operation	7.5
helprpt	Scans a file or directory for help	7.0
hex2num	Converts number from IEEE hexadecimal format	7.0
hgexport	Exports a figure	7.0
hggroup	Creates a Handle Graphics group object	7.0
hgtransform	Creates graphics transformation object	7.0
hidegui	(Obsolete) Set the *figure* HandleVisibility property instead	7.4
hidegui	(Depreciated) Set the *figure's* handlevisibility property	7.6
hisrtc	(Change) Significant performance improvements	7.10
hist	(Change) Now calls MEX function histc	5.3
hist	(Change) Changes to the data tips in the histogram display	7.7
histc	Histogram binning given bin edges	5.3
histc	(Change) Bug fixed	6.1
hold	(Change) Now supports all option, which holds the plot so that subsequent plots do not reset the color and linestyle order	7.0
horzcat	(Change) Better error checking for sparse matrix functions	7.10
hthelp	(Obsolete) Use web instead	7.1
hthelp	(Removed) Use web instead	7.5
hypot	Square root of the sum of squares	7.1
ichol	New incomplete Cholesky factorization function; preferred replacement for the cholinc function	7.12
icubic	(Removed) No longer supported	7.0
idivide	Integer division similar to A./B; fractional quotients are rounded to integers using a specified rounding mode	7.2
ifft	(Change) FFT functions are now multithreaded	7.8

`ifft`	(Change) Now supports larger input arrays (size in one dimension greater than $2^{31}-1$) on 64-bit platforms	7.9
`ifft2`	(Change) FFT functions are now multithreaded	7.8
`ifft2`	(Change) Now supports larger input arrays (size in one dimension greater than $2^{31}-1$) on 64-bit platforms	7.9
`ifftn`	(Change) FFT functions are now multithreaded	7.8
`ifftn`	(Change) Now supports larger input arrays (size in one dimension greater than $2^{31}-1$) on 64-bit platforms	7.9
`ifftshift`	Inverse FFT shift	5.2
`ilu`	Performs the sparse incomplete LU factorization	7.4
`imfinfo`	(Change) Now returns Exif data for JPEG or TIFF files. New `DigitalCamera` and `GPSInfo` fields	7.6
`imformats`	Eases the task of adding read and write support for new file formats	6.5
`immapprox`	Approximates indexed image by one with fewer colors	7.8
`import`	Adds to the current Java packages import list	6.0
`importdata`	(Change) Now supports XLS, XLSX, XLSB, and XLSM formats on Windows platforms with an appropriate version of Excel installed	7.8
`imread`	Reads image data	5.0
`imread`	Added ability to read a specified portion of a TIFF image	7.0
`imread`	(Change) Improved support for TIFF files. Improved performance in some instances	7.5
`imread`	(Change) Now supports rewriting portions of a TIFF file rather than replacing the entire file. New Tiff object allows access to functions in the LibTIFF library	7.9
`imread`	(Change) Now supports n-channel J2C JPEG 2000 files	7.11
`imwrite`	Writes image data	5.0
`imwrite`	(Change) Now supports GIF format	7.0.4
`imwrite`	(Change) New `RowsPerStrip` parameter to specify number of image rows per strip written to TIFF files. Default <= 8KB	7.6
`imwrite`	(Change) Now supports rewriting portions of a TIFF file rather than replacing the entire file. New Tiff object allows access to functions in the LibTIFF library	7.9
`imwrite`	Added support for JPEG 2000 format files	7.10

Function	Description	Release
`imwrite`	(Change) Now supports n-channel J2C JPEG 2000 files	7.11
`ind2sub`	Subscripts from single linear index	5.0
`inmem`	Functions in memory	5.0
`inmem`	Now reports path information	7.0
`inpolygon`	Detects points inside a polygonal region	5.0
`inputdlg`	Displays input dialog	5.0
`inputname`	Input argument name	5.0
`inputParser`	New OOP class available for parsing and validating inputs to functions	7.4
`inspect`	The Property Inspector has a new interface but no new functionality	7.2
`instrfindall`	Finds all serial port objects with specified property values	7.0
`int16`	Conversion to 16-bit signed integer data types	5.3
`int16`	(Change) Now rounds noninteger inputs rather than truncating	7.0
`int16`	(Change) Multithreading extended to integer conversion and arithmetic	7.10
`int32`	Conversion to 32-bit signed integer data types	5.3
`int32`	(Change) Now rounds noninteger inputs rather than truncating	7.0
`int32`	(Change) Multithreading extended to integer conversion and arithmetic	7.10
`int64`	Creates signed 64-bit integer array	6.5
`int64`	(Change) Now rounds noninteger inputs rather than truncating	7.0
`int64`	(Change) Multithreading extended to integer conversion and arithmetic	7.10
`int8`	Conversion to 8-bit signed integer data types	5.3
`int8`	(Change) Now rounds noninteger inputs rather than truncating	7.0
`int8`	(Change) Multithreading extended to integer conversion and arithmetic	7.10
`interp1`	(Change) `'cubic'` option now calls `pchip`	6.0
`interp1`	(Change) Now enables data extrapolation	6.1
`interp1`	(Change) Now supports multidimensional arrays for Y	7.0

`interp1`	(Change) Now optionally returns a pp-form for evaluation with `ppval`	7.0
`interp3`	3-D data interpolation	5.0
`interp4`	Replaced by `interp2`	6.0
`interp4`	(Removed) No longer supported	7.0
`interp5`	Replaced by `interp2`	6.0
`interp5`	(Removed) No longer supported	7.0
`interp6`	Replaced by `interp2`	6.0
`interp6`	(Removed) No longer supported	7.0
`interpn`	n-D data interpolation	5.0
`interpstreamspeed`	Interpolates streamline vertices from speed	6.0
`intersect`	Set intersection	5.0
`intmax`	Maximum integer value given integer data type	7.0
`intmin`	Minimum integer value given integer data type	7.0
`intwarning`	Controls state of the integer data type warnings	7.0
`intwarning`	To be removed. Use `warning` instead	7.10
`intwarning`	Depreciated and will be removed in a future release	7.10
`intwarning`	(Removed) No replacement	7.11
`intwarning`	(Removed) No replacement	7.12
`ipermute`	Inverse permute dimensions of n-D array	5.0
`isa`	(Change) Now returns consistent results for objects with base class of logical	7.11
`iscell`	True for cell array	5.0
`iscolumn`	True if argument is a column vector	7.11
`iscom`	True for COM/ActiveX objects	7.0
`isdir`	Replaced with `exist`	6.0
`isempty`	(Change) Now supports `containers.Map` objects	7.8
`isequal`	True if arrays are equal	5.0
`isequal`	(Change) When used to compare structures, input argument field creation order no longer has an impact on equality	6.5
`isequal`	(Change) Now ignores class: `isequal(double(3),single(3))` now returns `True`	7.9
`isequal`	(Fix) Now returns correct values when inputs are arrays of objects containing NaN values	7.11

Function	Description	Release
isequalwithequalnans	True if arrays are equal with NaN s considered equal	6.5
isequalwithequalnans	(Fix) Now returns correct values when inputs are arrays of objects containing NaN values	7.11
isevent	True if event of object	7.0
isfield	(Change) Now supports cell array input	7.1
isfinite	True for finite elements	5.0
isfloat	True for floating-point data	7.0
ishghandle	True for Handle Graphics Object handle	7.0
isieee	Obsolete. All MATLAB platforms use IEEE arithmetic	6.0
isinteger	True for integer data	7.0
isinterface	True for COM interface	7.0
isjava	True for Java Object	6.0
iskeyword	Generates or tests if argument is a MATLAB keyword	6.0
islogical	True for logical arrays	5.0
islogical	(Change) Now returns consistent results for objects with base class of logical	7.11
ismac	Returns true on OSX platforms	7.4
ismatrix	True if argument is a (2-D) matrix	7.11
ismember	Detects members of a set	5.0
ismember	Now calls the MEX function ismemc to maximize speed	5.1
ismember	(Change) [tf, idx] = ismember(...) now returns indices idx of located members; in this case, MEX function ismemc2 is called to maximize speed	6.5
isnumeric	True for numeric array	5.0
isocaps	Computes isosurface end-cap geometry	5.3
isocolors	Computes colors of isosurface vertices	6.0
isonormals	Computes normals at isosurface vertices	5.3
isosurface	Extracts isosurface data from volume data	5.3
isosurface	Extracts isosurface	6.0
isprime	True for prime numbers	5.0
ispuma	True for computers running Mac OS X 10.1.x	7.0
ispuma	(Depreciated) OS X 10.1 is no longer supported on the Mac platform	7.4

`isrow`	True if argument is a row vector	7.11
`isscalar`	True if argument is a scalar	7.0
`issorted`	True if array is sorted	6.5
`issorted`	Now works on cell arrays of strings	7.2
`isstr`	Replaced by `ischar`	6.0
`isstr`	To be removed. Use `ischar` instead	7.10
`isstrprop`	True for string elements matching a variety of specifications	7.0
`isstruct`	True for structure	5.0
`isvarname`	True if input string is a valid variable name	6.0
`isvector`	True if argument is a row or column vector	7.0
`javaaddpath`	Adds directories to the dynamic Java path	7.0
`javaArray`	Creates Java array	6.0
`javaclasspath`	Gets and sets Java path	7.0
`javacomponent`	Creates a Java AWT Component and puts in a figure	7.0
`javaMethod`	Invokes Java method	6.0
`javaObject`	Constructs Java object	6.0
`javarmpath`	Removes directory from dynamic Java path	7.0
`lasterr`	(Depreciated) Use `MException` objects instead	7.9
`lasterror`	Returns last error message and related information	6.5
`lasterror`	(Change) Now returns stack information from the last error	7.1
`lasterror`	(Depreciated) Use `MException` objects instead	7.9
`lastwarn`	Returns last warning string	5.2
`layout`	(Obsolete) No replacement	7.1
`layout`	(Removed) No replacement	7.5
`ldivide`	(Change) Multithreaded support added	7.6
`ldivide (.\)`	(Change) Now supports `int64` and `uint64` classes natively	7.11
`ldl`	Enhancements to improve control, performance, and memory usage, as well as access to upper and lower triangular factors and lower symbolic factor	7.3
`ldl`	Full `ldl` factorization and solving for Hermitian matrices	7.3
`ldl`	(Change) Now supplies factorization and solving for an additional output argument, the scaling matrix, when the input matrix is real, sparse, and symmetric	7.5
`ldl`	(Change) New algorithms	7.6

Function	Description	Release
le	No longer accepts complex integer inputs	7.3
legend	(Change) The v6 option for high-level plotting functions is obsolete	7.5
libfunctions	Generic DLL Interface function	7.0
libfunctionsview	Generic DLL Interface function	7.0
libisloaded	Generic DLL Interface function	7.0
libpointer	Generic DLL Interface function	7.0
libstruct	Generic DLL Interface function	7.0
lightangle	Creates or moves *light* object in spherical coordinates	5.2
lines	Colormap that follows axes colororder property	5.0
linkaxes	Synchronizes limits of specified *axes* objects	7.0
linkdata	Turns data linking on or off for a *figure*	7.6
linkprop	Maintains same value for corresponding Handle Graphics properties	7.0
linsolve	Solves $Ax = y$, given specific structure of A	7.0
loadlibrary	Generic DLL Interface function	7.0
log	(Change) Multithreaded support added	7.6
log1p	Computes $\log(1+x)$ accurately	7.0
log2	(Change) Multithreaded support added	7.6
logical	Converts (casts) to logical array	5.0
loglog	(Change) The v6 option for high-level plotting functions is obsolete	7.5
logm	(Change) New algorithms	7.6
lscov	Can now specify either Cholesky or orthogonal decomposition	7.0
lsqnonneg	(Change) New name and calling sequence for function nnls	5.3
lsqnonneg	(Change) More efficient; now accepts sparse matrices as input and maintains sparsity throughout	7.7
lsqnonneg	(Change) No longer uses the optional input x0 as a starting point	7.10
lsqr	LSQR implementation of Conjugate Gradients on Normal Equations	6.0
lt	No longer accepts complex integer inputs	7.3

`ltitr`	(Change, previously undocumented) Linear time-invariant time response kernel	7.0
`lu`	(Change) Now uses UMFPACK for sparse matrices	6.5
`lu`	Enhancements to improve control, performance, and memory usage, as well as access to upper and lower triangular factors and lower symbolic factor	7.3
`lu`	(Change) Additional output helps improve numerical stability of sparse `lu` factorization	7.3
`luinc`	Incomplete LU factorization	5.0
`luinc`	Enhancements to improve control, performance, and memory usage, as well as access to upper and lower triangular factors and lower symbolic factor	7.3
`luinc`	(Obsolete) Use `ilu` instead	7.12
`makecontentsfile`	Makes a new `Contents.m` file	7.0
`makehgtransform`	Creates graphical transformation matrix	7.0
`makemcode`	Makes M-file for regenerating object and its children	7.0
`mat2cell`	Breaks matrix up into a cell array of matrices; previously part of the *Neural Networks Toolbox*	6.5
`mat2str`	Converts matrix to string	5.0
`mat2str`	(Change) Now converts nondouble data types	7.0.1
`mat2str`	(Fix) Now returns correct values for character array input	7.4
`matq2ws`	(Obsolete) No replacement	7.1
`matq2ws`	(Removed) No replacement	7.5
`matqdlg`	(Obsolete) No replacement	7.1
`matqdlg`	(Removed) No replacement	7.5
`matqparse`	(Obsolete) No replacement	7.1
`matqparse`	(Removed) No replacement	7.5
`matqueue`	(Obsolete) No replacement	7.1
`matqueue`	(Removed) No replacement	7.5
`max([]) = []`	(Change) Defined output for empty inputs	5.0
`max`	(Change) Now returns an error when input arguments are complex integers	7.0.4
`max`	(Change) Now uses the phase angle in the case of equal magnitude complex input. No longer returns warning messages for inputs of different data types	7.3

Function	Description	Release
max	(Change) Multithreading supported	7.8
maxNumCompThreads	Gets/sets the maximum number of complex threads	7.5
maxNumCompThreads	(Change) Ability to adjust the maximum number of computational threads has been removed; multithreading support is either on or off	7.8
maxNumCompThreads	(Depreciated) To be discontinued with no replacement planned	7.9
meditor	(Removed) Stand-alone editor removed; use the Editor/ Debugger instead	7.6
memory	(Change) Displays memory usage and availability information on Windows platforms. Formerly provided help text describing how to free additional memory for MATLAB	7.6
menubar	(Obsolete) Set the MenuBar property of the *figure* to none	7.3
menubar	(Removed) Set the menubar property of the *figure* to none	7.5
menuedit	(Obsolete) Use guide instead	7.1
menuedit	(Removed) Use guide instead	7.5
menulabel	(Obsolete) Use '&' and the 'Accelerator' property instead	7.1
menulabel	(Depreciated) No replacement available	7.4
mesh	(Change) The v6 option for high-level plotting functions is obsolete	7.5
meshdom	Replaced by meshgrid	6.0
meshdom	(Removed) No longer supported	7.0
methodsview	Displays information on all methods implemented by a Java or MATLAB class	6.0
MException	New OOP class available for exception handling in function M-files	7.5
mexext	MEX-file extension	5.0
mexext	(Change) Now returns mexw32 on 32-bit Windows platforms	7.1
mexext	MEX-files on 64-bit Windows now use the .mexw64 extension	7.2
mexext	(Change) New MEX-file extensions for Intel Mac (mexmaci) and Solaris 64-bit (mexs64) platforms	7.4
mfilename	Name of the currently running M-file	5.0

`migrateistimefirst`	Time series object `isTimeFirst` property behavior is changing in a future release. `migrateistimefirst` is a temporary function to help migrate this property to conform to the newer usage; it will be removed in future releases	7.10
`mimofr`	(Change, previously undocumented) Linear time-invariant frequency response	7.0
`min([]) = []`	(Change) Defined output for empty inputs	5.0
`min`	(Change) Now returns an error when input arguments are complex integers	7.0.4
`min`	(Change) Now uses the phase angle in the case of equal magnitude complex input. No longer returns warning messages for inputs of different data types	7.3
`min`	(Change) Multithreading supported	7.8
`minres`	Solves system of equations by using Minimal Residual Method	6.0
`minus (-)`	(Change) Now supports `int64` and `uint64` classes natively	7.11
`mislocked`	True if M-file cannot be cleared	5.2
`mldivide`	(Change) Now supports complex regular sparse input matrices. Performance improvements for sparse rectangular matrix input	7.9
`mldivide (\)`	(Change) Now supports `int64` and `uint64` classes natively	7.11
`mlint`	Examines M-file or a directory of M-files for potential problems and makes suggestions for possible improvements	7.0
`mlint`	(Change) New `-notok` option; hyperlink to Editor/Debugger in output	7.1
`mlint`	(Change) The `mlint` functionality has been built into the Editor/Debugger. Some of the message text has changed as well	7.2
`mlint`	(Change) Enhanced preferences and message text changes	7.3
`mlint`	(Change) New `cyc` option to determine McCabe (cyclomatic) complexity of M-files	7.5
`mlint`	(Change) New `-config` option to override preference settings	7.6
`mlintrpt`	Examines M-file or a directory of M-files for potential problems and makes suggestions for possible improvements	7.0
`mlintrpt`	(Change) New option applies preferences from a settings file	7.3

Function	Description	Release
mlock	Prevents M-file clearing	5.2
mmcompinfo	Multimedia compressor information	7.0
mmfileinfo	Gets information about a multimedia file (PC only)	7.0
mmfileinfo	(Change) Now reads files on the MATLAB path as well as the current directory. New Path field added to the mmfileinfo output struct	7.5
mmreader	Video file reader for Windows platforms	7.5
mmreader	(Change) Now supported on Linux platforms	7.8
mmreader	(Change) Now supports Motion JPEG 2000 (.mj2) files on all platforms except Solaris	7.9
mmreader. getFileFormats	Returns a platform-specific list of video file formats mmreader supports	7.10
mmreader. isPlatformSupported	(Change) Always returns True. Use mmreader.getFileFormats instead	7.10
mmreader	To be removed. Use VideoReader instead	7.11
mod	Modulus	5.0
mod	(Change) Now supports int64 and uint64 classes natively	7.11
mode	Finds most frequent values in a sample. mode is now a reserved word	7.1
mode	(Change) Returns NaN for an empty array input	7.4
movefile	(Change) Now also renames a file or directory	6.5
movie	(Change) No longer a built-in function; the syntax is unchanged	7.11
movie2avi	(Change) Now generates uncompressed AVI files on Windows platforms unless the Indio5 codec is installed	7.5
mpower (^)	(Change) Now supports int64 and uint64 classes natively	7.11
mrdivide	(Change) Significant performance improvements	7.10
mrdivide (/)	(Change) Now supports int64 and uint64 classes natively	7.11
msgbox	Displays message dialog	5.0
mtimes (*)	(Change) Now supports int64 and uint64 classes natively	7.11
multibandread	Supports reading data from raw files	6.5
multibandwrite	Supports writing data to raw files	6.5
munlock	Permits M-file clearing	5.2
mwIndex	New platform-independent index type declaration	7.3

`mwSize`	New platform-independent size type declaration for values such as array dimensions and number of elements	7.3
`namelengthmax`	Returns maximum variable and function name lengths	6.5
`nargin`	(Change) Now accepts either a function name or function handle as an input argument	7.0.1
`nargout`	(Change) Now accepts either a function name or function handle as an input argument	7.0.1
`nargoutchk`	Validates number of output arguments	6.0
`native2unicode`	Converts a character string from a native character set to Unicode	7.0.1
`nccreate`	(New) Creates variable in a NetCDF file	7.12
`ncdisp`	(New) Displays contents of a NetCDF file	7.12
`nchoosek`	All possible combinations of n elements taken k at a time	5.0
`ncinfo`	(New) Returns information about a NetCDF file	7.12
`ncread`	(New) Reads data and attributes of a NetCDF file	7.12
`ncreadatt`	(New) Reads an attribute from a NetCDF file	7.12
`ncwrite`	(New) Writes data to a NetCDF file	7.12
`ncwriteatt`	(New) Writes an attribute to a NetCDF file	7.12
`ncwriteschema`	(New) Adds NetCDF schema definitions to a NetCDF file	7.12
`ndgrid`	Generates arrays for n-D functions and interpolation	5.0
`ndims`	Number of array dimensions	5.0
`NearestNeighbor/` `(DelaunayTri)`	(Change) The two-output form returns the corresponding Eculidean distances between the query points and their nearest neighbors	7.11
`netcdf`	A number of new functions support MATLAB interface to the NetCDF libraries. Use `help netcdf` for overviews of the available functionality	7.11
`nextpow2`	Due to change to element-by-element calculation in a future release	7.8
`nextpow2`	(Change) Now returns a vector the same size as the input	7.10
`nnls`	Replaced by `lsnonneg`	6.0
`nnls`	(Removed) No longer supported	7.0
`normest`	2-norm estimate	5.0
`notebook`	(Change) The `-setup` switch no longer accepts arguments	7.3
`now`	Current date and time	5.0

Function	Description	Release
nthroot	n^{th} real root	7.0
num2cell	Converts matrix to cell array	5.0
num2hex	Converts number to IEEE hexadecimal format	7.0
numel	Number of elements in an array	6.0
numel	(Change) numel(A,varargin) returns the number of subscripted elements in A(varargin{:})	6.1
numel	(Fix) Now returns consistent results for built-in classes and subclasses	7.9
ode*	(Change) ODE solvers can now solve problems without using an ODE file	6.0
ode*	(Change) Now optionally returns solution structure for use by deval	6.1
ode113	Solves nonstiff differential equations, variable order method	5.0
ode15i	Ordinary differential equation solver for implicit equations	7.0
ode15s	Solves stiff differential equations, variable order method	5.0
ode23	Solves nonstiff differential equations, low order method	5.0
ode23s	Solves stiff differential equations, low order method	5.0
ode23t	Solves moderately stiff differential equations	5.2
ode23tb	Solves stiff differential equations by using crude error tolerances	5.2
ode45	Solves nonstiff differential equations, medium order method	5.0
odefile	Problem definition file for ODE solvers	5.0
odeget	Extracts options from ODE options structure	5.0
odeset	Creates or edits options structure for ODE solvers	5.0
odeset	(Change) Nonnegativity constraints can be imposed on computed solutions	7.1
odextend	Extends solution of ordinary differential equations	7.0
onCleanup	Specifies task(s) to perform just before exiting the current function	7.6
ones	(Change) Now accepts a final argument specifying numeric data type of result	7.0
ones	(Change) No longer accepts complex dimension arguments	7.5
opengl	Allows switching between hardware and software-based OpenGL® rendering	7.0.1

`optimget`	Gets optimization options	5.3
`optimset`	Sets or modifies optimization options	5.3
`ordeig`	Returns the vector of eigenvalues of a quasitriangular matrix or matrix pair	7.0.1
`orderfields`	Order fields of a structure	6.5
`ordqz`	Reorders QZ factorization	7.0
`ordschur`	Reorders Shur factorization	7.0
`pagedlg`	Open Page Layout dialog	5.1
`pagedlg`	(Obsolete) Use `pagesetupdlg` instead	7.3
`pagedlg`	(Removed) Use `guide` instead	7.5
`pagesetupdlg`	(Depreciated) Use `printpreview` instead	7.12
`pan`	Interactively pans the plot view	7.0
`parfor`	New reserved word	7.3
`path2rc`	(Obsolete) Replaced by `savepath`	7.0
`pause`	(Change) Fractional second argument is accepted	5.3
`pause`	(Change) New `query` argument to return pause state	7.7
`pbaspect`	Sets or gets plot box aspect ratio and plot box aspect mode	5.2
`pcg`	Preconditioned conjugate gradients method	5.0
`pchip`	Piecewise cubic Hermite interpolating polynomial interpolation	6.0
`pcode`	Creates pseudocode	5.0
`pcode`	(Change) Internal format of P-files change	6.5
`pcode`	(Change) Internal format of P-files change	7.5
`pdepe`	Solves partial differential equations in one dimension	6.0
`pdeval`	Evaluates solution computed by `pdepe`	6.0
`perl`	Calls Perl script using platform executable	6.5
`perms`	All possible permutations	5.0
`permute`	Permutes dimensions of n-D array	5.0
`persistent`	Declares a variable persistent	5.2
`pie`	Pie chart	5.0
`pie3`	3-D pie chart	5.0
`playshow`	(Obsolete) Replaced by the `echodemo` helper function	7.1
`plot`	(Change) The v6 option for high-level plotting functions is obsolete	7.5

Function	Description	Release
plot3	(Change) The v6 option for high-level plotting functions is obsolete	7.5
plotbrowser	Shows or hides the plot browser for a *figure*	7.0
plottools	Shows or hides the plot-editing tools for a *figure*	7.0
plotyy	Plots with *y*-axis labels on both the left and right	5.0
plus (+)	(Change) Now supports int64 and uint64 classes natively	7.11
poleig	Can now return a vector of condition numbers for the eigenvalues	7.0
polyarea	Area of a polygon	5.0
polyeig	(Change) Can now return only the eigenvalues	6.1
polyfit	(Change) Now supports centering and scaling of data	6.0
polyval	(Change) Now supports centering and scaling of data	6.0
popupstr	(Depreciated) No replacement available	7.4
power (.^)	(Change) Now supports int64 and uint64 classes natively	7.11
ppval	(Change) Now supports ppval(xx,pp) to permit use of ppval with function functions	6.1
ppval	(Change) Now supports multidimensional arrays for Y	7.0
primes	Generates a list of prime numbers	5.0
print	(Change) The -dln03 option has been removed	7.4
print	(Depreciated) Certain printer and graphics format option strings are depreciated and will be removed in future versions	7.7
print	(Change) The -adobecset option and -dill device are depreciated and will be removed	7.9
print	(Change) The -dmfile option to print will be removed in a future release	7.11
printdlg	Opens Print dialog	5.1
printdlg	Prints dialog box	7.0
printdmfile	(Change) The printdmfile function will be removed in a future release	7.11
printpreview	Displays preview of figure to be printed	7.0
prod([]) = 1	(Change) Defined output for empty inputs	5.0
prod	(Change) Multithreading supported	7.8
profile	Profiles M-file execution	5.0

`profile`	No longer supports the `-detail` flag's `builtin` option	7.0
`profile`	(Change) New `-nohistory` option added	7.2
`profreport`	(Obsolete) Replaced by `profsave`	7.0
`profview`	Produces graphical profile report	6.5
`propedit`	The `-v6` option is no longer supported. The Version 6 Property Editor has been removed	7.3
`propertyeditor`	Shows or hides the property editor for a *figure*	7.0
`psi`	Evaluates Digamma function	6.5
`publish`	Runs a script M-file and saves the results	7.0
`publish`	(Change) New `catchError` option available. Italic text now supported	7.2
`publish`	(Change) New `maxOutputLines` field added; default value is `Inf`	7.4
`publish`	(Change) New `codetoEvaluate` option; `stopOnError` option removed	7.6
`publish`	(Change) New `figureSnapMethod` option to specify figure details be included in the snapshot. Now supports in-line LaTeX math symbols	7.7
`publish`	(Change) New `figureSnapMethod` options are `entireGUIWindow` (default) and `entireFigureWindow` which include the title bar and all other window decorations in the snapshot	7.8
`pwch`	Piecewise cubic Hermite interpolation	7.0
`qmr`	Quasi-minimal residual method	5.0
`qr`	(Change) Now supports complex sparse input matrices with a third output argument containing a fill-reducing permutation for sparse matrix input	7.9
`qr`	(Change) The upper triangular matrix produced by the factorization routine in `qr` always contains real, nonnegative diagonal elements	7.10
`qr`	Reverts to pre-v7.10 behavior; the diagonal of R may contain complex or negative elements	7.12
`qrdelete`	(Change) Can now delete rows as well as columns	6.5
`qrinsert`	(Change) Can now insert rows as well as columns	6.5
`quad`	(Change) Function sampling bug fixed	6.1
`quad2d`	Provides additional quadrature functionality for nonrectangular areas of integration	7.8

Function	Description	Release
quad8	Replaced by quadl	6.0
quad8	(Depreciated) Use quadl instead	7.0
quad8	(Obsolete) Use quadl instead	7.3
quad8	(Removed) Use quadl instead	7.6
quadgk	Numerically evaluates the integral, adaptive Gauss–Kronrod quadrature	7.5
quadl	Numerically computes integral by using Lobatto quadrature; replaces quad8	6.0
quadv	Vectorized quad function	7.0
questdlg	Displays question dialog	5.0
quiver	(Change) The v6 option for high-level plotting functions is obsolete	7.5
quiver3	3-D quiver plot	5.0
quiver3	(Change) The v6 option for high-level plotting functions is obsolete	7.5
rand	(Change) Now supports the Mersenne twister algorithm	7.1
rand	(Change) Now uses the Mersenne twister algorithm as default rather than Subtract-with-Borrow	7.4
rand	(Change) No longer accepts complex dimension arguments	7.5
randi	(New) Returns random integers from a uniform discrete distribution	7.7
randn	(Change) No longer accepts complex dimension arguments	7.5
randn	(Change) New longer period random number algorithm is the default	7.7
RandStream object	(Changes) RandnAlg property replaced by NormalTransform property, setDefaultStream method replaced by setGlobalStream method, getDefaultStream method replaced by getGlobalStream method	7.12
rbbox	Rubberband box	5.0
rdivide	(Change) Multithreaded support added	7.6
rdivide (./)	(Change) Now supports int64 and uint64 classes natively	7.11
recycle	Sets option to move deleted files to a recycle folder	7.0
recycle	Determines if deleted files go to Recycle Bin	7.0
reducepatch	Reduces number of patch faces	5.3

`reducevolume`	Reduces number of volume data elements	5.3
`refreshdata`	Refreshes data in plot	7.0
`regexp`	Matches regular expression	6.5
`regexp`	(Change) Calling `regexp` with `'tokenExtents'` and `'once'` options now returns a double array	7.2
`regexp`	(Change) New ?%cmd operator to help with debugging regular expressions	7.3
`regexpi`	Matches regular expression, ignoring case	6.5
`regexpi`	(Change) Calling `regexpi` with `'tokenExtents'` and `'once'` options now returns a double array	7.2
`regexpi`	(Change) New ?%cmd operator to help with debugging regular expressions	7.3
`regexprep`	Replaces string, using regular expression	6.5
`regexprep`	(Change) Now supports the use of character representations (like \t or \n) in replacement strings	7.0.1
`regexptranslate`	Returns a regular expression for a literal string containing wildcard or metacharacters	7.2
`registerevent`	Registers events for a specified control at runtime	7.0
`rehash`	Refreshes function and file system caches	6.0
`rem`	(Change) Multithreaded support added	7.6
`rem`	(Change) Now supports `int64` and `uint64` classes natively	7.11
`repmat`	Replicates and tiles an array	5.0
`reshape`	Changes the shape of an array	5.0
`reshape`	(Change) `reshape(A,...,[],...)` now calculates the size required for the empty dimension	6.1
`restoredefaultpath`	Restores default MATLAB path	7.0
`rethrow`	Reissues error	6.5
`rethrow`	(Change) Now accepts stack information as input	7.1
`rethrow`	(Change) The `rethrow(lasterror)` usage is depreciated. Use the `rethrow(MException)` or `rethrow(MException.last)` instead	7.9
`rexgrep`	(Change) New `'split'` option to split an input string into sections	7.5
`rgb2ind`	Converts RGB image to indexed image	7.8

Function	Description	Release
ribbon	Draws lines as 3-D strips	5.0
rmdir	Removes directory and, optionally, contents as well	6.5
rmfield	Removes field from structure	5.0
rng	New function to control the random number generator. Use rng rather than the 'seed', 'state', or 'twister' inputs to the rand or randn functions	7.12
save	(Change) Now supports compressing MAT-files	7.0
save	(Change) Now supports -v7.3 option to save MAT-files in uncompressed HDF5 format permitting file sizes greater than 2 GB. Compressed HDF5 format will be the default MAT-file format in a future version	7.3
save	No longer accepts -compress, -uncompress, -unicode, or -nounicode options. Compression and Unicode are the defaults. Use the -v6 argument to disable compression and Unicode	7.3
save	(Fix) The -rexgrep argument is no longer taken to be a file name if no file name is supplied	7.4
save	(Change) Data items over 2 GB stored in a MAT-file using the -v7.3 option are now compressed	7.6
saveas	(Depreciated) Certain printer and graphics format option strings are depreciated and will be removed in future versions	7.7
saveas	(Change) The mmat format option is depreciated and will be removed in a future release	7.11
savepath	Saves current MATLAB path; replaces path2rc function	7.0
saxis	(Obsolete) No longer used	6.0
saxis	(Removed) No longer supported	7.0
scatter	2-D scatter plot	5.1
scatter	(Change) The v6 option for high-level plotting functions is obsolete	7.5
scatter3	3-D scatter plot	5.1
scatter3	(Change) The v6 option for high-level plotting functions is obsolete	7.5
secd	Trigonometric function with arguments in degrees	7.0
secd	(Change) Significant performance improvements	7.11
selectmoveresize	Interactively selects, moves, or resizes objects	5.0

`semilogx`	(Change) The v6 option for high-level plotting functions is obsolete	7.5
`semilogy`	(Change) The v6 option for high-level plotting functions is obsolete	7.5
`sendmail`	Sends e-mail	6.5
`serial`	(Change) Now supported on all platforms	7.8
`set`	(Change) Do not use `get` or `set` to manage properties of a Java object; this usage will generate an error in future releases	7.6
`setdiff`	Set difference	5.0
`setenv`	Sets an environment variable in the underlying operating system	7.2
`setfield`	Sets field in structure	5.0
`setpixelposition`	Sets position of object in pixels	7.0
`setstatus`	(Depreciated) No replacement available	7.4
`setstr`	Replaced by `char`	6.0
`setstr`	To be removed. Use `char` instead	7.10
`setuprop`	(Obsolete) Use `setappdata` instead	7.1
`setuprop`	(Removed) Use `setappdata` instead	7.5
`setxor`	Set exclusive OR	5.0
`shiftdim`	Shifts dimensions	5.0
`showplottool`	Shows or hides one of the plot-editing components for a figure	7.0
`shrinkfaces`	Reduces size of patch faces	5.3
`sign`	(Change) Now supports `int64` and `uint64` classes natively	7.11
`sind`	Trigonometric function with arguments in degrees	7.0
`sind`	(Change) Significant performance improvements	7.11
`single`	Conversion to single-precision data type	5.3
`smooth3`	Smoothes 3-D data	5.3
`snapnow`	Includes a snapshot of output in a published document	7.6
`sort`	(Change) Now works on data types other than double precision	6.0
`sort`	(Change) Now supports an optional last argument that specifies the sort direction	7.0
`sort`	No longer accepts complex integer inputs	7.3

Function	Description	Release
sortrows	Sorts rows in ascending order	5.0
sortrows	(Change) Now calls MEX function sortrowsc to maximize speed. With cell array of strings input, now calls MEX function sortcellchar to maximize speed	6.1
sortrows	(Change) Significant performance improvements	7.10
spfun	(Change) Better error checking for sparse matrix functions	7.10
spline	(Change) Now supports multidimensional arrays for Y	7.0
spones	(Change) Better error checking for sparse matrix functions	7.10
spparms	Enhancements to improve control, performance, and memory usage, as well as access to upper and lower triangular factors and lower symbolic factor	7.3
sprand	Random uniformly distributed sparse arrays	5.0
sprand	(Change) Better error checking for sparse matrix functions	7.10
sprandn	(Change) Better error checking for sparse matrix functions	7.10
sprandsym	(Change) Better error checking for sparse matrix functions	7.10
spring	Colormap of magenta and yellow	5.0
sprintf	(Change) Additional numeric positional arguments available for formatted string format specifiers	7.4
squeeze	Eliminates singleton dimensions	5.0
stairs	(Change) The v6 option for high-level plotting functions is obsolete	7.5
std	(Change) std([]) now returns NaN rather than empty	6.0
stem	(Change) Stem tips can be filled or unfilled	5.0
stem	(Change) The v6 option for high-level plotting functions is obsolete	7.5
stem3	3-D stem plot	5.0
stem3	(Change) The v6 option for high-level plotting functions is obsolete	7.5
str2double	Converts character string to double-precision value	5.3
str2func	Constructs function handle from function name string	6.0
str2func	(Change) Can now convert an anonymous function definition to a function handle	7.8
str2mat	Replaced by char	6.0
str2mat	To be removed. Use char instead	7.10
strcat	String concatenation	5.0

`strcmpi`	Compares strings, ignoring case	5.2
`stream2`	Computes 2-D streamlines	5.3
`stream3`	Computes 3-D streamlines	5.3
`streamline`	Draws streamlines	5.3
`streamparticles`	Draws stream particles	6.0
`streamribbon`	Draws stream ribbons	6.0
`streamslice`	Draws streamlines	6.0
`streamtube`	Draws stream tubes	6.0
`strfind`	Searches for occurrence of second string argument in first string argument	6.1
`strfind`	(Change) Now supports cell array of strings as input	7.0
`strjust`	(Change) Now does right, left, and center justification	5.2
`strmatch`	Finds matches for a string	5.0
`strncmp`	Compares first n characters	5.0
`strncmpi`	Compares first n characters in strings ignoring case	5.2
`strread`	To be removed. Use `textscan` instead	7.10
`strtok`	(Change) Now supports cell array of strings as input	7.0
`strtrim`	Removes leading and trailing white space from a string	7.0
`struct`	Creates structure array	5.0
`struct2cell`	Converts structure to cell array	5.0
`structfun`	Applies a function to each field of a structure	7.1
`strvcat`	Vertical string concatenation	5.0
`strvcat`	To be removed. Use `char` instead	7.10
`sub2ind`	Single index from subscripts	5.0
`subplot`	(Change) The v6 option for high-level plotting functions is obsolete	7.5
`subvolume`	Extracts subset of volume data set	5.3
`sum([]) = 0`	(Change) Defined output for empty inputs	5.0
`sum`	(Change) Can now be used with all integer data types	5.3
`sum`	(Change) Multithreading supported	7.8
`sum`	(Change) Now supports `int64` and `uint64` classes natively	7.11
`summer`	Colormap of green and yellow	5.0
`surf`	(Change) The v6 option for high-level plotting functions is obsolete	7.5

Function	Description	Release
surf2patch	Converts surface data to patch data	5.3
svd	(Change) Can now return only the first two outputs, U and S	6.1
svd	(Change) Adds support for economy decomposition on matrices having few rows and many columns using svd(A, 'econ') (Documented in MATLAB 7.1.)	7.0
svd	(Change) Now supports economy decomposition (introduced but undocumented in v7.0)	7.1
svd	(Change) Significant performance improvement for the three-output form of svd	7.11
svds	A few singular values	5.0
swapbytes	Swap byte-ordering	7.1
symamd	Symmetric approximate minimum degree permutation	6.0
symbfact	Enhancements to improve control, performance, and memory usage, as well as access to upper and lower triangular factors and lower symbolic factor	7.3
symmlq	Solves system of equations by using symmetric LQ method	6.0
symmmd	(Depreciated) Use symamd instead	7.0
symmmd	(Obsolete) Use symamd instead	7.6
symmmd	(Removed) Use symamd instead	7.7
table1	Replaced by interp1	6.0
table1	(Depreciated) Use interp1 instead	7.3
table1	(Obsolete) Use interp1 or interp1g instead	7.6
table2	Replaced by interp2	6.0
table2	(Obsolete) Use interp2 instead	7.6
table8	(Removed) No replacement	7.3
tand	Trigonometric function with arguments in degrees	7.0
tand	(Change) Significant performance improvements	7.11
tar	Archive files to a tar file	7.0.4
tar	Partial paths and wildcards (e.g., '~' and '*') are accepted in filename arguments	7.2
tempname	(Change) Now generates a longer and more unique filename	7.5
terminal	(Removed) No longer supported	7.0
tetramesh	Tetrahedron mesh plot for use with delaunayn	6.1

`texlabel`	Creates the TeX format from a character string	5.3
`textread`	To be removed. Use `textscan` instead	7.10
`textscan`	Reads text file into a cell array; has more features than `textread`	7.0
`textscan`	(Change) Now reads data from strings in addition to files	7.0.4
`textscan`	(Change) New `CollectOutput` switch to return like values in the same cell array	7.4
`tfqmr`	Implements a transpose-free quasi-minimal-residual method for solving systems of linear equations	7.8
`tic`	(Change) Now supports multiple consecutive timings	7.7
`timer`	Creates and controls timer objects to schedule execution of MATLAB code	6.5
`timerfindall`	Finds all timer objects with specified property values	7.0
`times (.*)`	(Change) Now supports `int64` and `uint64` classes natively	7.11
`timeseries`	New functions for time series analysis. Must be manually enabled on Linux 64-bit platforms	7.1
`title`	(Change) Now accepts *axes* handle as first argument	7.0
`toc`	(Change) Now supports multiple consecutive timings	7.7
`toolboxdir`	Returns the absolute path to the specified toolbox	7.2
`trimesh`	Triangular mesh plot	5.0
`triplequad`	Evaluates triple integral	6.5
`triplot`	2-D triangular plot for use with `delaunay`	6.1
`TriRep`	New OOP class provides enhanced computational geometry tools	7.8
`TriScatteredInterp`	New OOP class provides enhanced computational geometry tools	7.8
`trisurf`	Triangular surface plot	5.0
`true`	Creates array of logical True	6.5
`true`	(Change) No longer accepts complex dimension arguments	7.5
`tscollection`	New functions for time series analysis. Must be manually enabled on Linux 64-bit platforms	7.1
`tsearch`	Searches for enclosing Delaunay triangle	5.0
`tsearch`	(Depreciated) Use `DelaunayTri/pointLocation` instead	7.9
`tsearch`	(Obsolete) Use `DelaunayTri/pontLocation` methods instead	7.10

Function	Description	Release
typecast	Converts data types without changing the underlying data	7.1
uibuttongroup	Creates *buttongroup* object	7.0
uicontainer	Creates *container* object	7.0
uicontrol	(Change) uicontrol(h) now transfers focus to the *uicontrol* having handle h. Multiline 'edit' style *uicontrol* objects now have a vertical scroll bar. *uicontrol* objects now have a 'KeyPressFcn' callback	7.0
uigetfile	(Change) Now permits selection of multiple files	7.0
uigetfile	(Change) Now supports meta directories such as '.', '..', and '/'	7.6
uigettoolbar	(Depreciated) No replacement yet available	7.4
uint16	Conversion to 16-bit unsigned integer data types	5.3
uint16	(Change) Now rounds noninteger inputs rather than truncating	7.0
uint16	(Change) Multithreading extended to integer conversion and arithmetic	7.10
uint32	(Change) Now rounds noninteger inputs rather than truncating	7.0
uint32	(Change) Multithreading extended to integer conversion and arithmetic	7.10
uint64	(Change) Now rounds noninteger inputs rather than truncating	7.0
uint64	(Change) Multithreading extended to integer conversion and arithmetic	7.10
uint8	Conversion to 8-bit unsigned integer data types	5.3
uint8	(Change) Now rounds noninteger inputs rather than truncating	7.0
uint8	(Change) Multithreading extended to integer conversion and arithmetic	7.10
uipanel	Creates *uipanel* container object	7.0
uipushtool	Creates pushbutton in *uitoolbar* object	7.0
uiputfile	(Change) Now supports meta directories such as '.', '..', and '/'	7.6
uiresume	Resumes suspended M-file execution	5.0
uitab	(Change) Changes are being made to this undocumented function	7.11

`uitabgroup`	(Change) Changes are being made to this undocumented function	7.11
`uitable`	Creates a *uitable* object	7.0
`uitable`	Creates a new graphic table component	7.6
`uitoggletool`	Creates togglebutton in *uitoolbar* object	7.0
`uitoolbar`	Creates *uitoolbar* object	7.0
`uitree`	Creates *uitree* object	7.0
`uitreenode`	Creates a *node* object in a *uitree* component	7.0
`uiwait`	Blocks M-file execution	5.0
`uiwait`	(Change) `uiwait(handle,t)` now times out after time t has elapsed	7.0
`uminus (-)`	(Change) Now supports `int64` and `uint64` classes natively	7.11
`umtoggle`	(Obsolete) Set the `Checked` property of the `uimenu` object	7.3
`umtoggle`	(Removed) Set the `Checked` property of the *uimenu* object	7.5
`unicode2native`	Converts a character string from Unicode to a native character set	7.0.1
`union`	Union of two sets	5.0
`unique`	Unique elements of a vector	5.0
`unique`	(Change) New `'first'` and `'last'` options	7.3
`unit32`	Conversion to 32-bit unsigned integer data types	5.3
`unit64`	Creates unsigned 64-bit integer array	6.5
`unloadlibrary`	Generic DLL Interface function	7.0
`unregisteral levents`	Unregisters all events for a specified control at runtime	7.0
`unregisterevent`	Unregisters events for a specified control at runtime	7.0
`untar`	Extracts files from a tar file or URL	7.0.4
`unzip`	Uncompresses files and directories	6.5
`unzip`	(Change) Argument can be a file or a URL	7.0.4
`unzip`	Now preserves original write attribute of all extracted files	7.10
`uplus (+)`	(Change) Now supports `int64` and `uint64` classes natively	7.11
`urlread`	Reads content using URL	6.5
`urlwrite`	Writes content using URL	6.5
`userpath`	Views/sets/clears the user directory from the top of the MATLAB search path	7.6

Function	Description	Release
validateattributes	Checks the validity (e.g., numeric, nonempty) of an input array	7.5
validateattributes	(Change) Now supports checking size and range of input values	7.8
validatestring	Checks the validity (e.g., character, nonempty) of text string input	7.5
varargin	Passes a variable number of function arguments	5.0
varargout	Returns a variable number of function arguments	5.0
ver	(Change) Now returns more detailed information, and hostid information is no longer provided	6.5
verLessThan	Compares specified toolbox version with currently running version	7.4
vertcat	(Change) Better error checking for sparse matrix functions	7.10
VideoReader	Replacement for mmreader; both return identical VideoReader objects	7.11
VideoWriter	Improvement over avifile supports files larger than 2 GB	7.11
VideoWriter	(Change) Now supports Motion JPEG 2000 files	7.12
view	(Change) No longer supports 4-by-4 transformation matrices as inputs. Use view([az el]) instead	7.9
volumebounds	Gets coordinate and color limits for volume data	6.0
voronoi	Voronoi diagram	5.0
voronoi	(Change) Now makes use of Qhull	6.1
voronoi	(Change) No longer supports the QHULL or QHULL options arguments	7.9
voronoi	(Change) No longer uses the options argument	7.10
voronoin	n-D Voronoi diagram	6.0
voronoin	(Change) Now supports user-settable options	7.0
waitfor	Blocks execution until condition is satisfied	5.0
warning	Displays warning message	5.0
warning	(Change) Additional numeric positional arguments available for formatted string format specifiers	7.4
wavplay	To be removed. Use audioplayer and play instead	7.11
wavrecord	To be removed. Use audioplayer and record instead	7.11

web	(Change) Use `-browser` parameter to specify which system browser to use (default is Firefox.) Any `docopt.m` browser specification is now ignored	7.8
weekday	Day of week	5.0
weekday	New output options	7.0
what	(Change) New `package` information available	7.7
who	(Change) Now displays information for nested functions separately	7.1
whos	(Change) Now displays information for nested functions separately	7.1
whos	(Change) Modifications to the output format	7.3
winopen	On Windows platforms, opens a file in its appropriate application	6.5
winter	Colormap of blue and green	5.0
wizard	(Obsolete) No replacement	7.1
wizard	(Removed) Use `guide` instead	7.5
wk1finfo	To be removed. Use `xlsinfo` instead	7.10
wk1finfo	To be removed. Use `xlsinfo` instead	7.11
wk1read	To be removed. Use `xlsread` instead	7.10
wk1read	To be removed. Use `xlsread` instead	7.11
wk1write	To be removed. Use `xlswrite` instead	7.10
wk1write	To be removed. Use `xlswrite` instead	7.11
ws2matq	(Obsolete) No replacement	7.1
ws2matq	(Removed) No replacement	7.5
xlabel	(Change) Now accepts *axes* handle as first argument	7.0
xlim	Sets or gets *x*-axis limits and limit mode	5.2
xlsfinfo	Output format change	7.0
xlsinfo	Now supports Excel files in formats other than XLS	7.2
xlsread	Date import enhanced	7.0
xlsread	(Change) Now operates on function handles; date formats changed	7.0.4
xlsread	Now supports Excel files in formats other than XLS	7.2
xlsread	(Change) Now supports XLS, XLSX, XLSB, and XLSM formats on Windows platforms with an appropriate version of Excel installed	7.8

Function	Description	Release
xlswrite	Writes Matrix to Excel spreadsheet	7.0
xlswrite	(Change) Now supports XLS, XLSX, XLSB, and XLSM formats on Windows platforms with an appropriate version of Excel installed	7.8
xmlread	Reads XML document	6.5
xmlwrite	Writes XML document	6.5
xslt	Transforms XML document using XSLT engine	6.5
ylabel	(Change) Now accepts *axes* handle as first argument	7.0
ylim	Sets or gets *y*-axis limits and limit mode	5.2
zeros	(Change) Now accepts a final argument specifying numeric data type of result	7.0
zeros	(Change) No longer accepts complex dimension arguments	7.5
zip	Compresses files and directories	6.5
zip	Partial paths and wildcards (e.g., '~' and '*') are accepted in filename arguments	7.2
zlabel	(Change) Now accepts *axes* handle as first argument	7.0
zlim	Sets or gets *z*-axis limits and limit mode	5.2

Index